A *Very* Applied First Course in Partial Differential Equations

MICHAEL K. KEANE
Department of Mathematical Sciences
United States Air Force Academy

Prentice Hall
Upper Saddle River, New Jersey 07458

Library of Congress Cataloging-in-Publication Data

Keane, Michael, K.

A very applied first course in partial differential equations / Michael K. Keane

p. cm.

ISBN: 0-13-030417-4

1. Differential equations, Partial. I. Title

QA377.K38 2002

515'.353–dc21   2001040032

Acquisition Editor: George Lobell
Editor-in-Chief: Sally Yagan
Vice President/Director of Production and Manufacturing: David W. Riccardi
Executive Managing Editor: Kathleen Schiaparelli
Senior Managing Editor: Linda Mihatov Behrens
Production Editor: Bob Walters
Manufacturing Buyer: Alan Fischer
Manufacturing Manager: Trudy Pisciotti
Marketing Manager: Angela Battle
Marketing Assistant: Vince Jansen
Director of Marketing: John Tweeddale
Editorial Assistant: Melanie VanBenthuysen
Art Director: Jayne Conte
Cover Design: Bruce Kenselaar
Cover Illustration: Michael K. Keane

©2002 Prentice-Hall, Inc.
Upper Saddle River, New Jersey 07458

Printed in the United States of America

10   9   8   7   6   5   4   3   2   1

ISBN: 0-13-030417-4

Pearson Education LTD., *London*
Pearson Education Australia PTY, Limited, *Sydney*
Pearson Education Singapore, Pte. Ltd.
Pearson Education North Asia Ltd., *Hong Kong*
Pearson Education Canada, Ltd., *Toronto*
Pearson Educaciûn de Mexico S.A. de C. V.
Pearson Education—Japan, *Tokyo*
Pearson Education Malaysia, Pte. Ltd.

To my loving wife, Jean,

and

in memory of my friend, Sonya

Contents

List of Figures

Preface

This text is designed for a one-semester course in partial differential equations for the undergraduate student of engineering, physics, applied mathematics, social science, biology, and other sciences, for example, economics. The text covers the method of separation of variables, Fourier series, classical problems of physics and engineering, Sturm–Liouville eigenvalue problems, power series solutions of variable coefficient ordinary differential equations, and transform methods. Wherever possible, mathematical topics are motivated by physical laws or problems. As such, mathematical modeling of physical data and applications are stressed. Throughout the text, completely worked examples/counterexamples are used to develop mathematical concepts. This reduces the potential for the student to see mathematics as a set of magical steps, and it allows the student time to develop his/her own methodology for solving problems based on comprehension of the mathematical process. When a purely mathematical topic is developed, such as Fourier series, the approach taken is constructive in methodology by building on material the student has encountered in other courses. This provides the student a framework of connections allowing easy comprehension of the material, and it assists the instructor in developing the student's insight into higher mathematics.

Mathematical texts can be very intimidating to many students. Therefore, this text is designed to be truly readable and "student friendly." Whenever the text or parts of the text have been used in class, student end-of-course critiques indicated that the readability and usability of the text is in the 99th percentile.

The text is motivated by applications, which help the student in his/her studies in other areas of engineering and science. Many topics are introduced by using a physical model as opposed to a purely theoretical approach. For example, the section on the method of characteristics for first-order partial differential equations with constant coefficients is introduced by the physical example of a surfer catching a wave. Another example is the uniqueness of solution for the one-dimensional wave equation, which is developed by first considering conservation of energy for a vibrating string, a concept that most students should understand from either their first physics or calculus courses. Theoretical topics, such as Fourier series, are introduced by first discussing real vector spaces and the fact that different basis can be developed for n-dimensional space by considering an $n \times n$ matrix with n distinct eigenvalues and their corresponding eigenvectors.

The prerequisites for a student in a course using this text are the calculus sequence and elementary ordinary differential equations. An introduction to linear algebra would be helpful, but not necessary.

I have included a review of ordinary differential equations in the appendices. I have found this extremely valuable for many students. Also, for a more theoretical approach an appendix with proofs of selected theorems is provided.

Course Outline

A possible outline for a one semester course is the following:

Chapters 1 through 8, which is the core material. This provides for the development of the three classes of linear second-order partial differential equations, elliptic, parabolic and hyperbolic and the three types of boundary conditions, Dirichlet, Neumann, and Robin. Additionally, Chapters 1 through 8 gives a thorough discussion of the separation of variables technique, coverage of the relevant theorems of Fourier series and an introduction to the Sturm-Liouville boundary value problem. Once Chapters 1 through 8 are covered, there are several options. For a complete development of classical solution methods of second-order linear partial differential equations, I would suggest including Chapter 11, which develops the Fourier and Laplace transforms. For a wider set of applications, I would suggest including Chapters 9 and 10. It is also possible to chose selected topics from Chapters 9, 10, and 11 for a broad discussion of applications and technique.

Although the text is not directly tied to a mathematical software package, such as Mathematica, many of the exercises require the student to find partial sums of Fourier series. Also, students are required in the exercises to graph both the Fourier series representation of a function and a three-dimensional view of the solution of a partial differential equation for various partial sums. Thus, students should be familiar with some type of mathematical software package.

You may contact the author directly for Mathematica files and other complementary material related to the text by email at

keanemj@gateway.net

Acknowledgments

In writing this text, I have drawn from my classroom experience, and I have been influenced by many sources. I benefited from discussions with colleagues and students. I thank everyone who has made comments and suggestions. Their interest in improving the book encouraged me.

It is with great pleasure that I thank George Lobell, my Prentice Hall Editor, for his support, valuable suggestions, and encouragement. In addition, I would like to thank Bob Walters, my Production Editor for his assistance and advice and Adam Lewenberg for his assistance and work on the graphics.

I want to take this opportunity to express my gratitude to the following mathematicians who reviewed earlier versions of this text and offered many valuable suggestions and insights, which certainly improved the content, readability, and accuracy of the text:

Dr. Daniel C. Biles, Western Kentucky University;

Dr. Christine M. Guenther, Pacific University;

Dr. Jose Barrionuevo, University of South Alabama;

Dr. Mark Kon, Boston University;

Dr. William Margulies, California State University;

Dr. Mikhail Shvartsman, University of Saint Thomas;

Dr. Johnny Henderson, Auburn University.

I would like to express my thanks to the following individuals who gave valuable support, advice, and suggestions during the development of the text:

Dr. Keith Bergeron, United States Air Force Academy;

Dr. Patricia Egleston, United States Air Force Academy;

Dr. Ali Haghighat, Nuclear Engineering Program University Park;

Colonel and Dr. Daniel W Litwhiler, United States Air Force Academy;

Dr. Glenn E. Sjoden United States Air Force Academy;

Dr. Dawn L. Stewart United State Air Force Academy.

Finally, I want to thank Dr. Michael Round, United States Air Force Academy, for his time in reviewing several drafts of this text for grammatical accuracy.

I am particularly grateful to my wife, Jean, and to a friend of mine, Sonya, for their support during this long project. It is with great pleasure that I dedicate this book to them.

Michael K. Keane

Chapter 1

Introduction

The theory of ordinary differential equations (ODEs) was well established in the early part of the 18th Century. However, the theory of partial differential equations (PDEs) was still in its infancy and only studied by a few pioneers. Then, in 1747, the famous mathematician Jean Le Rond D'Alembert, while studying the problem of vibrating strings, developed the following form of the wave equation:

$$\frac{\partial^2 u(x,t)}{\partial t^2} = \frac{\partial^2 u(x,t)}{\partial x^2}. \tag{1.1}$$

He also published, in the *Memoirs* of the Berlin Academy, the solution to Equation (1.1), i.e.,

$$u(x,t) = f(x+t) + g(x-t),$$

where f and g are arbitrary functions.

The field was advanced further by another famous mathematician, Leonhard Euler. He developed the solution

$$u(x,t) = f(x+ct) + g(x-ct),$$

which is the solution to a more general wave equation

$$\frac{\partial^2 u(x,t)}{\partial t^2} = c^2 \frac{\partial^2 u(x,t)}{\partial x^2}. \tag{1.2}$$

Also in 1752, while studying hydrodynamics, Euler developed the three-dimensional equation

$$\frac{\partial^2 u(x,y,z)}{\partial x^2} + \frac{\partial^2 u(x,y,z)}{\partial y^2} + \frac{\partial^2 u(x,y,z)}{\partial z^2} = 0. \tag{1.3}$$

However, it wasn't until 1782 that P. S. Laplace formalized Equation (1.3) while working in celestial mechanics. Equation (1.3) is now known as Laplace's equation.

Next, the mathematical theory of heat, which is characterized by the equation

$$\frac{\partial u(x,t)}{\partial t} = k\frac{\partial^2 u(x,t)}{\partial x^2},\tag{1.4}$$

was studied extensively by J. B. Fourier. In 1822, his paper on the solutions of the heat equation, *Theorie analytique de la chaleur* set forth his idea that any function $y = f(x)$ could be represented by a trigonometric series, now known as a Fourier series. The Fourier series is one of the main topics of this text because it provides a solution of second-order PDEs that are linear and homogeneous with linear and homogeneous boundary conditions.

Careful inspection of Equations (1.2, 1.3, and 1.4) reveals something that they have in common: they are all second-order PDEs.

Although the study of second order PDEs is two and a half centuries old, it is far from obsolete. In fact, you can develop a second order PDE for almost anything that occurs in nature or is constructed by man. They have been used traditionally in physics, celestial mechanics, and meteorology. However, second order PDEs are being developed and applied to problems in economics, mathematical physiology, nuclear transport theory, aerospace industry, geophysics (particularly in the areas of lava flow and plate tectonics), car design, electrical engineering, forestry, industrial and community pollution, oceanography, and a host of other areas. With these diverse fields researching second-order PDEs, one might get the impression that there is no common thread to all the different fields. However, this impression is quite wrong, second-order PDEs can be broken down into three major classes, known as elliptic, hyperbolic, and parabolic. These names may seem strange; however, they are rooted in the study of quadratic equations.

A form of the quadratic equation is $ax^2 + 2bx + c = 0$, which may be solved by using the quadratic formula

$$r = \frac{-b \pm \sqrt{b^2 - ac}}{a}.\tag{1.5}$$

In Equation (1.5), we call $b^2 - ac$ the discriminant. If $b^2 - ac < 0$, then the equation is called elliptic. If $b^2 - ac > 0$, then the equation is called hyperbolic. If $b^2 - ac = 0$, then the equation is called parabolic.

The quadratic form of an equation applies to a linear PDE in two variables. Consider the general linear second-order PDE in two variables:

$$a\frac{\partial^2 u(x,y)}{\partial x^2} + 2b\frac{\partial^2 u(x,y)}{\partial x \partial y} + c\frac{\partial^2 u(x,y)}{\partial y^2}$$

$$+d\frac{\partial u(x,y)}{\partial x} + e\frac{\partial u(x,y)}{\partial y} + fu(x,y) = g.\tag{1.6}$$

Mathematicians call

$$a\frac{\partial^2 u(x,y)}{\partial x^2} + 2b\frac{\partial^2 u(x,y)}{\partial x \partial y} + c\frac{\partial^2 u(x,y)}{\partial y^2}\tag{1.7}$$

the principal part of Equation (1.6). From the principal part of the equation, you can determine if a linear second-order PDE is elliptic, hyperbolic, or parabolic. For instance, Laplace's equation, Equation (1.3), has a discriminant equal to -1. Thus, it is representative of the elliptic class of PDEs. We first encounter this class of PDE in Chapter 5. The wave equation, Equation (1.1), has a discriminant equal to 1. Hence, it is representative of the hyperbolic class of PDEs. We encounter this class of PDE in Chapter 3. Finally, the heat equation, Equation (1.4), has a discriminant equal to 0. Therefore, it is representative of the parabolic class of PDEs, which is introduced in Chapter 2.

Laplace's equation and the heat and wave equations are the primary representatives of their particular classes. However, not all linear second order PDEs are readily classified. For example, Tricomi's equation,

$$y \frac{\partial^2 u(x, y)}{\partial x^2} + \frac{\partial^2 u(x, y)}{\partial y^2} = 0,$$

is elliptic for $y > 0$, parabolic for $y = 0$, and hyperbolic for $y < 0$. Thus, one must be very careful when classifying PDEs.

As we study second-order PDEs, it is very important to remember the three classes, since everything that we learn about the representative of each class of PDE applies to the entire class. Now, we move on to the study of PDEs by investigating the heat equation.

EXERCISES 1

1.1. For each of the following linear second order PDEs, identify in what regions of the two-dimensional plane the equation is elliptic, hyperbolic, or parabolic.

(a) $2\dfrac{\partial^2 u(x, y)}{\partial x^2} + 4\dfrac{\partial^2 u(x, y)}{\partial y^2} + 4\dfrac{\partial^2 u(x, y)}{\partial y \partial x} - u(x, y) = 0.$

(b) $\dfrac{\partial^2 u(x, y)}{\partial x^2} + 2y\dfrac{\partial^2 u(x, y)}{\partial y \partial x} + \dfrac{\partial^2 u(x, y)}{\partial y^2} + u(x, y) = 0.$

(c) $\sin(xy)\dfrac{\partial^2 u(x, y)}{\partial x^2} - 6\dfrac{\partial^2 u(x, y)}{\partial y \partial x} + \dfrac{\partial^2 u(x, y)}{\partial y^2} + \dfrac{\partial u(x, y)}{\partial y} = 0.$

(d) $\dfrac{\partial^2 u(x, y)}{\partial x^2} - \cos(x)\dfrac{\partial^2 u(x, y)}{\partial y \partial x} + \dfrac{\partial^2 u(x, y)}{\partial y^2} + \dfrac{\partial u(x, y)}{\partial y} - \dfrac{\partial u(x, y)}{\partial x} + 5u(x, y) = 0.$

Chapter 2

The One-Dimensional Heat Equation

2.1 INTRODUCTION

For over one million years, fire and its product, heat, have contributed significantly to the rise of man. Thus, naturally, the study of heat transfer has become very important.

At first, the physical properties of heat could only be guessed at through the senses (e.g., the object is too hot or too cold). Only after Sir Isaac Newton and Gottfried Leibniz[1] developed calculus in the seventeenth century could we mathematically model the physical properties of heat transfer. Today we know three types of heat transfer: conduction, convection, and radiation.

Although the mathematical formulation of all three types of heat transfer is important, this chapter concentrates on mathematical formulas for heat conduction in a one-dimensional rod. It begins with the derivation of the heat equation, then branches out to discuss various boundary conditions and their physical meanings. Next, we investigate uniqueness of solution for the heat equation. Finally, the steady-state (equilibrium) temperature distribution solution for heat conduction in a one-dimensional rod is discussed.

2.2 DERIVATION OF HEAT CONDUCTION IN A ONE-DIMENSIONAL ROD

Why is heat transfer in a one-dimensional rod called **heat conduction**? To answer this question, we define what we mean by **heat transfer**, investigate how it occurs in a one-dimensional rod, and examine the three different types: convection, radiation, and conduction.

---

[1]Sir Isaac Newton and Gottfried W. Leibniz were contemporary seventeenth century mathematicians who independently developed calculus.

Let's start by defining heat transfer. Simply put, heat transfer is the way heat is distributed throughout a material (sometimes steady-state, sometimes time dependent) when either an internal or external heat source is applied. Next, we discuss how it occurs in a one-dimensional rod.

Physically, every one-dimensional rod has a cross-sectional area, A. Thus, the one-dimensional rod is actually three-dimensional. See Figure (2.1). A rod is considered one-dimensional if the heat transfer occurs in a one directional manner, down the length of the rod. This means the heat transfer across the cross-sectional area is uniform. Thus, the heat moves in one direction through the length of the rod. For example, consider a lit match and two different rods. The first rod has the cross-sectional area equivalent to the cross-sectional area of one human hair; the second rod has the cross-sectional area of a large steel I-beam. When the end of the first rod is placed near the flame of the lit match, the end is uniformly heated, and the heat moves uniformly down the length of the rod. When the end of the second rod is placed near the flame of the lit match the flame doesn't even cover the entire cross-sectional area of the rod. Therefore, the cross-sectional area of the second rod is not uniformly heated, and the flow of heat through the rod is not in one direction. Now, we'll investigate the three types of heat transfer, starting with convection.

Convection of heat is the transfer of heat by the actual motion of the material being heated. For example, if you hold your hand above the surface of boiling water your hand feels the heat. Some of this heat is from the movement of the air molecules, initially against the surface of the boiling water. As the air heats above the boiling water, the air molecules rise, taking heat with them. As these molecules rise, other cooler air molecules replace them. The rising molecules can even be seen as steam. If you think of a rod as a thin piece of copper wire, do the molecules of the wire move away from the wire as it is heated? No. Thus, heat transfer in a one-dimensional rod is not convection. Having eliminated convection, we proceed and consider radiation as a possible reason for heat transfer in a rod.

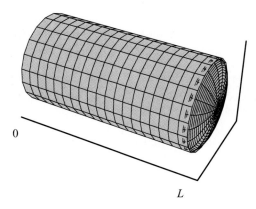

0

L

Figure 2.1: One-dimensional rod of length L.

Radiation refers to the continual emission of energy from the surface of all bodies. For example, hold your hand near the side of a radiator. (Note: do not put your hand near the top of the radiator, because then heat convection also occurs). You feel the heat. This is due to heat coming out of the radiator in a wave motion, in the infrared electromagnetic spectrum similar to electromagnetic waves. In a one-dimensional rod, heat transfer doesn't move in a wave-like motion down the length of the rod. Therefore, just as in convection, it seems radiation of heat doesn't account for heat transfer in the rod.

The last type of heat transfer is conduction. Conduction comes from collisions of neighboring molecules, transferring heat by kinetic energy. Here, the molecules move slightly, but do not actually move out of position like they do in convection; nor do they move like an electromagnetic wave as in radiation. Now their movement is more like that of tennis balls in a cylindrical container that is a little wider than a tennis ball, and long enough to hold several tennis balls. Imagine the tennis balls are molecules. When heat is applied, they can move slightly back and forth, bouncing against each other and against the side of the can. But they remain essentially in the same position. This is what happens when heat is applied to one end of a solid rod. The molecules get excited, but they do not move away from the heat, as in convection, nor do they move in a wave similar to an electromagnetic wave, as in radiation; instead, they collide with one another and retain their relative position.

Using these brief explanations of heat transfer, we can see that heat transfer in a one-dimensional rod is best described by conduction. Why is this so important? Once we understand that heat transfer in a one-dimensional rod is conduction, we may use three very important basic properties of conduction, called laws, which are attributed to Joseph Fourier. These **heat conduction laws** are as follows:

Law 1. *Heat flow is from points of higher temperature to points of lower temperature.*

Law 2. *Conduction can only take place in a body when different parts of the body, including the ends, are at different temperatures.*

Law 3. *Heat flow changes depending on the material. Different materials, even at the same temperature have different heat flows.*

These three laws form the basis for the mathematical model of heat conduction in a one-dimensional rod that we'll now develop.

2.2.1   Derivation of the Mathematical Model

One equation governing heat conduction in a one-dimensional nonuniform rod without lateral insulation may be written in words as: rate of change with respect to time of thermal energy in the rod equals the rate of flow of thermal energy due to conduction, minus thermal energy loss via convection, plus thermal energy produced within the rod. This word equation is translated into mathematics as

$$c(x)\rho(x)\frac{\partial u(x,t)}{\partial t} = \frac{\partial}{\partial x}\left(K_0(x)\frac{\partial u(x,t)}{\partial x}\right)$$

$$-\beta(x)\left[u(x,t) - v(x,t)\right] + Q(x,t). \tag{2.1}$$

This is quite an intimidating equation, and its solutions aren't easy to find. Eventually, you will be able to solve Equation (2.1) for particular cases. However, we will simplify Equation (2.1) to develop a careful approach to its solution. First, we must identify the terms, then make reasonable assumptions about them. This allows us to derive a new equation. Mathematicians use this method all the time. Thus, viewing Equation (2.1), we see that there are three double-variable functions $u(x,t)$, $Q(x,t)$, and $v(x,t)$ and four single variable functions $c(x)$, $\rho(x)$, $K_0(x)$, and $\beta(x)$.

The first double-variable function, dependent on distance and time, is $u(x,t)$: a continuous function that describes the temperature distribution in the rod. This function depends on distance because the temperature, $u(x,t)$, usually is not the same at every point x along the length of the rod. It depends on time because the temperature, $u(x,t)$, is expected to change with time due to some external or internal source.

The second double-variable function, dependent on distance and time, is $Q(x,t)$: a function representing internal heat source, called "source" for short. It describes the heat energy generated inside the rod. This is an interesting function because not all materials that can be used to make a rod generate heat; some materials will only generate heat after an external heat source is applied. For example, a rod made of steel does not generate heat even when an external heat source is applied, but a rod composed of a radioactive material generates its own heat, regardless of external source. Generally, the source function is either known or determined experimentally.

The last double-variable function is $v(x,t)$, the temperature of the surrounding medium. It depends on position and time, since the rod could connect different temperature areas that may change over time. This function plays a big part in Equation (2.1) if there is no lateral insulation.

Moving into the single-variable functions, we have $c(x)$, the specific heat of the rod. **Specific heat** is the heat energy required to raise the temperature of one unit of material mass one degree. Specific heat is a spatial function, since the material composition of the rod depends on position. If the rod is composed of only one material, or the mixture of materials is uniform, the specific heat, $c(x)$, becomes the constant, c.

The next single-variable function is $\rho(x)$. It is the **mass density** of the rod. It also depends on position in the rod because the rod could be composed of several materials, each having a different mass density. Mass density is usually measured as mass per unit volume. As in specific heat, if the rod is composed of one material, or the mixture of materials is uniform, the mass density, $\rho(x)$, becomes a constant, ρ.

The third single-variable function is $K_0(x)$, the **thermal conductivity** of the material. It is usually determined experimentally. Each different material, or mixture of materials, has a different thermal conductivity. For instance, the thermal conductivity of copper at 273 Kelvin is 390 watts per meter per degree Kelvin; whereas, the thermal conductivity of asbestos at 273 Kelvin is 0.15 watts per meter per degree Kelvin. Thus, thermal conductivity depends on the material composition of the rod, and it may also depend on the temperature. However, if the range of temperature is not allowed to have large swings (the case we will consider), then thermal conductivity will primarily depend on the material at each position, x, like specific heat, $c(x)$, and mass density, $\rho(x)$. Hence, like specific heat and mass density, if the rod is composed of one material or has a uniform mixture of various materials, the thermal conductivity, $K_0(x)$, becomes a constant K_0.

The last single-variable function in Equation (2.1) is $\beta(x)$. It appears in the heat-conduction equation multiplied by $[u(x,t) - v(x,t)]$, which is a convection term. Hence, $\beta(x)$ is named the convection proportionality function. In many cases, $\beta(x)$ is actually a constant. However, in our equation we treat it as a function.

In the table below the terms of Equation (2.1) are listed with a short description:

Function	Description
$u(x,t)$	temperature distribution of the rod
$Q(x,t)$	internal heat source
$v(x,t)$	temperature of surrounding medium
$c(x)$	specific heat of material
$\rho(x)$	mass density of material
$K_0(x)$	thermal conductivity of material
$\beta(x)$	convection proportionality function

Now, let's make some intelligent assumptions to simplify Equation (2.1).

First, let's make the assumption that the rod is made of a uniform material, then $c(x)$, $\rho(x)$, and $K_0(x)$ all become constants c, ρ, and K_0. When we impose a second assumption, the rod has perfect lateral insulation, which means convection can't take place. This assumption forces $\beta(x)$ to equal 0 in Equation (2.1). Using these two assumptions, we change Equation (2.1) to

$$c\rho\frac{\partial u(x,t)}{\partial t} = K_0\frac{\partial^2 u(x,t)}{\partial x^2} + Q(x,t). \qquad (2.2)$$

Equation (2.2) is the mathematical model for heat conduction in a one-dimensional uniform rod with perfect lateral insulation.

Now, we are going to derive Equation (2.2). Consider the physical model of a one-dimensional uniform rod with perfect lateral insulation. Apply the conservation of thermal energy law, which basically states that the rate of change with respect to time of thermal energy in the rod must equal the rate of flow of thermal energy across the boundaries, plus the thermal energy produced within the rod. However, when conservation of thermal energy is applied, the problem must be approached from a calculus point of view. In other words, consider only a thin slice first and then expand it mathematically to the full rod.

Shown in Figure (2.2) is a one-dimensional rod oriented in the positive x direction. That is, the x-axis runs down the center of the rod and the rod has length L, so that $0 \leq x \leq L$.

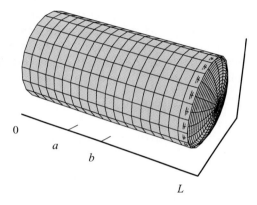

Figure 2.2: One-dimensional rod of length L.

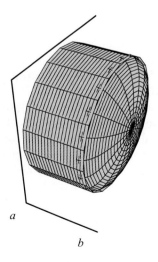

Figure 2.3: Arbitrary slice $[a, b]$ of one-dimensional rod.

Now, applying the conservation of thermal energy to any arbitrary slice $[a, b]$ of our rod (see Figure (2.3)) we can claim that

Rate of change with respect to time of total heat inside the slice

$=$ Rate of flow of heat across the boundaries of the slice

$+$ Total heat generated inside the slice. $\hspace{2cm}$ (2.3)

By using the previously defined functions, variables, axioms, and some calculus, the foregoing equation can be expressed mathematically. (*Note:* In the following equations, A is the cross-sectional area of the rod).

First,

$$\text{Total heat inside the slice of rod } [a, b] = \int_a^b c \, \rho \, A \, u(x, t) \, dx.$$

Next, using differential calculus on this equation, we may express the left side of Equation (2.3) as

$$\frac{d}{dt} \left[\int_a^b c \, \rho \, A \, u(x, t) \, dx \right]. \hspace{2cm} (2.4)$$

We will keep the constants c, ρ, and A inside the integral for the moment. This makes the calculations easier.

The second term in Equation (2.3), rate of flow of heat across the boundaries of the slice $[a, b]$, is somewhat more complicated. Since there is a rate, we know from differential calculus that a derivative is somehow involved. This derivative must relate temperature distribution at the boundaries of the slice to heat flow at these same boundaries. The correct solution lies in understanding heat flow as described by the three heat Conduction laws:

Law one says heat flow is from points of higher temperature to points of lower temperature. This law describes *how* heat flows.

Law two says conduction can only take place in a body when different parts of the body, including the ends, are at different temperatures. Therefore, Law two describes *when* heat flow takes place.

Law three says heat flow changes depending on the material. Different materials, even at the same temperature, have different heat flows. This law indicates heat flow *depends on material*. Hence, we must introduce a function describing the thermal conductivity of the material. At present our rod is uniform, so the function describing thermal conductivity becomes a constant.

Now, suppose the temperature in the rod is hotter to the left of the slice $[a, b]$ than it is to the right of the slice. See Figure (2.4).

By Law 1, this means that heat flow is in the positive x direction. The equation for heat flow across a cross-sectional area A is

$$\text{heat flow } = -K_0 \frac{\partial u(x, t)}{\partial x} A. \hspace{2cm} (2.5)$$

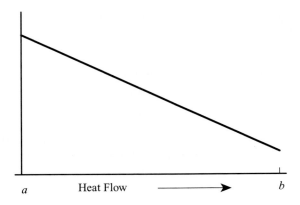

a   Heat Flow $\longrightarrow$ b

Figure 2.4: Temperature distribution of slice $[a, b]$.

The constant K_0 is the thermal conductivity of the material. The partial derivative, $\dfrac{\partial u(x, t)}{\partial x}$, is negative since it indicates the slope of the derivative in the x direction; therefore, the negative sign in front of K_0 must be introduced to show that heat flow is to the right, the positive x direction. The relationship between the three heat conduction laws to heat flow, expressed in Equation (2.5), is **Fourier's[2] law of heat conduction**.

Using Equation (2.5), the slice $[a, b]$ gains thermal energy at the boundary a and loses thermal energy at the boundary b. Therefore, the second term in Equation (2.3), rate of flow of heat across the boundaries of the slice $[a, b]$, is mathematically modeled as

$$-K_0 A \left[\frac{\partial u(a, t)}{\partial x} - \frac{\partial u(b, t)}{\partial x} \right]. \tag{2.6}$$

Again using calculus, we can model the third and final term in Equation (2.3), total heat generated inside the slice $[a, b]$ per unit time t as

$$\int_a^b A\, Q(x, t)\, dx. \tag{2.7}$$

Therefore, utilizing Equations (2.4, 2.6, and 2.7), we can rewrite Equation (2.3) as

$$\frac{d}{dt} \left[\int_a^b c\, \rho\, A\, u(x, t)\, dx \right] = -K_0 A \left[\frac{\partial u(a, t)}{\partial x} - \frac{\partial u(b, t)}{\partial x} \right]$$

$$+ \int_a^b A\, Q(x, t)\, dx. \tag{2.8}$$

---

[2] J. B. Fourier (1768–1830), a very influential mathematician, first postulated the heat conduction laws, and developed what is known as Fourier's law of heat conduction.

The constant cross-sectional area, A, occurs in every term of Equation (2.8). Thus, it is easily canceled, yielding

$$\frac{d}{dt}\left[\int_a^b c\,\rho\,u(x,t)\,dx\right] = -K_0\left[\frac{\partial u(a,t)}{\partial x} - \frac{\partial u(b,t)}{\partial x}\right]$$

$$+ \int_a^b Q(x,t)\,dx. \qquad (2.9)$$

Equation (2.9) can also be simplified by using calculus on

$$\frac{d}{dt}\left[\int_a^b c\,\rho\,u(x,t)\,dx\right]$$

and

$$-K_0\left[\frac{\partial u(a,t)}{\partial x} - \frac{\partial u(b,t)}{\partial x}\right].$$

Applying the fundamental theorem of integral calculus to the second term of Equation (2.9), namely,

$$-K_0\left[\frac{\partial u(a,t)}{\partial x} - \frac{\partial u(b,t)}{\partial x}\right],$$

we get

$$-K_0\left[\frac{\partial u(a,t)}{\partial x} - \frac{\partial u(b,t)}{\partial x}\right] = K_0 \int_a^b \frac{\partial^2 u(x,t)}{\partial x^2}\,dx,$$

provided that we assume that $\dfrac{\partial^2 u(x,t)}{\partial x^2}$ is continuous. This assumption is really not out of line as the function $\dfrac{\partial^2 u(x,t)}{\partial x^2}$ is the rate of change of the heat flow. We expect it to be continuous, since any kind of discontinuity implies a sharp change in heat flow, violating the three heat conduction laws.

For the first term of Equation (2.9), a new calculus formula is required. It is called **Leibniz's**[3] **formula**. The theorem is stated as follows and a proof may be found in Appendix B.

Theorem 1. *Suppose $f(x,t)$ and the partial derivative $\dfrac{\partial f(x,t)}{\partial t}$ are continuous in some region of the xt-plane where $a \leq x \leq b$ and $c \leq t \leq d$, then*

$$\frac{d}{dt}\left[\int_a^b f(x,t)\,dx\right] = \int_a^b \frac{\partial f(x,t)}{\partial t}\,dx.$$

---

[3]G. W. Leibniz (1646–1716), a contemporary of Sir Isaac Newton, developed the modern notation for calculus in 1676.

The first term of Equation (2.9) satisfies Leibniz's formula because $u(x,t)$ is continuous and $\dfrac{\partial u(x,t)}{\partial t}$ is expected to be continuous. Therefore,

$$\frac{d}{dt}\left[\int_a^b c\,\rho\; u(x,t)\,dx\right] = \int_a^b c\,\rho\,\frac{\partial u(x,t)}{\partial t}\,dx.$$

Using the two previous simplifications, we can rewrite equation (2.9) as

$$\int_a^b c\,\rho\,\frac{\partial u(x,t)}{\partial t}\,dx = \int_a^b K_0\frac{\partial^2 u(x,t)}{\partial x^2}\,dx + \int_a^b Q(x,t)\,dx. \tag{2.10}$$

For equation (2.10) to be true, we must have

$$\int_a^b \left(c\rho\,\frac{\partial u(x,t)}{\partial t} - K_0\frac{\partial^2 u(x,t)}{\partial x^2} - Q(x,t)\right) dx = 0. \tag{2.11}$$

Since the choice of a and b was arbitrary, Equation (2.11) must be true for all choices of a and b within the length of the rod. Using proof by contradiction, we can show that Equation (2.11) is true only if the integrand is zero. Thus, we can omit the integration and consider only the integrand—that is,

$$c\rho\,\frac{\partial u(x,t)}{\partial t} - K_0\frac{\partial^2 u(x,t)}{\partial x^2} - Q(x,t) = 0$$

or

$$c\rho\,\frac{\partial u(x,t)}{\partial t} = K_0\frac{\partial^2 u(x,t)}{\partial x^2} + Q(x,t). \tag{2.12}$$

Equation (2.12) is identical to Equation (2.2). Thus, using the conservation of thermal energy law and calculus, a one-dimensional uniform rod, with perfect lateral insulation, has been correctly modeled mathematically. In Equation (2.12), the source heat energy $Q(x,t)$ is usually given. Therefore, the only unknown is $u(x,t)$. If $Q(x,t) = 0$, then Equation (2.12) becomes

$$\frac{\partial u(x,t)}{\partial t} = k\frac{\partial^2 u(x,t)}{\partial x^2}, \tag{2.13}$$

where $k = \dfrac{K_0}{c\,\rho}$ and is called the thermal diffusivity. Appendix E contains a table of the thermal diffusivity of common materials, which will be used in this text.

2.2.2   Initial Temperature

Whenever an experiment is conducted, the initial state for that experiment is determined beforehand. In the case of heat conduction in a one-dimensional rod, we have the initial temperature distribution of the rod, known as the initial condition (IC) of the rod. Since we are talking about temperature distribution, the IC is a function of x. It is the condition that must exist before the experiment starts (at the start of time for the experiment). Therefore, at $t = 0$, we have $u(x,0) = f(x)$, $0 \leq x \leq L$. Remember, even if the IC is a constant c throughout the rod, the IC may still be thought of as a function of x.

The following equation, Equation (2.14), describes heat conduction in a one-dimensional rod with the constraints of perfect thermal lateral insulation, constant mass density, constant thermal conductivity, constant specific heat, no heat source, and an IC:

$$\frac{\partial u(x,t)}{\partial t} = k\frac{\partial^2 u(x,t)}{\partial x^2}$$

$$u(x,0) = f(x), \ 0 \leq x \leq L.$$

(2.14)

Also, remember that Equation (2.14) is the primary representative of the parabolic class of linear second-order PDEs.

EXERCISES 2.2

2.2.1. Suppose you are given a thin slice of a uniform one-dimensional rod with perfect lateral insulation. State the word equation for conservation of thermal energy.

2.2.2. Given specific heat, $c(x)$; mass density, $\rho(x)$; thermal conductivity, $K_0(x)$; and temperature, $u(x,t)$, state Fourier's law of heat conduction.

2.2.3. Briefly explain the basic idea of Fourier's law of heat conduction.

2.2.4. Suppose we have heat conduction in a rod with perfect lateral insulation, no internal heat sources, and specific heat, mass density, and thermal conductivity as functions of x, that is, $c(x)$, $\rho(x)$, and $K_0(x)$. Starting with the conservation of thermal energy law, derive a new form of the heat conduction equation.

2.2.5. Suppose we have heat conduction in a uniform rod with an internal source of heat energy, but there is no lateral insulation. Thus, the heat flows freely in and out across the lateral boundary at a rate proportional to the difference between the temperature, $u(x,t)$, in the rod and the surrounding medium, $\beta(x,t)$. Starting with the conservation of thermal energy law, derive a new form of the heat conduction equation. *Hint:* This problem requires a new formulation of the conservation of thermal energy law.

2.2.6. Suppose you have a perfect laterally insulated uniform rod, but instead of a heat source it has a heat sink (a sink is where heat is absorbed). Starting with the conservation of thermal energy law, derive the heat conduction equation.

2.2.7. Suppose you have a perfect laterally-insulated uniform rod, but the cross-sectional area is a function of x (that is, $A(x)$). Starting with the conservation of thermal energy law, derive a new form of the heat conduction equation.

2.2.8. Show that $u(x,t) = e^{-t} \sin x$ is a solution of

$$\frac{\partial u(x,t)}{\partial t} = \frac{\partial^2 u(x,t)}{\partial x^2},$$

subject to

$$u(x,0) = \sin x.$$

2.2.9. Show that $u(x,t) = e^{-3t} \cos 2x$ is a solution of

$$\frac{\partial u(x,t)}{\partial t} = \frac{3}{4} \frac{\partial^2 u(x,t)}{\partial x^2},$$

subject to

$$u(x,0) = \cos 2x.$$

2.2.10. Show that $u(x,t) = 1 + e^{2t}(4e^{3x} + 5e^{-3x})$ is a solution of

$$\frac{\partial u(x,t)}{\partial t} = \frac{2}{9} \frac{\partial^2 u(x,t)}{\partial x^2},$$

subject to

$$u(x,0) = 1 + \frac{4e^{6x} + 5}{e^{3x}}.$$

As time goes to infinity, what do you expect to happen to the one-dimensional rod?

2.2.11. Solve $\dfrac{\partial^2 u(x,y)}{\partial x \partial y} = xy.$

Exercises (2.2.12–2.2.14) involve first-order ordinary differential equations (ODEs). A review of this material may be found in Appendix C.

2.2.12. Solve $u'(t) = atu(t)$ where $a \in \mathbb{R}$. Graph several members of the family of solution curves.

2.2.13. Solve $u'(t) = atu(t) + \sin t$ where $a \in \mathbb{R}$. Graph several members of the family of solution curves.

2.2.14. Solve the following first-order ODEs

$$(1) \qquad g'(t) + 6g(t) = t, \qquad g(0) = 1.$$

$$(2) \qquad y' = 18y + xy, \qquad y(1) = 8.$$

$$(3) \qquad h'(x) - h(x) = \sin x, \qquad h(0) = \pi.$$

$$(4) \qquad 3g'(z) = \cos y - 9g(z), \qquad g\left(\frac{\pi}{2}\right) = 0.$$

2.3 BOUNDARY CONDITIONS FOR A ONE-DIMENSIONAL ROD

In Section 2.2, we derived Equation (2.12) and Equation (2.13), commonly referred to as the heat equations, for a one-dimensional rod. Also, we discussed the initial temperature distribution in the rod, then modeled it as an initial condition. However, when you model physical phenomena, it is good practice to model as much as possible; to satisfy this demand, we must discuss the ends of the rod. The ends of the rod are usually called the boundaries; when their state is modeled, they are called boundary conditions (BCs). In this section, we cover the mathematical models of various types of BCs.

2.3.1 Boundary Conditions of the First Kind

A name for BCs of the first kind is **Dirichlet**[4] **conditions**. When the boundaries of a one-dimensional rod take on the temperature of surrounding mediums, the boundaries are said to have **specified temperatures**. For example, if the rod's boundary at $x = 0$ is held in a hot water bath where the temperature changes with time, then that boundary may be mathematically modeled as

$$u(0, t) = g_1(t). \tag{2.15}$$

Also, suppose that the rod's boundary at $x = L$ is held in a different bath of hot water where the temperature changes in time, but differently than at $x = 0$. Then that boundary may be mathematically modeled as

$$u(L, t) = g_2(t).$$

(*Note:* $g_1(t)$ and $g_2(t)$ describe the temperature of the different baths as they change with time. Some texts use $u_{b_1}(t)$ and $u_{b_2}(t)$ to describe the temperature in different baths).

Sometimes, the temperatures of baths do not depend on time and are therefore constants, T_1 and T_2. If T_1 and T_2 equal zero, then we have $u(0, t) = 0$ and

---

[4]Peter Gustav Lejune Dirichlet (1805–1859) was a Prussian born mathematician who was highly influenced by Fourier in the early 19th century

$u(L, t) = 0$. In this case, the boundaries being described are in contact with zero-degree baths, and they are known as homogeneous boundary conditions of the first kind. Homogeneous boundary conditions are required to develop the **separation of variables solution technique**. This technique is one of the major topics covered in this book.

2.3.2   Boundary Conditions of the Second Kind

BCs of the second kind are also known as **Neumann**[5] **conditions**, and they describe rate of heat flow across the boundaries. Here, ends of the rod are covered by insulation material. For example, consider the rod's boundary at $x = L$ where there is insulation material between the end of the rod and the surrounding medium. Now, suppose heat is flowing outward from the end of the rod at $x = L$. That is,

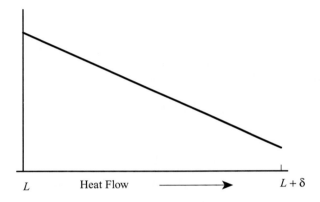

Figure 2.5: Heat flow from end $x = L$.

heat flow is from left to right in the positive x direction. This means the rod is hotter than the surrounding medium. See Figure (2.5).

From Fourier's law of heat conduction, the equation that models this example is

$$-K_0(L)\frac{\partial u(L, t)}{\partial x} = \psi(t), \qquad (2.16)$$

where $\psi(t)$ is given. (*Note:* $K_0(L)$ is the thermal conductivity of the insulation. Also note, Equation (2.16) describes the value of the derivative at the point L. Thus, we cannot simply integrate to make it a specified temperature. A very special case of Equation (2.16) occurs when $\psi(t) = 0$. Then Equation (2.16) can be written as

$$\frac{\partial u(L, t)}{\partial x} = 0,$$

meaning we have the case of perfect insulation. (This may seem impossible, but it is a standard case to model mathematically).

---

[5]Franz Neumann, (1798–1895) was a mathematical physicist who worked in Königsberg.

2.3.3   Boundary Conditions of the Third Kind

A name given to BCs of the third kind is **Robin's**[6] **conditions**. These conditions are based on Newton's law of cooling, which states heat loss by convection from one body to another is proportional to the temperature difference between the two bodies. How can heat be lost by convection when we are discussing conduction? Consider the left boundary of a rod of length L. That is the boundary at $x = 0$. Suppose the rod is hotter than the surrounding medium—a liquid, which is being stirred rapidly so that it maintains a constant temperature throughout, which is changing with time t. Let the temperature of the medium be $g_1(t)$. However, the liquid adjacent to the end of the rod is slightly hotter than the liquid just a little farther away. Hence, it heats up slightly, moving away from the end of the rod in a convective manner. Once away from the end of the rod, the slightly hotter liquid is rapidly mixed in with the rest of the medium. The mathematical model for this is

$$-K_0(0)\frac{\partial u(0,t)}{\partial x} = -h[u(0,t) - g_1(t)]. \qquad (2.17)$$

Here h is a constant of proportionality and is assumed to be positive. It is called the **coefficient of convective heat transfer**. The minus signs in Equation (2.17) describe the physical situation of the rod being hotter than the liquid medium. Thus, heat is leaving the rod on the left side, which is in the negative direction. See Figure (2.6).

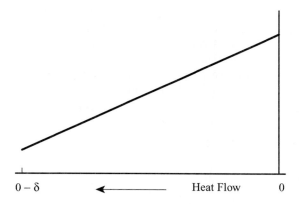

Figure 2.6: Heat flow at end of rod $x = 0$.

Robin's conditions can be thought of as the generalization of Dirichlet and Neumann boundary conditions. For example, if $h \to 0$ in Equation (2.17), then Equation (2.17) mathematically models Neumann's Condition of perfect insulation. If $h \to \infty$ in Equation (2.17), then Equation (2.17) mathematically models Dirichlet's Condition of a time-dependent specified temperature.

This completes our discussion of the three types of BCs. Using what we learned

---

[6]Victor G. Robin (1855–1897) was a French mathematician

in Sections (2.2.1 and 2.3), we may now mathematically model a physical situation involving heat conduction. This is done in the following example.

EXAMPLE 2.1. Mathematically model heat conduction in a one-dimensional uniform rod of length L with no internal heat source, thermal diffusivity of k, perfect lateral insulation, and initial condition as a function of x. Also, the left boundary, at $x = 0$, is in direct contact with a zero-degree bath, and the left boundary, at $x = L$, is perfectly insulated.

Solution: The equation is

$$\frac{\partial u(x,t)}{\partial t} = k\frac{\partial^2 u(x,t)}{\partial x^2},$$

subject to the IC

$$u(x,0) = f(x)$$

and the BCs

$$u(0,t) = 0,$$

$$\frac{\partial u(L,t)}{\partial x} = 0. \qquad \blacksquare$$

Notice, both the initial condition and the boundary conditions must be given for a complete physical description. When both are given, we can generate an accurate mathematical model of the physical description.

Knowing how to mathematically model a physical situation involving heat conduction is important. However, just as important is the capability of giving a physical description from a mathematical model. This is done in the next example.

EXAMPLE 2.2. Given that

$$\frac{\partial u(x,t)}{\partial t} = k\frac{\partial^2 u(x,t)}{\partial x^2} - \alpha Q(x,t),$$

subject to the BCs

$$\left\{ \begin{array}{l} \dfrac{\partial u(0,t)}{\partial x} = 6\,\text{watts}, \\[2ex] \dfrac{\partial u(L,t)}{\partial x} = 0, \end{array} \right.$$

and IC

$$u(x,0) = f(x),$$

write a short paragraph developing a possible physical model.

Solution: A possible physical model for the equations in this example is heat conduction in a perfect laterally insulated uniform one-dimensional rod of length L, with thermal diffusivity of k. Also, the source term is a sink, and it is being modified by a proportionality constant, α. This means heat energy is being withdrawn from the rod. The left boundary, at $x = 0$, has a rate of heat flow of a constant six watts. This may be due to imperfect insulation. The right boundary, at $x = L$, is perfectly insulated. Finally, the initial condition is a function of x. Please note, the initial condition does not have to match the boundary conditions at $x = 0$ and $x = L$. ∎

EXERCISES 2.3

2.3.1. Write a short paragraph (five sentences or less) that describes the physical problem modeled by the equations

$$\frac{\partial u(x,t)}{\partial t} = \frac{\partial^2 u(x,t)}{\partial x^2},$$

subject to IC

$$u(x, 0) = -x,$$

and the BCs

$$\frac{\partial u(0,t)}{\partial x} = 0 \text{ and } u(L, t) = \alpha(t).$$

2.3.2. Write a short paragraph (five sentences or less) that describes the physical problem modeled by the equations

$$\frac{\partial u(x,t)}{\partial t} = \frac{1}{4}\frac{\partial^2 u(x,t)}{\partial x^2} - 15[u(x,t) - (8x - 2e^{-t} + 2)],$$

subject to IC

$$u(x, 0) = 8x,$$

and the BCs

$$u(0, t) = 0 \text{ and } u(\pi, t) = 16\pi \cos\frac{\pi}{3}.$$

2.3.3. Suppose a nonuniform metal rod of length π, with perfect lateral insulation, has an initial temperature distribution of $\sin x$. Initially, one end of the rod is fixed at a temperature of $0°C$, while the rest of the rod is placed in liquid nitrogen. What would be a possible mathematical model that describes this problem? Explain your answer.

2.3.4. Consider a one-dimensional rod of length L. Assume that heat energy is flowing into the rod at $x = 0$ proportional to the temperature difference between the end temperature of the rod and the known external temperature. Develop the mathematical model for this condition. Briefly justify your answer.

2.3.5. Consider a one-dimensional rod of length L. Assume heat energy is flowing into the rod at $x = L$ proportional to the temperature difference between the end temperature of the rod and the known external temperature. Develop the mathematical model for this condition. Briefly justify your answer.

2.3.6. Given Equation (2.17), show that if $h \longrightarrow 0$, then we get Neumann's condition of perfect insulation. If $h \longrightarrow \infty$, then we get Dirichlet's condition.

2.3.7. Consider a one-dimensional rod of length H. Assume the rod has no lateral insulation, is nonuniform, and an internal heat source doesn't exist. Also, assume at the boundary $x = 0$, the rod is held at a constant temperature of $15°C$, and at the boundary $x = H$, the rod is imperfectly insulated, allowing a heat energy flow of a constant -8 watts. Develop the mathematical model that includes a possible initial temperature-distribution equation. Briefly explain your choice of initial temperature-distribution equation.

2.3.8. Show that $u(x, t) = e^{-2t} \sin x$ is a solution of

$$\frac{\partial u(x, t)}{\partial t} = 2\frac{\partial^2 u(x, t)}{\partial x^2},$$

subject to the BCs

$$u(0, t) = 0 \text{ and } u(2\pi, t) = 0,$$

and IC

$$u(x, 0) = \sin x.$$

2.3.9. Show that $u(x, t) = e^{-5t} \cos 3x$ is a solution of

$$\frac{\partial u(x, t)}{\partial t} = \frac{5}{9}\frac{\partial^2 u(x, t)}{\partial x^2},$$

subject to the BCs

$$\frac{\partial u(0, t)}{\partial x} = 0 \text{ and } \frac{\partial u\left(\frac{\pi}{3}, t\right)}{\partial x} = 0,$$

and IC

$$u(x, 0) = \cos 3x.$$

2.3.10. Solve $\dfrac{\partial^2 u(x, y)}{\partial x \partial y} = xy$.

Exercises (2.3.11 through 2.3.15) involve first-order ODEs. A review of this material may be found in Appendix C.

2.3.11. Carbon 14 obeys the law of radioactive decay. Determine k if the half-life of carbon 14 is 5568 years. Next, suppose the initial amount of carbon 14 is 9 gm. Determine the amount of carbon 14 left after 256 years.

2.3.12. Suppose a mothball loses volume by sublimation at a rate proportional to its surface area. Write an ODE to describe this phenomenon, making sure to define each term in your equation. Now, suppose the mothball's initial radius is 4 cm. If it takes 30 days for the radius to decrease 2 cm, how long will it take for the radius to decrease to 1 mm?

2.3.13. Suppose in a population of yeast cells, growing exponentially, the initial population of cells is 1200. Fifteen minutes later it is 1700. Find the growth rate for the population.

2.3.14. A cup of coffee is initially at boiling point, 100^0C. The temperature of the room is 20°C. Find the temperature of the coffee as a function of time. (*Hint:* use Newton's law of cooling.)

2.3.15. In a furnace, the temperature of the inner wall of an area 2m^2 is 450°C. The temperature of the outer wall is 80°C. There is a 0.5m of brick insulation (thermal conductivity of brick is 0.38) between the walls. How much heat escapes in three minutes? (*Hint:* Assume steady state has been reached across the walls and remember Fourier's law of heat conduction:
heat flow $= -K_0 \dfrac{du(x)}{dx} A$.)

2.4 THE MAXIMUM PRINCIPLE AND UNIQUENESS

So far, we can develop the heat equation for a one-dimensional rod, and we realize that we must have ICs and BCs to fully state the problem. However, we have no idea if we have enough initial data to have a solution for any time, t, or if the solution is unique, that is, is the problem well-posed (see Definition (2)). In this section, we address these issues and answer them. Since this text is primarily application oriented, we do not go into great depth.

Definition 2. *(Well-posed) A problem involving a partial differential equation is said to be well-posed if there exists a unique solution, and the solution depends continuously on the data of the problem.*

Suppose we are given

$$\frac{\partial u(x,t)}{\partial t} = k\frac{\partial^2 u(x,t)}{\partial x^2}, \tag{2.18}$$

subject to Dirichlet BCs

$$\begin{cases} u(0, t) = g(t) \\[2mm] u(L, t) = h(t) \end{cases} \tag{2.19}$$

and to IC

$$u(x, 0) = f(x). \tag{2.20}$$

We want to determine if a unique solution exists. Our intuition tells us that there is a unique solution. However, proving it is a slightly different matter. We start by stating a maximum–minimum theorem for diffusion of heat in a one-dimensional rod. Then we state and prove an immediate result of the maximum–minimum theorem, which shows that the problem given in Equations (2.18, 2.19, and 2.20) has an unique solution. Finally, we state another result of the maximum–minimum theorem, that of continuous dependence of the solution on the initial data. Thus, the problem is well-posed.

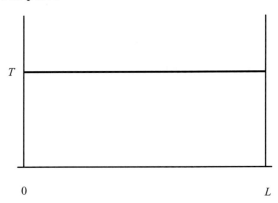

Figure 2.7: The domain of $u(x, t)$ in the rectangle $0 \le x \le L$ and $0 \le t \le T$.

Figure (2.7) is a two-dimensional diagram of heat flow in a one-dimensional uniform rod with perfect insulation and Dirichlet BCs. The vertical sides of the rectangle $0 \le x \le L$ and $0 \le t \le T$ refer to the BCs of the one-dimensional rod, which may be changing with respect to some function of time, t, throughout the experiment. The lower base of the rectangle indicates the IC of the one-dimensional rod at the start of the experiment.

Theorem 3. *(Maximum–Minimum theorem) Let T be an element of the real numbers such that $T > 0$. Suppose the function $u(x, t)$ is continuous in a closed rectangle, R, given by*

$$0 \le x \le L \text{ and } 0 \le t \le T,$$

as shown in Figure (2.7), and satisfies the heat equation given in Equation (2.18) in the interior of the rectangle. Then, $u(x, t)$ attains its maximum or minimum on *the base of the rectangle $t = 0$ or on the vertical sides of the rectangle $x = 0$ or $x = L$.*

A proof of the maximum–minimum theorem may be found in Appendix B. However, a common sense approach may give us some understanding of the theorem.

Fourier's heat conduction axioms tell us that heat flows from high to low. In the absence of a source or sink in the rod, the heat must flow from any hot spots to cooler spots within the rod or out of the ends. If the rod is being burned at one end and is in a freezer at the other end, the heat will flow from the burning end to the end in the freezer. However, the end that is being burned will still always be hotter than any other point of the rod and the end in the freezer will always be cooler than any other point on the rod.

We now proceed to Corollary (4), which tells us that if a solution exists, then it is unique.

Corollary 4. *(Uniqueness) There is at most one solution to the problem given in Equations (2.18, 2.19, and 2.20).*

Proof. Suppose $u_1(x, t)$ and $u_2(x, t)$ are both solutions to the problem given in Equations (2.18, 2.19, and 2.20). Let $w(x, t) = u_1(x, t) - u_2(x, t)$. Then, $w(x, t)$ satisfies Equation (2.18) with $w(x, 0) = 0$, and $w(0, t) = w(L, t) = 0$. Letting T be any number greater than zero, we have $w(x, t) = 0$ throughout the entire rectangle given in the maximum–minimum theorem. Thus, since T is arbitrary, we have $w(x, t) = u_1(x, t) - u_2(x, t) = 0$, which implies $u_1(x, t) = u_2(x, t)$. $\square$

Using a method called the **energy integral method**, it can be shown that

$$\frac{\partial u(x, t)}{\partial t} = k \frac{\partial^2 u(x, t)}{\partial x^2},$$

subject to Neumann BCs

$$\begin{cases} \dfrac{\partial u(0, t)}{\partial x} = g(t), \\[2mm] \dfrac{\partial u(L, t)}{\partial x} = h(t), \end{cases}$$

and to IC

$$u(x, 0) = f(x),$$

has a unique solution. However, that method is beyond the scope of this text.

Another result of the maximum–minimum theorem is continuous dependence of the solution on the data (both the boundary conditions and the initial conditions).

Corollary 5. *(Continuous dependence of the solution on the initial data) The solution of the problem given in Equations (2.18, 2.19, and 2.20) depends continuously on the initial and boundary conditions in the following way: let $u_1(x, t)$ and*

$u_2(x, t)$ *be solutions to the problem with initial data* $g_1(t)$, $h_1(t)$, *and* $f_1(x)$ *and* $g_2(t)$, $h_2(t)$, *and* $f_2(x)$, *respectively. Let* T *and* ε *be any positive real numbers. If*

$$\max_{0 \le x \le L} |f_1(x) - f_2(x)| \le \varepsilon,$$

$$\max_{0 \le t \le T} |g_1(t) - g_2(t)| \le \varepsilon, \quad and$$

$$\max_{0 \le t \le T} |h_1(t) - h_2(t)| \le \varepsilon,$$

then

$$\max_{0 \le x \le L, \, 0 \le t \le T} |u_1(x, t) - u_2(x, t)| \le \varepsilon.$$

Therefore, the maximum–minimum theorem and its two immediate corollaries provide for the well-posedness of the problem given in Equations (2.18, 2.19, and 2.20).

Equations (2.18, 2.19, and 2.20) describe the heat equation, sometimes called the diffusion equation, in a one-dimensional rod. From the preceeding discussion we know that the heat equation is well-posed. At this time, it is convenient to state the heat equation for multiple spatial dimensions for the Cartesian coordinate and in different coordinate systems. The heat equation in two and three spatial dimensions for Cartesian coordinates are given by Equations (2.21 and 2.22) respectively.

$$\frac{\partial u(x, y, t)}{\partial t} = k \left(\frac{\partial^2 u(x, y, t)}{\partial x^2} + \frac{\partial^2 u(x, y, t)}{\partial y^2} \right). \tag{2.21}$$

$$\frac{\partial u(x, y, z, t)}{\partial t} = k \left(\frac{\partial^2 u(x, y, z, t)}{\partial x^2} + \frac{\partial^2 u(x, y, z, t)}{\partial y^2} + \frac{\partial^2 u(x, y, z, t)}{\partial z^2} \right). \tag{2.22}$$

The heat equation in polar, cylindrical, and spherical coordinate systems are given by Equations (2.23, 2.24, and 2.25) respectively.

$$\frac{\partial u(r, \theta, t)}{\partial t} = k \left(\frac{1}{r} \frac{\partial}{\partial r} \left[r \frac{\partial u(r, \theta, t)}{\partial r} \right] + \frac{1}{r^2} \frac{\partial^2 u(r, \theta, t)}{\partial \theta^2} \right). \tag{2.23}$$

$$\frac{\partial u(r, \theta, z, t)}{\partial t} = k \left(\frac{1}{r} \frac{\partial}{\partial r} \left[r \frac{\partial u(r, \theta, z, t)}{\partial r} \right] + \frac{1}{r^2} \frac{\partial^2 u(r, \theta, z, t)}{\partial \theta^2} \right.$$

$$\left. + \frac{\partial^2 u(r, \theta, z, t)}{\partial z^2} \right). \tag{2.24}$$

$$\frac{\partial u(r, \theta, \phi, t)}{\partial t} = k \left(\frac{1}{r^2} \frac{\partial}{\partial r} \left[r^2 \frac{\partial u}{\partial r} \right] + \frac{1}{r^2 \sin \theta} \left[\sin \theta \frac{\partial^2 u}{\partial \theta^2} \right] \right.$$

$$\left. + \frac{1}{r^2 \sin^2 \theta} \frac{\partial^2 u}{\partial \phi^2} \right). \tag{2.25}$$

The right side of Equations (2.21, 2.22, 2.23, 2.24, and 2.25) is called the Laplacian, and it is given in many equations as $\nabla^2 u$ where the arguments of u are understood from the nature of the problem.

We will state and use Equations (2.21, 2.22, 2.23, 2.24, and 2.25) in later chapters in the text. However, you should know how to derive each of the equations from the Cartesian counterpart. For instance, the heat equation in polar coordinates, Equation (2.23), may be derived from the Cartesian two spatial dimension heat equation, Equation (2.21). A demonstration of the derivation is given in Example (2.3).

EXAMPLE 2.3. Derive the polar form,

$$\frac{\partial u(r,\theta,t)}{\partial t} = k \left(\frac{1}{r} \frac{\partial}{\partial r} \left[r \frac{\partial u(r,\theta,t)}{\partial r} \right] + \frac{1}{r^2} \frac{\partial^2 u(r,\theta,t)}{\partial \theta^2} \right),$$

of the heat equation from the two spatial dimension Cartesian form,

$$\frac{\partial u(x,y,t)}{\partial t} = k \left(\frac{\partial^2 u(x,y,t)}{\partial x^2} + \frac{\partial^2 u(x,y,t)}{\partial y^2} \right).$$

Solution: Remembering that $x = r\cos\theta$ and $y = r\sin\theta$ we have $r = \sqrt{x^2 + y^2}$ and $\theta = \tan^{-1}\left(\frac{y}{x}\right)$. Therefore,

$$\frac{\partial r}{\partial x} = \frac{x}{\sqrt{x^2+y^2}} = \cos\theta, \text{ which implies } \frac{\partial^2 r}{\partial x^2} = \frac{y^2}{(x^2+y^2)^{3/2}} = \frac{\sin^2\theta}{r}.$$

Similarly,

$$\frac{\partial r}{\partial y} = \sin\theta, \text{ which implies } \frac{\partial^2 r}{\partial y^2} = \frac{\cos^2\theta}{r}.$$

Also,

$$\frac{\partial \theta}{\partial x} = -\frac{\sin\theta}{r}. \text{ Therefore, } \frac{\partial^2 \theta}{\partial x^2} = \frac{2\cos\theta\sin\theta}{r^2},$$

and

$$\frac{\partial \theta}{\partial y} = \frac{\cos\theta}{r}. \text{ Thus, } \frac{\partial^2 \theta}{\partial y^2} = -\frac{2\cos\theta\sin\theta}{r^2}.$$

We are now ready to find the polar coordinate form of the heat equation. Using the chain rule, we have

$$\frac{\partial u}{\partial x} = \frac{\partial u}{\partial r}\frac{\partial r}{\partial x} + \frac{\partial u}{\partial \theta}\frac{\partial \theta}{\partial x},$$

which means

$$\frac{\partial^2 u}{\partial x^2} = \frac{\partial}{\partial x}\left[\frac{\partial u}{\partial r}\frac{\partial r}{\partial x} + \frac{\partial u}{\partial \theta}\frac{\partial \theta}{\partial x}\right] = \left(\frac{\partial^2 u}{\partial r^2}\frac{\partial r}{\partial x} + \frac{\partial^2 u}{\partial \theta r}\frac{\partial \theta}{\partial x}\right)\frac{\partial r}{\partial x} + \frac{\partial u}{\partial r}\frac{\partial^2 r}{\partial x^2}$$

$$+ \left(\frac{\partial^2 u}{\partial \theta^2}\frac{\partial \theta}{\partial x} + \frac{\partial^2 u}{\partial r\theta}\frac{\partial r}{\partial x}\right)\frac{\partial \theta}{\partial x} + \frac{\partial u}{\partial \theta}\frac{\partial^2 \theta}{\partial x^2}$$

$$= \frac{\partial^2 u}{\partial r^2}\cos^2\theta - \frac{\partial^2 u}{\partial r\theta}\frac{2\sin\theta\cos\theta}{r} + \frac{\partial^2 u}{\partial \theta^2}\frac{\sin^2\theta}{r^2} + \frac{\partial u}{\partial r}\frac{sin^2\theta}{r} + \frac{\partial u}{\partial \theta}\frac{2\cos\theta\sin\theta}{r^2}.$$

Similarly,

$$\frac{\partial^2 u}{\partial y^2} = \frac{\partial^2 u}{\partial r^2}\sin^2\theta + \frac{\partial^2 u}{\partial r\theta}\frac{2\sin\theta\cos\theta}{r} + \frac{\partial^2 u}{\partial \theta^2}\frac{\cos^2\theta}{r^2} + \frac{\partial u}{\partial r}\frac{cos^2\theta}{r} - \frac{\partial u}{\partial \theta}\frac{2\cos\theta\sin\theta}{r^2}.$$

Thus,

$$\frac{\partial^2 u}{\partial x^2} + \frac{\partial^2 u}{\partial y^2} = \frac{\partial^2 u}{\partial r^2}\cos^2\theta - \frac{\partial^2 u}{\partial r\theta}\frac{2\sin\theta\cos\theta}{r} + \frac{\partial^2 u}{\partial \theta^2}\frac{\sin^2\theta}{r^2} + \frac{\partial u}{\partial r}\frac{\sin^2\theta}{r}$$

$$+ \frac{\partial u}{\partial \theta}\frac{2\cos\theta\sin\theta}{r^2} + \frac{\partial^2 u}{\partial r^2}\sin^2\theta + \frac{\partial^2 u}{\partial r\theta}\frac{2\sin\theta\cos\theta}{r}$$

$$+ \frac{\partial^2 u}{\partial \theta^2}\frac{\cos^2\theta}{r^2} + \frac{\partial u}{\partial r}\frac{\cos^2\theta}{r} - \frac{\partial u}{\partial \theta}\frac{2\cos\theta\sin\theta}{r^2} = \frac{\partial^2 u}{\partial r^2} + \frac{1}{r}\frac{\partial u}{\partial r} + \frac{1}{r^2}\frac{\partial^2 u}{\partial \theta^2}$$

$$= \frac{1}{r}\frac{\partial}{\partial r}\left[r\frac{\partial u}{\partial r}\right] + \frac{1}{r^2}\frac{\partial^2 u}{\partial \theta^2}.$$

Hence, we have

$$\frac{\partial u(r,\theta,t)}{\partial t} = k\left(\frac{1}{r}\frac{\partial}{\partial r}\left[r\frac{\partial u(r,\theta,t)}{\partial r}\right] + \frac{1}{r^2}\frac{\partial^2 u(r,\theta,t)}{\partial \theta^2}\right).$$

In the exercises, you must derive the cylindrical and spherical form of the Laplacian.

EXERCISES 2.4

2.4.1. Prove Corollary (5).

2.4.2. Show that $A\cos\omega_0 t + B\sin\omega_0 t$ can be written in the form $r\sin(\omega_0 t + \theta)$. Determine r and θ in terms of A and B. (*Hint:* Use the trigonometric angle–sum relations.)

2.4.3. The Laplacian in the Cartesian coordinate system is defined as

$$\nabla^2 u(x,y,z) = \frac{\partial^2 u}{\partial x^2} + \frac{\partial^2 u}{\partial y^2} + \frac{\partial^2 u}{\partial z^2}.$$

(1) If $x = r\cos\theta$, $y = r\sin\theta$, and $z = z$, show that the Laplacian can be written in the cylindrical coordinate system as

$$\nabla^2 u(r, \theta, z) = \frac{1}{r}\frac{\partial}{\partial r}\left(r\frac{\partial u}{\partial r}\right) + \frac{1}{r^2}\frac{\partial^2 u}{\partial \theta^2} + \frac{\partial^2 u}{\partial z^2}.$$

(2) If $x = r\sin\theta\cos\phi$, $y = r\sin\theta\sin\phi$, and $z = r\cos\theta$, show that the Laplacian can be written in the spherical coordinate system as

$$\nabla^2 u(r, \theta, \phi) = \frac{1}{r^2}\frac{\partial}{\partial r}\left(r^2\frac{\partial u}{\partial r}\right) + \frac{1}{r^2\sin\theta}\left(\sin\theta\frac{\partial^2 u}{\partial \theta^2}\right)$$

$$+\frac{1}{r^2\sin^2\theta}\frac{\partial^2 u}{\partial \phi^2}.$$

2.4.4. (1) Given the cylindrical coordinates

$$x = \rho\cos\phi,$$

$$y = \rho\sin\phi, \text{ and}$$

$$z = z;$$

$\rho^2 = x^2 + y^2$ and $\phi = \tan^{-1}(\frac{y}{x})$, show that

(a) $\dfrac{\partial \rho}{\partial x} = \cos\phi$.

(b) $\dfrac{\partial \rho}{\partial y} = \sin\phi$.

(c) $\dfrac{\partial \phi}{\partial x} = \dfrac{-\sin\phi}{\rho}$.

(d) $\dfrac{\partial \phi}{\partial y} = \dfrac{\cos\phi}{\rho}$.

(2) Given $\vec{r} = \rho\cos\phi\vec{i} + \rho\sin\phi\vec{j} + z\vec{k}$. Show that the square of the element of arc length is

$$(ds)^2 = d\vec{r}d\vec{r} = (d\rho)^2 + \rho^2(d\phi)^2 + (dz)^2.$$

2.4.5. Given the cylindrical coordinates

$$x = \rho\cos\phi,$$

$$y = \rho\sin\phi, \text{ and}$$

$$z = z;$$

$\rho^2 = x^2 + y^2$ and $\phi = \tan^{-1}(\frac{y}{x})$, prove that the cylindrical coordinate system is orthogonal.

2.4.6.   (1) Given the spherical coordinates

$$x = \rho \sin \theta \cos \phi,$$

$$y = \rho \sin \theta \sin \phi, \text{ and}$$

$$z = \rho \cos \theta;$$

$\rho^2 = x^2 + y^2 + z^2$ and $\phi = \tan^{-1}\left(\dfrac{y}{x}\right)$, show that

(a) $\dfrac{\partial \rho}{\partial x} = \sin \theta \cos \phi.$

(b) $\dfrac{\partial \rho}{\partial y} = \sin \theta \sin \phi.$

(c) $\dfrac{\partial \rho}{\partial z} = \cos \theta.$

(d) $\dfrac{\partial \phi}{\partial x} = \dfrac{-\sin \phi}{\rho \sin \theta}.$

(e) $\dfrac{\partial \phi}{\partial y} = \dfrac{\cos \phi}{\rho \sin \theta}.$

(2) Given $\vec{r} = \rho \sin \theta \cos \phi \vec{i} + \rho \sin \theta \sin \phi \vec{j} + \rho \cos \theta \vec{k}$, show that the square of the element of arc length is

$$(ds)^2 = d\vec{r} d\vec{r} = (d\rho)^2 + \rho^2 (d\theta)^2 + \rho^2 \sin^2 \theta (d\phi)^2.$$

2.4.7. Given the parabolic cylindrical coordinates

$$x = \tfrac{1}{2}(\alpha^2 - \beta^2),$$

$$y = \alpha\beta, \text{ and}$$

$$z = z;$$

show that the square of the element of arc length is

$$(ds)^2 = (\alpha^2 + \beta^2)(d\alpha)^2 + (\alpha^2 + \beta^2)(d\beta)^2 + (dz)^2.$$

2.5   STEADY-STATE TEMPERATURE DISTRIBUTION

From the previous two sections we know how to set up the mathematical model for heat conduction in a uniform rod with perfect lateral insulation, no internal sources, specified temperatures at both ends, and length L. This mathematical model is

$$\frac{\partial u(x,t)}{\partial t} = k \frac{\partial^2 u(x,t)}{\partial x^2}, \tag{2.26}$$

subject to the BCs

$$\begin{cases} u(0,t) = g(t) \\ u(L,t) = h(t) \end{cases} \tag{2.27}$$

and to IC

$$u(x,0) = f(x). \tag{2.28}$$

Now, suppose $g(t)$ and $h(t)$ are the fixed constants a and b, respectively. Then, the solution to this model will simplify to a steady-state temperature distribution, independent of time t. This solution is known as the equilibrium solution.

To understand what we mean by steady-state, think of a uniform rod with perfect lateral insulation and no internal sources. Now, apply a constant temperature forever to both ends. Since neither end is insulated, eventually the temperature in the rod will adjust to the temperature distribution specified by the heat sources at both ends.

Definition 6. *A **steady-state** temperature distribution is a temperature distribution that does not depend on time.*

From Definition (6), $u(x,t)$ from Equation (2.26) becomes $u(x)$. Thus, $\dfrac{\partial u(x,t)}{\partial t} = \dfrac{\partial u(x)}{\partial t} = 0$ and $k\dfrac{\partial^2 u(x,t)}{\partial x^2} = k\dfrac{d^2 u(x)}{dx^2} = 0$ or

$$\frac{d^2 u(x)}{dx^2} = 0, \tag{2.29}$$

with BCs

$$u(0) = a \text{ and } u(L) = b. \tag{2.30}$$

Since the IC is concerned with the temperature distribution at time $t = 0$ and steady-state is time independent, the IC is generally ignored.

Equation (2.29) is a rather simple second-order ordinary differential equation. By integrating it twice, we arrive at the general solution

$$u(x) = C_1 x + C_2. \tag{2.31}$$

Applying the BC $u(0) = a$ in Equation (2.31), we get $C_2 = a$. Applying the second boundary condition, $u(L) = b$, we get $C_1 = \dfrac{b-a}{L}$. Thus our specific solution is

$$u(x) = \frac{b-a}{L}x + a. \tag{2.32}$$

In this problem, the steady-state temperature distribution is a straight line with slope $\dfrac{b-a}{L}$ and $u(x)$ intercept of a. Figure (2.8) graphically illustrates $u(x)$.

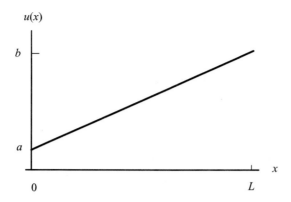

Figure 2.8: Steady-state temperature distribution with $u(0) = a$ and $u(L) = b$.

What would happen if we changed Equation (2.26) so that it had a time-independent source term, $Q(x)$, and we still wanted to solve for the steady-state solution? Equation (2.26) becomes

$$\frac{\partial u(x,t)}{\partial t} = k\frac{\partial^2 u(x,t)}{\partial x^2} + Q(x).$$

Would the solution, Equation (2.32), change? How would this effect Figure (2.8)? These questions are important to answer for a fuller understanding of this steady-state problem; they'll be asked again in the exercises at the end of the section.

Next, suppose we want to find the steady-state temperature distribution in a uniform one-dimensional rod with no internal source, perfect lateral insulation, and perfect insulation on the boundaries. We again start with Equation (2.29)

$$\frac{d^2 u(x)}{dx^2} = 0.$$

However, the BCs change to

$$\frac{du(0)}{dx} = 0 = \frac{du(L)}{dx}.$$

Again, the general solution to the ODE is

$$u(x) = C_1 x + C_2.$$

Applying both the first and second boundary condition yields

$$\frac{du(0)}{dx} = 0 = C_1.$$

This implies $u(x) = C_2$. It appears that steady-state temperature in a uniform rod, with perfect insulation everywhere, is an arbitrary constant. But is it arbitrary? No.

In this case, it turns out that the initial condition of the rod, $u(x,0) = f(x)$, plays an important part. The energy associated with the initial temperature distribution cannot escape because of the perfect insulation. Since temperature distribution in the rod must follow the three heat conduction axioms, the temperature distribution levels out to a constant. But what constant? This question can only be answered by going back to Equation (2.8), which is the mathematical representation of conservation of thermal energy. It states that for any thin slice of rod $[a, b]$,

$$\frac{d}{dt}\left[\int_a^b c\rho A u(x,t)dx\right] = -K_0 A\left[\frac{\partial u(a,t)}{\partial x} - \frac{\partial u(b,t)}{\partial x}\right]$$

$$+ \int_a^b AQ(x,t)dx.$$

This equation is valid for the entire rod, not just the thin slice $[a, b]$; therefore, it can be written as

$$\frac{d}{dt}\left[\int_0^L c\rho A u(x,t)dx\right] = -K_0 A\left[\frac{\partial u(0,t)}{\partial x} - \frac{\partial u(L,t)}{\partial x}\right]$$

$$+ \int_0^L AQ(x,t)dx. \tag{2.33}$$

Because there is no internal source, $Q(x,t) = 0$. Also, we have a uniform rod, which means c, ρ, and K_0 are constant and can be written as $k = \dfrac{K_0}{c\rho}$. Thus, Equation (2.33) may be written as

$$\frac{d}{dt}\left[\int_0^L u(x,t)dx\right] = -k\left[\frac{\partial u(0,t)}{\partial x} - \frac{\partial u(L,t)}{\partial x}\right]. \tag{2.34}$$

We can further reduce Equation (2.34) by remembering that both ends are perfectly insulated, which means $\dfrac{\partial u(0,t)}{\partial x} = 0$ and $\dfrac{\partial u(L,t)}{\partial x} = 0$. Thus, Equation (2.34) becomes

$$\int_0^L u(x,t)dx = \text{constant}.$$

This tells us that the total initial heat energy inside the rod must equal the total final heat energy inside the rod. We did **not** say that the initial temperature distribution is the same as the final temperature distribution. Since $u(x,0) = f(x)$, the total initial heat energy is

$$\int_0^L u(x,0)dx = \int_0^L f(x)dx.$$

Also, since $u(x) = C_2$, the total steady-state heat energy in the rod (the total final heat energy) is

$$\int_0^L C_2 dx = C_2 L.$$

Setting total initial heat energy equal to total final heat energy and solving for C_2, we arrive at

$$C_2 = \frac{1}{L} \int_0^L f(x) \, dx,$$

which is the average of the initial temperature distribution.

As a final example, consider the following problem:

EXAMPLE 2.4. Consider a one-dimensional uniform rod with perfect lateral insulation, internal heat source of $25 \cos x$, thermal diffusivity of 5, and initial temperature distribution of $25 \cos x$. At the boundary $x = 0$, the rod is in direct contact with a bath held at the constant temperature of $25°C$, and at the boundary $x = \pi$, the rod has perfect insulation. Determine and graph the steady-state temperature distribution.

Solution: First, describe the above physical problem as a mathematical model. We have

$$\frac{\partial u(x,t)}{\partial t} = 5 \frac{\partial^2 u(x,t)}{\partial x^2} + 25 \cos x,$$

subject to the IC

$$u(x,0) = 25 \cos x$$

and BCs

$$\begin{cases} u(0,t) = 25 \\ \dfrac{\partial u(\pi,t)}{\partial x} = 0. \end{cases}$$

Second, state and solve the steady-state problem. The steady-state problem is

$$5 \frac{d^2 u(x)}{dx^2} + 25 \cos x = 0,$$

subject to BCs

$$\begin{cases} u(0) = 25 \\ \dfrac{du(\pi)}{dx} = 0. \end{cases}$$

The steady-state problem has the general solution

$$u(x) = 5\cos x + C_1 x + C_2.$$

Applying the BC $u(0) = 25$, implies $C_2 = 20$. Applying the BC $\dfrac{du(\pi)}{dx} = 0$, implies $C_1 = 0$. Thus, the specific steady-state solution is

$$u(x) = 5\cos x + 20.$$

Figure (2.9) is the graph of the steady-state temperature distribution.　■

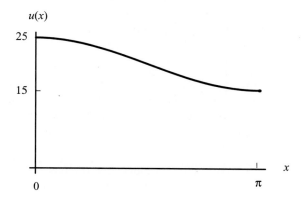

Figure 2.9: Graph of $5\cos x + 20$.

EXERCISES 2.5

2.5.1. Given a one-dimensional uniform rod with perfect lateral insulation, determine and graph the steady-state heat distribution given the following boundary conditions, source, and thermal diffusivity:

(1)　　$u(0) = 0,$　　　$u(L) = a,$　　　$Q(x) = 0,$　　$k = 2.$

(2)　　$u(0) = b,$　　　$u(L) = 0,$　　　$Q(x) = x,$　　$k = 1.$

(3)　　$u'(0) = 0,$　　　$u(2\pi) = a,$　　　$Q(x) = 0,$　　$k = 1.$

(4)　　$u(0) = 0,$　　　$u'(L) = 0,$　　　$Q(x) = x^2,$　　$k = 5.$

(5)　　$u(0) = u'(0),$　　$u(\pi) = 15,$　　　$Q(x) = 0,$　　$k = 1.$

(6)　　$u(0) = 2,$　　　$u(L) = u'(L),$　　$Q(x) = -x,$　　$k = 2.$

2.5.2. Consider the mathematical model

$$\frac{\partial u(x,t)}{\partial t} = k\frac{\partial^2 u(x,t)}{\partial x^2} + \alpha u(x,t); \quad \alpha > 0, \quad k > 0,$$

subject to the BCs

$$u(0,t) = 1 \text{ and } \frac{\partial u(\pi,t)}{\partial x} = -1$$

and IC

$$u(x,0) = \cos x - x.$$

(1) Write a short paragraph (five sentences or less) describing a possible physical model.

(2) Determine the steady-state temperature distribution.

2.5.3. Determine the steady-state solution for a uniform rod with perfect lateral insulation if the boundary at $x = 0$ is kept at a constant temperature of $-10^0 C$, the boundary at $x = 100$ cm is kept at a constant temperature of $15^0 C$, and there exists a time-independent heat source, which is a linear function based on the position in the rod.

2.5.4. Find the steady-state solution for a uniform one-dimensional rod with no internal source and perfect lateral insulation that satisfies the radiation condition

$$\frac{\partial u(0,t)}{\partial x} - u(0,t) = 0$$

at the end $x = 0$ and is kept at a constant temperature T_2 at the end $x = L$.

2.5.5. Find the steady-state solution for a uniform one-dimensional rod with an internal source of $Q(x,t) = x$, perfect lateral insulation, that is kept at a constant temperature T_1 at the end $x = 0$, and satisfies the radiation condition

$$\frac{\partial u(L,t)}{\partial x} - u(L,t) = 0$$

at the end $x = L$.

2.5.6. Consider the mathematical model

$$(\cos \pi x)\frac{\partial u(x,t)}{\partial t} = e^x \frac{\partial^2 u(x,t)}{\partial x^2} + \frac{t^2 e^{-x}}{t^2 + 1},$$

subject to the BCs

$$u(0,t) = 0 \text{ and } \frac{\partial u(1,t)}{\partial x} = 1$$

and IC

$$u(x,0) = x^2.$$

(1) Write a short paragraph (five sentences or less) describing a possible physical model.

(2) Determine the steady-state temperature distribution.

2.5.7. Consider

$$\frac{\partial u(x,t)}{\partial t} = \frac{\partial^2 u(x,t)}{\partial x^2} + x - \beta,$$

subject to the BCs

$$\frac{\partial u(0,t)}{\partial x} = 0 \text{ and } \frac{\partial u(2,t)}{\partial x} = 0$$

and IC

$$u(x,0) = \cos \frac{\pi x}{2}.$$

(1) Find the equilibrium temperature distribution.

(2) For what values of β does the equilibrium temperature distribution exist? Explain physically.

2.5.8. Suppose you were given

$$\frac{\partial u(x,t)}{\partial t} = 9 \frac{\partial^2 u(x,t)}{\partial x^2} + 9x,$$

subject to the BCs

$$u(0,t) = 0 \text{ and } u(\pi,t) = 0$$

and IC

$$u(x,0) = e^{-x} \sin 3x.$$

Write a technical report, including explanations and mathematical details, which contains at the least the following:

(1) a physical interpretation of the previous problem and

(2) for very large time:

    i. the solution,

    ii. the heat energy generated per unit time inside the rod,

    iii. the heat energy flowing out of the rod per unit time at each end, and

    iv. the relationship between parts (b) and (c).

2.5.9. Suppose you were given

$$\frac{\partial u(x,t)}{\partial t} = \frac{\partial^2 u(x,t)}{\partial x^2} + \frac{\pi^2 u(x,t)}{4},$$

subject to the BCs

$$u(0,t) = 0 \text{ and } u(6,t) = 0$$

and IC

$$u(x,0) = \pi \left(6x - x^2\right).$$

Write a technical report, including explanations and mathematical details, which contains at the least the following:

(1) a physical interpretation of the above problem and

(2) for very large time:

   i. the solution,

   ii. the heat energy generated per unit time inside the rod,

   iii. the heat energy flowing out of the rod per unit time at each end, and

   iv. the relationship between parts (b) and (c).

2.5.10. Suppose you were given

$$\frac{\partial u(x,t)}{\partial t} = \frac{\partial^2 u(x,t)}{\partial x^2} + e^x,$$

subject to the BCs

$$u(0,t) = 0 \text{ and } u(\pi,t) = 0$$

and IC

$$u(x,0) = \pi \sin x.$$

Write a technical report, including explanations and mathematical details, which contains, at the least the following:

(1) a physical interpretation of the above problem and

(2) for very large time:

   i. the solution,

   ii. the heat energy generated per unit time inside the rod,

   iii. the heat energy flowing out of the rod per unit time at each end, and

   iv. the relationship between parts (b) and (c).

2.5.11. Consider a uniform one-dimensional rod with no internal source and perfect lateral insulation. Assume that $u(x, 0) = f(x)$, and suppose the following boundary conditions are given:

$$u(\pi, t) = u(-\pi, t) \text{ and } \frac{\partial u(\pi, t)}{\partial x} = \frac{\partial u(-\pi, t)}{\partial x}.$$

(1) State a possible physical explanation. Briefly justify your answer.

(2) Does a steady-state solution exist? If so, what is the steady state solution? Explain your answer. If a steady-state solution does not exist, explain why.

2.5.12. Given the Cauchy-Riemann equations

$$\begin{cases} \dfrac{\partial u}{\partial x} - \dfrac{\partial v}{\partial y} = 0 \\[2mm] \dfrac{\partial u}{\partial y} + \dfrac{\partial v}{\partial x} = 0, \end{cases}$$

show that both u and v satisfy Laplace's equation

$$\frac{\partial^2 z}{\partial x^2} + \frac{\partial^2 z}{\partial y^2} = 0,$$

which may be considered as the multi-dimensional steady-state equation. Laplace's equation is discussed in Chapter 5.

2.5.13. Fick's law of diffusion for chemical species in mathematical physiology is given by

$$Q = -D\nabla c,$$

where Q is the flow of a chemical across a membrane, D is the diffusion coefficient and is dependent on the solute and the fluid in which the chemical is dissolved, and c is the amount of chemical. Fick's law can be used to derive an analogue of Ohm's law for a membrane of thickness, L, with different chemical concentrations on each side of the membrane. If the medium is isotropic (diffusion occurs the same regardless of the direction of the measurement), then we get

$$\frac{\partial c}{\partial t} = D\frac{\partial^2 c}{\partial x^2},$$

subject to

$c(0, t) = C_l$, the chemical concentration on the left of the membrane,

and

$c(L, t) = C_r$, the chemical concentration on the right of the membrane.

Find the steady-state solution.[7]

---

[7] James Keener and James Sneyd, *Mathematical Physiology*, ©1998 by Springer-Verlag, New York, pp. 36-38. Reprinted by permission.

Chapter 3

The One-Dimensional Wave Equation

3.1 INTRODUCTION

The study of wave equations covers a wide range of physical problems. For instance, the wave equation that governs vibration of a microphone diaphragm is

$$\frac{\partial^2 u}{\partial x^2} + \frac{1}{x}\frac{\partial u}{\partial x} - \left(a\frac{\partial^2 u}{\partial t^2} + b\frac{\partial u}{\partial t} \right) = 0,$$

where $u(x,t)$ is the displacement of a diaphragm in a capacitor microphone. Other physical examples are the electromagnetic waves in a transmission line, wave motion in an ocean, and vibrations in a beam. In later chapters, we will develop equations and solutions for some of these physical examples. However in this chapter, we consider the most basic wave equation: vibration of a one-dimensional string.

We derive the wave equation for vibration in a one-dimensional string in Section (3.2). Boundary conditions (BCs) for the one-dimensional wave equation are discussed in Section (3.3). Section (3.4) covers conservation of energy for the wave equation and uniqueness of solution when using Neumann boundary conditions. We conclude this chapter with the method of characteristics for first-order PDEs and d'Alembert's[1] solution for the one-dimensional wave equation.

3.2 DERIVATION OF THE ONE-DIMENSIONAL WAVE EQUATION

When you think of vibrations in a one-dimensional string, you should ask yourself, "Is the string vertical or horizontal, tightly stretched or loose?" There may be a difference in the derivation of the equations. In fact, there are no vibrations in

---

[1] Jean Le Rond d'Alembert (1717–1783) was the first to develop a form of the wave equation

a string that is not tightly stretched. Gravity could play a completely different role in a string that is vertical versus a string that is horizontal. In this section, we restrict ourselves to a tightly stretched horizontal string. Also, we make the following assumptions:

- Vibrations of the string are small. This means that as the string vibrates, we have a very small change in the slope of the string from the "at rest" position, which is horizontal.

- Only vertical vibrations will be considered—horizontal vibrations can be neglected because of the small slope.

- Vertical displacement then depends on the position on the string, along with time, and can be modeled as $y = u(x,t)$. Thus, the slope is represented by $\dfrac{dy}{dx} = \dfrac{\partial u}{\partial x} = \tan[\theta(x,t)]$.

- The string is perfectly flexible. It offers no resistance to bending.

- Newton's law of motion, $F = ma$ (which we will write as $ma = F$), is applied to a small section of the string (x to $x + \triangle x$).

Consider Figure (3.1), which is out of proportion for labeling purposes:

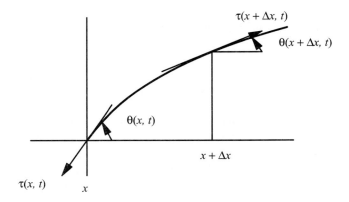

Figure 3.1: Finite string segment.

Here τ, a tangential force, represents the tension in the string, and θ is the angle of the displacement of the string from the horizontal at the ends x and $x + \triangle x$.

It is unknown whether the string is uniform. Therefore, we represent the mass of the string as a function. However, we only expect the mass of the string to change with position, x, not with time, t. Thus, we chose $\rho(x)$ to model mass, m, per unit length and assume it is a known quantity.

In addition to mass, we must consider other forces on the string. One example is the **restoring force.** This is the force that tries to return the string to its at-rest position. This force models the added tension on the string caused by vertical displacement. Another example is the **resistance force** to the string's velocity. The medium (air, water, oil, etc.) that the string vibrates in causes this force, which attempts to slow any motion.

Using our assumptions, plus the discussion on forces and mass, we can now give a fairly good mathematical model of the one-dimensional wave equation. Using Newton's law of motion we arrive at

$$(\triangle x)\rho(x)\frac{\partial^2 u}{\partial t^2} = \tau(x + \triangle x, t) \sin\left[\theta(x + \triangle x, t)\right] - \tau(x, t) \sin\left[\theta(x, t)\right]$$

$$-(\triangle x)\alpha\frac{\partial u}{\partial t} - (\triangle x)\beta u + (\triangle x)Q(x, t). \qquad (3.1)$$

Mass times acceleration is represented by the term $(\triangle x)\rho(x)\dfrac{\partial^2 u}{\partial t^2}$. The right side of Equation (3.1) is the sum of forces. We have two terms for tension: one for the left end at x and one for the right end at $x + \triangle x$. Since only vertical vibrations are considered, the terms $\tau(x + \triangle x, t) \sin\left[\theta(x + \triangle x, t)\right]$ and $\tau(x, t) \sin\left[\theta(x, t)\right]$ are the vertical components of the tensile force. Assuming that $\alpha > 0$, the term $(\triangle x)\alpha\dfrac{\partial u}{\partial t}$ models the resistance force. $(\triangle x)\beta u$ with $\beta > 0$ models the restoring force. Other possible external forces, such as gravity, are modeled as $(\triangle x)Q(x, t)$.

Dividing Equation (3.1) by $\triangle x$ and taking the limit as $\triangle x \to 0$, we get

$$\rho(x)\frac{\partial^2 u}{\partial t^2} = \frac{\partial}{\partial x}\left(\tau(x, t) \sin\left[\theta(x, t)\right]\right) - \alpha\frac{\partial u}{\partial t} - \beta u + Q(x, t). \qquad (3.2)$$

In our assumptions, the slope was

$$\frac{dy}{dx} = \frac{\partial u}{\partial x} = \tan\left[\theta(x, t)\right].$$

When the angle θ is small ($\theta \approx 0$), which is another of our assumptions, we have

$$\frac{dy}{dx} = \frac{\partial u}{\partial x} = \tan\left[\theta(x, t)\right] = \frac{\sin\left[\theta(x, t)\right]}{\cos\left[\theta(x, t)\right]} \approx \sin\left[\theta(x, t)\right]$$

or

$$\frac{dy}{dx} = \frac{\partial u}{\partial x} \approx \sin\left[\theta(x, t)\right]. \qquad (3.3)$$

Replacing $\sin\left[\theta(x, t)\right]$ by $\dfrac{\partial u}{\partial x}$ in Equation (3.2) yields

$$\rho(x)\frac{\partial^2 u}{\partial t^2} = \frac{\partial}{\partial x}\left(\tau(x, t)\frac{\partial u}{\partial x}\right) - \alpha\frac{\partial u}{\partial t} - \beta u + Q(x, t). \qquad (3.4)$$

Equation (3.4) is the mathematical model for small vibrations in a small piece of perfectly flexible string that is horizontal and tightly stretched.

If we make the assumption that the string is perfectly elastic (a valid assumption for most strings), then the tensile force $\tau(x, t)$ may be approximated by the constant

τ, which is the initial tension on the unperturbed string. Also, if we assume the string is made of a uniform material, then the mass density $\rho(x)$ becomes the constant ρ and Equation (3.4) becomes

$$\rho\frac{\partial^2 u}{\partial t^2} = \tau\frac{\partial^2 u}{\partial x^2} - \alpha\frac{\partial u}{\partial t} - \beta u + Q(x,t).$$

Another assumption we can make is that the term $Q(x,t)$ only models gravity. If we know the tension force is high compared to the force of gravity, we may neglect gravity. Doing so yields

$$\rho\frac{\partial^2 u}{\partial t^2} = \tau\frac{\partial^2 u}{\partial x^2} - \alpha\frac{\partial u}{\partial t} - \beta u, \tag{3.5}$$

which is a form of the well-known Telegrapher's equation.

As its name implies, the Telegrapher's equation models electromagnetic wave transmission in a wire. Also, it is an important equation in analyzing the time-dependent Boltzmann's equation in the theory of neutron transport. We discuss solutions to the Telegrapher's equation in a later chapter.

To simplify Equation (3.5) we may, for the moment, neglect the forces of friction and restoration. Mathematically, this means we assume that α and β are zero. Thus, the mathematical model for vibrations in a one-dimensional, uniform, perfectly flexible, highly stretched string is

$$\rho\frac{\partial^2 u}{\partial t^2} = \tau\frac{\partial^2 u}{\partial x^2} \tag{3.6}$$

or in its more usual form

$$\frac{\partial^2 u}{\partial t^2} = c^2\frac{\partial^2 u}{\partial x^2}, \tag{3.7}$$

where $c^2 = \dfrac{\tau}{\rho}$. Because τ is tension and ρ is mass per unit length, c has the dimension of length/time, known as velocity. Actually, c is the specific velocity of wave propagation along the string.

The one-dimensional wave equation applies to many different physical systems. For example,

$$\frac{\partial^2 u}{\partial t^2} = k\frac{\partial^2 u}{\partial x^2}$$

is the mathematical model of longitudinal or torsional vibrations in a rod. Here, the constant k is a physical parameter known as the Young's modulus.

Another example is electrical current along a wire. Kirchoff's[2] laws give us

$$\frac{\partial i}{\partial x} + C\frac{\partial v}{\partial t} + Gv = 0 \tag{3.8}$$

---

[2]Gustav Kirchoff (1824–1887) was a physicist who obtained significant results in the study of PDEs

and

$$\frac{\partial v}{\partial x} + L\frac{\partial i}{\partial t} + Ri = 0, \tag{3.9}$$

where the variables have the following meaning:

- x is the location along the wire.

- t is time.

- $i(x,t)$ is the current along the wire.

- $v(x,t)$ is the potential along the wire.

- C is the capacitance.

- G is the leakage conductance.

- R is the resistance.

- L is the self-inductance.

Using Equations (3.8 and 3.9), we can derive a form of the wave equation by differentiating Equation (3.8) with respect to x and differentiating Equation (3.9) with respect to t. This yields

$$\frac{\partial^2 i}{\partial x^2} + C\frac{\partial^2 v}{\partial x \partial t} + G\frac{\partial v}{\partial x} = 0 \tag{3.10}$$

and

$$\frac{\partial^2 v}{\partial t \partial x} + L\frac{\partial^2 i}{\partial t^2} + R\frac{\partial i}{\partial t} = 0. \tag{3.11}$$

Multiplying Equation (3.11) by C, then subtracting it from Equation (3.10), yields

$$\frac{\partial^2 i}{\partial x^2} + C\left(\frac{\partial^2 v}{\partial x \partial t} - \frac{\partial^2 v}{\partial t \partial x}\right) + G\frac{\partial v}{\partial x} - CL\frac{\partial^2 i}{\partial t^2} - CR\frac{\partial i}{\partial t} = 0. \tag{3.12}$$

In Equation (3.12) the terms in the parenthesis, $\dfrac{\partial^2 v}{\partial x \partial t} - \dfrac{\partial^2 v}{\partial t \partial x}$, will equal zero if, in the function $v(x,t)$, the first derivatives, $\dfrac{\partial v}{\partial t}$ and $\dfrac{\partial v}{\partial x}$, and the mixed partials, $\dfrac{\partial^2 v}{\partial x \partial t}$ and $\dfrac{\partial^2 v}{\partial t \partial x}$, are all continuous. This condition is expected in an electrical system. Therefore, Equation (3.12) becomes

$$\frac{\partial^2 i}{\partial x^2} + G\frac{\partial v}{\partial x} - CL\frac{\partial^2 i}{\partial t^2} - CR\frac{\partial i}{\partial t} = 0. \tag{3.13}$$

Now, using Equation (3.9) in the form $\dfrac{\partial v}{\partial x} = -L\dfrac{\partial i}{\partial t} - Ri$, we can rewrite Equation (3.13) as

$$\frac{\partial^2 i}{\partial x^2} - GL\frac{\partial i}{\partial t} - CL\frac{\partial^2 i}{\partial t^2} - CR\frac{\partial i}{\partial t} - GRi = 0$$

or in the more familiar wave form,

$$CL\frac{\partial^2 i}{\partial t^2} = \frac{\partial^2 i}{\partial x^2} - (GL + CR)\frac{\partial i}{\partial t} - GRi. \tag{3.14}$$

In a similar fashion, we can derive a wave equation for the potential:

$$CL\frac{\partial^2 v}{\partial t^2} = \frac{\partial^2 v}{\partial x^2} - (GL + CR)\frac{\partial v}{\partial t} - GRv. \tag{3.15}$$

If, in Equations (3.14 and 3.15), $G = R = 0$, then we arrive at the equations

$$\frac{\partial^2 i}{\partial t^2} = \frac{1}{CL}\frac{\partial^2 i}{\partial x^2}$$

and

$$\frac{\partial^2 v}{\partial t^2} = \frac{1}{CL}\frac{\partial^2 v}{\partial x^2},$$

which are easily recognized as the one-dimensional wave equation. Please remember that the wave equation is the primary representative of the hyperbolic class of linear second-order PDEs.

Other examples of physical systems that use the wave equation as a mathematical model include sound waves, water waves, probability waves of quantum mechanics, and vibrations in solids. However, to get a more complete picture of the wave equation, we must know what conditions were present when we started the experiment. These are called initial conditions; we discuss them next.

When we derived the heat equation we also discussed an initial condition (IC). This IC described the heat distribution in the rod at the start of time. In the wave equation, Equation (3.7), we have a second partial derivative with respect to time. Thus, we must have two initial conditions (ICs).

One IC of the wave equation describes the starting position of the string. The starting position of the string is the location of the string when a stopwatch starts ($t = 0$). Therefore, the starting position depends only on x, and we model it as $u(x, 0) = f(x)$.

For a second IC, we will consider a plucked violin string. Suppose after the violin string is plucked, we allow it to vibrate for a few seconds before we start a stopwatch. We then determine the starting position. We also notice the string is moving. This movement is the instantaneous velocity. It must be mathematically modeled for a more accurate picture of the experiment. We know velocity is always the first derivative with respect to time of our function. Thus, we have $\dfrac{\partial u}{\partial t}$. Initial velocity implies $t = 0$ in the function, resulting in $\dfrac{\partial u(x, 0)}{\partial t} = g(x)$.

Thus, our two ICs are

$$
\begin{cases}
u(x,0) = f(x) \\
\dfrac{\partial u(x,0)}{\partial t} = g(x).
\end{cases}
$$

EXERCISES 3.2

3.2.1. Given the one-dimensional wave equation

$$
\frac{\partial^2 u}{\partial t^2} = c^2 \frac{\partial^2 u}{\partial x^2} - \alpha \frac{\partial u}{\partial t} - \beta u + Q(x,t),
$$

explain in two sentences or less the physical meaning of the terms

$$
\alpha \frac{\partial u}{\partial t},
$$

$$
\beta u, \text{ and}
$$

$$
Q(x,t).
$$

3.2.2. Show that for all positive integers m, each of the following functions satisfies

$$
\frac{\partial^2 u}{\partial t^2} = c^2 \frac{\partial^2 u}{\partial x^2}.
$$

(1) $u_1(x,t) = \sin[m\pi x]\sin[m\pi ct]$.
(2) $u_2(x,t) = \sin[m\pi x]\cos[m\pi ct]$.
(3) $u_3(x,t) = \cos\left[\left(m + \dfrac{1}{2}\right)\pi x\right]\sin\left[\left(m + \dfrac{1}{2}\right)\pi ct\right]$.

3.2.3. Given

$$
\frac{\partial i}{\partial x} + C\frac{\partial v}{\partial t} + Gv = 0
$$

and

$$
\frac{\partial v}{\partial x} + L\frac{\partial i}{\partial t} + Ri = 0.
$$

Derive

$$
CL\frac{\partial^2 v}{\partial t^2} = \frac{\partial^2 v}{\partial x^2} - (GL + CR)\frac{\partial v}{\partial t} - GRv.
$$

3.2.4. Show that

$$u(x,t) = 2\frac{1}{\cosh^2(x-4t)}$$

is a solution of the Korteweg-deVries equation

$$\frac{\partial u}{\partial t} + 6u\frac{\partial u}{\partial x} + \frac{\partial^3 u}{\partial x^3}.$$

3.2.5. A linear approximation of one-dimensional isentropic flow of an ideal gas (a gas in which the only stress across any element of area is normal to it) is given by

$$\left\{ \begin{array}{l} \dfrac{\partial u}{\partial t} + \dfrac{\partial \rho}{\partial x} = 0 \\[3mm] \dfrac{\partial u}{\partial x} + c^2\dfrac{\partial \rho}{\partial t} = 0 \end{array} \right.,$$

where $u = u(x,t)$ is the velocity of the gas and $\rho = \rho(x,t)$ is the density of the gas. Show that u and ρ satisfy the wave equation.

The following exercises involve second-order ODEs. A review of this material may be found in Appendix C.

3.2.6. Solve the following second-order homogeneous ODEs:

(1) $u''(t) + 6u'(t) + 3u(t) = 0,$ $u(0) = 1,$ $u'(0) = 0.5.$

(2) $y'' + 6y = 0.$

(3) $s''(t) - 6s(t) = 0,$ $s(1) = 0,$ $s'(0) = 1.$

(4) $g''(x) + 4g'(x) = 0.$

(5) $h''(t) - 3h'(t) + 2h(t) = 0,$ $h(0) = 4,$ $h'(3) = 1.$

3.2.7. Consider a weight of 5 pounds attached to a steel spring that has a natural length of 1 foot. The mass stretches the spring 0.25 foot. Suppose the system is started in motion by stretching the spring an additional 0.1 foot in the downward direction, then released. Determine and solve the resulting equation of motion neglecting air resistance.

3.2.8. Solve the following second-order nonhomogeneous ODEs.

(1) $y''(x) + 3y'(x) - 4y(x) = 2e^{-4x}.$

(2) $z''(y) + 2z'(y) - 3z(y) = 3e^{-3y}.$

(3) $x''(t) + 2x'(t) + 4x(t) = 3\sin t.$

(4) $u''(t) + \alpha u'(t) + \omega_0 u(t) = \cos \omega t$, $\alpha^2 - 4\omega_0^2 < 0$.

(5) $\dfrac{d^2 y}{dt} - 2\dfrac{dy}{dt} + y = te^t + 4$.

3.2.9. Determine the solution of the differential equation

$$mu''(t) + cu'(t) + ku(t) = F \cos \omega t$$

satisfying the following ICs. Assume $c^2 - 4km < 0$.

(1) $u(0) = \alpha$, $u'(0) = 0$.

(2) $u(0) = 0$, $u'(0) = \alpha$.

(3) $u(0) = \alpha$, $u'(0) = \beta$.

3.2.10. Solve

$$u''(t) + u(t) = \cos \omega t, \ \omega \neq 1$$

subject to ICs

$$\begin{cases} u(0) = 0 \\ u'(0) = 0. \end{cases}$$

Show that the solution may be written as

$$u(t) = \frac{2}{1 - \omega^2} \sin\left(\frac{(1+\omega)\,t}{2}\right) \sin\left(\frac{(1-\omega)t}{2}\right).$$

Using your favorite mathematical software, graph the solution for at least three different values of ω.

3.3 BOUNDARY CONDITIONS

In the previous section, we derived Equation (3.6) and Equation (3.7), which are commonly referred to as the equations that govern wave motion in a one-dimensional string. Also, initial position, $u(x, 0) = f(x)$, and initial velocity, $\dfrac{\partial u(x, 0)}{\partial t} = g(x)$, the ICs, were discussed. However, when we model physical phenomena, we must try to model as many constraints as possible. We would fall short of this goal if we did not discuss the ends of the string, called boundaries, as we did in Chapter 1. In this section, we discuss the mathematical model of various boundary conditions for a one-dimensional string.

3.3.1   Boundary Conditions of the First Kind

For a one-dimensional string, Dirichlet conditions describe where boundaries are attached. A fixed attachment would be considered a fixed boundary condition. For example, consider a one-dimensional string fixed at some constant displacement S, then $u(0,t) = S$ and $u(L,t) = S$. A special case of constant displacement is $u(0,t) = 0$ and $u(L,t) = 0$. These describe boundaries fixed with zero displacement from the horizontal axis, homogeneous BCs.

Alternatively, we can use Dirichlet conditions to describe how boundaries of the one-dimensional string are controlled or specified. That is, a function describes the physical displacement of the end of the string over time from the horizontal axis. For example, if the boundary $x = 0$ of the string moves with time in an up and down motion only, then that boundary may be mathematically modeled as

$$u(0,t) = g_1(t). \tag{3.16}$$

3.3.2   Boundary Conditions of the Second Kind

Neumann conditions describe the tensile force on a one-dimensional string at the boundaries. *Note:* A mathematical model of tensile force (from the derivation of the wave equation) is $\tau \dfrac{\partial u(x,t)}{\partial x}$. For example, suppose at the boundary $x = L$, the tensile force applied changes with time. The equation governing this is

$$\tau \frac{\partial u(L,t)}{\partial x} = g_2(t). \tag{3.17}$$

The tensile force does not have to change with time. It could be a constant. In this case, the mathematical model for the boundary at $x = L$ is

$$\frac{\partial u(L,t)}{\partial x} = S.$$

A very special case of constant tensile force is when $S = 0$, and we have at the boundary $x = L$, as in

$$\frac{\partial u(L,t)}{\partial x} = 0.$$

This particular boundary condition means that at the boundary $x = L$ the string is attached to a frictionless sleeve which moves vertically. This may seem impossible, but it is a standard mathematically modeled case. A more interesting case is boundary conditions of the third kind.

3.3.3   Boundary Conditions of the Third Kind

Robin's conditions for a one-dimensional string describe some type of an elastic attachment at both ends of the string. For example, consider Figure (3.2). Here, the ends of the string are attached to a spring. The spring has its other end fixed.

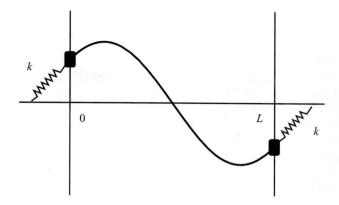

Figure 3.2: String attached to elastic ends with their ends fixed.

The spring constant is assumed to be positive. Also, we will assume the spring constant is the same at both ends, and we will denote it as k.

The mathematical model for this physical condition is

$$\tau \frac{\partial u(0,t)}{\partial x} = ku(0,t)$$

for the end $x = 0$. For the end, $x = L$, we have

$$\tau \frac{\partial u(L,t)}{\partial x} = -ku(L,t).$$

Note: The spring constant may not be the same at both ends. This condition would indicate different springs attached to either end of the string.

Another example of Robin's conditions is shown in Figure (3.3). Here, the end

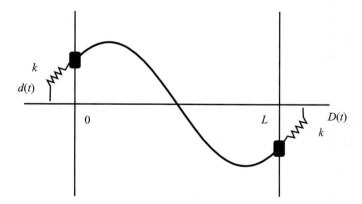

Figure 3.3: String attached to elastic ends with their ends displaced.

$x = 0$ is attached to a spring where the end of the spring can move in a vertical direction. The displacement of the end of the string is described by $u(0,t)$. The displacement of the left end of the spring can be described as $d(t)$. The tension on the end of the spring at its attachment to the string is described by $\tau\dfrac{\partial u(0,t)}{\partial x}$. By setting the vertical tension of the spring equal to the difference of the displacements, we arrive at the mathematical model

$$\tau\frac{\partial u(0,t)}{\partial x} = k[u(0,t) - d(t)]. \tag{3.18}$$

Assuming the spring obeys Hooke's[3] law, k is the spring constant. The signs in Equation (3.18) are identical to Newton's law of cooling, which was developed in Chapter 1. The end $x = L$ has a similar mathematical model, and it is left as an exercise.

EXERCISES 3.3

3.3.1. Explain in your own words what happens to the Robin's condition

$$\tau\frac{\partial u(0,t)}{\partial x} = h[u(0,t) - \gamma_1(t)]$$

when

(1) $h \longrightarrow \infty$.

(2) $h \longrightarrow 0$.

3.3.2. State the mathematical model for the given information:

    (a) Small vertical vibrations in a uniform tightly-stretched string.

    (b) Fixed left end at 4.

    (c) Free right end.

    (d) String length of 3π.

    (e) Tension of 4.

    (f) Mass density of 3.

    (g) Initial displacement of $x^2 + 2x + 4$.

    (h) Initial velocity of 0.

3.3.3. State the mathematical model for the given information:

    (a) Small vertical vibrations in a nonuniform tightly-stretched string.

    (b) Free left end.

---

[3]R. Hooke (1638–1703) was professor of geometry at Gresham College and secretary of the Royal Society.

(c) Fixed right end at -2.

(d) String length of $\dfrac{\pi}{2}$.

(e) Tension of 4.

(f) Mass density of $2x + 1$.

(g) Initial displacement of 0.

(h) Initial velocity of $-2 \sin x$.

3.3.4. Consider a one-dimensional tightly stretched string in the horizontal position. Suppose the mass density of the string is constant; the left end, at $x = 0$, is fixed at a height of 3.5 off the horizontal axes; the right end, at $x = \pi$, is allowed to move freely; the initial position of the string is given by $f(x) = x^2 - 2\pi x + 3.5$, and the initial velocity is 0. State the mathematical model.

3.3.5. Given the following equation, describe the physical situation:

$$(x + 5)^3 \frac{\partial^2 u}{\partial t^2} = \frac{\partial}{\partial x}\left((2x + 1) \frac{\partial u}{\partial x} \right) + \left(\frac{-x^2}{4\pi} + x \right) e^{-t},$$

subject to BCs:

$$\begin{cases} \dfrac{\partial u(0, t)}{\partial x} = t^2 + 2t + 1 \\[2mm] u(4, t) = 0 \end{cases}$$

and ICs:

$$\begin{cases} u(x, 0) = \cos x + \sin x - 1 \\[2mm] \dfrac{\partial u(x, 0)}{\partial t} = \ln(x + 1) - 2.60759. \end{cases}$$

3.3.6. Given the following equation, describe the physical situation:

$$(x + 1)^2 \frac{\partial^2 u}{\partial t^2} = \frac{\partial}{\partial x}\left((3x + 5) \frac{\partial u}{\partial x} \right) + \sin xt,$$

subject to BCs:

$$\begin{cases} u(0, t) = 0 \\[2mm] \dfrac{\partial u(L, t)}{\partial x} = 3 \cos t \end{cases}$$

and ICS:

$$\begin{cases} u(x, 0) = \ln(x + 1) \\[2mm] \dfrac{\partial u(x, 0)}{\partial t} = 0. \end{cases}$$

3.3.7. Develop the mathematical model for Robin's conditions at the boundary $x = L$ described in Figure (3.3).

3.3.8. A uniform string with fixed ends has an initial displacement of $2x$ for $0 \leq x < \pi$ and $3\pi - x$ for $\pi \leq x \leq 3\pi$. It is known that the initial velocity is 0 and the string is vibrating in a medium that resists the vibrations. (The medium produces a resistance proportional to the velocity.) Suppose the resistance constant of proportionality is 0.01. State the mathematical model.

3.3.9. A uniform string with a fixed end at 0 and free end at 2π has an initial displacement of $-x$ for $0 \leq x < \dfrac{3\pi}{2}$ and $3x - 6\pi$ for $\dfrac{3\pi}{2} \leq x \leq 2\pi$. It is known that the initial velocity is 0 and the string is vibrating in a medium that resists the vibrations. (The medium produces a resistance proportional to the velocity.) Suppose the resistance constant of proportionality is 0.03. State the mathematical model.

3.3.10. A uniform string with free ends has an initial displacement of

$$
u(x,0) = \begin{cases} -x, \ 0 \leq x \leq 1 \\[2mm] \dfrac{3}{2}x - \dfrac{5}{2}, \ 1 \leq x < 3 \\[2mm] -2x + 8, \ 3 \leq x \leq 4. \end{cases}
$$

It is known that the initial velocity is 0 and the string is vibrating in a medium that resists the vibrations. (The medium produces a resistance proportional to the velocity.) Suppose the resistance constant of proportionality is 0.13. State the mathematical model.

3.4   CONSERVATION OF ENERGY FOR A VIBRATING STRING

The **principle of conservation of energy** states that if no energy is lost due to friction or other possible forces, then in a mechanical system the sum of the instantaneous kinetic and potential energy is equal to a constant. From physics, we know kinetic energy of a point with mass m is $E_k = \dfrac{1}{2}mv^2$, where v is the velocity of the mass point at any time t. Also, potential energy is given by $E_p = \dfrac{1}{2}kx^2$, where k is a force constant of proportionality and x is the coordinate of the body.

Consider the equations for a uniform tightly stretched vibrating string with free ends and initial conditions,

$$
\frac{\partial^2 u}{\partial t^2} = c^2 \frac{\partial^2 u}{\partial x^2}, \tag{3.19}
$$

subject to BCs

$$\begin{cases} \dfrac{\partial u(0, t)}{\partial x} = 0 \\[3mm] \dfrac{\partial u(L, t)}{\partial x} = 0 \end{cases} \tag{3.20}$$

and ICs

$$\begin{cases} u(x, 0) = f(x) \\[3mm] \dfrac{\partial u(x, 0)}{\partial t} = g(x). \end{cases} \tag{3.21}$$

Here, the kinetic energy is

$$E_k = \frac{1}{2} \int_0^L \left(\frac{\partial u(x, t)}{\partial t} \right)^2 dx, \tag{3.22}$$

and the potential energy is

$$E_p = \frac{c^2}{2} \int_0^L \left(\frac{\partial u(x, t)}{\partial x} \right)^2 dx. \tag{3.23}$$

Equations (3.22 and 3.23) are different than the single-point mass equations because, for a string, we must sum over the entire length. Thus, we have

$$\begin{aligned} E &= E_k + E_p \\ &= \frac{1}{2} \int_0^L \left(\frac{\partial u(x, t)}{\partial t} \right)^2 dx + \frac{c^2}{2} \int_0^L \left(\frac{\partial u(x, t)}{\partial x} \right)^2 dx. \end{aligned} \tag{3.24}$$

We need the sum of the instantaneous kinetic and potential energy, which means taking the derivative with respect to time, t. Hence, the equation

$$\frac{dE}{dt} = \frac{d}{dt} \left[\frac{1}{2} \int_0^L \left(\frac{\partial u(x, t)}{\partial t} \right)^2 dx + \frac{c^2}{2} \int_0^L \left(\frac{\partial u(x, t)}{\partial x} \right)^2 dx \right]$$

becomes

$$\frac{dE}{dt} = \frac{d}{dt} \left[\frac{1}{2} \int_0^L \left(\frac{\partial u(x, t)}{\partial t} \right)^2 + c^2 \left(\frac{\partial u(x, t)}{\partial x} \right)^2 dx \right]. \tag{3.25}$$

Using Leibniz's formula on Equation (3.25) yields

$$\frac{dE}{dt} = \left[\frac{1}{2} \int_0^L \frac{\partial}{\partial t} \left(\frac{\partial u(x, t)}{\partial t} \right)^2 + c^2 \frac{\partial}{\partial t} \left(\frac{\partial u(x, t)}{\partial x} \right)^2 dx \right],$$

which reduces to

$$\frac{dE}{dt} = \left[\int_0^L \frac{\partial^2 u(x,t)}{\partial t^2} \left(\frac{\partial u(x,t)}{\partial t} \right) + c^2 \frac{\partial^2 u(x,t)}{\partial x \partial t} \left(\frac{\partial u(x,t)}{\partial x} \right) dx \right]. \quad (3.26)$$

From Equation (3.19), Equation (3.26) becomes

$$\frac{dE}{dt} = \left[\int_0^L c^2 \frac{\partial^2 u(x,t)}{\partial x^2} \left(\frac{\partial u(x,t)}{\partial t} \right) + c^2 \frac{\partial^2 u(x,t)}{\partial x \partial t} \left(\frac{\partial u(x,t)}{\partial x} \right) dx \right]. \quad (3.27)$$

The integrand of Equation (3.27) may be recognized as the derivative of a product, which is

$$\frac{\partial}{\partial x} \left[\frac{\partial u(x,t)}{\partial t} \frac{\partial u(x,t)}{\partial x} \right] = \frac{\partial^2 u(x,t)}{\partial x^2} \left(\frac{\partial u(x,t)}{\partial t} \right) + \frac{\partial^2 u(x,t)}{\partial x \partial t} \left(\frac{\partial u(x,t)}{\partial x} \right).$$

Therefore, Equation (3.27) becomes

$$\frac{dE}{dt} = \left[c^2 \int_0^L \frac{\partial}{\partial x} \left[\frac{\partial u(x,t)}{\partial t} \frac{\partial u(x,t)}{\partial x} \right] dx \right]$$

$$= c^2 \left[\frac{\partial u(x,t)}{\partial t} \frac{\partial u(x,t)}{\partial x} \right]_0^L,$$

which means the sum of the instantaneous kinetic and potential energy is equal to a constant.

We may now proceed and prove that if Equations (3.19, 3.20, and 3.21) have a solution, then the solution is unique.

Proof. Let $u_1(x,t)$ and $u_2(x,t)$ be solutions to the given equations. Then, $v(x,t) = u_1(x,t) - u_2(x,t)$. Thus, $\frac{\partial v(0,t)}{\partial x} = \frac{\partial u_1(0,t)}{\partial x} - \frac{\partial u_2(0,t)}{\partial x} = 0$ and $\frac{\partial v(L,t)}{\partial x} = \frac{\partial u_1(L,t)}{\partial x} - \frac{\partial u_2(L,t)}{\partial x} = 0$. Also, $v(x,0) = u_1(x,0) - u_2(x,0) = 0$ and $\frac{\partial v(x,0)}{\partial t} = \frac{\partial u_1(x,0)}{\partial t} - \frac{\partial u_2(x,0)}{\partial t} = 0$. Thus, the sum of the instantaneous kinetic and potential energies of $v(x,t)$ must equal a constant. We form the energy equation

$$E = \frac{1}{2} \int_0^L \left(\frac{\partial v(x,t)}{\partial t} \right)^2 dx + \frac{c^2}{2} \int_0^L \left(\frac{\partial v(x,t)}{\partial x} \right)^2 dx. \quad (3.28)$$

Taking the derivative with respect to t yields

$$\frac{dE}{dt} = \frac{1}{2} \frac{d}{dt} \left[\int_0^L \left(\frac{\partial v(x,t)}{\partial t} \right)^2 dx + \frac{c^2}{2} \int_0^L \left(\frac{\partial v(x,t)}{\partial x} \right)^2 dx \right], \quad (3.29)$$

which becomes

$$\frac{dE}{dt} = \left[\frac{\partial v(x,t)}{\partial t} \frac{\partial v(x,t)}{\partial x} \right]_0^L = 0.$$

This means that $E(t) = 0$, which in turn implies that the integrands of Equation (3.28) are zero. Thus, $\frac{\partial v(x,t)}{\partial t}$ and $\frac{\partial v(x,t)}{\partial x}$ must be identically zero, which implies $v(x,t) \equiv 0$. Thus, $u_1(x,t) = u_2(x,t)$. $\qquad\square$

EXERCISES 3.4

3.4.1. Given a uniform tightly stretched vibrating string with a fixed end at $x = 0$, a free end at $x = L$, and initial conditions, determine what happens to the total energy E.

3.4.2. at the end $x = L$, and the springs other end fixed, and initial conditions, determine what happens to the total energy E.
Note: The spring constant is assumed to be a positive.

3.4.3. Given the equations for a uniform tightly stretched vibrating string with homogeneous fixed ends and initial conditions,

$$\frac{\partial^2 u}{\partial t^2} = c^2 \frac{\partial^2 u}{\partial x^2}, \tag{3.30}$$

subject to BCs:

$$\begin{cases} u(0,t) = 0 \\ u(L,t) = 0 \end{cases} \tag{3.31}$$

and ICs:

$$\begin{cases} u(x,0) = f(x) \\ \dfrac{\partial u(x,0)}{\partial t} = g(x). \end{cases} \tag{3.32}$$

Show that if this problem has a solution, then the solution is unique.

3.5 FIRST-ORDER PDES: METHOD OF CHARACTERISTICS

In this section, we use the method of characteristics to analyze first-order constant coefficient PDEs. A **characteristic** of a first-order PDE is a curve where the PDE becomes an ODE. A common example of this is a surfer catching a wave. The surfer moves along the wave with a velocity equal to the wave's. The wave may move faster or slower depending on any number of variables. But, from the surfer's point of view, the wave does not change.

We begin our investigation with the simple first-order PDE

$$\frac{\partial z}{\partial t} - c\frac{\partial z}{\partial x} = 0 \tag{3.33}$$

with IC

$$z(x,0) = f(x), \tag{3.34}$$

where $-\infty < x < \infty$ and $0 < t < \infty$. Unlike the second-order wave equation, which has two initial conditions, a first-order PDE will only have one initial condition. This initial condition describes the initial position at time $t = 0$. Remember, initial conditions depend on the number of partial derivatives with respect to time that are in the equation. A first-order PDE only has one partial derivative with respect to time. Also, there are no boundary conditions because the variable, x, varies from negative infinity to positive infinity.

One way to solve this problem is to consider the rate of change of $z(x(t), t)$ as measured by a moving observer, $x = x(t)$. The chain rule implies

$$\frac{d}{dt}z(x(t), t) = \frac{\partial z}{\partial x}\frac{dx}{dt} + \frac{\partial z}{\partial t}. \tag{3.35}$$

Comparing these quantities with equation (3.33), we see that if the observer moves along at velocity $-c$, then $\frac{dx}{dt} = -c$ and $\frac{d}{dt}z(x(t), t) = 0$. That is, z is a constant.

The term $\frac{\partial z}{\partial t}$ in Equation (3.35) represents the change in z with respect to time at the fixed position x. The first term $\frac{\partial z}{\partial x}\frac{dx}{dt}$, in Equation (3.35), represents the change that the observer sees as the observer moves into different regions of the domain of z.

Using Equation (3.35), we have reduced the PDE in Equation (3.33) to two first-order ordinary differential equations: $\frac{dx}{dt} = -c$ and $\frac{dz}{dt} = 0$. If we consider that

$$\frac{dx}{dt} = -c,$$

we see that the solution is

$$x(t) = -ct + a, \tag{3.36}$$

where a represents the initial point on the characteristic curve when $t = 0$. Using the IC (3.34), we have $z(x(t), t) = z(a, 0) = f(a)$.

Solving for a in Equation (3.36) yields $a = x(t) + ct$, or simply $a = x + ct$. Therefore,

$$z(x, t) = f(x + ct). \qquad (3.37)$$

Equation (3.37) is known as the general solution of Equation (3.33).

EXAMPLE 3.1. Consider

$$\frac{\partial z}{\partial t} - 3\frac{\partial z}{\partial x} = 0$$

with initial condition

$$z(x, 0) = \sin x,$$

where $-\infty < x < \infty$ and $0 < t < \infty$. Letting

$$z(x, t) = z(x(t), t).$$

We find the derivative of z with respect to t, which is

$$\frac{dz}{dt} = \frac{\partial z}{\partial x}\frac{dx}{dt} + \frac{\partial z}{\partial t}.$$

Thus,

$$\frac{dx}{dt} = -3,$$

which implies

$$x(t) = -3t + a.$$

Also,

$$\frac{dz}{dt} = 0,$$

which implies

$$z(x(t), t) = c.$$

Thus,

$$z(x(0), 0) = \sin(x(0)).$$

Therefore, using Equation (3.37), we find that the general solution is

$$z(x, t) = \sin(x + 3t). \qquad \blacksquare$$

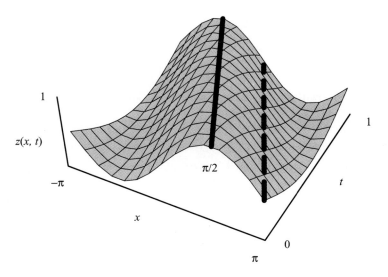

Figure 3.4: The solution sketch for $\dfrac{\partial z}{\partial t} - 3\dfrac{\partial z}{\partial x} = 0$.

Figure (3.4) shows a sketch of the solution for values $0 < t \le 1$ and $-\pi < x < \pi$. Two of the parallel characteristics $x = a - 3t$ are clearly evident in the solution sketch. They are represented by a solid line for $a = \dfrac{\pi}{2}$ and a dashed line for $a = \pi$.

We will now consider the more general first-order constant coefficient PDE

$$\frac{\partial u}{\partial t} + \alpha \frac{\partial u}{\partial x} + \beta u = 0 \tag{3.38}$$

with IC

$$u(x,0) = f(x), \tag{3.39}$$

where $-\infty < x < \infty$, $0 < t < \infty$, and $\beta > 0$.

As introduced in the beginning of this section, one way to solve this problem is to consider the rate of change of $u(x(t), t)$ as measured by a moving observer, $x = x(t)$. Applying the chain rule yields

$$\frac{du}{dt} = \frac{\partial u}{\partial x}\frac{dx}{dt} + \frac{\partial u}{\partial t}. \tag{3.40}$$

If we assume that $\dfrac{dx}{dt} = \alpha$, the left side of Equation (3.40) can be substituted into Equation (3.38). Then, we have

$$\frac{du}{dt} + \beta u = 0 \text{ or } \frac{du}{dt} = -\beta u,$$

which is a first-order ODE with solution

$$u(x,t) = c(x)e^{-\beta t}.$$

Applying the IC (3.39) yields a general solution for $u(x,t)$ of

$$u(x,t) = f(x)e^{-\beta t}.$$

Note: The solution of the first-order PDE is the solution of a first-order ODE, where the arbitrary constant of the ODE is replaced by an arbitrary function.

To determine a more specific solution for $u(x,t)$ based on our assumption $\dfrac{dx}{dt} = \alpha$, we must solve for x and place that solution in the function $f(x)$. The solution to $\dfrac{dx}{dt} = \alpha$ is $x = \alpha t + a$, where a is the constant representing the initial position of the characteristic curve when $t = 0$. Therefore, when $x = a$ and $t = 0$, we have $u(x,t) = u(a,0)$. Thus, the complete solution of Equation (3.38) with IC (3.39) is

$$u(x,t) = f(x - \alpha t)e^{-\beta t}. \tag{3.41}$$

Because of the term $e^{-\beta t}$, we know as time goes to infinity, $u(x,t)$ tends to zero. This seems quite reasonable if we assume Equation (3.38) is a first-order wave equation. In this case, the term βu is a damping function that tends to flatten out the wave as time passes.

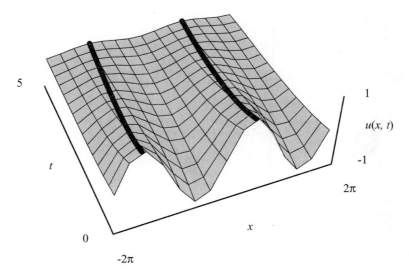

Figure 3.5: Solution sketch for $u(x,t) = \sin(x - 0.25t)e^{-0.5t}$.

EXAMPLE 3.2. Consider

$$\frac{\partial u}{\partial t} + 0.25\frac{\partial u}{\partial x} + 0.5u = 0 \tag{3.42}$$

with IC

$$u(x, 0) = \sin x, \tag{3.43}$$

where $-\infty < x < \infty$ and $0 < t < \infty$. Since $\alpha = 0.25$ and $\beta = 0.5$, we know from Equation (3.41),

$$u(x, t) = f(x - \alpha t)e^{-\beta t},$$

that the specific solution is

$$u(x, t) = \sin(x - 0.25t)e^{-0.5t}.$$

Figure (3.5) shows a sketch for $-2\pi \leq x \leq 2\pi$ and $0 < t < 5$. Two parallel characteristics are shown on the graph. However, remember these characteristics take into consideration the exponential $e^{-0.5t}$. ∎

EXERCISES 3.5

3.5.1. Solve and graph the following using the method of characteristics.

(1) $\dfrac{\partial z}{\partial t} + 5\dfrac{\partial z}{\partial x} = 0$ with IC $z(x, 0) = 2x$.

(2) $\dfrac{\partial z}{\partial t} - 3\dfrac{\partial z}{\partial x} = 0$ with IC $z(x, 0) = x^2 - 2x + 1$.

(3) $\dfrac{\partial w}{\partial t} - \dfrac{\partial w}{\partial x} + 6w = 0$ with IC $w(x, 0) = \cos x$.

(4) $\dfrac{\partial u}{\partial t} + 4\dfrac{\partial u}{\partial x} - 2u = 0$ with IC $u(x, 0) = \sin x$.

3.5.2. Consider the simple concentration problem

$$\frac{\partial u}{\partial t} + \frac{\partial u}{\partial x} = 0, \quad 0 < t < \infty, \quad -\infty < x < \infty,$$

with IC

$$u(x, 0) = \cos x.$$

Solve and graph the solution using your choice of mathematical software. Does the solution satisfy the PDE and the IC?

3.5.3. Solve

$$\frac{\partial v}{\partial t} + x\frac{\partial v}{\partial x} = 0, \quad 0 < t < \infty, \quad -\infty < x < \infty,$$

with IC

$$v(x, 0) = 0.5x^2.$$

Graph the solution using your choice of mathematical software. Does the solution satisfy the PDE and the IC?

3.5.4. Solve

$$\frac{\partial u}{\partial t} + \frac{\partial u}{\partial x} + tu = 0, \quad 0 < t < \infty, \quad -\infty < x < \infty,$$

with IC

$$u(x, 0) = x^2.$$

Graph the solution using your choice of mathematical software. Does the solution satisfy the PDE and the IC?

3.5.5. Solve the surface wave problem

$$\frac{\partial u}{\partial t} + \alpha \frac{\partial u}{\partial x} + \beta \frac{\partial u}{\partial y} + \gamma u = 0, \quad 0 < t < \infty, \quad -\infty < x < \infty, \quad -\infty < y < \infty,$$

with IC

$$u(x, y, 0) = \sin(x + y),$$

where α, β, and γ are constants.

3.5.6. Solve the surface wave problem

$$\frac{\partial u}{\partial t} + \alpha \frac{\partial u}{\partial x} + \beta \frac{\partial u}{\partial y} + \gamma u = 0, \quad 0 < t < \infty, \quad -\infty < x < \infty, \quad -\infty < y < \infty,$$

with initial condition

$$u(x, y, 0) = \sin(x + y),$$

where $\alpha = 2$, $\beta = -1$, and $\gamma = 1$.

3.5.7. Given the first-order semilinear PDE

$$a(x, y, u(x, u)) \frac{\partial u(x, y)}{\partial x} + b(x, y, u(x, y)) \frac{\partial u(x, y)}{\partial x} = c(x, y, u(x, y)),$$

show that the method of characteristics yields

$$\frac{dx}{a(x, y, u(x, y))} = \frac{dy}{b(x, y, u(x, y))} = \frac{du(x, y)}{c(x, y, u(x, y))}.$$

When an IC is given, Equation (3.5) is sometimes called the first-order Cauchy[4] problem.

3.5.8. Given $\dfrac{\partial^2 u}{\partial x^2} - \dfrac{1}{c^2} \dfrac{\partial^2 u}{\partial t^2} + \alpha \dfrac{\partial u}{\partial t} + \beta u = 0$, which is another form of the Telegraph

---

[4]Augustin-Louis Cauchy (1789–1857) a very famous French mathematician who derived the name of determinant

Equation, let $v = u$, $w = \dfrac{\partial u}{\partial x}$, and $z = \dfrac{\partial u}{\partial t}$. Show that v, w, and z must satisfy the following system of three equations,

$$\frac{\partial v}{\partial t} - z = 0,$$

$$\frac{\partial w}{\partial t} - \frac{\partial z}{\partial x} = 0$$

and

$$\frac{\partial z}{\partial t} - c^2 \left(\frac{\partial w}{\partial x} + \alpha z + \beta v \right) = 0.$$

3.6 D'ALEMBERT'S SOLUTION TO THE ONE-DIMENSIONAL WAVE EQUATION

In the last section, we considered the method of characteristics as a solution to first-order PDEs. In this section, we show the method of characteristics may be applied to solving the one-dimensional wave equation. Here, the chain rule has an important role. This was first published by Jean Le Rond d'Alembert in 1747. We'll start with the one-dimensional wave equation with no BCs, That is, the string is considered infinite:

$$\frac{\partial^2 u}{\partial t^2} = c^2 \frac{\partial^2 u}{\partial x^2} \tag{3.44}$$

with ICs

$$\left. \begin{array}{l} u(x,0) = f(x) \\[2mm] \dfrac{\partial u(x,0)}{\partial t} = g(x) \end{array} \right\}, \quad -\infty < x < \infty. \tag{3.45}$$

Remembering our calculus, if $u(x,t)$ is a real-valued function (which it is) such that $\dfrac{\partial u}{\partial x}$, $\dfrac{\partial u}{\partial t}$, and $\dfrac{\partial^2 u}{\partial t \partial x}$ or $\dfrac{\partial^2 u}{\partial x \partial t}$ (called the mixed partials) are continuous at every point in the domain, then at every point in the domain the derivatives $\dfrac{\partial^2 u}{\partial t \partial x}$ or $\dfrac{\partial^2 u}{\partial x \partial t}$ exist and $\dfrac{\partial^2 u}{\partial t \partial x} = \dfrac{\partial^2 u}{\partial x \partial t}$. To use this theorem, we must assume that the mixed partials are continuous at every point in the domain. Thus, we can rewrite Equation (3.44) as

$$\frac{\partial^2 u}{\partial t^2} + c \frac{\partial^2 u}{\partial t \partial x} - c \frac{\partial^2 u}{\partial x \partial t} - c^2 \frac{\partial^2 u}{\partial x^2} = 0. \tag{3.46}$$

This can be "factored" in the following way:

$$\left(\frac{\partial}{\partial t} - c \frac{\partial}{\partial x} \right) \left(\frac{\partial u}{\partial t} + c \frac{\partial u}{\partial x} \right) = 0.$$

Letting $z = \left(\dfrac{\partial u}{\partial t} + c \dfrac{\partial u}{\partial x} \right)$ yields the first-order PDE

$$\frac{\partial z}{\partial t} - c\frac{\partial z}{\partial x} = 0,$$

which we know from the previous section has the general solution

$$z(x,t) = \frac{\partial u}{\partial t} + c\frac{\partial u}{\partial x} = Q(x + ct), \tag{3.47}$$

where Q is some arbitrary function. *Note:* In Equation (3.37), $z(x,t) = f(x + ct)$, where the function $f(x)$ was the initial condition. In Equation (3.47), we do not know the function explicitly. Therefore, we assign it the arbitrary function Q.

Similarly, if we "factored" Equation (3.46), that is, we found that

$$\left(\frac{\partial}{\partial t} + c\frac{\partial}{\partial x} \right)\left(\frac{\partial u}{\partial t} - c\frac{\partial u}{\partial x} \right) = 0$$

and let $v = \left(\dfrac{\partial u}{\partial t} - c \dfrac{\partial u}{\partial x} \right)$, we would have another first-order PDE

$$\frac{\partial v}{\partial t} + c\frac{\partial v}{\partial x} = 0,$$

which has the solution

$$v(x,t) = \frac{\partial u}{\partial t} - c\frac{\partial u}{\partial x} = P(x - ct) \tag{3.48}$$

where P is some arbitrary function.

Adding Equation (3.47) to Equation (3.48), we obtain

$$\frac{\partial u}{\partial t} = \frac{1}{2}\left[Q(x + ct) + P(x - ct) \right].$$

If we let $-c\dfrac{\partial F(x - ct)}{\partial t} = \dfrac{1}{2}P(x - ct)$ and $c\dfrac{\partial G(x + ct)}{\partial t} = \dfrac{1}{2}Q(x + ct)$, then use some calculus, we obtain

$$u(x,t) = G(x + ct) + F(x - ct) \tag{3.49}$$

as the general solution of Equation (3.44). It is interesting to note that Equation (3.49) expresses the solution of the one-dimensional wave equation as the sum of two moving waves, moving in opposite directions with velocity c. This makes sense if you remember that we are dealing with an infinite string. You pluck the string, and the wave moves off in both directions.

Applying ICs (3.45), we obtain a specific solution. The first IC says that

$$u(x,0) = G(x) + F(x) = f(x). \tag{3.50}$$

To apply the second IC, we use the chain rule. Consider the term $G(x + ct)$ in Equation (3.49). It is actually a composition of the functions $G(X)$ and $X(x, t) = x + ct$. Thus, $\dfrac{\partial G}{\partial t} = \dfrac{dG}{dX}\dfrac{\partial X}{\partial t}$. The fact that $\dfrac{\partial X}{\partial t} = c$ implies $\dfrac{\partial G}{\partial t} = c\dfrac{\partial G}{\partial X} = c\dfrac{dG}{dX}$, since G is a function of X only. Therefore, $c\dfrac{dG}{dX} = cG'(X) = cG'(x + ct)$. Similarly, $\dfrac{\partial F}{\partial t} = -cF'(x - ct)$.

Using the method of the previous paragraph and applying the second IC, we obtain

$$\frac{\partial u(x, 0)}{\partial t} = c\left[G'(x) - F'(x)\right] = g(x). \tag{3.51}$$

To find a solution for $G(x)$, we take the derivative of Equation (3.50) with respect to x, multiply by c, and add it to Equation (3.51), which yields

$$\frac{1}{2}\left[\frac{g(x)}{c} + f'(x)\right] = G'(x).$$

By integrating, we see that

$$G(x) = \frac{1}{2}f(x) + \frac{1}{2c}\int_0^x g(s)ds + k_1. \tag{3.52}$$

A similar calculation yields for $F(x)$

$$F(x) = \frac{1}{2}f(x) - \frac{1}{2c}\int_0^x g(s)ds + k_2. \tag{3.53}$$

In Equations (3.52 and 3.53), k_1 and k_2 are constants of integration. However, since $G(x) + F(x) = f(x)$, we must have $k_1 + k_2 = 0$. Therefore, the general solution of Equation (3.44) is the sum of Equations (3.52 and 3.53), each shifted a distance of ct; $u(x, t)$ can be written in the form

$$u(x, t) = \frac{1}{2}\left[f(x + ct) + f(x - ct)\right] + \frac{1}{2c}\left[\int_0^{x+ct} g(s)ds - \int_0^{x-ct} g(s)ds\right]$$

or

$$u(x, t) = \frac{1}{2}\left[f(x + ct) + f(x - ct)\right] + \frac{1}{2c}\int_{x-ct}^{x+ct} g(s)ds. \tag{3.54}$$

This is known as d'Alembert's solution to the one-dimensional wave equation. Physically, we combine two sets of characteristic curves (sometimes called two families of characteristic curves) to form the solution of the one-dimensional wave equation.

EXAMPLE 3.3. Consider

$$\frac{\partial^2 u(x, t)}{\partial t^2} = c^2\frac{\partial^2 u(x, t)}{\partial x^2}, \quad -\infty < x < \infty,$$

subject to

$$u(x,0) = \begin{cases} h, & \text{if } |x| < a \\ 0, & \text{if } |x| > a \end{cases}$$

$$\frac{\partial u(x,0)}{\partial t} = 0.$$

Determine the solution using d'Alembert's method and graph the solution for several values of time t.

Solution: The general form of d'Alembert's solution is

$$u(x,t) = \frac{1}{2}\left[f(x+ct) + f(x-ct)\right] + \frac{1}{2c}\int_{x-ct}^{x+ct} g(s)\,ds.$$

Since $g(x) = 0$, we have

$$u(x,t) = \frac{1}{2}\left[f(x+ct) + f(x-ct)\right]$$

and know that $F(x) = G(x) = \frac{1}{2}f(x)$. This means

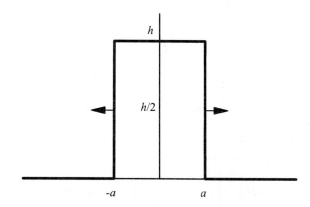

Figure 3.6: Graph of d'Alembert's solution for $t = 0$.

$$F(x) = G(x) = \begin{cases} \dfrac{h}{2}, & \text{if } |x| < a \\ 0, & \text{if } |x| > a. \end{cases} \qquad \blacksquare$$

In Figure (3.6), we have $t = 0$ and $u(x,t) = \frac{1}{2}\left[f(x) + f(x)\right] = f(x) = h$. Since $|x| < a$, we have the endpoints of the graph at $-a$ and a. Remember, the full pulse is made up of two smaller pulses. Each smaller pulse has a height of $\dfrac{h}{2}$, and as soon

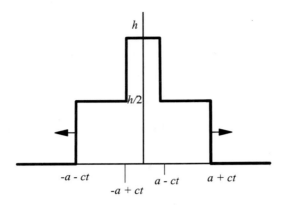

Figure 3.7: Graph of d'Alembert's solution for $0 < t < \dfrac{a}{c}$.

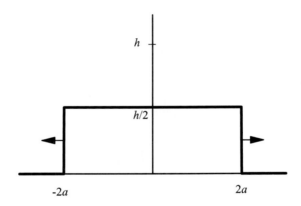

Figure 3.8: Graph of d'Alembert's solution for $t = \dfrac{a}{c}$.

as $t \neq 0$ the pulses will start to separate, as shown in Figure (3.7).

In Figure (3.7), $0 < t < \dfrac{a}{c}$, we have half of the pulse going to the left and the other half of the pulse going to the right, both with a height of $\dfrac{h}{2}$. However, since $t < \dfrac{a}{c}$, we still have the height of h where the pulses overlap. We determine the endpoints for each change of the pulses by again considering $|x| < a$. Here, x actually is $x+ct$ and $x-ct$. For $x+ct$, we have $-a < x+ct < a$ or $-a-ct < x < a-ct$. For $x - ct$, we have $-a < x - ct < a$ or $-a + ct < x < a + ct$. Also, note the pulse going to the left has endpoints of $-a - ct$ and $a - ct$, and the pulse going to the right has endpoints of $-a + ct$ and $a + ct$.

In Figure (3.8), $t = \dfrac{a}{c}$, both pulses are now at the height of $\dfrac{h}{2}$, and the pulses no longer overlap. Using the formula $|x| < a$ where $x = x + ct = x + a$, we obtain the endpoints for the left pulse, which are $-2a$ and 0. The right endpoints are obtained similarly by letting $x = x - ct = x - a$. Thus, the right endpoints become 0 and $2a$. Finally, for Figure (3.9), $t > \dfrac{a}{c}$, and we have two separate pulses, each of height

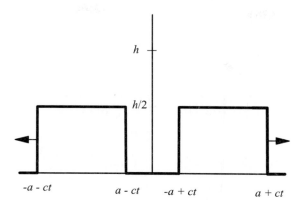

Figure 3.9: Graph of d'Alembert's solution for $t > \dfrac{a}{c}$.

$\dfrac{h}{2}$. One pulse is traveling to the left and the other pulse is traveling to the right. The pulse traveling to the left has endpoints of $-a - ct$ and $a - ct$. These endpoints are determined by letting $x = x + ct$ while remembering $|x| < a$. Thus, we have $|x + ct| < a$ or $-a < x + ct < a$, which becomes $-a - ct < x < a - ct$. Similarly, using $x = x - ct$, the right endpoints are found to be $-a + ct$ and $a + ct$.

EXERCISES 3.6

3.6.1. What is the solution to the initial-value problem

$$\frac{\partial^2 u}{\partial t^2} = 2\frac{\partial^2 u}{\partial x^2}; \quad 0 < t < \infty; \quad -\infty < x < \infty,$$

subject to

$$u(x, 0) = \begin{cases} 1, & |x| < 1 \\ 0, & |x| > 1 \end{cases}$$

$$\frac{\partial u(x, 0)}{\partial t} = 0?$$

Sketch the solution for several values of time, t.

3.6.2. What is the solution to the initial-value problem

$$\frac{\partial^2 u}{\partial t^2} = 3\frac{\partial^2 u}{\partial x^2}; \quad 0 < t < \infty; \quad -\infty < x < \infty,$$

subject to

$$u(x, 0) = 0$$

and

$$\frac{\partial u(x, 0)}{\partial t} = \begin{cases} 1, & |x| < 1 \\ \\ 0, & |x| > 1? \end{cases}$$

Sketch the solution for several values of time, t.

3.6.3. What is the solution to the initial-value problem

$$\frac{\partial^2 u}{\partial t^2} = \frac{\partial^2 u}{\partial x^2}; \quad 0 < t < \infty; \quad -\infty < x < \infty,$$

subject to

$$u(x, 0) = \begin{cases} e^{-x}, & |x| < 1 \\ \\ 0, & |x| > 1 \end{cases}$$

and

$$\frac{\partial u(x, 0)}{\partial t} = 0?$$

Sketch the solution for several values of time, t.

3.6.4. What is the solution to the initial-value problem

$$\frac{\partial^2 u}{\partial t^2} = 4\frac{\partial^2 u}{\partial x^2}; \quad 0 < t < \infty; \quad -\infty < x < \infty,$$

subject to

$$u(x, 0) = 0$$

and

$$\frac{\partial u(x, 0)}{\partial t} = \begin{cases} e^{-x}, & |x| < 1 \\ \\ 0, & |x| > 1? \end{cases}$$

Sketch the solution for several values of time, t.

3.6.5. What is the solution to the initial-value problem

$$\frac{\partial^2 u}{\partial t^2} = 9\frac{\partial^2 u}{\partial x^2}; \quad 0 < t < \infty; \quad -\infty < x < \infty,$$

subject to

$$u(x,0) = \begin{cases} -\sin x, \ |x| < \pi \\ \\ 0, \ |x| > \pi \end{cases}$$

and

$$\frac{\partial u(x,0)}{\partial t} = 0?$$

Sketch the solution for several values of time, t.

3.6.6. What is the solution to the initial-value problem

$$\frac{\partial^2 u}{\partial t^2} = \frac{\partial^2 u}{\partial x^2}; \quad 0 < t < \infty; \quad -\infty < x < \infty,$$

subject to

$$u(x,0) = \begin{cases} -\cos x, \ |x| < \pi \\ \\ 0, \ |x| > \pi \end{cases}$$

and

$$\frac{\partial u(x,0)}{\partial t} = 0?$$

Sketch the solution for several values of time, t.

3.6.7. What is the solution to the initial-value problem

$$\frac{\partial^2 u}{\partial t^2} = \frac{1}{4}\frac{\partial^2 u}{\partial x^2}; \quad 0 < t < \infty; \quad -\infty < x < \infty,$$

subject to

$$u(x,0) = 0$$

and

$$\frac{\partial u(x,0)}{\partial t} = \begin{cases} \sin x, \ |x| < \pi \\ \\ 0, \ |x| > \pi? \end{cases}$$

Sketch the solution for several values of time, t.

3.6.8. What is the solution to the initial-value problem

$$\frac{\partial^2 u}{\partial t^2} = \frac{1}{9}\frac{\partial^2 u}{\partial x^2}; \quad 0 < t < \infty; \quad -\infty < x < \infty,$$

subject to

$$u(x, 0) = 0$$

and

$$\frac{\partial u(x,0)}{\partial t} = \begin{cases} \cos x, & |x| < \dfrac{\pi}{2} \\[2mm] 0, & |x| > \dfrac{\pi}{2}? \end{cases}$$

Sketch the solution for several values of time, t.

3.6.9. Solve for propagation of electrical vibrations in an infinite conductor for the condition

$$GL = CR,$$

where G, L, C, and R are the leakage conductance, self-inductance, capacitance, and resistance, respectively, per unit length of the conductor. At the start of the experiment, the current is $i(x, 0) = f(x)$ and the voltage is $v(x, 0) = g(x)$.

3.6.10. Given

$$\frac{\partial^2 u}{\partial t^2} = c^2 \frac{\partial^2 u}{\partial x^2}, \tag{3.55}$$

subject to

$$\left. \begin{array}{l} u(x, 0) = f(x) \\[3mm] \dfrac{\partial u(x,0)}{\partial t} = g(x) \end{array} \right\}, \quad -\infty < x < \infty, \tag{3.56}$$

consider the coordinate transformations $\xi = x - ct$ and $\eta = x + ct$. Using the chain rule on $v(\xi, \eta)$, develop d'Alembert' solution for Equations (3.55 and 3.56). *Hint:* Let $u(x, t) = v(\xi, \eta)$ and determine $\dfrac{\partial^2 u}{\partial t^2}$ and $\dfrac{\partial^2 u}{\partial x^2}$ in terms of $v(\xi, \eta)$.

Chapter 4

The Essentials of Fourier Series

4.1 INTRODUCTION

In previous chapters, we discussed the derivation of the heat and wave equations. Also, we discussed the solutions to the steady-state problem for the heat equation, and d'Alembert's method for the wave equation in an infinite string. Although these solution methods are important in their own right, they do not tell us the entire story of the partial differential equations governing heat conduction in a rod, or motion of a tightly stretched finite string. In fact, the steady-state method tells us nothing about heat conduction in a rod at time $t = 5$ seconds, or, for that matter, at any finite time t. Although there are ways of developing d'Alembert's method for a finite string, they aren't always the best or the preferred solutions, nor is there an equivalent d'Alembert's method for the vibrations in a drumhead. However, other methods can solve wave motion in a string and temperature distribution in a rod, as well as multispatial dimensions; these methods are based on Fourier series.

Traditionally, Fourier series is introduced as the closed form solution to a second-order PDE with boundary conditions. Although this method is completely correct, and was originally introduced by Fourier, it sometimes leaves you with the feeling that something is missing. The effect is that students must rely on brute memory to get through the material. In this chapter, we introduce Fourier series by making use of an earlier concept, which you've already used or have been introduced to, linear algebra.

Linear algebra is central to a fuller understanding of mathematics. It lays the foundation for the higher mathematics used in physics, engineering, and applied mathematics. The framework of linear algebra plays a crucial role in the development of quantum mechanics, solving systems of ODEs in engineering, and the advanced mathematical tools used by applied mathematicians.

Using linear algebra as an introduction to Fourier series may seem to you like the blind leading the blind. Many engineering and physics students have never been

formally introduced to linear algebra. But, all students who have taken the calculus
sequence, ODEs, and some type of engineering mathematics course have worked
with linear algebra through most of their college career. Therefore, this chapter
builds on this material to develop the ideas from linear algebra that are needed
to understand the connection with Fourier series in the section titled "Elements of
Linear Algebra." The chapter then develops the mechanics of Fourier series. Finally,
even and odd functions and their Fourier series representation are discussed.

The linear algebra portion of this chapter should not be construed as a replacement for a linear algebra course. We cover only those theorems and ideas relevant
to developing Fourier series. Also, we do not present the proofs nor much of the
mechanical manipulations that are necessary for a better understanding of linear
algebra. If you enjoy the material covered in the linear algebra portion, I would
suggest you take a linear algebra course specifically designed for applications. Also,
the notation of this chapter may be new to you. If this is the case, I suggest you
read Appendix D on mathematical notation.

4.2   ELEMENTS OF LINEAR ALGEBRA

First, linear algebra needs to be defined. This is not an easy task because, quite
literally, it means the algebra of lines or the algebraic study of lines. Therefore, we
need to think of a more informative approach. I'll start with an explanation of the
term, "linear," in linear algebra. Then, I'll explain how the word "algebra" fits in.

You have a geometric and algebraic understanding of linear. For example, the
equation $y = mx + b$ geometrically describes a line in the xy-plane called $\mathbb{R}^2$. In
algebra the same equation $y = mx + b$ is called a linear equation. However, in linear
algebraic vocabulary this equation is known as an **affine equation**. An affine
equation is basically a shift of a linear equation. In linear algebra, a linear equation
must go through the origin: This means that $b = 0$ in the equation $y = mx + b$.
This concept of "linear" in linear algebra forces a mathematical assumption. That
assumption is that a line going through the origin contains all the information about
all the lines parallel to it in the plane, except the y intercept, which is nothing more
than the shift of the line. In $\mathbb{R}^3$, this concept is a little more complicated, but the
idea is the same.

Now that we understand what "linear" means in linear algebra, the concept of
"algebra" in linear algebra must also be explained. In high school, you learned
about algebra on the real numbers. This is basically the study of multiplication
and addition on general real numbers, using equations to explain simple physical
phenomena. In the equations, scalars (commonly called constants) came from the
real numbers. This really does not change much in linear algebra. You still work
with equations. Also, scalars are still real numbers. However, the general numbers
in the equations have changed to vectors. Thus, $\mathbb{R}^2$ is a space in linear algebra.

Linear algebra, as in all subjects, has evolved past its humble beginnings of
$\mathbb{R}^2$ and has been applied to other spaces. To discuss these more evolved spaces,
development of a more general space structure was needed. The general space
structure was called a vector space. Don't let the words "vector space" scare you.

The definition is just a mathematical formalization describing something that occurs in nature. In fact, you have been living in a vector space all your life. For example, if you consider where you are standing as the origin, then an axis running through your head into the ground could be the z-axis and the xy-plane would be the ground you are standing on. This is actually a rough copy of $\mathbb{R}^3$, which is a vector space.

Understanding the definition of a vector space is fundamental to the study of linear algebra. We define vector space in the next subsection.

4.2.1 Vector Space

Definition 7. *A **vector space** $\mathcal{V}$ is a non-empty set of elements x and y (called vectors), such that the algebraic operations, vector addition and scalar multiplication, hold.*

This is a very concise definition, but what does it mean? We will start with vector addition.

Vector addition includes several ideas. First, when adding two vectors in a vector space $\mathcal{V}$, the sum is also in $\mathcal{V}$. This concept is known as closure of the space. Second, vector addition is commutative: If $\mathbf{x}$ and $\mathbf{y}$ are vectors in $\mathcal{V}$, then $\mathbf{x} + \mathbf{y} = \mathbf{y} + \mathbf{x}$. Third, vector addition is associative. This means that if $\mathbf{x}$, $\mathbf{y}$, and $\mathbf{z}$ are vectors in $\mathcal{V}$, then $\mathbf{x} + (\mathbf{y} + \mathbf{z}) = (\mathbf{x} + \mathbf{y}) + \mathbf{z}$. Last, there exists a unique vector $\mathbf{0}$ in $\mathcal{V}$ such that if $\mathbf{x}$ is any vector in $\mathcal{V}$, then $\mathbf{x} + \mathbf{0} = \mathbf{x}$. Also, there exists the vector $-\mathbf{x}$ in $\mathcal{V}$, such that $\mathbf{x} + (-\mathbf{x}) = \mathbf{0}$. Next, we consider scalar multiplication.

Scalar multiplication also includes several ideas. First, a scalar is nothing more than a number, and sometimes you will read or hear the phrase "a real vector space." This means that the scalars are all real numbers. If you are reading a math text and you see the phrase "complex vector space," then the scalars come from the complex plane and have the form $a + bi$. In this text, we will work exclusively with real vector spaces. Also, a vector space $\mathcal{V}$ is closed under scalar multiplication. Second, scalar multiplication distributes over vector addition. This means if $\mathbf{x}$ and $\mathbf{y}$ are vectors in $\mathcal{V}$, and a is any real number, then $a(\mathbf{x} + \mathbf{y}) = a\mathbf{x} + a\mathbf{y}$. Third, scalar multiplication is associative. That is, if $\mathbf{x}$ is any vector in $\mathcal{V}$ and a and b are real numbers, then $a(b\mathbf{x}) = (ab)\mathbf{x}$. Finally, there exists an identity element for scalar multiplication. Since we are working exclusively with real vector spaces, you can clearly see that 1 is our identity element.

In many texts, Definition (7) is written in the following form:

Definition 8. *A **real vector space** $\mathcal{V}$ is a non-empty set of elements x and y (called vectors), such that:*

1. For any pair of vectors x and y in $\mathcal{V}$, there exists a unique vector in $\mathcal{V}$ that is the sum of x and y.

2. Vector addition is commutative. For any pair of vectors x and y in $\mathcal{V}$, then $x + y = y + x$.

3. Vector addition is associative. For any vectors x, y, and z in $\mathcal{V}$, then $x+(y+z) = (x+y)+z$.

4. There exists a unique vector 0 in $\mathcal{V}$, such that for all vectors x in $\mathcal{V}$, $x+0=x$.

5. For every vector x in $\mathcal{V}$, there exists the vector $-x$ in $\mathcal{V}$, such that $x+(-x) = 0$.

6. For all vectors x in $\mathcal{V}$ and all real numbers, a, the unique vector ax is in $\mathcal{V}$.

7. Scalar multiplication distributes over vector addition. For any pair of vectors x and y in $\mathcal{V}$ and real number a, we have $a(x+y) = ay+ax$.

8. Scalar multiplication is associative. For all vectors x in $\mathcal{V}$ and the real numbers a and b, $a(bx) = (ab)x$.

9. The scalar 1 is the identity element for scalar multiplication.

You may use either Definition (7) or Definition (8). Examples of real vector spaces are $\mathbb{R}^2$ (the xy-plane) and $\mathbb{R}^3$ (three-dimensional space).

The next subsection increases our knowledge of a vector space $\mathcal{V}$ by introducing terms you already know from previous courses. These terms are linear dependence, linear independence, and basis.

4.2.2   Linear Dependence, Linear Independence, and Basis

Central to the concepts of linear dependence and linear independence is the concept of linear combination of vectors in a vector space. A linear combination of vectors is the term for the addition of two or more vectors in a vector space. In general, we use the following definition:

Definition 9. *Let $\mathcal{V}$ be a vector space. Suppose the vectors u_1, u_2, u_3, ..., u_n are elements of $\mathcal{V}$ and a_1, a_2, a_3, ..., a_n are scalars. Then, a linear combination of the vectors u_1, u_2, u_3, ..., u_n is*

$$a_1 u_1 + a_2 u_2 + a_3 u_3 + \ldots + a_n u_n.$$

Note: A linear combination of vectors in a vector space yields a vector. Remember, the zero vector, written 0, is a valid vector.

EXAMPLE 4.1. Consider the following vectors in standard form in $\mathbb{R}^3$.

$$\mathbf{u} = \begin{bmatrix} 1 \\ -2 \\ 3 \end{bmatrix}, \mathbf{v} = \begin{bmatrix} -2 \\ 0 \\ 1 \end{bmatrix}, \text{ and } \mathbf{w} = \begin{bmatrix} 5 \\ -6 \\ 8 \end{bmatrix}.$$

A possible linear combination is

$$\begin{bmatrix} 1 \\ -2 \\ 3 \end{bmatrix} + 2\begin{bmatrix} -2 \\ 0 \\ 1 \end{bmatrix} - 5\begin{bmatrix} 5 \\ -6 \\ 8 \end{bmatrix}.$$ ∎

Also, at first glance, the three vectors in the above example seem to be unrelated. But, a careful inspection shows $3\mathbf{u} - \mathbf{v} = \mathbf{w}$ or

$$3 \begin{bmatrix} 1 \\ -2 \\ 3 \end{bmatrix} - \begin{bmatrix} -2 \\ 0 \\ 1 \end{bmatrix} = \begin{bmatrix} 5 \\ -6 \\ 8 \end{bmatrix}.$$

This means the vector $\mathbf{w}$ can be written as a linear combination of the other two vectors. Thus, the vectors $\mathbf{u}$, $\mathbf{v}$, and $\mathbf{w}$ are linearly dependent. The formal definition of linear dependence in a general vector space follows:

Definition 10. *Let $\mathcal{V}$ be a vector space. Suppose the vectors $\mathbf{u}_1$, $\mathbf{u}_2$, $\mathbf{u}_3$, ..., $\mathbf{u}_n$ are elements of $\mathcal{V}$. If a_1, a_2, a_3, ..., a_n are scalars and a linear combination of the vectors $\mathbf{u}_1$, $\mathbf{u}_2$, $\mathbf{u}_3$, ..., $\mathbf{u}_n$,*

$$a_1\mathbf{u}_1 + a_2\mathbf{u}_2 + a_3\mathbf{u}_3 + \ldots + a_n\mathbf{u}_n = \mathbf{0},$$

(where a_i, $1 \leq i \leq n$) are not all equal to zero, then the vectors $\mathbf{u}_1$, $\mathbf{u}_2$, $\mathbf{u}_3$, ..., $\mathbf{u}_n$ are linearly dependent.

To explain this definition, consider the three vectors $\mathbf{u}$, $\mathbf{v}$, and $\mathbf{w}$ given in Example (4.1). Since $\mathbf{w} = 3\mathbf{u} - \mathbf{v}$, we have $\mathbf{w} - 3\mathbf{u} + \mathbf{v} = \mathbf{0}$. Here $a_1 = 1$, $a_2 = -3$, and $a_3 = 1$. Now that we know what linearly dependent vectors are, we need to define linearly independent vectors.

Definition 11. *Let $\mathcal{V}$ be a vector space. Suppose the vectors $\mathbf{v}_1$, $\mathbf{v}_2$, $\mathbf{v}_3$, ..., $\mathbf{v}_n$ are elements of $\mathcal{V}$ and a_1, a_2, a_3, ..., a_n scalars. Then, the linear combination of the vectors $\mathbf{v}_1$, $\mathbf{v}_2$, $\mathbf{v}_3$, ..., $\mathbf{v}_n$ is given by*

$$a_1\mathbf{v}_1 + a_2\mathbf{v}_2 + a_3\mathbf{v}_3 + \ldots + a_n\mathbf{v}_n = \mathbf{0}.$$

The vectors $\mathbf{v}_1$, $\mathbf{v}_2$, $\mathbf{v}_3$, ..., $\mathbf{v}_n$ are said to be linearly independent only when $a_i = 0$ for all i such that $1 \leq i \leq n$.

To demonstrate this definition clearly, consider the following example.

EXAMPLE 4.2. Let

$$\mathbf{v}_1 = \begin{bmatrix} 1 \\ 1 \\ -1 \end{bmatrix}, \mathbf{v}_2 = \begin{bmatrix} -1 \\ 1 \\ -1 \end{bmatrix}, \text{ and } \mathbf{v}_3 = \begin{bmatrix} 0 \\ 1 \\ 1 \end{bmatrix}.$$

We will show that these vectors are linearly independent. Suppose we form the linear combination of the vectors $\mathbf{v}_1$, $\mathbf{v}_2$, and $\mathbf{v}_3$ as

$$a_1 \begin{bmatrix} 1 \\ 1 \\ -1 \end{bmatrix} + a_2 \begin{bmatrix} -1 \\ 1 \\ -1 \end{bmatrix} + a_3 \begin{bmatrix} 0 \\ 1 \\ 1 \end{bmatrix} = \begin{bmatrix} 0 \\ 0 \\ 0 \end{bmatrix},$$

where we assume at least one of the scalars a_1, a_2, or a_3 does not equal 0. From the first line across all three vectors, we find that $a_1 = a_2$ because $a_1 - a_2 = 0$. Also, from the second line across all three vectors, $a_3 = -2a_1$ since $a_1 + a_2 + a_3 = 0$. Using substitution we have $2a_1 + a_3 = 0$. However, from the third line across all three lines across all three vectors, $-a_1 - a_2 + a_3 = 0$, implying $a_3 = 2a_1$. Thus, $2a_1 = a_3 = -2a_1$. This in turn implies a contradiction to our assumption, since $2a_1 = a_3 = -2a_1$ implies $a_1 = 0$, which means $a_2 = 0$ and $a_3 = 0$.   ∎

It is easy to see that the vectors

$$\begin{bmatrix} 1 \\ 0 \end{bmatrix} \text{ and } \begin{bmatrix} 0 \\ 1 \end{bmatrix}$$

in $\mathbb{R}^2$ and

$$\begin{bmatrix} 1 \\ 0 \\ 0 \end{bmatrix}, \begin{bmatrix} 0 \\ 1 \\ 0 \end{bmatrix}, \text{ and } \begin{bmatrix} 0 \\ 0 \\ 1 \end{bmatrix}$$

in $\mathbb{R}^3$ are linearly independent. These vectors are referred to as the **standard basis** vectors for the vector spaces $\mathbb{R}^2$ and $\mathbb{R}^3$, respectively. Considering the format of the standard basis in $\mathbb{R}^2$ and $\mathbb{R}^3$, we could guess that the standard basis in $\mathbb{R}^n$ would have n vectors, each vector with n components, one in the nth position and all other positions with a zero. This would be a good guess and quite correct. In fact, the usual way to state the standard basis in $\mathbb{R}^n$ is

$$\begin{bmatrix} 1 \\ 0 \\ 0 \\ 0 \\ \vdots \\ 0 \end{bmatrix}, \begin{bmatrix} 0 \\ 1 \\ 0 \\ 0 \\ \vdots \\ 0 \end{bmatrix}, \begin{bmatrix} 0 \\ 0 \\ 1 \\ 0 \\ \vdots \\ 0 \end{bmatrix}, \cdots, \begin{bmatrix} 0 \\ 0 \\ 0 \\ \vdots \\ 0 \\ 1 \end{bmatrix},$$

where it is understood that there are n vectors with n components in each vector.

There are other bases for $\mathbb{R}^2$, $\mathbb{R}^3$, and $\mathbb{R}^n$. These will be discussed in a later subsection. Next, we'll discuss orthogonality and inner product. These topics may sound new to you, but they really are not.

4.2.3   Orthogonality and Inner Product

$\mathbb{R}^2$ and $\mathbb{R}^3$ are the vector spaces most familiar to you. The following is a good geometric definition of orthogonality: If two vectors intersect and the intersection forms a right angle, then the two vectors are orthogonal to each other. This means the two vectors are perpendicular to each other. A more mathematical definition is as follows:

Definition 12. *Two vectors are orthogonal if and only if their dot product is zero.*

Since we know how to perform the dot product, Definition (12) gives us a way to determine if two random vectors are orthogonal. This removes the uncertainty of guessing or trying to geometrically determine orthogonality, which may require the use of a straight edge and protractor.

You have determined in many previous courses, by using the dot product, that the standard bases in $\mathbb{R}^2$ and $\mathbb{R}^3$ is an orthogonal set, and you can see that the standard basis in $\mathbb{R}^n$ is also an orthogonal set. Also, it is standard practice in calculus courses to show two or more random vectors are orthogonal. Therefore, I will not run through an example.

The question that should come to your mind is "Why use the term orthogonal if it just means perpendicular in a geometric sense?" In $\mathbb{R}^2$ and $\mathbb{R}^3$, it is easy to understand when two vectors are perpendicular, thus orthogonal. In $\mathbb{R}^n$ for $n > 3$, it is impossible to see, but you can somewhat imagine perpendicular lines by relating them to $\mathbb{R}^2$ or $\mathbb{R}^3$. Thus, you have an idea of the concept of orthogonality here, also. However, a vector space was defined in general terms and real vector spaces—$\mathbb{R}^2$, $\mathbb{R}^3$, and $\mathbb{R}^n$—are not the only vector spaces, they are just the vector spaces you are most familiar with. For instance, another vector space is the vector space composed of all 2×2 matrices with real-number components. It is called the matrix space of 2×2 matrices, and it is denoted $\mathbf{M}_{2,2}$. In the matrix space, $\mathbf{M}_{2,2}$, perpendicular matrices have no meaning. Thus, "orthogonal" has become the word most preferred. Also, the term "orthogonal" has become tied to another term, "inner product."

The standard inner product of two vectors in $\mathbb{R}^2$, $\mathbb{R}^3$, or $\mathbb{R}^n$ is our old friend the dot product. However, the mathematical notation has changed. The dot product between two vectors $\mathbf{u}$ and $\mathbf{v}$ was always denoted as $\mathbf{u} \bullet \mathbf{v}$. Now the inner product between the same two vectors $\mathbf{u}$ and $\mathbf{v}$ is denoted as $\langle \mathbf{u}, \mathbf{v} \rangle$, and when we discuss the standard inner product on $\mathbb{R}^2$, $\mathbb{R}^3$, or $\mathbb{R}^n$, the notation is $\langle \mathbf{u}, \mathbf{v} \rangle = \mathbf{u} \bullet \mathbf{v}$.

Since the dot product is now an inner product, the definition for orthogonality can be restated:

Definition 13. *Two vectors are orthogonal if and only if their inner product is zero.*

We use the term inner product because the dot product only works on the vector spaces $\mathbb{R}^n$, for $n \in \mathbb{N}$. In other vector spaces, the dot product would make no sense. For instance, the matrix space $\mathbf{M}_{2,2}$ cannot use the dot product as its inner product. This naturally leads us to the general definition of an inner product.

Definition 14. *Let $\mathcal{V}$ be any real vector space. An inner product on $\mathcal{V}$ is a function that assigns a real number to each pair of vectors $\boldsymbol{u}$ and $\boldsymbol{v}$ of $\mathcal{V}$, written $\langle \boldsymbol{u}, \boldsymbol{v} \rangle = a \in \mathbb{R}$, satisfying:*

1. $\langle \boldsymbol{u}, \boldsymbol{u} \rangle > 0$ for $\boldsymbol{u} \neq 0$, and $\langle \boldsymbol{u}, \boldsymbol{u} \rangle = 0$ if and only if $\boldsymbol{u} = 0$.

2. $\langle \boldsymbol{u}, \boldsymbol{v} \rangle = \langle \boldsymbol{v}, \boldsymbol{u} \rangle$ for any $\boldsymbol{u}$ and $\boldsymbol{v}$ in $\mathcal{V}$.

3. $\langle \boldsymbol{u} + \boldsymbol{v}, \boldsymbol{w} \rangle = \langle \boldsymbol{u}, \boldsymbol{w} \rangle + \langle \boldsymbol{v}, \boldsymbol{w} \rangle$ for any $\boldsymbol{u}$, $\boldsymbol{v}$, and $\boldsymbol{w}$ in $\mathcal{V}$.

4. $\langle b\boldsymbol{u}, \boldsymbol{v} \rangle = b \langle \boldsymbol{u}, \boldsymbol{v} \rangle$ for $\boldsymbol{u}$ and $\boldsymbol{v}$ in $\mathcal{V}$ and b a scalar in $\mathbb{R}$.

It is left as an exercise for you to show that the inner product (dot product) in the vector spaces $\mathbb{R}^2$ and $\mathbb{R}^3$ is, in fact, an inner product. Also, it should be noted, any vector space where an inner product has been defined is known as an inner product space.

An example of an inner product space, that is not one of the $\mathbb{R}^n$'s is the inner product space composed of all 2×2 matrices with real-number components. If A and B are elements of $\mathbf{M}_{2,2}$, then the inner product is defined as $\langle A, B \rangle = tr(B^T A)$. In the next example, we prove the space of all 2×2 matrices with real-number components is an inner product space.

EXAMPLE 4.3. We must show that the space of all 2×2 matrices with real-number components, $\mathbf{M}_{2,2}$, is an inner product space. We are using the inner product, $\langle A, B \rangle = tr(B^T A)$. We start with the vector A, where

$$A = \left[\begin{array}{cc} a & b \\ c & d \end{array} \right].$$

First, show that

$$\langle A, A \rangle > 0 \text{ for } A \neq 0 \text{ and } A \in \mathbf{M}_{2,2}.$$

Thus, we have

$$\langle A, A \rangle = tr(A^T A) = tr\left(\left[\begin{array}{cc} a & c \\ b & d \end{array} \right] \left[\begin{array}{cc} a & b \\ c & d \end{array} \right] \right)$$

$$= tr\left(\left[\begin{array}{cc} a^2 + c^2 & ab + cd \\ ba + dc & b^2 + d^2 \end{array} \right] \right)$$

$$= a^2 + c^2 + b^2 + d^2 > 0 \text{ for } a, b, c, \text{ and } d \neq 0.$$

If

$$A = \left[\begin{array}{cc} 0 & 0 \\ 0 & 0 \end{array} \right],$$

then

$$\langle A, A \rangle = 0 \text{ for } A \in \mathbf{M}_{2,2}.$$

Also, if

$$A = \left[\begin{array}{cc} a & b \\ c & d \end{array} \right]$$

and we have

$$\langle A, A \rangle = tr(A^T A) = tr \left(\begin{bmatrix} a & c \\ b & d \end{bmatrix} \begin{bmatrix} a & b \\ c & d \end{bmatrix} \right)$$

$$= tr \left(\begin{bmatrix} a^2 + c^2 & ab + cd \\ ba + dc & b^2 + d^2 \end{bmatrix} \right)$$

$$= a^2 + c^2 + b^2 + d^2 = 0$$

then a, b ,c, and $d = 0$. Second, we show that

$$\langle A, B \rangle = \langle B, A \rangle \text{ for any } A \text{ and } B \text{ in } \mathbf{M}_{2,2}.$$

Let

$$A = \begin{bmatrix} a & b \\ c & d \end{bmatrix} \text{ and } B = \begin{bmatrix} e & f \\ g & h \end{bmatrix}.$$

Then,

$$\langle A, B \rangle = tr(B^T A) = tr \left(\begin{bmatrix} e & g \\ f & h \end{bmatrix} \begin{bmatrix} a & b \\ c & d \end{bmatrix} \right)$$

$$= tr \left(\begin{bmatrix} ea + gc & eb + gd \\ fa + hc & fb + hd \end{bmatrix} \right)$$

$$= ea + gc + fb + hd,$$

and

$$\langle B, A \rangle = tr(A^T B) = tr \left(\begin{bmatrix} a & c \\ b & d \end{bmatrix} \begin{bmatrix} e & f \\ g & h \end{bmatrix} \right)$$

$$= tr \left(\begin{bmatrix} ae + cg & af + ch \\ be + dg & bf + dh \end{bmatrix} \right)$$

$$= ae + cg + bf + dh.$$

Therefore,

$$\langle A, B \rangle = \langle B, A \rangle \text{ for any } A \text{ and } B \text{ in } \mathbf{M}_{2,2}.$$

Next, we show that

$$\langle A + B, C \rangle = \langle A, C \rangle + \langle B, C \rangle \text{ for any } A, B, \text{ and } C \text{ in } \mathbf{M}_{2,2}.$$

Let

$$A = \begin{bmatrix} a & b \\ c & d \end{bmatrix}, \ B = \begin{bmatrix} e & f \\ g & h \end{bmatrix}, \text{ and } C = \begin{bmatrix} m & n \\ o & p \end{bmatrix}.$$

Then,

$$\langle A + B, C \rangle = tr\left(C^T(A+B)\right) = tr\left(\begin{bmatrix} m & o \\ n & p \end{bmatrix} \begin{bmatrix} a+e & b+f \\ c+g & d+h \end{bmatrix}\right)$$

$$= tr\left(\begin{bmatrix} m(a+e) + o(c+g) & m(b+f) + o(d+h) \\ n(a+e) + p(c+g) & n(b+f) + p(d+h) \end{bmatrix}\right)$$

$$= m(a+e) + o(c+g) + n(b+f) + p(d+h),$$

and

$$\langle A, C \rangle + \langle B, C \rangle = tr(C^T A) + tr(C^T B)$$

$$= tr\left(\begin{bmatrix} m & o \\ n & p \end{bmatrix} \begin{bmatrix} a & b \\ c & d \end{bmatrix}\right) + tr\left(\begin{bmatrix} m & o \\ n & p \end{bmatrix} \begin{bmatrix} e & f \\ g & h \end{bmatrix}\right)$$

$$= tr\left(\begin{bmatrix} ma + oc & mb + od \\ na + pc & nb + pd \end{bmatrix}\right) + tr\left(\begin{bmatrix} me + og & mf + oh \\ ne + pg & nf + ph \end{bmatrix}\right)$$

$$= ma + oc + nb + pd + me + og + nf + ph$$

$$= m(a+e) + o(c+g) + n(b+f) + p(d+h).$$

Therefore,

$$\langle A + B, C \rangle = \langle A, C \rangle + \langle B, C \rangle \text{ for any } A, B, \text{ and } C \text{ in } \mathbf{M}_{2,2}.$$

Finally, we show that

$$\langle \beta A, B \rangle = \beta \langle A, B \rangle \text{ for any } A \text{ and } B \text{ in } \mathbf{M}_{2,2} \text{ and } \beta \text{ in } \mathbb{R}.$$

Let

$$A = \begin{bmatrix} a & b \\ c & d \end{bmatrix} \text{ and } B = \begin{bmatrix} e & f \\ g & h \end{bmatrix}.$$

Then,

$$\langle \beta A, B \rangle = tr\left(B^T \beta A\right) = tr\left(\begin{bmatrix} e & g \\ f & h \end{bmatrix}\begin{bmatrix} \beta a & \beta b \\ \beta c & \beta d \end{bmatrix}\right)$$

$$= tr\left(\begin{bmatrix} e\beta a + g\beta c & e\beta b + g\beta d \\ f\beta a + h\beta c & f\beta b + h\beta d \end{bmatrix}\right)$$

$$= e\beta a + g\beta c + f\beta b + h\beta d = \beta\left(ea + gc + fb + hd\right),$$

and

$$\beta\langle A, B \rangle = \beta tr\left(\begin{bmatrix} e & g \\ f & h \end{bmatrix}\begin{bmatrix} a & b \\ c & d \end{bmatrix}\right)$$

$$= \beta tr\left(\begin{bmatrix} ea + gc & eb + gd \\ fa + hc & fb + hd \end{bmatrix}\right)$$

$$= \beta\left(ea + gc + fb + hd\right).$$

Thus,

$$\langle \beta A, B \rangle = \beta\langle A, B \rangle \text{ for any } A \text{ and } B \text{ in } \mathbf{M}_{2,2} \text{ and } \beta \text{ in } \mathbb{R}.$$

Therefore, we have shown that $\mathbf{M}_{2,2}$ is an inner product space with inner product defined by $\langle A, B \rangle = tr(B^T A)$. ∎

Later in this chapter, a completely different inner product in a completely different space will be introduced. This new inner product is crucial to our understanding of how Fourier series works with PDEs. For now though, we move onto topics you should be familiar with, eigenvalues and eigenvectors.

4.2.4   Eigenvalues and Eigenvectors

In this subsection, the one definition is stated for the general case $n \times n$. However, all discussions and examples will be with respect to the vector space $\mathbb{R}^2$. This vector space is quite familiar, and it facilitates the comprehension of the ideas underlying the material of **eigenvalues** and **eigenvectors**.

The first time you encountered eigenvalues and eigenvectors may have been in an ODE course. Eigenvalues and eigenvectors played an important part in the solution of systems of first-order linear ODEs. However, the emphasis was on the solution to the system, not on eigenvalues and eigenvectors. Here, we are primarily concerned with eigenvalues and eigenvectors themselves. Let's start with a definition.

Definition 15. *Let A be an $n \times n$ matrix. Then,, a real number λ is called an eigenvalue of the matrix A, if and only if for some nonzero $n \times 1$ vector $\boldsymbol{x}$, the equation*

$$A\boldsymbol{x} = \lambda\boldsymbol{x}$$

is true. If the vector $\boldsymbol{x}$ exists, then the vector $\boldsymbol{x}$ is called the eigenvector corresponding to the eigenvalue λ.

The definition does not indicate how to find the eigenvalue and the associated eigenvector. However, we determine the eigenvalues by solving the equation

$$|A - \lambda I| = 0.$$

Remember the symbol $|\;\;|$ means to find the determinant of the matrix.

EXAMPLE 4.4. Consider the matrix

$$A = \begin{pmatrix} 5 & 3 \\ 3 & 5 \end{pmatrix}.$$

Determine the eigenvalues and associated eigenvectors for the matrix A. First, form the matrix $A - \lambda I$. This matrix is

$$A - \lambda I = \begin{pmatrix} 5 & 3 \\ 3 & 5 \end{pmatrix} - \lambda \begin{pmatrix} 1 & 0 \\ 0 & 1 \end{pmatrix} = \begin{pmatrix} 5 - \lambda & 3 \\ 3 & 5 - \lambda \end{pmatrix}.$$

Next, find the determinant of $A - \lambda I$, and set $|A - \lambda I| = 0$. That is,

$$0 = \begin{vmatrix} 5 - \lambda & 3 \\ 3 & 5 - \lambda \end{vmatrix} = (5 - \lambda)(5 - \lambda) - 9 = \lambda^2 - 10\lambda + 25 - 9$$

$$= \lambda^2 - 10\lambda + 16.$$

The equation $\lambda^2 - 10\lambda + 16 = 0$ is called the **characteristic polynomial**, and it can be solved for roots by factoring. Once factored, the roots are $\lambda = 2$ and $\lambda = 8$. These are the eigenvalues. Generally, they are labeled as $\lambda_1 = 2$ and $\lambda_2 = 8$.

To determine the associated eigenvector for each of the eigenvalues, form the equation $(A - \lambda I)\,\mathbf{x} = \mathbf{0}$ for each eigenvalue. Thus, for the eigenvector corresponding to the first eigenvalue, we solve

$$\mathbf{0} = (A - \lambda_1 I)\,\mathbf{x} = \begin{pmatrix} 5 - \lambda_1 & 3 \\ 3 & 5 - \lambda_1 \end{pmatrix} \begin{pmatrix} x_1 \\ x_2 \end{pmatrix}.$$

This equation can be solved using the Gauss–Jordan reduction method and the eigenvector $\mathbf{x}$ is given by

$$\begin{pmatrix} x_1 \\ x_2 \end{pmatrix} = \begin{pmatrix} 1 \\ -1 \end{pmatrix}.$$

We now have the corresponding eigenvector to the eigenvalue $\lambda_1 = 2$. In a similar fashion, the corresponding eigenvector $\mathbf{x}$ to the eigenvalue $\lambda_2 = 8$ is

$$\begin{pmatrix} x_1 \\ x_2 \end{pmatrix} = \begin{pmatrix} 1 \\ 1 \end{pmatrix}. \qquad\qquad \blacksquare$$

It is interesting to note that the eigenvectors

$$\begin{pmatrix} 1 \\ -1 \end{pmatrix} \text{ and } \begin{pmatrix} 1 \\ 1 \end{pmatrix}$$

are orthogonal to each other.

The previous example shows that eigenvalues and eigenvectors can be determined in a systematic way. Also, it is important to note an $n \times n$ matrix generally has n distinct eigenvalues, each of which has a corresponding eigenvector. In the next subsection, we tie all the pieces of linear algebra that are analogous to the development of Fourier series together.

4.2.5   Significance

In the previous subsections, vector spaces, linear combinations, linear independence, inner product, orthogonality, and eigenvalues and eigenvectors were discussed, but they were not related. Here, hopefully, we will fit the pieces of the puzzle together. All of the examples will be centered on the real vector spaces, $\mathbb{R}^2$ and $\mathbb{R}^3$, since you are most familiar with them.

Everything that occurs in this section occurs in a vector space. This is an important concept. It means mathematics and reality are tied together in some fashion and that fashion is the development of mathematical objects that act the same as three space, $\mathbb{R}^3$.

In a vector space, a basis is extremely important and necessary. A basis is composed of linearly independent vectors. Another important point in the real vector spaces $\mathbb{R}^n$, $n \in \mathbb{N}$, is that the n in the superscript position tells us how many vectors are in the basis. For instance, $\mathbb{R}^3$ has three vectors in the standard basis, and they are

$$\begin{bmatrix} 1 \\ 0 \\ 0 \end{bmatrix}, \begin{bmatrix} 0 \\ 1 \\ 0 \end{bmatrix}, \text{ and } \begin{bmatrix} 0 \\ 0 \\ 1 \end{bmatrix}.$$

Also, the n in the superscript position identifies the vector space to be finite dimensional. This does not mean the space does not go on to infinity, it just means there are only a finite number of copies of the real axis being used to make up the space. Are there infinite dimensional spaces? The answer is yes, and in the next section you will be introduced to one, the function space of piecewise smooth functions. But, before we discuss infinite dimensional space, let's tie together some of the tools we need from the real vector spaces.

Basis vectors are linearly independent. This means any vector in the space can be expressed as a linear combination of the basis vectors.

EXAMPLE 4.5. Express the vector

$$\begin{bmatrix} -3 \\ 18 \\ 7 \end{bmatrix}$$

as a linear combination of the standard basis vectors. The solution is

$$\begin{bmatrix} -3 \\ 18 \\ 7 \end{bmatrix} = -3 \begin{bmatrix} 1 \\ 0 \\ 0 \end{bmatrix} + 18 \begin{bmatrix} 0 \\ 1 \\ 0 \end{bmatrix} + 7 \begin{bmatrix} 0 \\ 0 \\ 1 \end{bmatrix}. \qquad \blacksquare$$

Although Example (4.5) is easy to complete and, in fact, can be completed almost unconsciously, it really should bring up two very important questions. First, is the standard basis the only basis? Second, if another basis exists, how do we determine the multiplicative constants of the basis to form a linear combination as in the standard basis? We will answer each of these questions because an understanding of the underlying concepts is necessary.

The standard basis is not the only basis for $\mathbb{R}^3$, or for that matter any vector space. The standard basis is the easiest one to work with because in most vector spaces the standard basis is known to be orthonormal, meaning the vectors that make up the basis have unit length and are orthogonal to each other.

Another orthonormal basis for $\mathbb{R}^3$ is

$$\begin{bmatrix} \dfrac{1}{\sqrt{3}} \\ \dfrac{1}{\sqrt{3}} \\ \dfrac{1}{\sqrt{3}} \end{bmatrix}, \quad \begin{bmatrix} \dfrac{-1}{\sqrt{6}} \\ \dfrac{2}{\sqrt{6}} \\ \dfrac{-1}{\sqrt{6}} \end{bmatrix}, \text{ and } \begin{bmatrix} \dfrac{-1}{\sqrt{2}} \\ 0 \\ \dfrac{1}{\sqrt{2}} \end{bmatrix}. \qquad (4.1)$$

If you want to express

$$\begin{bmatrix} -3 \\ 18 \\ 7 \end{bmatrix}$$

as a linear combination of the orthonormal basis in Equation (4.1), you must determine the constant multipliers. It does not seem to be an easy task. However, the inner product is the tool that makes it happen. If you compute the inner product of the vector with each vector in the basis in Equation (4.1), then the answers are the respective constants.

EXAMPLE 4.6. Express the vector

$$\begin{bmatrix} -3 \\ 18 \\ 7 \end{bmatrix}$$

as a linear combination of the orthonormal basis

$$\begin{bmatrix} \dfrac{1}{\sqrt{3}} \\ \dfrac{1}{\sqrt{3}} \\ \dfrac{1}{\sqrt{3}} \end{bmatrix}, \quad \begin{bmatrix} \dfrac{-1}{\sqrt{6}} \\ \dfrac{2}{\sqrt{6}} \\ \dfrac{-1}{\sqrt{6}} \end{bmatrix}, \text{ and } \begin{bmatrix} \dfrac{-1}{\sqrt{2}} \\ 0 \\ \dfrac{1}{\sqrt{2}} \end{bmatrix}.$$

Solution: First, compute the inner products of the vector and the orthonormal basis. That is, find

$$\left\langle \begin{bmatrix} -3 \\ 18 \\ 7 \end{bmatrix}, \begin{bmatrix} \dfrac{1}{\sqrt{3}} \\ \dfrac{1}{\sqrt{3}} \\ \dfrac{1}{\sqrt{3}} \end{bmatrix} \right\rangle = \frac{22}{\sqrt{3}},$$

$$\left\langle \begin{bmatrix} -3 \\ 18 \\ 7 \end{bmatrix}, \begin{bmatrix} \dfrac{-1}{\sqrt{6}} \\ \dfrac{2}{\sqrt{6}} \\ \dfrac{-1}{\sqrt{6}} \end{bmatrix} \right\rangle = \frac{32}{\sqrt{6}},$$

and

$$\left\langle \begin{bmatrix} -3 \\ 18 \\ 7 \end{bmatrix}, \begin{bmatrix} \dfrac{-1}{\sqrt{2}} \\ 0 \\ \dfrac{1}{\sqrt{2}} \end{bmatrix} \right\rangle = \frac{10}{\sqrt{2}}.$$

Note: Here the inner product yields the contribution of the vector in the direction of each unit vector; this is because the basis vectors are orthonormal. Next, form the linear combination, which is

$$\frac{22}{\sqrt{3}} \begin{bmatrix} \dfrac{1}{\sqrt{3}} \\ \dfrac{1}{\sqrt{3}} \\ \dfrac{1}{\sqrt{3}} \end{bmatrix} + \frac{32}{\sqrt{6}} \begin{bmatrix} \dfrac{-1}{\sqrt{6}} \\ \dfrac{2}{\sqrt{6}} \\ \dfrac{-1}{\sqrt{6}} \end{bmatrix} + \frac{10}{\sqrt{2}} \begin{bmatrix} \dfrac{-1}{\sqrt{2}} \\ 0 \\ \dfrac{1}{\sqrt{2}} \end{bmatrix} = \begin{bmatrix} -3 \\ 18 \\ 7 \end{bmatrix}. \qquad \blacksquare$$

Example (4.6) shows us how to find the constants of the linear combination when we have an orthonormal basis.

Since we know other bases exist for the real vector spaces, how do we find them and how do we make them orthonormal? The answer lies in two theorems stated here, with proofs in Appendix B and a process, and it involves topics discussed in the previous subsections. The topics are eigenvalues and eigenvectors.

Theorem 16. *Let λ_1, λ_2, $\lambda_3, \ldots \lambda_n$ be distinct eigenvalues of an $n \times n$ matrix. Then, the corresponding eigenvectors x_1, x_2, $x_3, \ldots x_n$ form a linearly independent set of vectors. That is, $c_1 x_1 + c_2 x_2 + c_3 x_3 + \ldots + c_n x_n = 0$ if and only if $c_i = 0$ for $1 \le i \le n$.*

Theorem 17. *If an $n \times n$ matrix A has n distinct eigenvalues, then the corresponding eigenvectors form a basis for $\mathbb{R}^n$.*

Remember, eigen means characteristic. Thus, eigenvalues are the characteristic values of an $n \times n$ matrix, and the corresponding eigenvectors are the characteristic vectors of the $n \times n$ matrix. Knowing that characteristic means distinguishing trait or feature tells us that the eigenvalues and corresponding eigenvectors are a distinguishing feature of the $n \times n$ matrix and the corresponding n-space. What could be a more distinguishing feature of n-space than the basis. Using a basis in n-space allows one to describe every vector in that space by using a linear combination of the basis and Theorem (17) tells us that eigenvectors are a basis.

Both theorems are necessary. Theorem (16) tells us the eigenvectors are linearly independent. This is required for any basis. Theorem (17) informs us the eigen-vectors form a basis for the real vector space, which has the same superscript as the size of the matrix. But how do we know if the eigenvectors are orthonormal? Generally, the eigenvectors are not orthogonal and usually do not have unit length. This is why a process is needed: The Gram–Schmidt orthonormalization process, which makes the eigenvectors orthonormal. However, the process is rather lengthy to explain and not needed for this course. What is important is that eigenvalues and eigenvectors have a role in determining a basis in vector spaces. This idea will show up again in the next section.

EXERCISES 4.2

4.2.1. Which of the following sets of vectors in $\mathbb{R}^3$ are linearly dependent, and which are linearly independent:

(1) $\left\{ \begin{bmatrix} 1 \\ 1 \\ 0 \end{bmatrix}, \begin{bmatrix} 0 \\ 2 \\ 3 \end{bmatrix}, \begin{bmatrix} 1 \\ 2 \\ 3 \end{bmatrix}, \begin{bmatrix} 3 \\ 6 \\ 6 \end{bmatrix} \right\}$,

(2) $\left\{ \begin{bmatrix} 1 \\ 2 \\ 3 \end{bmatrix}, \begin{bmatrix} -1 \\ 2 \\ -5 \end{bmatrix}, \begin{bmatrix} 0 \\ 1 \\ 7 \end{bmatrix} \right\}$, or

(3) $\left\{ \begin{bmatrix} 1 \\ 1 \\ 1 \end{bmatrix}, \begin{bmatrix} 1 \\ 2 \\ 3 \end{bmatrix}, \begin{bmatrix} 0 \\ 1 \\ 0 \end{bmatrix} \right\}$.

4.2.2. Find the characteristic polynomial, eigenvalues, and corresponding eigenvec-tors for the following matrices:

(1) $\begin{bmatrix} 2 & 1 \\ -1 & 3 \end{bmatrix}$,

(2) $\begin{bmatrix} 2 & -2 & 3 \\ 0 & 3 & -2 \\ 0 & -1 & 2 \end{bmatrix}$, and

(3) $\begin{bmatrix} 2 & 2 & 3 \\ 1 & 2 & 1 \\ 2 & -2 & 1 \end{bmatrix}$.

4.2.3. Given the matrix

$$\begin{bmatrix} 1 & 0 & 2 \\ 0 & 2 & 1 \\ -1 & 0 & 6 \end{bmatrix},$$

find the eigenvalues and eigenvectors. Then, perform the following process:

(1) Name the eigenvectors α_1, α_2, and α_3 (order doesn't matter).

(2) Let vector $\beta_1 = \alpha_1$.

(3) Find the vector β_2 by using the formula

$$\beta_2 = \alpha_2 - \frac{\langle \alpha_2, \beta_1 \rangle}{\langle \beta_1, \beta_1 \rangle} \beta_1,$$

where $\langle a, b \rangle$ is the standard inner product for $\mathbb{R}^3$.

(4) Find the vector β_3 by using the formula

$$\beta_3 = \alpha_3 - \frac{\langle \alpha_3, \beta_1 \rangle}{\langle \beta_1, \beta_1 \rangle} \beta_1 - \frac{\langle \alpha_3, \beta_2 \rangle}{\langle \beta_2, \beta_2 \rangle} \beta_2.$$

(5) The vectors β_1, β_2, and β_3 are an orthogonal basis for $\mathbb{R}^3$. Make them an orthonormal basis.

(6) This process is known as the Gram–Schmidt process. Prove you have produced an orthonormal basis for $\mathbb{R}^3$, which can be used to describe every vector in $\mathbb{R}^3$.

4.2.4. Show that the dot product in the vector spaces $\mathbb{R}^2$ and $\mathbb{R}^3$ is an inner product.

4.2.5. Show that the space of all 2×3 matrices with real-number components, $\mathbf{M}_{2,3}$, is an inner product space. *Note:* Use the inner product $\langle A, B \rangle = tr\left(B^T A\right)$, where A and B are elements of $\mathbf{M}_{2,3}$.

4.2.6. Show that the space of all 3×2 matrices with real number components, $\mathbf{M}_{3,2}$, is an inner product space. *Note:* Use the inner product $\langle A, B \rangle = tr\left(B^T A\right)$, where A and B are elements of $\mathbf{M}_{3,2}$.

4.2.7. If w_1 and w_2 are positive real numbers, show that the definition

$$(x_1, x_2) \bullet (y_1, y_2) = w_1 x_1 y_1 + w_2 x_2 y_2$$

yields a different inner product on $\mathbb{R}^2$. (*Hint:* Show all four properties of an inner product hold.)

4.3   A NEW SPACE: THE FUNCTION SPACE OF PIECEWISE SMOOTH FUNCTIONS

The function space of piecewise smooth (PWS) functions is an inner product space where the elements are PWS functions. Therefore, a new inner product must be defined, because the inner product on PWS functions can't be the dot product. Also, unlike the vector spaces $\mathbb{R}^n$, orthogonality in a function space may not easily be recognized geometrically. But, we are getting ahead of ourselves. Inner product and orthogonality are discussed in the next subsection. First, we must define the elements of the function space of PWS functions.

For our purposes, the function space of PWS functions is composed of PWS functions, each of which exists on the closed finite interval $[a, b]$. Also, a PWS function seems to imply the function can be in pieces. This implication is true to a certain extent. To understand piecewise smooth functions, we first need to define the term **jump discontinuity.**

Definition 18. *A function $f(x)$ has a jump discontinuity at the point $x = x_0$, if the limit of $f(x_0)$ from the left, written $\lim_{x \to x_0^-} f(x)$, exists and the limit of $f(x_0)$ from the right, written $\lim_{x \to x_0^+} f(x)$, exists, but $\lim_{x \to x_0^-} f(x) \neq \lim_{x \to x_0^+} f(x)$.*

It is important to note that the word "exist" in the definition of jump discontinuity means the limit must be finite. Below, in Figure (4.1) and Figure (4.2), are two graphical examples of functions with at least one jump discontinuity. Figure (4.1) is $f(x) = [|x + 1|]$, the step function on the interval $[-2, 2]$. Whereas, Figure (4.2) is a function with random jump discontinuities on the interval $[-L, L]$. Also,

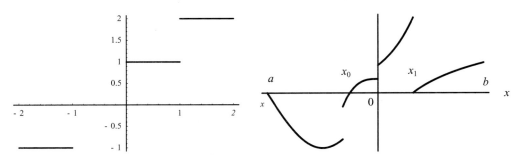

Figure 4.1: $f(x) = [|x + 1|]$ on the interval $[-2, 2]$.

Figure 4.2: A function, $f(x)$, with several jump discontinuities.

a good counterexample of the definition of jump discontinuity is the function $\tan x$ on the interval $[0, \pi]$, which is graphed in Figure (4.3). It is a good counterexample, because at $x = \dfrac{\pi}{2}$, $\tan x$ has a discontinuity, but it doesn't fit our definition of a jump discontinuity.

With a good understanding of jump discontinuity, we can define a piecewise smooth function.

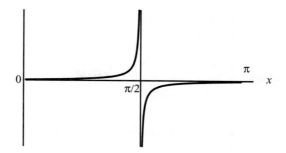

Figure 4.3: $f(x) = \tan x$ on the interval $[0, \pi]$.

Definition 19. *A function $f(x)$ on the interval $[a, b]$ is piecewise smooth if the interval $[a, b]$ can be broken up into a finite number of subintervals such that the function, $f(x)$, and its derivative, $\dfrac{df(x)}{dx}$, are continuous on each of the subintervals.*

This definition of a piecewise smooth function given implies that the function $f(x)$ may not be continuous on the interval $[a, b]$, but the discontinuities are only jump discontinuities. Thus, the function $f(x)$ is bounded on the interval $[a, b]$. Both Figures (4.1 and 4.2) are good examples of piecewise smooth functions. Another example is $f(x) = |x|$, graphed in Figure (4.4) on the interval $[-2, 2]$. *Note:* The

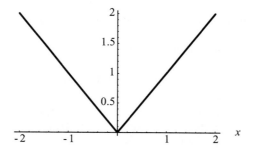

Figure 4.4: $f(x) = |x|$ on the interval $[-2, 2]$.

function $f(x) = |x|$ is continuous on the interval $[-2, 2]$, but at $x = 0$ the derivative does not exist. Thus, we break the function into the intervals $[-2, 0)$ and $(0, 2]$ where the function and the derivative are both continuous, and we recognize it as a PWS function.

Summarizing the properties of the function space of piecewise smooth functions we have the following:

- The function space exists on the closed interval $[a, b]$.

- The elements are functions.

- The functions and their first derivatives are continuous except at possibly a finite number of jump discontinuities.

Understanding the setting for the function space of piecewise smooth functions is only part of the story. The other pieces of the story are discovering reasonable definitions for inner product, orthogonality, and basis.

4.3.1   Inner Product, Orthogonality, and Basis in a Function Space

In the vector space $\mathbb{R}^n$, $n \in \mathbb{N}$, the dot product is the most familiar inner product. However, in the function space of PWS functions on the interval $[a, b]$ the dot product would make no sense. Therefore, a new inner product must be determined. To aid us, the definition for an inner product is restated below for a general vector space.

Definition 20. *Let $\mathcal{V}$ be any real vector space. An inner product on $\mathcal{V}$ is a function that assigns to each pair of vectors $\boldsymbol{u}$ and $\boldsymbol{v}$ of $\mathcal{V}$ a real number, written $\langle \boldsymbol{u}, \boldsymbol{v} \rangle = c \in \mathbb{R}$, satisfying the following:*

1. $\langle \boldsymbol{u}, \boldsymbol{u} \rangle > 0$ for $\boldsymbol{u} \neq 0$, and $\langle \boldsymbol{u}, \boldsymbol{u} \rangle = 0$ if and only if $\boldsymbol{u} = 0$.

2. $\langle \boldsymbol{u}, \boldsymbol{v} \rangle = \langle \boldsymbol{v}, \boldsymbol{u} \rangle$ for any $\boldsymbol{u}$ and $\boldsymbol{v}$ in $\mathcal{V}$.

3. $\langle \boldsymbol{u} + \boldsymbol{v}, \boldsymbol{w} \rangle = \langle \boldsymbol{u}, \boldsymbol{w} \rangle + \langle \boldsymbol{v}, \boldsymbol{w} \rangle$ for any $\boldsymbol{u}$, $\boldsymbol{v}$, and $\boldsymbol{w}$ in $\mathcal{V}$.

4. $\langle b\boldsymbol{u}, \boldsymbol{v} \rangle = b \langle \boldsymbol{u}, \boldsymbol{v} \rangle$ for $\boldsymbol{u}$ and $\boldsymbol{v}$ in $\mathcal{V}$ and b a scalar in $\mathbb{R}$.

Remember, the words "real vector space" mean the associated scalar field to the vector space is the real numbers.

Let's look at the different operations we can perform with functions. First, there is addition. In general, when two functions are added together, we get another function not a real number. Next, we could consider multiplying two functions together. But, multiplication of two functions usually yields another function. Moving into the realm of calculus, differentiation could be considered. However, differentiation is really defined at a point and does not apply to functions that are piecewise smooth because of the jump discontinuity. Therefore, differentiation would not be a good candidate for the inner product. Finally, we can consider integration. A definite integral can be applied to a function on the entire interval and produces an answer that is a real number. Also, a function that has jump discontinuities, as defined previously, can be integrated, and we can easily integrate the product of two functions. Thus, integration is a good candidate for the inner product. Therefore, we define a possible new inner product as

$$\langle u, v \rangle = \int_a^b u(x)v(x) \; dx.$$

Note: u and v are no longer boldfaced since they are functions. Thus, not vectors in the traditional sense.

Now, it must be determined if this definition meets all the criteria of the definition of an inner product.

First, we must ask, "is

$$\langle u, u \rangle = \int_a^b u(x)u(x) \, dx = \int_a^b [u(x)]^2 \, dx \geq 0$$

on the interval $[a, b]$?" We know, from calculus, the square of a function is always nonnegative. The integral of a nonnegative function is also nonnegative, since the integral is always nonnegative when the curve of the entire function is above the x-axis. Therefore, $\langle u, u \rangle \geq 0$ for all functions $u(x)$ in the function space of PWS functions. Also, if $u = 0$, then

$$\langle u, u \rangle = \langle 0, 0 \rangle = \int_a^b 0 \, dx = \int_a^b 0 \, dx = 0.$$

If

$$\langle u, u \rangle = \int_a^b u(x)u(x) \, dx = \int_a^b [u(x)]^2 \, dx = 0,$$

then we have $u = 0$.

Second, we must show $\langle u, v \rangle = \langle v, u \rangle$ for any u and v in the vector space. Using the general theorems of integration, we have

$$\langle u, v \rangle = \int_a^b u(x)v(x) \, dx = \int_a^b v(x)u(x) \, dx = \langle v, u \rangle$$

for any u and v in the function space of PWS functions.

Third, we must show $\langle u + v, w \rangle = \langle u, w \rangle + \langle v, w \rangle$ for any u, v, and w in the function space of PWS functions. Again, using the general theorems of integration, we have

$$\langle u + v, w \rangle = \int_a^b [u(x) + v(x)] \, w(x) \, dx = \int_a^b u(x)w(x) + v(x)w(x) \, dx$$

$$= \int_a^b u(x)w(x) \, dx + \int_a^b v(x)w(x) \, dx = \langle u, w \rangle + \langle v, w \rangle$$

for any u, v, and w in the function space of PWS functions.

Finally, we must show $\langle cu, v \rangle = c \langle u, v \rangle$ for u and v in the function space of PWS functions and c a scalar in $\mathbb{R}$. The theorems on general integration provide us with the solution

$$\langle cu, v \rangle = \int_a^b cu(x)v(x) \, dx = c \int_a^b u(x)v(x) \, dx = c \langle u, v \rangle$$

for any u and v in the function space of PWS functions and c a scalar in $\mathbb{R}$.

We have shown the definition

$$\langle u, v \rangle = \int_a^b u(x)v(x) \, dx$$

is an inner product on the function space of PWS functions. In fact, it is the standard inner product for the function space of PWS functions, which is analogous to the dot product in the vector spaces $\mathbb{R}^n$, for any $n \in \mathbb{N}$.

It is convenient to change the notation sightly as in

$$\langle f, g \rangle = \int_a^b f(x)g(x) \, dx.$$

This change actually makes the inner product easier to recognize. Also, in the text's study of PDEs, the interval of greatest importance to us is the interval $[-L, L]$, instead of the general interval $[a, b]$. Therefore, we change our inner product notation to

$$\langle f, g \rangle = \int_{-L}^L f(x)g(x) \, dx$$

in the function space of PWS functions.

Having an inner product for the function space of PWS functions implies a concept of orthogonality. By the definition of orthogonality, two functions are orthogonal if the inner product is zero. This means, if $f(x)$ is orthogonal to the function $g(x)$, then

$$\langle f, g \rangle = \int_{-L}^L f(x)g(x) \, dx = 0.$$

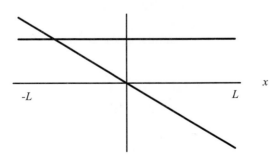

Figure 4.5: The graph of $f(x) = -x$ and $g(x) = 2$ on the interval $[-L, L]$.

EXAMPLE 4.7. Let $f(x) = -x$ and $g(x) = 2$. Then,

$$\int_{-L}^L (-x)(2) \, dx = \int_{-L}^L -2x \, dx = -x^2 \, \big|_{-L}^L = -(L^2 - (-L)^2) = 0.$$

Figure (4.5) shows the graph of $f(x)$ and $g(x)$. ■

In Example (4.7), the functions $f(x) = -x$ and $g(x) = 2$ are shown to be orthogonal in the function space of PWS functions. We now must find the analogous orthogonal basis in the function space of PWS functions.

In the vector space $\mathbb{R}^3$, the natural basis, axes, can be defined by the three vectors

$$\begin{bmatrix} 1 \\ 0 \\ 0 \end{bmatrix}, \begin{bmatrix} 0 \\ 1 \\ 0 \end{bmatrix}, \text{ and } \begin{bmatrix} 0 \\ 0 \\ 1 \end{bmatrix},$$

which are perpendicular, orthogonal, to each other. An analogous definition of the basis of the vector space $\mathbb{R}^3$ would involve elements from the function space of PWS functions, and these elements must be functions defined on the interval $[-L, L]$.

The functions that are analogous to the standard basis in $\mathbb{R}^n$, $n \in \mathbb{N}$, are orthogonal to each other on the interval $[-L, L]$. However, this is where the analogies stop. The standard basis in $\mathbb{R}^n$, $n \in \mathbb{N}$ for a given n (say $n = 15$) means there are 15 vectors in the basis. Thus, we know we have a finite dimensional vector space. The standard basis in the function space of PWS functions is infinite. Thus, our function space is infinitely dimensional. The questions to be answered are "What are the functions?" and "How do we get them?"

To answer the first question, the functions a where a is a constant and the trigonometric functions, sine and cosine form the standard basis for the function space of PWS functions. However, the argument of the functions is not a simple x. The argument is $\dfrac{n\pi x}{L}$, $n \in \mathbb{N}$. Thus, the standard basis is a, $\sin \dfrac{n\pi x}{L}$, and $\cos \dfrac{n\pi x}{L}$, $n \in \mathbb{N}$.

The second question is answered by considering the general ODE

$$\varphi''(x) = -\lambda \varphi(x), \tag{4.2}$$

subject to the BCs

$$\varphi(L) = \varphi(-L)$$

$$\frac{d\varphi(L)}{dx} = \frac{d\varphi(-L)}{dx} \tag{4.3}$$

on the interval $[-L, L]$. We must solve Equation (4.2) for $\lambda < 0$, $\lambda = 0$, and $\lambda > 0$.

For $\lambda < 0$, we let $-s = \lambda$, $s > 0$. Then, Equation (4.2) becomes

$$\varphi''(x) = s\varphi(x).$$

From your previous courses you should recognize this ODE and know that the solution is dependent on λ. First, for $\lambda < 0$, we have the solution

$$\varphi(x) = c_1 \cosh\left(\sqrt{s}x\right) + c_2 \sinh\left(\sqrt{s}x\right). \tag{4.4}$$

Using the first BC from Equation (4.3) yields

$$\varphi(L) = c_1 \cosh\left(\sqrt{s}L\right) + c_2 \sinh\left(\sqrt{s}L\right)$$

$$= \varphi(-L) = c_1 \cosh\left(-\sqrt{s}L\right) + c_2 \sinh\left(-\sqrt{s}L\right). \tag{4.5}$$

Since the hyperbolic cosine function is an even function and the hyperbolic sine function is an odd function, we rewrite Equation (4.5) as

$$c_1 \cosh\left(\sqrt{s}L\right) + c_2 \sinh\left(\sqrt{s}L\right) = c_1 \cosh\left(\sqrt{s}L\right) - c_2 \sinh\left(\sqrt{s}L\right),$$

which becomes

$$2c_2 \sinh\left(\sqrt{s}L\right) = 0. \tag{4.6}$$

Equation (4.6) implies $c_2 = 0$. Therefore, Equation (4.4) reduces to

$$\varphi(x) = c_1 \cosh\left(\sqrt{s}x\right). \tag{4.7}$$

Using the second BC from Equation (4.3) on Equation (4.7) yields

$$\frac{d\varphi(L)}{dx} = \sqrt{s}c_1 \sinh\left(\sqrt{s}L\right) = \frac{d\varphi(-L)}{dx} = \sqrt{s}c_1 \sinh\left(-\sqrt{s}L\right),$$

which can be rewritten as

$$2\sqrt{s}c_1 \sinh\left(\sqrt{s}L\right) = 0,$$

implying $c_1 = 0$. Thus, for $\lambda < 0$, we have only the trivial solution.

Next, we assume $\lambda = 0$. Then, Equation (4.2) becomes

$$\varphi''(x) = 0.$$

Integrating twice yields

$$\varphi(x) = d_1 x + d_2. \tag{4.8}$$

Applying the first BC from Equation (4.3) yields

$$\varphi(L) = d_1 L + d_2 = \varphi(-L) = -d_1 L + d_2. \tag{4.9}$$

Equation (4.9) can be rewritten as

$$2d_1 L = 0,$$

which implies $d_1 = 0$. Thus, Equation (4.8) becomes

$$\varphi(x) = d_2. \tag{4.10}$$

The second BC from Equation (4.3) indicates that we must find the derivative of $\varphi(x) = d_2$. Since d_2 is a constant, the second BC is trivially true. Thus, for $\lambda = 0$ we have $\varphi(x)$ equal to the constant d_2.

Finally, we assume that $\lambda > 0$. The solution, in this case, is

$$\varphi(x) = a\cos\sqrt{\lambda}x + b\sin\sqrt{\lambda}x. \tag{4.11}$$

Applying the first BC from Equation (4.3) yields

$$\varphi(L) = a\cos\left(\sqrt{\lambda}L\right) + b\sin\left(\sqrt{\lambda}L\right)$$

$$= \varphi(-L) = a\cos\left(-\sqrt{\lambda}L\right) + b\sin\left(-\sqrt{\lambda}L\right). \tag{4.12}$$

Since the cosine function is an even function and the sine function is an odd function, we rewrite Equation (4.12) as

$$a\cos\left(\sqrt{\lambda}L\right) + b\sin\left(\sqrt{\lambda}L\right) = a\cos\left(\sqrt{\lambda}L\right) - b\sin\left(\sqrt{\lambda}L\right),$$

which becomes

$$2b\sin\left(\sqrt{\lambda}L\right) = 0. \tag{4.13}$$

Applying the second BC from Equation (4.3) yields

$$\frac{d\varphi(L)}{dx} = \sqrt{\lambda}\left(b\cos\left(\sqrt{\lambda}L\right) - a\sin\left(\sqrt{\lambda}L\right)\right)$$

$$= \frac{d\varphi(-L)}{dx} = \sqrt{\lambda}\left(b\cos\left(-\sqrt{\lambda}L\right) - a\sin\left(-\sqrt{\lambda}L\right)\right). \tag{4.14}$$

Again, since the cosine function is an even function and the sine function is an odd function, we rewrite Equation (4.14) as

$$\sqrt{\lambda}\left(b\cos\left(\sqrt{\lambda}L\right) - a\sin\left(\sqrt{\lambda}L\right)\right) = \sqrt{\lambda}\left(b\cos\left(\sqrt{\lambda}L\right) + a\sin\left(\sqrt{\lambda}L\right)\right),$$

which, after some algebraic manipulations, becomes

$$2a\sin\left(\sqrt{\lambda}L\right) = 0. \tag{4.15}$$

Equations (4.13 and 4.15) are very similar. Dividing out the $2a$ and $2b$ yields

$$\sin\left(\sqrt{\lambda}L\right) = 0$$

for both equations. Solving for λ, we find $\lambda = \left(\frac{n\pi}{L}\right)^2$, $n = 1, 2, 3, \ldots$ Remember, the sine function equals zero when the argument of the sine function is an integer multiple of π. Thus, when $\lambda > 0$, the solution for Equation (4.2) is

$$\varphi(x) = a\cos\frac{n\pi x}{L} + b\sin\frac{n\pi x}{L}. \tag{4.16}$$

We have all the values of λ, called eigenvalues, which allow for a nontrivial solution to Equation (4.2) subject to the BCs given in Equation (4.3). These eigenvalues are analogous to the eigenvalues in Theorem (16) and Theorem (17). Also, Equations (4.10 and 4.16) make up the general solution of Equation (4.2). The solution is made up of three different types of functions. The three different functions are a constant, $\cos \dfrac{n\pi x}{L}$, and $\sin \dfrac{n\pi x}{L}$, called eigenfunctions. They are analogous to the corresponding eigenvectors of Theorems (16 and 17). The eigenfunctions are often referred to as orthogonal sequences. The orthogonal sequences are $\cos \dfrac{n\pi x}{L}$, and $\sin \dfrac{n\pi x}{L}$. Remember, when $n = 0$, $\sin \dfrac{n\pi x}{L} = 0$ and $\cos \dfrac{n\pi x}{L} = 1$. Thus, the constant may be included in the $\cos \dfrac{n\pi x}{L}$ sequence. However, in your previous course work, the orthogonal sequences were referred to as the fundamental solution set. We know the functions of a fundamental solution set are linearly independent, and linear independence is a basic property of any basis, and since they are also known as orthogonal sequences, we know that they are orthogonal. However, we will not take for granted that they are orthogonal. This will be proved.

As the previous paragraph suggests, solving the ODE, Equation (4.2), is analogous to solving an $n \times n$ matrix for the distinct eigenvalues and the corresponding eigenvectors. Remember, eigenvalues are found when you solve the matrix equation $A\mathbf{x} = \lambda\mathbf{x}$ or $|A - \lambda I|\, x = 0$. Also, when eigenvalues for an $n \times n$ matrix are found, the eigenvectors may be determined. By Theorem (17), Eigenvectors form a basis for the n-dimensional space that is represented by the $n \times n$ matrix. The space we are considering is a function space. Therefore, instead of eigenvectors, we have eigenfunctions, for the function space formed on the interval $[-L, L]$. It remains for us to show they are orthogonal.

To show that the eigenfunctions found are orthogonal on the interval $[-L, L]$, we must show several things. You must remember there is one constant and an infinite number of both sine and cosine eigenfunctions, all of which must be orthogonal to each other. We start by showing that

$$\int_{-L}^{L} \sin \frac{n\pi x}{L} \cos \frac{m\pi x}{L} \; dx = 0, \; n, \, m \in \mathbb{N}.$$

First, the trigonometry identities

$$\sin(a + b) = \sin a \cos b + \cos a \sin b$$

and

$$\sin(a - b) = \sin a \cos b - \cos a \sin b$$

must be used. We have

$$\sin \left(\frac{n\pi x}{L} + \frac{m\pi x}{L} \right) = \sin \frac{n\pi x}{L} \cos \frac{m\pi x}{L} + \cos \frac{n\pi x}{L} \sin \frac{m\pi x}{L} \qquad (4.17)$$

and

$$\sin \left(\frac{n\pi x}{L} - \frac{m\pi x}{L} \right) = \sin \frac{n\pi x}{L} \cos \frac{m\pi x}{L} - \cos \frac{n\pi x}{L} \sin \frac{m\pi x}{L}. \qquad (4.18)$$

Adding Equation (4.17) to Equation (4.18) yields

$$\sin\left(\frac{n\pi x}{L} + \frac{m\pi x}{L}\right) + \sin\left(\frac{n\pi x}{L} - \frac{m\pi x}{L}\right) = 2\sin\frac{n\pi x}{L}\cos\frac{m\pi x}{L}.$$

Dividing both sides by two and integrating from $-L$ to L, we find

$$\frac{1}{2}\int_{-L}^{L}\left\{\sin\left(\frac{n\pi x}{L} + \frac{m\pi x}{L}\right) + \sin\left(\frac{n\pi x}{L} - \frac{m\pi x}{L}\right)\right\}\,dx = \int_{-L}^{L}\sin\frac{n\pi x}{L}\cos\frac{m\pi x}{L}\,dx.$$

Integrating yields

$$\frac{1}{2}\int_{-L}^{L}\sin\left(\frac{n\pi x}{L} + \frac{m\pi x}{L}\right) + \sin\left(\frac{n\pi x}{L} - \frac{m\pi x}{L}\right)\,dx$$

$$= \frac{1}{2}\left[\int_{-L}^{L}\sin\left(\frac{(n+m)\pi}{L}\right)x\,dx + \int_{-L}^{L}\sin\left(\frac{(n-m)\pi}{L}\right)x\,dx\right]$$

$$= \frac{1}{2}\left[\left(\frac{-L}{(n+m)\pi}\right)\cos\left(\frac{(n+m)\pi}{L}\right)x + \left(\frac{-L}{(n-m)\pi}\right)\cos\left(\frac{(n-m)\pi}{L}\right)x\right]_{-L}^{L}$$

$$= \frac{1}{2}\left[\left(\frac{-L}{(n+m)\pi}\right)\cos\left(\frac{(n+m)\pi}{L}\right)L + \left(\frac{-L}{(n-m)\pi}\right)\cos\left(\frac{(n-m)\pi}{L}\right)L\right]$$

$$- \frac{1}{2}\left[\left(\frac{-L}{(n+m)\pi}\right)\cos\left(\frac{(n+m)\pi}{L}\right)(-L) + \left(\frac{-L}{(n-m)\pi}\right)\cos\left(\frac{(n-m)\pi}{L}\right)(-L)\right]$$

$$= 0.$$

Remember, the cosine of the angle equals the cosine of minus the angle.

The rest of the orthogonality integrals are

$$\int_{-L}^{L}\sin\frac{n\pi x}{L}\sin\frac{m\pi x}{L}\,dx = 0,\ n,\ m\in\mathbb{N},\ n\neq m;$$

$$\int_{-L}^{L}\cos\frac{n\pi x}{L}\cos\frac{m\pi x}{L}\,dx = 0,\ n,\ m\in\mathbb{N},\ n\neq m;$$

$$\int_{-L}^{L}d_2\cos\frac{n\pi x}{L}\,dx = 0,\ d_2\text{ a constant},\ n\in\mathbb{N},\ n\neq 0;$$

and

$$\int_{-L}^{L}d_2\sin\frac{n\pi x}{L}\,dx = 0,\ d_2\text{ a constant},\ n\in\mathbb{N},\ n\neq 0.$$

They are left as exercises.

We have shown that the eigenfunctions d_2, a constant, $\sin\dfrac{n\pi x}{L}$, and $\cos\dfrac{n\pi x}{L}$ are orthogonal eigenfunctions. In fact, they form the standard basis of the function space of PWS functions. Therefore, we are now ready to define a linear combination of our basis. *Note:* A linear combination of our basis must represent every function in the function space of PWS functions.

4.3.2   Definition of Trigonometric Fourier series

The eigenfunctions d_2, $\sin\dfrac{n\pi x}{L}$, and $\cos\dfrac{n\pi x}{L}$, $n \in \mathbb{N}$, were identified as the standard basis for the function space of PWS functions on the interval $[-L, L]$. This can only be true if every function in the function space of PWS functions can be represented in a linear combination of sine and cosine functions. Remember, a linear combination is the sum of the product of the eigenvectors, in this case eigenfunctions, and a scalar, a constant. Also, we know that there are an infinite number of eigenfunctions in our proposed standard basis. Therefore, a linear combination must include an infinite sum. Thus, if our standard basis was identified correctly, we have, for any function $f(x)$, in the function space of PWS functions, the definition

$$f(x) \approx a_0 + \sum_{n=1}^{\infty} \left[a_n \cos\left(\frac{n\pi x}{L}\right) + b_n \sin\left(\frac{n\pi x}{L}\right) \right], \qquad (4.19)$$

where a_0, a_n, and b_n are the scalar multipliers of our eigenfunctions and $\approx$ means approximately. *Note:* A scalar times a constant is just another constant. Thus, a_0 is the product of an arbitrary scalar and the constant d_2.

The right side of Equation (4.19) is known as a trigonometric Fourier series. It is called trigonometric, since sine and cosine functions are used in the sum. Also, Equation (4.19) is sometimes called an eigenfunction expansion, where sine and cosine functions are the eigenfunctions. There are other types of Fourier series. They also form a standard basis but for different intervals/coordinate systems. However, at this time, there is no need to go into detail about them. They will be discussed in Chapter 8, when generalized Fourier series are introduced, and used extensively in Chapter 10. For purposes of this text, the trigonometric Fourier series above will be called a Fourier series for short. When we introduce generalized Fourier series, appropriate names will be applied.

Equation (4.19) identifies the linear combination of our proposed standard basis. The constants a_0, a_n, and b_n, are known as Fourier coefficients, and we must determine them.

For each different function in the space of PWS function, the scalars a_0, a_n, and b_n are found in the same way the scalar multipliers were found in Example (4.6). In Example (4.6), the scalar multipliers were found by using the standard inner product for $\mathbb{R}^3$, called the dot product. The vector to be represented was used in the dot product with each eigenvector, yielding the scalar contribution in the direction of each eigenvector. When the eigenvectors and corresponding scalars were put in a linear combination, they formed another representation of the vector in question. Thus, it would seem natural to find the scalars a_0, a_n, and b_n in Equation

(4.19) for each function in the space of PWS functions by using the standard inner product for the space of PWS functions, integrating the product of the function to be represented and the eigenfunctions from $[-L, L]$. The obvious solution is not as easy as the question. How do we do it?

First, for convenience, the $\approx$ in Equation (4.19) will be changed to $=$. Thus, Equation (4.19) becomes

$$f(x) = a_0 + \sum_{n=1}^{\infty} \left[a_n \cos\left(\frac{n\pi x}{L}\right) + b_n \sin\left(\frac{n\pi x}{L}\right) \right]. \tag{4.20}$$

The inner product for the function space requires the integration over the interval $[-L, L]$ of the product of two functions. This means we have to multiply Equation (4.20) by some function, and the object of the multiplication and subsequent integration is the determination of a Fourier Coefficient. This implies using the orthogonality of the sine and cosine functions to our advantage. Thus, to find a_n, for any $n \in \mathbb{N}$, we would multiply Equation (4.20) by $\cos\left(\dfrac{m\pi x}{L}\right)$ for some particular $m \in \mathbb{N}$, $m \neq 0$. This yields

$$f(x) \cos\left(\frac{m\pi x}{L}\right) = a_0 \cos\left(\frac{m\pi x}{L}\right)$$

$$+ \sum_{n=1}^{\infty} \cos\left(\frac{m\pi x}{L}\right) \left[a_n \cos\left(\frac{n\pi x}{L}\right) + b_n \sin\left(\frac{n\pi x}{L}\right) \right]. \tag{4.21}$$

After some small algebraic manipulation on Equation (4.21), we get

$$f(x) \cos\left(\frac{m\pi x}{L}\right) = a_0 \cos\left(\frac{m\pi x}{L}\right)$$

$$+ \sum_{n=1}^{\infty} \cos\left(\frac{m\pi x}{L}\right) \left[a_n \cos\left(\frac{n\pi x}{L}\right) + b_n \sin\left(\frac{n\pi x}{L}\right) \right]. \tag{4.22}$$

We integrate Equation (4.22) from $-L$ to L. Thus, we have

$$\int_{-L}^{L} f(x) \cos\left(\frac{m\pi x}{L}\right) \, dx = a_0 \int_{-L}^{L} \cos\left(\frac{m\pi x}{L}\right) \, dx$$

$$+ \int_{-L}^{L} \sum_{n=1}^{\infty} \cos\left(\frac{m\pi x}{L}\right) \left[a_n \cos\left(\frac{n\pi x}{L}\right) + b_n \sin\left(\frac{n\pi x}{L}\right) \right] dx. \tag{4.23}$$

The first integral,

$$\int_{-L}^{L} f(x) \cos\left(\frac{m\pi x}{L}\right) \, dx,$$

in Equation (4.23) is in its final form unless we know the function, $f(x)$. The second integral,

$$a_0 \int_{-L}^{L} \cos\left(\frac{m\pi x}{L}\right) \, dx,$$

in Equation (4.23) is known to equal zero by orthogonality. The last integral,

$$\int_{-L}^{L} \sum_{n=1}^{\infty} \cos\left(\frac{m\pi x}{L}\right) \left[a_n \cos\left(\frac{n\pi x}{L}\right) + b_n \sin\left(\frac{n\pi x}{L}\right)\right] \, dx,$$

in Equation (4.23) presents a completely different problem. We must ask ourselves, "When does the integral of an infinite sum equal the infinite sum of the integrals?" This question is extremely important, and it will be discussed in more detail in Chapter 6. For the moment, we will accept as fact the integral can be taken inside the infinite sum, and we know that the infinite sum of sums is equal to the sum of the infinite sums. This yields

$$\sum_{n=1}^{\infty} a_n \int_{-L}^{L} \cos\left(\frac{n\pi x}{L}\right) \cos\left(\frac{m\pi x}{L}\right) \, dx$$

$$+ \sum_{n=1}^{\infty} b_n \int_{-L}^{L} \sin\left(\frac{n\pi x}{L}\right) \cos\left(\frac{m\pi x}{L}\right) \, dx. \tag{4.24}$$

The second integral in Equation (4.24),

$$b_n \int_{-L}^{L} \sin\left(\frac{n\pi x}{L}\right) \cos\left(\frac{m\pi x}{L}\right) \, dx,$$

is known to equal zero by orthogonality. The first integral in Equation (4.24),

$$a_n \int_{-L}^{L} \cos\left(\frac{n\pi x}{L}\right) \cos\left(\frac{m\pi x}{L}\right) \, dx,$$

will, by orthogonality, equal zero whenever $m \neq n$. However, when $m = n$, we have

$$a_n \int_{-L}^{L} \cos\left(\frac{n\pi x}{L}\right)^2 \, dx = a_n \int_{-L}^{L} \frac{1}{2} + \frac{1}{2} \cos\left(\frac{2n\pi x}{L}\right) \, dx = a_n L.$$

Thus, the entire infinite sum in Equation (4.24) becomes one term, and that term is $a_n L$. Therefore,

$$\int_{-L}^{L} f(x) \cos\left(\frac{n\pi x}{L}\right) \, dx = a_n L.$$

Solving for the Fourier Coefficient a_n yields

$$a_n = \frac{1}{L} \int_{-L}^{L} f(x) \cos\left(\frac{n\pi x}{L}\right) \, dx.$$

In a similar fashion, we find the Fourier coefficients

$$a_0 = \frac{1}{2L} \int_{-L}^{L} f(x)\ dx$$

and

$$b_n = \frac{1}{L} \int_{-L}^{L} f(x) \sin\left(\frac{n\pi x}{L}\right)\ dx.$$

In conclusion, we have the definition of the trigonometric Fourier series representation of a function in the space of PWS functions,

$$f(x) = a_0 + \sum_{n=1}^{\infty} \left[a_n \cos\left(\frac{n\pi x}{L}\right) + b_n \sin\left(\frac{n\pi x}{L}\right) \right], \tag{4.25}$$

and the integral equations for determining the Fourier coefficients, which are

$$a_0 = \frac{1}{2L} \int_{-L}^{L} f(x)\ dx, \tag{4.26}$$

$$a_n = \frac{1}{L} \int_{-L}^{L} f(x) \cos\left(\frac{n\pi x}{L}\right)\ dx, \tag{4.27}$$

and

$$b_n = \frac{1}{L} \int_{-L}^{L} f(x) \sin\left(\frac{n\pi x}{L}\right)\ dx. \tag{4.28}$$

Thus, for any function, $f(x)$, in the function space of PWS functions on $[-L, L]$, we can define a Fourier series and determine the Fourier Coefficients. It still must be determined if a Fourier series actually converges to the function, $f(x)$, in the function space of PWS functions. This important point is discussed in the next subsection.

4.3.3 Fourier series Representation of Piecewise Smooth Functions

This subsection starts with several examples of functions and their Fourier series representations. The examples are followed by the general theorem governing Fourier series convergence, called the Fourier's convergence theorem. We conclude this subsection with several theorems developing absolute and uniform convergence of a Fourier series to the function it represents. The definitions of absolute and uniform convergence will be provided at that time.

The first example is a Fourier series representation of a well-behaved polynomial,

$$f(x) = x^3 + 7x^2 - 4x - 10.$$

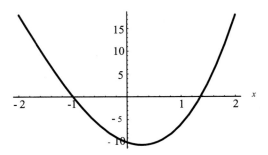

Figure 4.6: $f(x) = x^3 + 7x^2 - 4x - 10$ on the interval $[-2, 2]$.

EXAMPLE 4.8. Consider the function $f(x) = x^3 + 7x^2 - 4x - 10$ on the interval $[-2, 2]$ shown in Figure (4.6). The Fourier series representation of $f(x)$ is

$$x^3 + 7x^2 - 4x - 10 = a_0 + \sum_{n=1}^{\infty} \left[a_n \cos\left(\frac{n\pi x}{L}\right) + b_n \sin\left(\frac{n\pi x}{L}\right) \right],$$

where

$$a_0 = \frac{1}{2L} \int_{-L}^{L} f(x) \, dx = \frac{1}{4} \int_{-2}^{2} x^3 + 7x^2 - 4x - 10 \, dx = \frac{-2}{3}$$

and

$$a_n = \frac{1}{L} \int_{-L}^{L} f(x) \cos\left(\frac{n\pi x}{L}\right) \, dx$$

$$= \frac{1}{2} \int_{-2}^{2} \left(x^3 + 7x^2 - 4x - 10\right) \cos\left(\frac{n\pi x}{L}\right) \, dx$$

$$= \frac{1}{2} \left(\frac{224 \cos(n\pi)}{n^2 \pi^2} - \frac{224 \sin(n\pi)}{n^3 \pi^3} + \frac{72 \sin(n\pi)}{n\pi} \right)$$

$$= \frac{112(-1)^n}{n^2 \pi^2},$$

since $\cos(n\pi) = (-1)^n$ and $\sin(n\pi) = 0$ for all n. Also

$$b_n = \frac{1}{L} \int_{-L}^{L} f(x) \sin\left(\frac{n\pi x}{L}\right) \, dx$$

$$= \frac{1}{2} \int_{-2}^{2} \left(x^3 + 7x^2 - 4x - 10\right) \sin\left(\frac{n\pi x}{L}\right) \, dx$$

$$= \frac{1}{2}\left(\frac{192\cos(n\pi)}{n^3\pi^3} - \frac{1\sin(n\pi)}{n^4\pi^4} + \frac{64\sin(n\pi)}{n^2\pi^2}\right)$$

$$= \frac{96(-1)^n}{n^3\pi^3},$$

since $\cos(n\pi) = (-1)^n$ and $\sin(n\pi) = 0$ for all n. Thus, the Fourier series representation of the function $f(x)$ is

$$x^3 + 7x^2 - 4x - 10 = \frac{-2}{3} + \sum_{n=1}^{\infty}\left[\frac{112(-1)^n}{n^2\pi^2}\cos\left(\frac{n\pi x}{L}\right) + \frac{96(-1)^n}{n^3\pi^3}\sin\left(\frac{n\pi x}{L}\right)\right].$$

Figures (4.7, 4.8, 4.9 and 4.10) are four graphs of the Fourier series of the function $f(x)$ for the partial sums $S_1(x)$, $S_5(x)$, $S_{12}(x)$, and $S_{25}(x)$, respectively. *Note:* The Fourier series converges to the entire function, $f(x)$, rather quickly. ∎

$$S_1(x) = \frac{-2}{3} + \sum_{n=1}^{1}\left[\frac{112(-1)^n}{n^2\pi^2}\cos\left(\frac{n\pi x}{L}\right) + \frac{96(-1)^n}{n^3\pi^3}\sin\left(\frac{n\pi x}{L}\right)\right].$$

$$S_5(x) = \frac{-2}{3} + \sum_{n=1}^{5}\left[\frac{112(-1)^n}{n^2\pi^2}\cos\left(\frac{n\pi x}{L}\right) + \frac{96(-1)^n}{n^3\pi^3}\sin\left(\frac{n\pi x}{L}\right)\right].$$

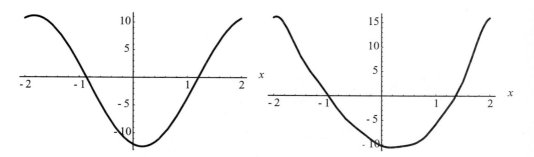

Figure 4.7: The Fourier series representation of $f(x) = x^3 + 7x^2 - 4x - 10$ for the partial sum $S_1(x)$.

Figure 4.8: The Fourier series representation of $f(x) = x^3 + 7x^2 - 4x - 10$ for the partial sum $S_5(x)$.

$$S_{12}(x) = \frac{-2}{3} + \sum_{n=1}^{12}\left[\frac{112(-1)^n}{n^2\pi^2}\cos\left(\frac{n\pi x}{L}\right) + \frac{96(-1)^n}{n^3\pi^3}\sin\left(\frac{n\pi x}{L}\right)\right].$$

$$S_{25}(x) = \frac{-2}{3} + \sum_{n=1}^{25}\left[\frac{112(-1)^n}{n^2\pi^2}\cos\left(\frac{n\pi x}{L}\right) + \frac{96(-1)^n}{n^3\pi^3}\sin\left(\frac{n\pi x}{L}\right)\right].$$

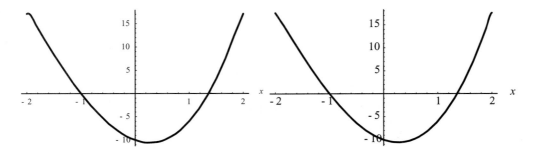

Figure 4.9: The Fourier series representation of $f(x) = x^3 + 7x^2 - 4x - 10$ for the partial sum $S_{12}(x)$.

Figure 4.10: The Fourier series representation of $f(x) = x^3 + 7x^2 - 4x - 10$ for the partial sum $S_{25}(x)$.

In the next two examples, the functions chosen demonstrate the action of the Fourier series at a jump discontinuity. An explanation is provided, then formalized, in Fourier's convergence theorem.

EXAMPLE 4.9. Consider the function $h(x) = x^2 + 2x - 1$ on the interval $[-5, 5]$ shown in Figure(4.11). The Fourier series representation of $h(x)$ is

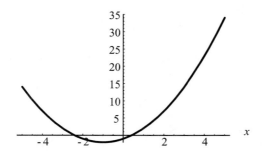

Figure 4.11: $h(x) = x^2 + 2x - 1$ on the interval $[-5, 5]$.

$$x^2 + 2x - 1 = a_0 + \sum_{n=1}^{\infty} \left[a_n \cos \frac{n\pi x}{5} + b_n \sin \frac{n\pi x}{5} \right]$$

where

$$a_0 = \frac{1}{2L} \int_{-L}^{L} h(x) \, dx = \frac{1}{10} \int_{-5}^{5} x^2 + 2x - 1 \, dx = \frac{22}{3},$$

$$a_n = \frac{1}{L} \int_{-L}^{L} h(x) \cos\left(\frac{n\pi x}{L}\right) dx = \frac{1}{5} \int_{-5}^{5} (x^2 + 2x - 1) \cos\left(\frac{n\pi x}{5}\right) dx$$

$$= \frac{100}{n^2\pi^2} \cos(n\pi) - \frac{100}{n^3\pi^3} \sin(n\pi) + \frac{48}{n\pi} \sin(n\pi) = \frac{100(-1)^n}{n^2\pi^2},$$

and

$$b_n = \frac{1}{L} \int_{-L}^{L} h(x) \sin\left(\frac{n\pi x}{L}\right) dx = \frac{1}{5} \int_{-5}^{5} (x^2 + 2x - 1) \sin\left(\frac{n\pi x}{5}\right) dx$$

$$= \frac{20}{n^2\pi^2} \sin(n\pi) - \frac{20}{n\pi} \cos(n\pi) = -\frac{20(-1)^n}{n\pi}.$$

Thus,

$$x^2 + 2x - 1 = \frac{22}{3} + \sum_{n=1}^{\infty} \left[\frac{100(-1)^n}{n^2\pi^2} \cos\frac{n\pi x}{5} - \frac{20(-1)^n}{n\pi} \sin\frac{n\pi x}{5} \right].$$

Figures (4.12, 4.13, 4.14, 4.15, 4.16, and 4.17) are six graphs of the Fourier series representation of $h(x)$ for the partial sums $S_1(x)$, $S_5(x)$, $S_{25}(x)$, $S_{50}(x)$, $S_{75}(x)$, and $S_{100}(x)$, respectively. *Note:* The partial sums converge to the function $h(x)$ much more slowly than the partial sums represented by Figures (4.7, 4.8, 4.9 and 4.10) of Example (4.8). The reason for the slower convergence of the partial sums are the x-values, $x = -5$ and $x = 5$. At these x-values $h(-5) \neq h(5)$. Thus, the Fourier series representation of the function does not seem to converge to the values of $h(-5)$ or $h(5)$. This example demonstrates the action of the Fourier series representation of a function with a jump discontinuity at the endpoints.

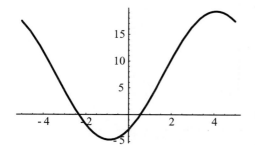

Figure 4.12: Fourier Series representation of $h(x) = x^2 + 2x - 1$ for the partial sum $S_1(x)$.

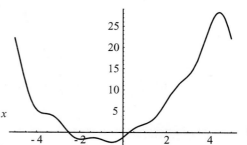

Figure 4.13: Fourier Series representation of $h(x) = x^2 + 2x - 1$ for the partial sum $S_5(x)$.

$$S_1(x) = \frac{22}{3} + \sum_{n=1}^{1} \left[\frac{100(-1)^n}{n^2\pi^2} \cos\frac{n\pi x}{5} - \frac{20(-1)^n}{n\pi} \sin\frac{n\pi x}{5} \right].$$

$$S_5(x) = \frac{22}{3} + \sum_{n=1}^{5} \left[\frac{100(-1)^n}{n^2\pi^2} \cos \frac{n\pi x}{5} - \frac{20(-1)^n}{n\pi} \sin \frac{n\pi x}{5} \right].$$

$$S_{25}(x) = \frac{22}{3} + \sum_{n=1}^{25} \left[\frac{100(-1)^n}{n^2\pi^2} \cos \frac{n\pi x}{5} - \frac{20(-1)^n}{n\pi} \sin \frac{n\pi x}{5} \right].$$

$$S_{50}(x) = \frac{22}{3} + \sum_{n=1}^{50} \left[\frac{100(-1)^n}{n^2\pi^2} \cos \frac{n\pi x}{5} - \frac{20(-1)^n}{n\pi} \sin \frac{n\pi x}{5} \right].$$

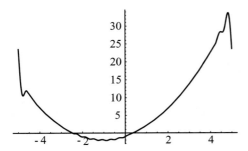

Figure 4.14: Fourier Series representation of $h(x) = x^2 + 2x - 1$ for the partial sum $S_{25}(x)$.

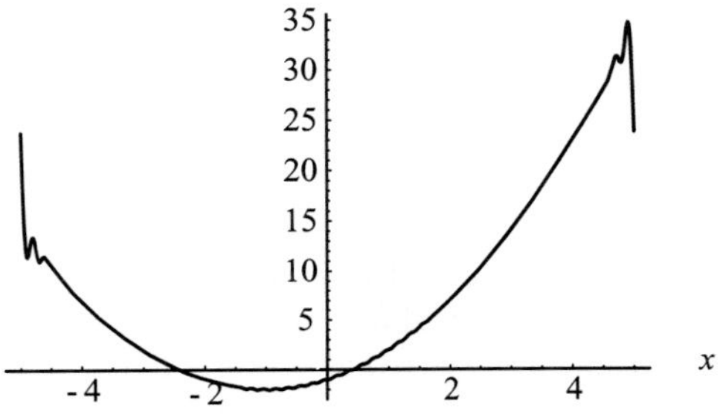

Figure 4.15: Fourier Series representation of $h(x) = x^2 + 2x - 1$ for the partial sum $S_{50}(x)$.

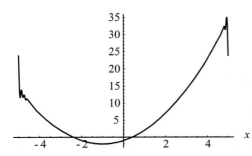

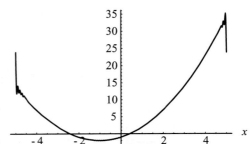

Figure 4.16: Fourier Series representation of $h(x) = x^2 + 2x - 1$ for the partial sum $S_{75}(x)$.

Figure 4.17: Fourier Series representation of $h(x) = x^2 + 2x - 1$ for the partial sum $S_{100}(x)$.

$$S_{75}(x) = \frac{22}{3} + \sum_{n=1}^{75} \left[\frac{100(-1)^n}{n^2\pi^2} \cos \frac{n\pi x}{5} - \frac{20(-1)^n}{n\pi} \sin \frac{n\pi x}{5} \right].$$

$$S_{100}(x) = \frac{22}{3} + \sum_{n=1}^{100} \left[\frac{100(-1)^n}{n^2\pi^2} \cos\frac{n\pi x}{5} - \frac{20(-1)^n}{n\pi} \sin\frac{n\pi x}{5} \right].$$

In Figure (4.18), the graph of the Fourier series representation of $f(x)$ over the interval $[-15, 15]$ is shown. We see the Fourier series duplicates itself three times on this new interval. This is known as the periodic extension of the Fourier series. Also, we see that the Fourier series representation of $f(x)$ does converge at $f(-5)$ and $f(5)$. Although, what it actually converges to is not yet known. ■

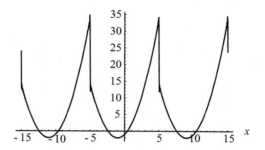

Figure 4.18: $h(x) = \frac{22}{3} + \sum_{n=1}^{100} \left[\frac{100(-1)^n}{n^2\pi^2} \cos\frac{n\pi x}{5} - \frac{20(-1)^n}{n\pi} \sin\frac{n\pi x}{5} \right]$ graphed on the interval $[-15, 15]$.

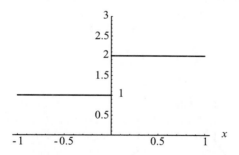

Figure 4.19: The graph of the function $g(x)$ on the interval $-1 \leq x \leq 1$.

Next, we will consider a function that has a jump discontinuity in the interior of the interval.

EXAMPLE 4.10. Consider the function

$$g(x) = \begin{cases} 1, & -1 \leq x < 0 \\ 2, & 0 \leq x \leq 1, \end{cases}$$

shown in Figure (4.19).

The Fourier series representation of $g(x)$ is

$$g(x) = a_0 + \sum_{n=1}^{\infty} \left[a_n \cos \frac{n\pi x}{1} + b_n \sin \frac{n\pi x}{1} \right],$$

where

$$a_0 = \frac{1}{2L} \int_{-L}^{L} g(x) \, dx = \frac{1}{2} \int_{-1}^{1} g(x) \, dx = \frac{3}{2},$$

$$a_n = \frac{1}{L} \int_{-L}^{L} g(x) \cos \left(\frac{n\pi x}{L} \right) \, dx = \frac{1}{1} \int_{-1}^{1} g(x) \cos \left(\frac{n\pi x}{1} \right) \, dx$$

$$= \frac{3}{n\pi} \sin(n\pi) = 0,$$

and

$$b_n = \frac{1}{L} \int_{-L}^{L} g(x) \sin \left(\frac{n\pi x}{L} \right) \, dx = \frac{1}{1} \int_{-1}^{1} g(x) \sin \left(\frac{n\pi x}{1} \right) \, dx$$

$$= \frac{1}{n\pi} - \frac{1}{n\pi} \cos(n\pi) = \frac{[1 - (-1)^n]}{n\pi}.$$

Thus,

$$g(x) = \frac{3}{2} + \sum_{n=1}^{\infty} \left[\frac{[1 - (-1)^n]}{n\pi} \sin (n\pi x) \right].$$

The Fourier series representation of the function $g(x)$ are graphed below for the partial sums $S_1(x)$, $S_{10}(x)$, $S_{25}(x)$, $S_{50}(x)$, $S_{100}(x)$, and $S_{1000}(x)$, respectively. Also, note that as n increases from 1 to 1000, the graphs of the Fourier series representation of the function $g(x)$ become increasingly more accurate. Again, please note that at the x values of -1 and 1, the Fourier series representation of $g(x)$ does not seem to converge to the points $g(-1)$ and $g(1)$. Also, at $x = 0$ the Fourier series representation of $g(x)$ seems to equal both 1 and 2. In Figure (4.26), the Fourier series representation of $g(x)$ is graphed over the interval $[-3, 3]$. Please note that the Fourier series representation of $g(x)$ repeats itself. Again, this is known as the periodic extension of the Fourier series. ∎

In Example (4.9), the function, $f(x) = x^2 + 2x - 1$, on the interval $[-5, 5]$ was a continuous function on that interval. However, $f(-5) \neq f(5)$. If we consider the graph of the Fourier series representation of $f(x)$ in Figure (4.18), we can see that at the point $(-5, f(-5))$ and the point $(5, f(5))$, the Fourier series converges to something.

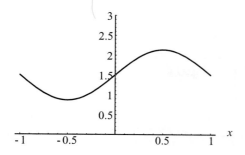

Figure 4.20: $g(x) = \frac{3}{2} + \sum_{n=1}^{1}\left[\frac{[1-(-1)^n]}{n\pi}\sin(n\pi x)\right]$ on the interval $[-1, 1]$.

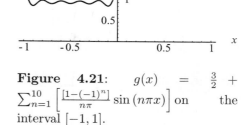

Figure 4.21: $g(x) = \frac{3}{2} + \sum_{n=1}^{10}\left[\frac{[1-(-1)^n]}{n\pi}\sin(n\pi x)\right]$ on the interval $[-1, 1]$.

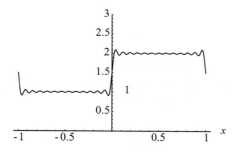

Figure 4.22: $g(x) = \frac{3}{2} + \sum_{n=1}^{25}\left[\frac{[1-(-1)^n]}{n\pi}\sin(n\pi x)\right]$ on the interval $[-1, 1]$.

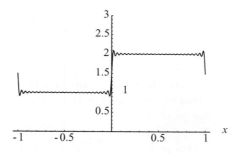

Figure 4.23: $g(x) = \frac{3}{2} + \sum_{n=1}^{50}\left[\frac{[1-(-1)^n]}{n\pi}\sin(n\pi x)\right]$ on the interval $[-1, 1]$.

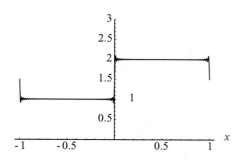

Figure 4.24: $g(x) = \frac{3}{2} + \sum_{n=1}^{100}\left[\frac{[1-(-1)^n]}{n\pi}\sin(n\pi x)\right]$ on the interval $[-1, 1]$.

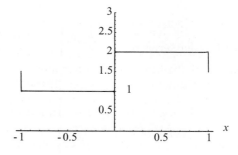

Figure 4.25: $g(x) = \frac{3}{2} + \sum_{n=1}^{1000}\left[\frac{[1-(-1)^n]}{n\pi}\sin(n\pi x)\right]$ on the interval $[-1, 1]$.

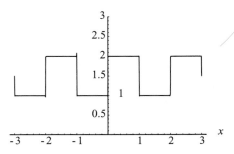

Figure 4.26: $g(x) = \frac{3}{2} + \sum_{n=1}^{1000} \left[\frac{[1-(-1)^n]}{n\pi} \sin{(n\pi x)} \right]$ on the interval $[-3, 3]$.

In Example (4.10), the function,

$$g(x) = \begin{cases} 1, & -1 \le x < 0 \\ 2, & 0 \le x \le 1, \end{cases}$$

on the interval $[-1, 1]$ has a jump discontinuity at $x = 0$ and $g(-1) \ne g(1)$. Again, if we consider the graph of the Fourier series representation of $g(x)$ in Figure (4.26), we see that at the points $(-1, g(-1))$, $(0, g(0))$, and $(1, g(1))$, the Fourier series does converge to something.

The convergent point of the Fourier series representation of a function $f(x)$ on the interval $[-L, L]$ with jump discontinuities within the interval is the average of the left and right limits of the function at the jump discontinuity. For instance, this means in Example (4.10) where

$$g(x) = \begin{cases} 1, & -1 \le x < 0 \\ 2, & 0 \le x \le 1, \end{cases}$$

at $x = 0$, the left limit of the $g(0^-) = 1$ and the right limit of $g(0^+) = 2$. Therefore, the Fourier series representation of $g(x)$ at $x = 0$ converges to the value

$$\frac{g(0^-) + g(0^+)}{2} = 1.5.$$

The Fourier series representation of a function $f(x)$ on the interval $[-L, L]$, where $f(-L) \ne f(L)$, converges to the average of the left and right limits of the endpoints on the periodic extension of the function. What is the periodic extension of a function? The easiest way to answer this question is to think of the cosine function on the interval $[-\pi, \pi]$ graphed in Figure (4.27). A periodic extension of the $\cos x$ on the interval $[-\pi, \pi]$ is the $\cos x$ graphed on the interval $[-3\pi, 3\pi]$ shown in Figure (4.28).

Figure (4.28) repeats the original curve shown in Figure (4.27), once on the left and once on the right. This is an example of a periodic extension. Basically, the curve is repeated on the left of the original curve as many times as

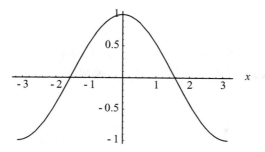

Figure 4.27: $\cos x$ on the interval $[-\pi, \pi]$.

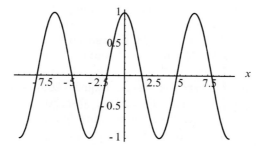

Figure 4.28: $\cos x$ on the interval $[-3\pi, 3\pi]$.

needed to reach negative infinity, and on the right of the original curve as many times as needed to reach positive infinity. Therefore, in Example (4.9), where $f(-5) \neq f(5)$, the Fourier series representation of $f(x)$ converges at $x = -5$ to $\dfrac{f(-5^-) + f(-5^+)}{2} = \dfrac{f(5) + f(-5^+)}{2} = \dfrac{34 + 14}{2} = 24$; at $x = 5$ the Fourier series converges to $\dfrac{f(5^-) + f(5^+)}{2} = \dfrac{f(5^-) + f(-5)}{2} = \dfrac{34 + 14}{2} = 24$.

Theorem (21) is Fourier's convergence theorem, which was proved by Peter Gustav Lejeune Dirichlet[1] in 1828 in the journal *Journal für die reine und angewandte Mathematik*.

Theorem 21. *(Fourier's Convergence Theorem) If $f(x)$ is an element of the function space of piecewise smooth functions on the interval $[-L, L]$, then the Fourier series representation of $f(x)$ converges to*

1. *the function $f(x)$, wherever $f(x)$ is continuous, and to*

2. *the average of the left and right limits of the function $f(x)$ at any jump discontinuity. That is, the Fourier series converges at a jump discontinuity $x = x_0$ to*

---

[1]P. G. L. Dirichlet (1805–1859) was a very famous mathematician from Prussia.

$$\frac{f(x_0^-) + f(x_0^+)}{2}.$$

3. if the endpoints, $f(-L)$ and $f(L)$ are not equal, then the Fourier Series converges, at $-L$ and at L, to $\dfrac{f(L^-) + f(-L^+)}{2}.$

The proof of Theorem (21) may be found in many different texts on Fourier series. In particular, *Fourier series*, by Georgi P. Tolstov, provides a complete derivation of this theorem and many of the other theorems governing convergence of Fourier series.

Since the Fourier series of a piecewise smooth function is guaranteed to converge to the function whenever the function is continuous, or to the average of the left- and right-hand limits wherever the function has a jump discontinuity, we do not have to graph the function or the periodic extension of the function before writing the Fourier series down and determining the Fourier coefficients. This concept is very important. We now develop the notion of uniform convergence of a Fourier series to a function.

Understanding how a Fourier series converges requires some knowledge of series convergence. Hence, we start with two definitions; absolute and uniform convergence. If you have not had a real analysis course, this definition and several other concepts may be difficult to understand. However, it gives you an idea of convergence and how it must be shown. So let's get started.

Definition 22. *An infinite series is said to be absolutely convergent if the series formed from it by replacing each term by its absolute value is convergent.*

Definition 23. *Let (f_n) be a sequence of functions defined on a subset $S \in \mathbb{R}$. Then, (f_n) converges uniformly on S to a function F defined on S, if for each $\varepsilon > 0$ there exists a number N such that*

$$|f_n(x) - f(x)| < \varepsilon$$

for all $x \in S$ and all $n > N$.

This definition gives us a precise way of determining uniform convergence. The following theorem, called the Weierstrass M-test,[2] provides a way to determine if a sequence of functions converges uniformly. The proof of this theorem may be found in *Analysis with an Introduction to Proof* by Steven R. Lay as well as many other texts on real analysis.

Theorem 24. *(Weierstrass M-test) Suppose that (f_n) is a sequence of functions defined on a subset $S \in \mathbb{R}$, and (M_n) is a sequence of nonnegative numbers such that*

$$|f_n(x)| \leq M_n$$

---

[2]Karl Weierstrass (1815–1897) was one of the greatest mathematics teacher of the mid-nineteenth century. His greatest contributions to mathematics were in the field of power series representation of a function.

for all $x \in S$ and all $n \in \mathbb{N}$. If $\displaystyle\sum_{n=1}^{\infty} M_n$ converges, then $\displaystyle\sum_{n=1}^{\infty} f_n$ converges uniformly on S.

It should be stressed that uniform convergence is much stronger than both pointwise or mean-square convergence. Although both pointwise and mean-square convergence are important concepts, pointwise convergence is not powerful enough for our purposes and mean-square convergence is beyond the scope of this text.

We now consider a simple and important fact. It is stated in a theorem and followed by a short proof.

Theorem 25. *Given a trigonometric Fourier series*

$$a_0 + \sum_{n=1}^{\infty} \left[a_n \cos\left(\frac{n\pi x}{L}\right) + b_n \sin\left(\frac{n\pi x}{L}\right) \right], \qquad (4.29)$$

which is not assumed to be the Fourier series of any function. If the series

$$\sum_{n=1}^{\infty} \left(|a_n| + |b_n| \right)$$

converges, the Fourier series, Equation (4.29), converges absolutely and uniformly.

Proof. Since

$$|a_n \cos\left(\frac{n\pi x}{L}\right) + b_n \sin\left(\frac{n\pi x}{L}\right)| \le |a_n \cos\left(\frac{n\pi x}{L}\right)| + |b_n \sin\left(\frac{n\pi x}{L}\right)|$$

by triangle inequality and

$$|a_n \cos\left(\frac{n\pi x}{L}\right)| + |b_n \sin\left(\frac{n\pi x}{L}\right)|| \le |a_n| + |b_n|,$$

because the sine and cosine functions are bounded by ±1. We have, by the Weierstrass M-test, that the Fourier series, Equation (4.29), converges uniformly. $\square$

The previous theorem and the Fourier convergence theorem, Theorem (21) imply Theorem (26).

Theorem 26. *The Fourier series representation of a continuous, piecewise smooth function $f(x)$ on $[-L, L]$ converges to $f(x)$ absolutely and uniformly.*

Theorem (26) is a very important theorem, and it implies that the Fourier series representation of a continuous piecewise smooth function $f(x)$ on $[L, L]$ is also continuous. This fact, with the necessary conditions, is stated in the next theorem.

Theorem 27. *If $f(x)$ is a piecewise smooth continuous function on the interval $[-L, L]$ and $f(-L) = f(L)$, then the Fourier series representation of the function $f(x)$ is continuous on $[-L, L]$.*

Similar theorems are stated for the Fourier sine and cosine series in the next section.

EXERCISES 4.3

4.3.1. Show that

$$\int_{-L}^{L} \sin\left(\frac{n\pi x}{L}\right) \sin\left(\frac{m\pi x}{L}\right) \, dx = 0, \ n \in \mathbb{N}, \ n \neq m$$

by direct computation.

4.3.2. Show that

$$\int_{-L}^{L} \cos\left(\frac{n\pi x}{L}\right) \, dx = 0, \ n \in \mathbb{N}$$

by direct computation.

4.3.3. Show that

$$\int_{-L}^{L} \sin\left(\frac{n\pi x}{L}\right) \, dx = 0, \ n \in \mathbb{N}$$

by direct computation.

4.3.4. Show that

$$\int_{-L}^{L} \cos\left(\frac{n\pi x}{L}\right) \cos\left(\frac{m\pi x}{L}\right) \, dx = 0, \ n \in \mathbb{N}, \ n \neq m$$

by direct computation.

4.3.5. Plot the functions $f(x) = x$ and $g(x) = 1$ on the interval $[-L, L]$. Are the functions $f(x)$ and $g(x)$ orthogonal? Show by direct computation

$$\int_{-L}^{L} f(x)g(x) \, dx = 0.$$

4.3.6. Plot the functions

$$u(x) = \frac{1}{2}\left(3x^2 - 1\right) \text{ and } v(x) = \frac{1}{2}\left(5x^3 - 3x\right)$$

on the interval $[-1, 1]$. Are the functions $u(x)$ and $v(x)$ orthogonal? Show by direct computation that

$$\int_{-1}^{1} u(x)v(x) \, dx = 0.$$

Also, show by direct computation that

$$\int_{-1}^{1} [u(x)]^2 \ dx = \frac{2}{5}$$

and

$$\int_{-1}^{1} [v(x)]^2 \ dx = \frac{2}{7}.$$

4.3.7. Plot the functions

$$u(x) = \frac{1}{8} \left(35x^4 - 30x^2 + 3\right) \text{ and } v(x) = \frac{1}{8} \left(63x^5 - 70x^3 + 15x\right)$$

on the interval $[-1, 1]$. Are the functions $u(x)$ and $v(x)$ orthogonal? Show by direct computation that

$$\int_{-1}^{1} u(x)v(x) \ dx = 0.$$

Also, show by direct computation that

$$\int_{-1}^{1} [u(x)]^2 \ dx = \frac{2}{9}$$

and

$$\int_{-1}^{1} [v(x)]^2 \ dx = \frac{2}{11}.$$

4.3.8. Some functions are orthogonal with a common weight function. For example, the Hermite polynomials are orthogonal with weight function e^{x^2}. In Chapter 8, we discuss weight functions. However, for the purpose of this exercise, we give you the general formula and ask you to show that the first four Hermite polynomials are orthogonal to each other. The general formula is

$$\int_{-\infty}^{\infty} e^{-x^2} H_n(x) H_m(x) \ dx.$$

The first four Hermite polynomials are

$$H_0(x) = 1,$$

$$H_1(x) = 2x,$$

$$H_2(x) = 4x^2 - 2, \text{ and}$$

$$H_3(x) = 8x^3 - 12x.$$

4.3.9. Show, by direct computation, that the Fourier coefficients a_0 and b_n have the formulas

$$a_0 = \frac{1}{2L} \int_{-L}^{L} f(x)\, dx$$

and

$$b_n = \frac{1}{L} \int_{-L}^{L} f(x) \sin\left(\frac{n\pi x}{L}\right)\, dx.$$

4.3.10. Determine the Fourier series for the following functions on the given bounds:

(1) 5 for $-2 \le x \le 2$.

(2) $2x^2 - 3x$ for $-\pi \le x \le \pi$.

(3) $4\cos 3x$ for $\dfrac{-\pi}{2} \le x \le \dfrac{\pi}{2}$.

(4) $1 + \cos\dfrac{2x}{3} + \sin\dfrac{5x}{3}$ for $-3\pi \le x \le 3\pi$.

4.3.11. Determine the Fourier series representation for the function $f(x) = 2x - 1$ on the interval $[-5, 5]$. Using your favorite mathematical software, plot the function $f(x)$ and the Fourier series representation of $f(x)$ for $n = 1$ to 5, $n = 1$ to 10, $n = 1$ to 50, and $n = 1$ to 200. In your own words, explain how the Fourier series representation of $f(x)$ converges at the points -5, 0, and 5.

4.3.12. Determine the Fourier series representation for the function $g(x) = 3\sin x$ on the interval $[-\pi, \pi]$. State in your own words any conclusions you may determine about the Fourier series representation for the function $g(x)$.

4.3.13. Let

$$h(x) = \begin{cases} 1, & -\dfrac{3\pi}{2} \le x < \dfrac{\pi}{2} \\[2mm] -1, & \dfrac{\pi}{2} \le x \le \dfrac{3\pi}{2} \end{cases}.$$

Determine the Fourier series representation for the function $h(x)$. Using your favorite mathematical software, plot the function $h(x)$ on the interval $\left[\dfrac{-3\pi}{2}, \dfrac{3\pi}{2}\right]$ and the Fourier series representation of $h(x)$ on the interval $[-3\pi, 3\pi]$ for $n = 1$ to 5, $n = 1$ to 10, $n = 1$ to 50, and $n = 1$ to 200. In your own words, explain how the Fourier series representation of $h(x)$ converges at the points $-\dfrac{3\pi}{2}$, $\dfrac{\pi}{2}$ and 0.

4.3.14. Do you need to solve for the Fourier coefficients before graphing the Fourier series representation of a function?

4.3.15. Suppose $f(x)$ are piecewise smooth functions on $[-L, L]$. Prove Bessel's inequality:

$$a_0^2 + \sum_{n=1}^{\infty} a_n^2 + b_n^2 \leq \frac{1}{L} \int_{-L}^{L} [f(x)]^2 \, dx.$$

Bessel's inequality implies the sum of the squares of the Fourier coefficients of any square integrable function always converge.

4.3.16. Show that the Fourier series

$$a_0 + \sum_{n=1}^{\infty} a_n \cos \frac{n\pi x}{L} + b_n \sin \frac{n\pi x}{L}$$

can be written in the form

$$A_0 + \sum_{n=1}^{\infty} A_n \cos \left(\frac{n\pi x}{L} + \theta_n \right),$$

where $A_n = \sqrt{a_n^2 + b_n^2}$. Also, find an expression for θ_n in terms of a_n and b_n.

4.3.17. Prove Riemann–Lebesgue's lemma: If $f(x)$ is a piecewise smooth function of period 2π, then

$$\lim_{n \to \infty} \int_{-\pi}^{\pi} f(x) \sin \left[\left(n + \frac{1}{2} \right) x \right] \, dx = 0.$$

Hint: You must use Bessel's inequality.

4.3.18. Prove Parseval's equality: If $f(x)$ is a piecewise smooth function on $[-L, L]$, then

$$\frac{1}{L} \int_{-L}^{L} [f(x)]^2 \, dx = 2a_0^2 + \sum_{n=1}^{\infty} \left(a_n^2 + b_n^2 \right),$$

where a_n and b_n are the Fourier coefficients of $f(x)$.

4.4 EVEN AND ODD FUNCTIONS AND FOURIER SERIES

The question to consider next is "How do Fourier series help us solve PDEs, since the heat and wave equations are developed on the interval $[0, L]$?" This question is answered in the following discussions.

In Chapters 2 and 3, the equation for heat conduction in a one-dimensional rod and the wave equation for transverse vibrations in a string were developed. The complete equations depended on developing sets of boundary conditions. All of the boundary conditions were stated on the interval $[0, L]$. Also, at the beginning of this chapter, Fourier series were described as useful for solving PDEs like the heat and wave equations. However, Fourier series were developed on the interval $[-L, L]$. Therefore, there must be some kind of mathematical translation. This translation is based on even and odd functions.

The definition for an even function is as follows:

Definition 28. *The function $f(x)$ is an even function on the interval $[-L, L]$ if*

$$f(-x) = f(x)$$

for all $x \in [-L, L]$.

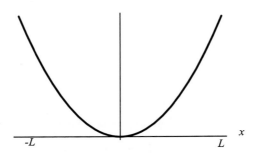

Figure 4.29: An even function on the interval $[-L, L]$.

The following definition is for an odd function:

Definition 29. *The function $f(x)$ is an odd function in the interval $[-L, L]$ if*

$$f(-x) = -f(x)$$

for all $x \in [-L, L]$.

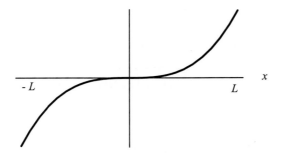

Figure 4.30: An odd function on the interval $[-L, L]$.

Definitions (28 and 29) explain how to show a function as even or odd. However, another way of describing a function as even or odd is graphically. The graph of an even function on the interval $[-L, L]$ is symmetric about the y axis. Figure (4.29) demonstrates this feature.

The graph of an odd function on the interval $[-L, L]$ is not symmetric about any axis. However, if you perform a 180^0 rotation of the first quadrant of the graph of a function, and it is identical to the third quadrant of the graph of the function, then the function is odd. Figure (4.30) demonstrates this feature. Please note, the graph of the function in the third quadrant is identical to a 180^0 rotation of the graph in the first quadrant.

When two even or two odd functions are multiplied together, the result is an even function. When an even and an odd function are multiplied together, the result is an odd function. Also, integration benefits from even and odd functions. If we integrate an even function, $f(x)$, from $-L$ to L, then we can simplify the integral by multiplying it by 2 and integrating from 0 to L, that is,

$$\int_{-L}^{L} f(x) \ dx = 2 \int_{0}^{L} f(x) \ dx, \ f(x) \text{ even}.$$

However, the integral of an odd function, $g(x)$, from $-L$ to L is 0.

In the function space of PWS functions on the interval $[-L, L]$, most functions are neither even nor odd. Therefore, knowing what even and odd functions are does not appear to help us. Also, as we said before, the heat and wave equations are usually given on the interval $[0, L]$. However, a mathematical technique of extending a function has been developed. Thus, if a function exists on the interval $[0, L]$, then we can extend it to the interval $[-L, L]$ by making an even or odd extension. This means we can choose which extension is convenient for our situation, then expand it in a Fourier series.

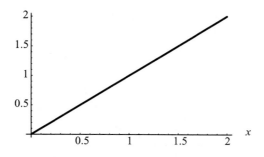

Figure 4.31: $f(x) = x$ on the interval $[0, 2]$.

EXAMPLE 4.11. Show how the function $f(x) = x$ on the interval $[0, 2]$ can be expanded in a Fourier series.

Solution: There are two choices and both are correct. First, Figure (4.31) shows the graph of $f(x) = x$ on the interval $[0, 2]$. Next, the function $f(x)$ is extended as an even function, which we will call $\overline{f}(x) = |x|$ on the interval $[-2, 2]$. Figure (4.32) graphically illustrates the even periodic extension of $\overline{f}(x) = |x|$ on the interval

$[-2, 2]$. The Fourier series representation of $\overline{f}(x) = |x|$ on the interval $[-2, 2]$ is then developed. The Fourier series representation of the function $\overline{f}(x) = |x|$ is

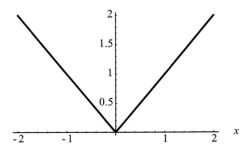

Figure 4.32: $\overline{f}(x) = |x|$ on the interval $[-2, 2]$.

$$\overline{f}(x) = |x| = a_0 + \sum_{n=1}^{\infty} \left[a_n \cos \frac{n\pi x}{L} + b_n \sin \frac{n\pi x}{L} \right]$$

$$= a_0 + \sum_{n=1}^{\infty} \left[a_n \cos \frac{n\pi x}{2} + b_n \sin \frac{n\pi x}{2} \right],$$

where

$$a_0 = \frac{1}{2L} \int_{-L}^{L} \overline{f}(x) \, dx = \frac{1}{4} \int_{-2}^{2} |x| \, dx.$$

However, $\overline{f}(x) = |x|$ on the interval $[-2, 2]$ is an even function. Therefore,

$$a_0 = \frac{1}{4} \int_{-2}^{2} |x| \, dx = \frac{1}{2} \int_{0}^{2} x \, dx = 1.$$

Also,

$$a_n = \frac{1}{L} \int_{-L}^{L} \overline{f}(x) \cos \frac{n\pi x}{L} \, dx = \frac{1}{2} \int_{-2}^{2} |x| \cos \frac{n\pi x}{2} \, dx.$$

$\overline{f}(x) \cos \dfrac{n\pi x}{2} = |x| \cos \dfrac{n\pi x}{2}$ on the interval $[-2, 2]$ is an even function; therefore,

$$a_n = \frac{1}{2} \int_{-2}^{2} |x| \cos \frac{n\pi x}{2} \, dx = \int_{0}^{2} x \cos \frac{n\pi x}{2} \, dx = \frac{4\left((-1)^n - 1\right)}{(n\pi)}.$$

Since $\overline{f}(x) \sin \dfrac{n\pi x}{2} = |x| \sin \dfrac{n\pi x}{2}$ on the interval $[-2, 2]$ is an odd function, the integral is zero. Thus,

$$\overline{f}(x) = |x| = 1 + \sum_{n=1}^{\infty} \frac{4\left((-1)^n - 1\right)}{(n\pi)^2} \cos \frac{n\pi x}{2}.$$

Figures (4.33, 4.34, 4.35, and 4.36) are four graphs of the Fourier cosine series of the function $\overline{f}(x) = |x|$ for the partial sums $S_1(x)$, $S_{10}(x)$, $S_{25}(x)$, and $S_{50}(x)$, respectively. Figure (4.37) is the Fourier cosine series representation of $\overline{f}(x) = |x|$ for the partial sum $S_{50}(x)$ on the interval $[-8, 8]$.

$$S_1(x) = 1 + \sum_{n=1}^{1} \frac{4\left((-1)^n - 1\right)}{(n\pi)^2} \cos \frac{n\pi x}{2}$$

and

$$S_{10}(x) = 1 + \sum_{n=1}^{10} \frac{4\left((-1)^n - 1\right)}{(n\pi)^2} \cos \frac{n\pi x}{2}.$$

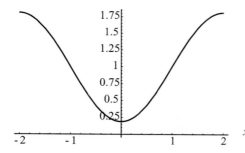

Figure 4.33: Fourier cosine series representation of $\overline{f}(x) = |x|$ for the partial sum $S_1(x)$.

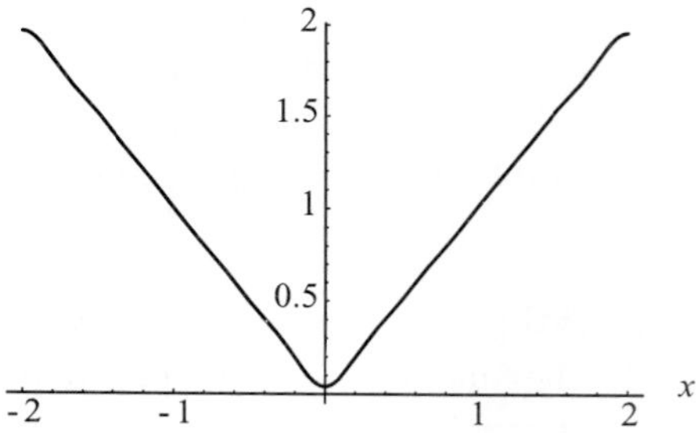

Figure 4.34: Fourier cosine series representation of $\overline{f}(x) = |x|$ for the partial sum $S_{10}(x)$.

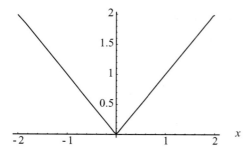

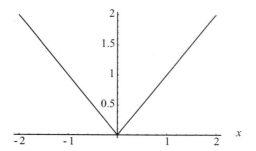

Figure 4.35: Fourier cosine series representation of $\overline{f}(x) = |x|$ for the partial sum $S_{25}(x)$.

Figure 4.36: Fourier cosine series representation of $\overline{f}(x) = |x|$ for the partial sum $S_{50}(x)$.

$$S_{25}(x) = 1 + \sum_{n=1}^{25} \frac{4\left((-1)^n - 1\right)}{(n\pi)^2} \cos \frac{n\pi x}{2}$$

and

$$S_{50}(x) = 1 + \sum_{n=1}^{50} \frac{4\left((-1)^n - 1\right)}{(n\pi)^2} \cos \frac{n\pi x}{2}.$$

The interval $[-8, 8]$ was chosen because it shows that the convergence is exact on

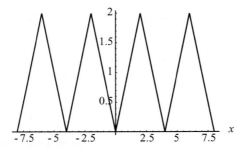

Figure 4.37: Fourier cosine series representation of $\overline{f}(x) = |x|$ on the interval $[-8, 8]$.

the interval $[-2, 2]$.

Finally, the function $f(x)$ is extended as an odd function, which we will call $\widetilde{f}(x) = x$ on the interval $[-2, 2]$. Figure (4.38) graphically illustrates $\widetilde{f}(x) = x$ on the interval $[-2, 2]$, and the Fourier series representation of $\widetilde{f}(x) = x$ on the interval $[-2, 2]$ is then developed. The Fourier series representation of the function $\widetilde{f}(x) = x$ is

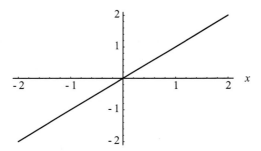

Figure 4.38: $\widetilde{f}(x) = x$ on the interval $[-2, 2]$.

$$\widetilde{f}(x) = x = a_0 + \sum_{n=1}^{\infty} \left[a_n \cos \frac{n\pi x}{L} + b_n \sin \frac{n\pi x}{L} \right]$$

$$= a_0 + \sum_{n=1}^{\infty} \left[a_n \cos \frac{n\pi x}{2} + b_n \sin \frac{n\pi x}{2} \right].$$

Here, $\widetilde{f}(x) = x$ on the interval $[-2, 2]$ is an odd function. Thus,

$$a_0 = \frac{1}{2L} \int_{-L}^{L} \widetilde{f}(x) \, dx = \frac{1}{4} \int_{-2}^{2} x \, dx = 0,$$

and

$$a_n = \frac{1}{L} \int_{-L}^{L} \widetilde{f}(x) \cos \frac{n\pi x}{L} \, dx = \frac{1}{2} \int_{-2}^{2} x \cos \frac{n\pi x}{2} \, dx = 0.$$

Whereas,

$$b_n = \frac{1}{L} \int_{-L}^{L} \widetilde{f}(x) \sin \frac{n\pi x}{L} \, dx = \frac{1}{2} \int_{-2}^{2} x \sin \frac{n\pi x}{2} \, dx = \int_{0}^{2} x \sin \frac{n\pi x}{2} \, dx$$

$$= \frac{-4 \, (-1)^n}{n\pi}.$$

Thus,

$$\widetilde{f}(x) = x = \sum_{n=1}^{\infty} \frac{-4 \, (-1)^n}{n\pi} \sin \frac{n\pi x}{2},$$

Figures (4.39, 4.40, 4.41, and 4.42) are four graphs of the Fourier sine series of

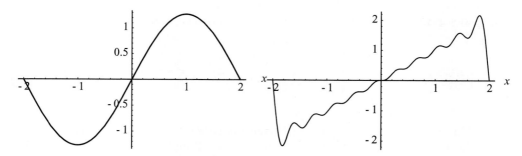

Figure 4.39: Fourier sine series representation of $\widetilde{f}(x) = x$ for the partial sum $S_1(x)$.

Figure 4.40: Fourier sine series representation of $\widetilde{f}(x) = x$ for the partial sum $S_{10}(x)$.

the function $\widetilde{f}(x) = x$ for the partial sums $S_1(x)$, $S_{10}(x)$, $S_{25}(x)$, and $S_{50}(x)$, respectively. Finally, Figure (4.43) is the Fourier series representation of $\widetilde{f}(x) = x$ for the partial sum $S_{50}(x)$ on the interval $[-8, 8]$.

$$S_1(x) = \sum_{n=1}^{1} \frac{-4 \, (-1)^n}{n\pi} \sin \frac{n\pi x}{2} \quad \text{and} \quad S_{10}(x) = \sum_{n=1}^{10} \frac{-4 \, (-1)^n}{n\pi} \sin \frac{n\pi x}{2}.$$

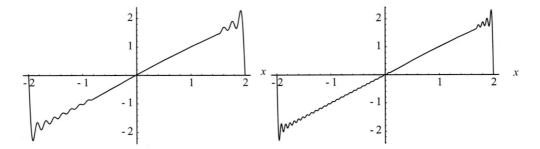

Figure 4.41: Fourier sine series representation of $\widetilde{f}(x) = x$ for the partial sum $S_{25}(x)$.

Figure 4.42: Fourier sine series representation of $\widetilde{f}(x) = x$ for the partial sum $S_{50}(x)$.

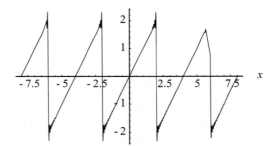

Figure 4.43: Fourier sine series representation of $\widetilde{f}(x) = x$ on the interval $[-8, 8]$.

$$S_{25}(x) = \sum_{n=1}^{25} \frac{-4\,(-1)^n}{n\pi} \sin\frac{n\pi x}{2} \text{ and } S_{50}(x) = \sum_{n=1}^{50} \frac{-4\,(-1)^n}{n\pi} \sin\frac{n\pi x}{2}.$$

The interval $[-8, 8]$ was chosen because it shows the convergence is not exact on the interval $[-2, 2]$. At the endpoints, $x = -2$ and $x = 2$, the Fourier series representation of the function, $\widetilde{f}(x) = x$, converges to 0. ∎

Note: In Figure (4.43), as x approaches -2 or 2 and -6 or 6 the Fourier series representation of the function, $f(x) = x$, seems to oscillate. This is known as the Gibbs phenomenon. The Gibbs phenomenon is discussed in greater detail at the end of Chapter 6.

Example (4.11) shows both an even and an odd extension of the function $f(x) = x$. The even extension is called the Fourier cosine series representation of the function, and it is given by

$$f(x) = a_0 + \sum_{n=1}^{\infty} a_n \cos\frac{n\pi x}{L},$$

where

$$a_0 = \frac{1}{L} \int_0^L f(x)\, dx$$

and

$$a_n = \frac{2}{L} \int_0^L f(x) \cos \frac{n\pi x}{L}\, dx.$$

The odd extension is called the Fourier sine series representation of the function, and it is given by

$$f(x) = \sum_{n=1}^{\infty} b_n \sin \frac{n\pi x}{L},$$

where

$$b_n = \frac{2}{L} \int_0^L f(x) \sin \frac{n\pi x}{L}\, dx.$$

However, which of the two different extensions is the right one to use? This is generally answered by the format of the problem. In most cases, either the Fourier sine series representation will be the correct one to use or the Fourier cosine series representation will be the correct one to use, but not both.

We conclude this section with two theorems on when the Fourier sine and cosine series representation of a function are continuous. As in Theorem (27), they follow directly from Theorem (26) stated at the end of the last section. These theorems are important and will be continually used throughout the rest of the text particularly in Chapter 6 and 8.

Theorem 30. *If $f(x)$ is a piecewise smooth continuous function on the interval $[0, L]$, then the Fourier cosine series representation of the function $f(x)$ is continuous.*

Theorem 31. *If $f(x)$ is a piecewise smooth continuous function on the interval $[0, L]$ and $f(0) = f(L) = 0$, then the Fourier sine series representation of the function $f(x)$ is continuous.*

EXERCISES 4.4

4.4.1. Determine the Fourier cosine series for the following functions on the given interval:

    (1)　7 for $0 \leq x \leq 2\pi$.

    (2)　$\sin x$ for $0 \leq x \leq 0.5$.

(3) $x^3 - 2x^2 + 1$ for $0 \le x \le \pi$.

(4) $\cos x$ for $0 \le x \le 3.1$.

4.4.2. Determine the Fourier sine series for the following functions on the given interval:

(1) 7 for $0 \le x \le 2\pi$.

(2) $\sin x$ for $0 \le x \le 0.5$.

(3) $x^3 - 2x^2 + 1$ for $0 \le x \le \pi$.

(4) $\cos x$ for $0 \le x \le 3.1$.

4.4.3. Given

$$\varphi''(x) = -\lambda\varphi(x),$$

subject to

$$\varphi(0) = 0 \text{ and } \varphi(L) = 0,$$

find all eigenvalues and eigenfunctions.

4.4.4. Given

$$\varphi''(x) = -\lambda\varphi(x),$$

subject to

$$\varphi'(0) = 0 \text{ and } \varphi'(L) = 0,$$

find all eigenvalues and eigenfunctions.

4.4.5. Given

$$\varphi''(x) = -\lambda\varphi(x)$$

subject to

$$\varphi'(0) = 0 \text{ and } \varphi(L) = 0,$$

find all eigenvalues and eigenfunctions.

4.4.6. Given

$$\varphi''(x) = -\lambda\varphi(x),$$

subject to

$$\varphi(0) = 0 \text{ and } \varphi'(L) = 0,$$

find all eigenvalues and eigenfunctions.

4.4.7. Find the Fourier cosine series representation for the function $f(x) = x^2$ on the interval $[0, 1]$. Using your favorite mathematical software, plot the function $f(x)$, the even extension of $f(x)$, and the Fourier cosine series representation of the function $f(x)$. Determine the necessary n, so that the Fourier cosine series representation of the function $f(x)$ is accurate.

4.4.8. Find the Fourier sine series representation for the function $f(x) = x^2$ on the interval $[0, 1]$. Using your favorite mathematical software, plot the function $f(x)$, the odd extension of $f(x)$, and the Fourier sine series representation of the function $f(x)$. Determine the necessary n, so that the Fourier sine series representation of the function $f(x)$ is accurate. Compare your results with those of the previous problem.

4.4.9. Is the Fourier cosine series representation of the function $g(x) = x^3$ on the interval $[0, 3]$, or the Fourier sine series representation of the function $g(x) = x^3$ on the interval $[0, 3]$, equivalent to the Fourier series representation of the function $h(x) = x^3$ on the interval $[-3, 3]$?

4.4.10. Given the fact that $f(x)$ is a continuous function on the interval $[-L, L]$,

(1) state the conditions when the Fourier series representation of the function $f(x)$ is equal to the function $f(x)$ for all x in the interval,

(2) state the conditions when the Fourier cosine series representation of the function $f(x)$ on the interval $[0, L]$ is equal to the function $f(x)$ for all x in the interval $[0, L]$, and

(3) state the conditions when the Fourier sine series representation of the function $f(x)$ on the interval $[0, L]$ is equal to the function $f(x)$ for all x in the interval $[0, L]$.

Chapter 5

Separation of Variables: The Homogeneous Problem

5.1   INTRODUCTION

So far, we have discussed both the steady-state temperature solution for the distribution of heat in a rod and d'Alembert's solution for the one-dimensional wave equation. In this chapter, we introduce a third method for solving PDEs: separation of variables.

Separation of variables is an important technique to master when studying solution methods of PDEs, because this technique leads to an infinite series solution, the Fourier Series solution. However, separation of variables does have its drawbacks. First, unlike the steady-state temperature solution, which makes no requirement on the homogeneity of the PDE or boundary conditions (BCs), separation of variables requires both the PDE and the BCs be homogeneous. Second, in general the spatial variable must have finite boundaries. If the spatial variable has semi-infinite or infinite boundaries, separation usually does not work. We tackle the problem of semi-infinite and infinite boundaries in Chapter 11. Finally, separation of variables requires the PDE and the BCs to be linear. Therefore, this chapter starts with a brief discussion of linear and homogeneous equations and BCs. To discuss this properly, we introduce the notion of an operator (don't get nervous, you've already seen lots of them). Operators are very important to mathematics and are used extensively by physicists and engineers.

Once we complete our discussion on operators and linear equations, we proceed with separation of variables. First, we cover separation of variables and the heat equation, followed immediately by separation of variables and the wave equation. The similarities between the two sections is quite evident and should be no surprise. Then, we apply separation of variables to multidimensional spatial problems. We complete the chapter with another application of separation of variables, Laplace's equation in Cartesian coordinates.

5.2   OPERATORS: LINEAR AND HOMOGENEOUS EQUATIONS

You have been working with operators all of your academic life. In first grade, when you learned about adding two numbers together, you began your study of operators. The operator in that case was addition, and addition on the real numbers could be considered a mapping written as $+ : \mathbb{R} \times \mathbb{R} \to \mathbb{R}$. Please refer to Appendix D. *Note:* $\mathbb{R}$ is a vector space. In calculus, you were introduced to two very important operators: They are the differential operator $\dfrac{d\_}{dx}$ and the integral operator $\displaystyle\int_a^b \_\, dx$. In both of the calculus operators, the underline, $\_$, holds the place for the function on which the operation occurs—for instance,

$$\frac{d}{dx}(x^2 + 2x - 1)$$

or

$$\int_a^b (x^2 + 2x - 1)dx.$$

In the study of ODEs, you learned that a differential operator can look like

$$\frac{d^2\_}{dx^2} + a\frac{d\_}{dx} + b,$$

which is the sum of other operators. Again, $\mathbb{R}$ is the vector space that is involved with the two calculus operators and the differential operator.

Judging from these three operators, it would seem reasonable to assume a partial differential operator could be

$$\frac{\partial\_}{\partial t} - k\frac{\partial^2\_}{\partial x^2}. \tag{5.1}$$

In fact, Equation (5.1) is a partial differential operator known as the heat operator.

Actually, an operator is like a function. However, it is more versatile. An operator is a mapping of vector spaces or function spaces. This is important to remember because the objects that operators operate on are, in this course, functions.

Now that we have looked at several different examples of operators, let's discuss a very special type of operator: the linear operator.

5.2.1   Linear Operators

Definition 32. *Let L be an operator. Then, if L has a vector space as it's domain and the range of L is a vector space over the same field, in our case $\mathbb{R}$, and L satisfies*

$$L\left[au(x,t) + bv(x,t)\right] = aL\left[u(x,t)\right] + bL\left[v(x,t)\right]$$

where a and b are arbitrary constants, then L is called a linear operator.

Note: Vector space in the previous definition may be replaced by function space. For example, we can show that $L = \int_a^b \_\, dx$ is a linear operator. This means that for two different functions $f(x)$ and $g(x)$ and two constants c and d, we have

$$L\left[cf(x) + dg(x)\right] = \int_a^b \left[cf(x) + dg(x)\right] dx.$$

By the property of definite integrals, we know the integral of the sum of two functions is equal to the sum of the integrals of each function. Thus, we have

$$\int_a^b \left[cf(x) + dg(x)\right] dx = \int_a^b cf(x) dx + \int_a^b dg(x) dx.$$

Applying the property of definite integrals again, we find that the integral of the product of a constant and a function is equal to the product of the constant and the integral of the function, resulting in

$$\int_a^b cf(x) dx + \int_a^b dg(x) dx = c\int_a^b f(x) dx + d\int_a^b g(x) dx.$$

This becomes

$$c\int_a^b f(x) dx + d\int_a^b g(x) dx = cL\left[f(x)\right] + dL\left[g(x)\right].$$

Therefore,

$$L\left[cf(x,t) + dg(x,t)\right] = cL\left[f(x,t)\right] + dL\left[g(x,t)\right],$$

proving that the integral operator $L = \int_a^b \_\, dx$ is a linear operator.

The heat operator, Equation (5.1), is also a linear operator. The following is a proof that the heat operator is linear. It is slightly more complicated than proving the integral operator is linear.

EXAMPLE 5.1. Consider the operator

$$L = \frac{\partial \_}{\partial t} - k\frac{\partial^2 \_}{\partial x^2} - g(x,t)\_,$$

where the underline (_) indicates where a function would be placed. To show L is a linear operator, we must show that

$$\frac{\partial(c_1 u_1 + c_2 u_2)}{\partial t} - k\frac{\partial^2(c_1 u_1 + c_2 u_2)}{\partial x^2} - g(x,t)(c_1 u_1 + c_2 u_2)$$

$$= c_1\frac{\partial(u_1)}{\partial t} - kc_1\frac{\partial^2(u_1)}{\partial x^2} - c_1 g(x,t)(u_1) +$$

$$c_2 \frac{\partial(u_2)}{\partial t} - kc_2 \frac{\partial^2(u_2)}{\partial x^2} - c_2 g(x,t)(u_2).$$

Let's start with the left side of this equation and work to the right side. We have the equation

$$\frac{\partial(c_1 u_1 + c_2 u_2)}{\partial t} - k \frac{\partial^2(c_1 u_1 + c_2 u_2)}{\partial x^2} - g(x,t)(c_1 u_1 + c_2 u_2)$$

$$= \frac{\partial(c_1 u_1 + c_2 u_2)}{\partial t} - k \frac{\partial^2(c_1 u_1 + c_2 u_2)}{\partial x^2} - c_1 g(x,t)u_1 - c_2 g(x,t)u_2$$

by distribution of multiplication over addition and commutative properties of multiplication. By the linearity of the differential operator, we obtain

$$\frac{\partial(c_1 u_1 + c_2 u_2)}{\partial t} - k \frac{\partial^2(c_1 u_1 + c_2 u_2)}{\partial x^2} - c_1 g(x,t)u_1 - c_2 g(x,t)u_2$$

$$= c_1 \frac{\partial(u_1)}{\partial t} + c_2 \frac{\partial(u_2)}{\partial t} - c_1 k \frac{\partial^2(u_1)}{\partial x^2} - c_2 k \frac{\partial^2(u_2)}{\partial x^2} - c_1 g(x,t)u_1 - c_2 g(x,t)u_2.$$

This equation can be rearranged by the commutative property of addition to produce

$$c_1 \frac{\partial(u_1)}{\partial t} + c_2 \frac{\partial(u_2)}{\partial t} - c_1 k \frac{\partial^2(u_1)}{\partial x^2} - c_2 k \frac{\partial^2(u_2)}{\partial x^2} - c_1 g(x,t)u_1 - c_2 g(x,t)u_2$$

$$= c_1 \frac{\partial(u_1)}{\partial t} - kc_1 \frac{\partial^2(u_1)}{\partial x^2} - c_1 g(x,t)(u_1) +$$

$$c_2 \frac{\partial(u_2)}{\partial t} - kc_2 \frac{\partial^2(u_2)}{\partial x^2} - c_2 g(x,t)(u_2),$$

which is the desired result. Thus,

$$L = \frac{\partial_-}{\partial t} - k \frac{\partial^2_-}{\partial x^2} - g(x,t)_-$$

is a linear operator.   ■

Understanding a new concept is often easier if a counterexample is given. Consider the following example, also based on the heat operator, a nonlinear heat operator.

EXAMPLE 5.2. Consider the operator

$$J = \frac{\partial()}{\partial t} - () \frac{\partial^2()}{\partial x^2} + \frac{\partial^3()}{\partial x^3}.$$

For the operator J, the parentheses, (), indicate a holding place for the function being operated on. To prove that J is a nonlinear operator, we show that

$$\frac{\partial(c_1 u_1 + c_2 u_2)}{\partial t} - (c_1 u_1 + c_2 u_2)\frac{\partial^2(c_1 u_1 + c_2 u_2)}{\partial x^2} + \frac{\partial^3(c_1 u_1 + c_2 u_2)}{\partial x^3}$$

$$\neq c_1 \frac{\partial(u_1)}{\partial t} - c_1 u_1 \frac{\partial^2(u_1)}{\partial x^2} + c_1 \frac{\partial^3(u_1)}{\partial x^3}$$

$$+ c_2 \frac{\partial(u_2)}{\partial t} - c_2 u_2 \frac{\partial^2(u_2)}{\partial x^2} + c_2 \frac{\partial^3(u_2)}{\partial x^3}.$$

Notice the *not equal to* sign in the previous equation. As in the last example, we start with the left side and work through the problem. By the distribution property of multiplication over addition, we find that

$$\frac{\partial(c_1 u_1 + c_2 u_2)}{\partial t} - (c_1 u_1 + c_2 u_2)\frac{\partial^2(c_1 u_1 + c_2 u_2)}{\partial x^2} + \frac{\partial^3(c_1 u_1 + c_2 u_2)}{\partial x^3}$$

$$= \frac{\partial(c_1 u_1 + c_2 u_2)}{\partial t} - (c_1 u_1)\frac{\partial^2(c_1 u_1 + c_2 u_2)}{\partial x^2}$$

$$- (c_2 u_2)\frac{\partial^2(c_1 u_1 + c_2 u_2)}{\partial x^2} + \frac{\partial^3(c_1 u_1 + c_2 u_2)}{\partial x^3}.$$

Applying the linearity of the differential operator once, twice, or three times where needed, and the commutative property, we obtain

$$\frac{\partial(c_1 u_1 + c_2 u_2)}{\partial t} - (c_1 u_1)\frac{\partial^2(c_1 u_1 + c_2 u_2)}{\partial x^2}$$

$$- (c_2 u_2)\frac{\partial^2(c_1 u_1 + c_2 u_2)}{\partial x^2} + \frac{\partial^3(c_1 u_1 + c_2 u_2)}{\partial x^3}$$

$$= c_1 \frac{\partial(u_1)}{\partial t} - c_1(c_1 u_1)\frac{\partial^2(u_1)}{\partial x^2} + c_2 \frac{\partial(u_2)}{\partial t} - c_2(c_1 u_1)\frac{\partial^2(u_2)}{\partial x^2}$$

$$- c_1(c_2 u_2)\frac{\partial^2(u_1)}{\partial x^2} - c_2(c_2 u_2)\frac{\partial^2(u_2)}{\partial x^2} + c_1 \frac{\partial^3(u_1)}{\partial x^3} + c_2 \frac{\partial^3(u_2)}{\partial x^3}.$$

No matter what algebraic method we apply, we can not rearrange this to be

$$c_1 \frac{\partial(u_1)}{\partial t} - c_1 u_1 \frac{\partial^2(u_1)}{\partial x^2} + c_1 \frac{\partial^3(u_1)}{\partial x^3}$$

$$+ c_2 \frac{\partial(u_2)}{\partial t} - c_2 u_2 \frac{\partial^2(u_2)}{\partial x^2} + c_2 \frac{\partial^3(u_2)}{\partial x^3}.$$

Thus, J is a nonlinear operator. ∎

Now, having the skill to identify a linear operator only comes in handy if you understand its placement in a linear equation.

5.2.2   Linear Equations

A **linear equation** contains a linear operator, L, operating on some function, $u(x)$, which is equal to another function, f. The function $f(x)$ can be constant or a function of x. Written mathematically, this translates to

$$L\left[u(x)\right] = f(x) \text{ or } L\left[u\right] = f. \qquad (5.2)$$

An example is the following heat equation with a source term $Q(x,t)$:

$$\frac{\partial u(x,t)}{\partial t} - k\frac{\partial^2 u(x,t)}{\partial x^2} = Q(x,t).$$

Here, the linear operator is a partial differential operator giving us a linear PDE. Another example is the wave equation with a damping function, $Q(x,t)$,

$$\frac{\partial^2 u(x,t)}{\partial t^2} - c^2\frac{\partial^2 u(x,t)}{\partial x^2} = Q(x,t).$$

The general linear second-order PDE in two spatial variables is

$$a(x,y)\frac{\partial^2 u(x,y)}{\partial x^2} + b(x,y)\frac{\partial^2 u(x,y)}{\partial y\partial x} + c(x,y)\frac{\partial^2 u(x,y)}{\partial y^2} + d(x,y)\frac{\partial u(x,y)}{\partial x}$$

$$+ e(x,y)\frac{\partial u(x,y)}{\partial y} + f(x,y)u(x,y) = g(x,y). \qquad (5.3)$$

A nonlinear PDE is one where one or more of the functions $a(x,y)$, $b(x,y)$, $c(x,y)$, $d(x,y)$, $e(x,y)$, or $f(x,y)$ in Equation (5.3) is a function of $u(x,y)$. For example, consider the equation

$$u(x,y)\frac{\partial^2 u(x,y)}{\partial x^2} + b(x,y)\frac{\partial^2 u(x,y)}{\partial y\partial x} + c(x,y)\frac{\partial^2 u(x,y)}{\partial y^2} + d(x,y)\frac{\partial u(x,y)}{\partial x}$$

$$+ e(x,y)\frac{\partial u(x,y)}{\partial y} + f(x,y)u(x,y) = g(x,y).$$

It is a nonlinear equation because $a(x,y)$ in Equation (5.3) has been changed to $u(x,y)$. For another example, consider

$$u(x,t)\frac{\partial^2 u(x,t)}{\partial t^2} - c^2\frac{\partial^2 u(x,t)}{\partial x^2} = Q(x,t).$$

This equation is nonlinear, since we have the product $u(x,t)\dfrac{\partial^2 u(x,t)}{\partial t^2}$. *Note:* The study of nonlinear PDEs is beyond the scope of this text. However, you still need to be able to identify a PDE as linear or nonlinear.

If $f = 0$ in Equation (5.2), then Equation (5.2) is called a **linear homogeneous equation.** The heat equation without a source term,

$$\frac{\partial u(x,t)}{\partial t} = k\frac{\partial^2 u(x,t)}{\partial x^2}, \tag{5.4}$$

is an example of a linear homogeneous partial differential equation. It should be clear that $u(x,t) \equiv 0$ is a solution for the linear homogeneous heat equation. This solution is called the **trivial solution**, a simple test to determine if a linear equation is homogeneous. If $u(x,t) \equiv 0$ is not a solution, then the equation is called **nonhomogeneous**. A solution technique for nonhomogeneous linear PDEs is discussed in Chapter 7.

Notice that in Equation (5.4) we have a second-order partial derivative with respect to the variable x. Recalling that a second-order linear homogeneous ODE has two unique solutions, it would not be surprising to find out that the linear homogeneous heat equation also has two or more unique solutions, in addition to the trivial solution. When you studied ODEs, you learned that if there are two solutions, then a linear combination of those solutions is also a solution. This is a property of linear operators and is known as the principle of superposition.

Definition 33. *Principle of superposition: If u_1 and u_2 are solutions to a linear homogeneous equation, then an arbitrary linear combination of them, $c_1u_1 + c_2u_2$, is also a solution of the same linear homogeneous equation.*

The principle of superposition is used extensively when solving PDEs, which will be pointed out later in this chapter. Also, in the previous definition, it is important to note that the word "homogeneous" is used. If we have a nonhomogeneous linear equation, then the principle of superposition does not apply. However, just as you learned in ODEs, there are methods to solve some nonhomogeneous linear PDEs that involve the principle of superposition. This subject will be covered in Chapter 7.

Boundary conditions can also be defined as linear and homogeneous. The special BCs studied in Chapters 2 and 3 where $u(0,t) = 0$ and $u(L,t) = 0$ or $\frac{\partial u(0,t)}{\partial x} = 0$ and $\frac{\partial u(L,t)}{\partial x} = 0$ are obviously linear and homogeneous. It is more difficult to determine whether Robin's conditions (where $\frac{\partial u(0,t)}{\partial x} = -hu(0,t)$ and $\frac{\partial u(L,t)}{\partial x} = hu(L,t)$, h is a positive constant) are linear and homogeneous. To determine if BCs are homogeneous, set $u(x,t) \equiv 0$ in the BCs. If you get $0 = 0$, then the BCs are homogeneous. To determine if BCs are linear, consider whether there is a product of the function $u(x,t)$ with the operator on the function $u(x,t)$. For instance,

$$\frac{\partial u(0,t)}{\partial x} = -u(0,t)\frac{\partial^2 u(0,t)}{\partial x^2}$$

is a nonlinear BC, whereas

$$\frac{\partial u(0,t)}{\partial x} = -hu(0,t),$$

when h is a positive constant, is linear. In fact, h can be a function of x and t, $h = h(x, t)$, and the BC will still be linear. In the next section, we will apply what we have learned about linearity.

EXERCISES 5.2

5.2.1. Show that

$$L = \frac{\partial}{\partial x} - k\frac{\partial^2}{\partial x^2}$$

is a linear operator.

5.2.2. Consider the equation

$$\frac{\partial^2 u}{\partial t^2} = k\frac{\partial^2 u}{\partial x^2} + \alpha(x, t)u + \beta(x, t).$$

(a) Identify the operator.

(b) Show that the operator is linear.

(c) If $\beta(x, t) = 0$, what can you say about the equation?

5.2.3. Let

$$L(u) = u\frac{\partial u}{\partial t} - k\frac{\partial u}{\partial x}.$$

Show that $L(u)$ is not a linear operator.

Note: Simply stating that there is a product between u and $\dfrac{\partial u}{\partial t}$ is not enough.

5.2.4. Show that

$$L = \frac{1}{r}\frac{\partial}{\partial r}\left[r\frac{\partial}{\partial r}\right] + \frac{1}{r^2}\frac{\partial^2}{\partial\theta^2}$$

is a linear operator.

5.2.5. Consider the equation

$$\frac{\partial^2 u}{\partial t^2} = \frac{1}{r}\frac{\partial}{\partial r}\left(r\frac{\partial u}{\partial r}\right) + \frac{1}{r^2}\frac{\partial^2 u}{\partial\theta^2} + \frac{\partial^2 u}{\partial z^2}.$$

(a) Identify the operator.

(b) Show that the operator is linear.

5.2.6. Consider the equation

$$\frac{\partial u}{\partial t} = \frac{1}{r^2}\frac{\partial}{\partial r}\left(r^2\frac{\partial u}{\partial r}\right) + \frac{1}{r^2\sin\theta}\left(\sin\theta\frac{\partial^2 u}{\partial\theta^2}\right) + \frac{1}{r^2\sin^2\theta}\frac{\partial^2 u}{\partial\phi^2}.$$

(a) Identify the operator.

(b) Show that the operator is linear.

5.2.7. Suppose that u_1 and u_2 are solutions to the linear homogeneous equation $L(u) = 0$. Prove that an arbitrary linear combination of u_1 and u_2 is a solution of $L(u)$. *Note:* Simply stating the principle of superposition is not enough.

5.3   SEPARATION OF VARIABLES: THE HEAT EQUATION IN A ONE-DIMENSIONAL ROD

Consider the heat equation for the uniform one-dimensional rod with perfect lateral insulation and no source term

$$\frac{\partial u}{\partial t} = k\frac{\partial^2 u}{\partial x^2}, \tag{5.5}$$

subject to BCs

$$\begin{cases} u(0,t) &= 0 \\ u(L,t) &= 0 \end{cases} \tag{5.6}$$

and IC

$$u(x,0) = f(x). \tag{5.7}$$

We want to have a solution for Equation (5.5) subject to the BCs, Equation (5.6), for any time t and $0 \leq x \leq L$. This is called solving the initial value problem. To do this, we first assure ourselves that Equations (5.5 and 5.6) are linear and homogeneous, using the tests described in the previous section. If Equations (5.5 and 5.6) are linear and homogeneous, we can apply the separation of variables technique, and we assume the function $u(x,t)$ is a product of two functions, one of time $G(t)$ and one of space $\varphi(x)$. In other words

$$u(x,t) = G(t)\varphi(x).$$

Then we find the appropriate derivatives. We have

$$\frac{\partial u}{\partial t} = G'(t)\varphi(x)$$

and

$$\frac{\partial^2 u}{\partial x^2} = G(t)\varphi''(x).$$

Substituting into Equation (5.5), we have

$$G'(t)\varphi(x) = kG(t)\varphi''(x). \tag{5.8}$$

Separating Equation (5.8) so that one side of the equation is in terms of t and the other side is in terms of x is called separation of variables. After doing so, we have

$$\frac{G'(t)}{kG(t)} = \frac{\varphi''(x)}{\varphi(x)}. \tag{5.9}$$

This may be very difficult to satisfy, since we are saying that a function of time is equal to a function of space. Or, more precisely, the left side of Equation (5.9) is a function of time, t, and does not vary with the variable x. However, it is equal to a function of x, which does not vary with time, t. Thus, both sides of Equation (5.9) must be equal to the same constant. The constant is called the separation constant, and we will denote the constant as $-\lambda$. Thus, we have the equation

$$\frac{G'(t)}{kG(t)} = \frac{\varphi''(x)}{\varphi(x)} = -\lambda. \tag{5.10}$$

The choice of $-\lambda$ as the separation constant is for convenience. We could have just as easily have chosen λ.

Having separated Equation (5.5) into time $\left(\dfrac{G'(t)}{kG(t)} = -\lambda \right)$ and space $\left(\dfrac{\varphi''(x)}{\varphi(x)} = -\lambda \right)$ equations, we now separate the BCs, Equation (5.6). Using our assumption for $u(x,t) = G(t)\varphi(x)$, we find that

$$u(0,t) = G(t)\varphi(0) = 0. \tag{5.11}$$

If $G(t) = 0$ for all time t, then the solution to Equation (5.11) is trivial. That is, $u(x,t) = 0$, which is not a very interesting problem. But since a BC usually refers to a spatial variable, it makes sense to assume that

$$\varphi(0) = 0.$$

Similarly

$$u(L,t) = G(t)\varphi(L) = 0$$

implies that

$$\varphi(L) = 0.$$

Thus, our BCs, Equation (5.6), have become

$$\left\{ \begin{array}{ccc} \varphi(0) & = & 0 \\ \varphi(L) & = & 0. \end{array} \right. \tag{5.12}$$

Using Equation (5.10), the equation in time is

$$G'(t) = -\lambda k G(t), \tag{5.13}$$

and the boundary value problem for the spatial variable is

$$\varphi''(x) = -\lambda\varphi(x), \tag{5.14}$$

subject to the BCs, Equations (5.12). Thus, we have completely separated Equation (5.5) subject to the BCs, Equation (5.6). *Note:* The initial condition for the time problem will be applied at a later stage of the solution process. Therefore, it does not need to be separated.

We will solve the spatial equation first, though, in general, either way will suffice. The second, more time-consuming, method would be to solve the time equation first, but then you'd have to fill in the variable λ after solving the spatial problem. Also, it is important that you develop a standard methodology that always works when solving PDEs using separation of variables. This helps when we start solving much more complicated problems using this method.

5.3.1 Spatial Problem Solution

Many texts assume $\lambda \geq 0$, since this is the only solution that relates to a physical situation for Equation (5.14) subject to the BCs, Equation (5.12). In this text, we do not make this assumption. I feel it is necessary for a student of physics, engineering, or mathematics to be aware of all solutions and to be able to work out all solutions for a given situation. Developing this awareness is not an easy task, but it is a necessary one. Thus, we will not make an assumption about λ but proceed in an orderly fashion to completely resolve Equation (5.14) subject to the BCs, Equation (5.12).

Case 1: $\lambda < 0$.

In this case it is convenient to assume that $\lambda = -s$, where $s > 0$. Thus, Equation (5.14) becomes

$$\varphi''(x) = s\varphi(x),$$

subject to

$$\begin{cases} \varphi(0) &= 0 \\ \varphi(L) &= 0. \end{cases}$$

We now have a second-order constant coefficient ODE where the coefficient is $s > 0$. The solution to this problem is

$$\varphi(x) = c_1 e^{\sqrt{s}x} + c_2 e^{-\sqrt{s}x}.$$

Using the noncomplex form of Euler's equations,

$$\sinh ax = \frac{e^{ax} - e^{-ax}}{2}$$

and

$$\cosh ax = \frac{e^{ax} + e^{-ax}}{2},$$

it is convenient to rewrite the solution as

$$\varphi(x) = c_3 \cosh \sqrt{s}x + c_4 \sinh \sqrt{s}x. \tag{5.15}$$

Following are the graphs of $\cosh x$, in Figure (5.1), and $\sinh x$, in Figure (5.2).

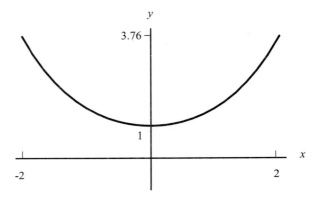

Figure 5.1: $y = \cosh x$.

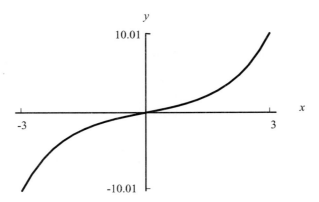

Figure 5.2: $y = \sinh x$.

The graphs point out the advantage of using the functions $\cosh x$ and $\sinh x$. Figure (5.1) indicates that $\cosh(0) = 1$, while Figure (5.2) indicates that only the $\sinh(0) = 0$. This helps us when we apply the BCs, Equation (5.12), to Equation (5.15). Applying $\varphi(0) = 0$ yields

$$\varphi(0) = 0 = c_3 \cosh(0) + c_4 \sinh(0).$$

Since $\sinh(0) = 0$ and $\cosh(0) = 1$, we must have $c_3 = 0$. Applying $\varphi(L) = 0$ yields

$$\varphi(L) = 0 = c_4 \sinh L;$$

since the $\sinh(x)$ is 0 only when $x = 0$, we must have $c_4 = 0$. Therefore, when $\lambda < 0$ there is only the trivial solution to Equation (5.14) subject to the BCs, Equation (5.12).

Case 2: $\lambda = 0$.

Here Equation (5.14) becomes

$$\varphi''(x) = 0.$$

This ODE is solved by integrating twice, which yields

$$\varphi(x) = d_1 x + d_2.$$

Applying the first BC, $\varphi(0) = 0$, we find that

$$\varphi(0) = 0 = d_2.$$

Applying the second BC, $\varphi(L) = 0$, we get

$$\varphi(L) = 0 = d_1 L,$$

which indicates that $d_1 = 0$, since $L \neq 0$. Therefore, when $\lambda = 0$ there is only the trivial solution to Equation (5.14) subject to the BCs, Equation (5.12).

Case 3: $\lambda > 0$.

Equation (5.14) is a second-order ODE with constant coefficients and a solution of

$$\varphi(x) = h_3 e^{i\sqrt{\lambda}x} + h_4 e^{-i\sqrt{\lambda}x}.$$

Using Euler's equations,

$$\sin ax = \frac{e^{iax} - e^{-iax}}{2i}$$

and

$$\cos ax = \frac{e^{iax} + e^{-iax}}{2},$$

this solution can be rewritten as

$$\varphi(x) = h_5 \cos \sqrt{\lambda}x + h_6 \sin \sqrt{\lambda}x.$$

Applying the first BC, $\varphi(0) = 0$, we get

$$\varphi(0) = 0 = h_5 \cos 0 + h_6 \sin 0.$$

Since $\sin 0 = 0$ and the $\cos 0 = 1$, we have $\varphi(0) = 0 = h_5$. Applying the second BC, $\varphi(L) = 0$, we have

$$\varphi(L) = 0 = h_6 \sin \sqrt{\lambda} L.$$

We could always claim that $h_6 = 0$, but this would only give us the unusable trivial solution again, $\varphi(x) = 0$. Hence, we must look for other possible solutions. The real clue comes from considering the function $y = \sin x$ shown in Figure (5.3). We see

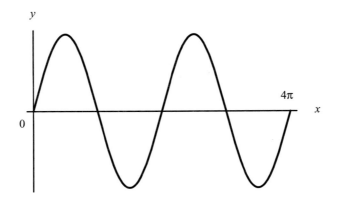

Figure 5.3: $y = \sin x$.

that there are numerous places that $\sin \sqrt{\lambda} L = 0$. For instance $\sqrt{\lambda} L = \pi$, $\sqrt{\lambda} L = 2\pi$, $\sqrt{\lambda} L = 3\pi, \ldots, \sqrt{\lambda} L = n\pi$. Solving for λ in general yields

$$\lambda_n = \left(\frac{n\pi}{L}\right)^2, \ n = 1, 2, 3, \ldots$$

We use the notation λ_n to indicate an infinite collection of solutions known as **eigenvalues**. Corresponding to each eigenvalue is an **eigenfunction** $\varphi_n(x) = h_n \sin \sqrt{\lambda_n} x = h_n \sin \left(\frac{n\pi x}{L}\right)$. Note that we have changed the coefficient from h_6 to h_n to indicate that each eigenfunction may have a different coefficient.

Thus, the complete solution to the spatial problem is given by the collection

$$\left. \begin{array}{l} \lambda_n = \left(\frac{n\pi}{L}\right)^2 \\[2mm] \varphi_n(x) = h_n \sin \left(\frac{n\pi x}{L}\right) \end{array} \right\}, \ n = 1, 2, 3, \ldots$$

Note: This is not a general solution to Equation (5.14) subject to the BCs, Equation (5.12). A general solution would be a linear combination of all possible solutions.

5.3.2   Time Problem Solution

From the spatial problem, we found that we could only solve the problem for selected values of λ denoted by $\lambda_n = \left(\frac{n\pi}{L}\right)^2$, $n = 1, 2, 3, \ldots$ Thus, Equation (5.13) becomes the family

$$G'_n(t) = -\lambda_n k G_n(t) = -\left(\frac{n\pi}{L}\right)^2 k G_n(t).$$

From your ODE course, you know that we have solutions

$$G_n(t) = c_n e^{-k\lambda_n t}, \; n = 1, 2, 3, \ldots$$

5.3.3   The Complete Solution

We now have the solutions for both the time ODE and the spatial ODE. From our assumed form of

$$u(x, t) = G(t)\varphi(x),$$

we find an infinite collection of solutions

$$u(x, t) = c_n e^{-k\lambda_n t} h_n \sin\left(\frac{n\pi x}{L}\right), \; n = 1, 2, 3, \ldots$$

Since c_n and h_n are arbitrary constants being multiplied together, we can replace them by B_n. Therefore,

$$u(x, t) = B_n e^{-k\lambda_n t} \sin\left(\frac{n\pi x}{L}\right), \; n = 1, 2, 3, \ldots$$

The question now remains as to which n gives the solution we really want. We know that for $n = 1$, we have

$$u(x, t) = B_1 e^{-k\lambda_1 t} \sin\left(\frac{\pi x}{L}\right).$$

When $n = 2$, we have

$$u(x, t) = B_2 e^{-k\lambda_2 t} \sin\left(\frac{2\pi x}{L}\right),$$

which is different from the case $n = 1$. When $n = m$, we have

$$u(x, t) = B_m e^{-k\lambda_m t} \sin\left(\frac{m\pi, x}{L}\right)$$

which is different from the previous two cases. In reality, we have an infinite number of independent solutions for $u(x, t)$. By the principle of superposition, we know that any sum of independent solutions is also a solution. Therefore, we have

$$u(x, t) = B_1 e^{-k\lambda_1 t} \sin\left(\frac{\pi x}{L}\right) + B_2 e^{-k\lambda_2 t} \sin\left(\frac{2\pi x}{L}\right)$$

$$+ \ldots + B_m e^{-k\lambda_m t} \sin\left(\frac{m\pi x}{L}\right) + \ldots,$$

which becomes, using summation notation,

$$u(x,t) = \sum_{n=1}^{\infty} B_n e^{-k\lambda_n t} \sin\left(\frac{n\pi x}{L}\right). \qquad (5.16)$$

From Chapter 4, we recognize this equation as similar to a Fourier sine series.

Having the general solution, we can apply the IC (Equation (5.7)), $u(x,0) = f(x)$, to arrive at the specific solution. We have

$$u(x,0) = f(x) = \sum_{n=1}^{\infty} B_n \sin\left(\frac{n\pi x}{L}\right).$$

Recall that, in Chapter 4, we learned that any piecewise smooth function can be represented by a Fourier series for $0 \le x \le L$. Therefore, we can solve for the coefficients B_n by using the orthogonality of the sine function. They are

$$B_n = \frac{\int_0^L f(x) \sin\left(\frac{n\pi x}{L}\right) dx}{\int_0^L \sin^2\left(\frac{n\pi x}{L}\right) dx} = \frac{2}{L} \int_0^L f(x) \sin\left(\frac{n\pi x}{L}\right) dx.$$

This completes the solution of Equation (5.5) subject to the BCs, Equation (5.6), and IC Equation (5.7). It describes temperature distribution in the one-dimensional rod for all time t. In fact, if we take the limit as time goes to infinity in Equation (5.16), we get the steady-state solution for Equation (5.5). We state this mathematically as

$$\lim_{t \to \infty} u(x,t) = \lim_{t \to \infty} \left[\sum_{n=1}^{\infty} B_n e^{-[k\lambda_n t]} \sin\left(\frac{n\pi x}{L}\right)\right] = 0.$$

A fully worked example of the separation of variables method for the heat equation is provided in Example (5.3).

EXAMPLE 5.3. Find the time-dependent solution for $u(x,t)$ when

$$\frac{\partial u}{\partial t} = 1.14 \frac{\partial^2 u}{\partial x^2}, \qquad (5.17)$$

subject to

$$\begin{cases} u(0,t) = 0 \\[2mm] \dfrac{\partial u(1,t)}{\partial x} = 0 \end{cases} \qquad (5.18)$$

and

$$u(x,0) = -x^2 + 2x. \qquad (5.19)$$

The PDE, Equation (5.17), and BCs, Equation (5.18), are linear and homogeneous. Therefore, separation of variables technique is valid. Letting $u(x,t) = G(t)\varphi(x)$, yields the time equation

$$G'(t) = -1.14\lambda G(t), \tag{5.20}$$

and the spatial equation

$$\varphi''(x) = -\lambda\varphi(x), \tag{5.21}$$

subject to the BCs

$$\varphi(0) = 0 \text{ and } \varphi'(1) = 0. \tag{5.22}$$

When solving the spatial equation, we must determine the valid values of λ. Thus, we first assume $\lambda < 0$. In this case, let $\lambda = -s$ and solve

$$\varphi''(x) = s\varphi(x),$$

subject to the BCs, Equation (5.22). This yields

$$\varphi(x) = c_1 \cosh \sqrt{s}x + c_2 \sinh \sqrt{s}x. \tag{5.23}$$

Applying the first BC, $\varphi(0) = 0$, to Equation (5.23) yields $c_1 = 0$. Applying the second BC, $\varphi'(1) = 0$, to Equation (5.23) yields $c_2 = 0$. Thus, there are no eigenvalues for $\lambda < 0$. Next, we consider $\lambda = 0$. Here, Equation (5.21) becomes

$$\varphi''(x) = 0,$$

which, after integrating twice, has as a solution

$$\varphi(x) = c_3 x + c_4. \tag{5.24}$$

Applying the first BC in Equation (5.22) to Equation (5.24) yields $c_4 = 0$. Applying the second BC, $\varphi'(1) = 0$, to Equation (5.24) yields $c_3 = 0$. Therefore, there are no eigenvalues for $\lambda = 0$. Finally, we assume $\lambda > 0$. In this case, the solution to Equation (5.21) is

$$\varphi(x) = c_5 \cos \sqrt{\lambda}x + c_6 \sin \sqrt{\lambda}x. \tag{5.25}$$

The first BC, $\varphi(0) = 0$, applied to Equation (5.25) yields $c_5 = 0$. The second BC, $\varphi'(1) = 0$, indicates either $c_6 = 0$, in which case we only have the trivial solution, or the eigenvalues

$$\lambda_n = \left[\frac{(2n-1)\pi}{2}\right]^2, \; n = 1, \, 2, \, 3, \, \dots \tag{5.26}$$

with corresponding eigenfunctions

$$\varphi_n(x) = c_n \sin \frac{(2n-1)\pi x}{2}. \tag{5.27}$$

Substituting the eigenvalues from Equation (5.26) into Equation (5.20), we determine the solution for the time equation, which is

$$G_n(t) = d_n e^{-1.14\lambda_n t}.$$

Thus, the solution for $u(x,t)$ is expressed as

$$u(x,t) = G_n(t)\varphi_n(x) = \sum_{n=1}^{\infty} b_n e^{-1.14\lambda_n t} \sin\frac{(2n-1)\pi x}{2}.$$

We determine the constant b_n by using the IC, Equation (5.19), and applying orthogonality. We have

$$u(x,0) = -x^2 + 2x = \sum_{n=1}^{\infty} b_n \sin\frac{(2n-1)\pi x}{2}.$$

Therefore, remembering $L = 1$, we find that

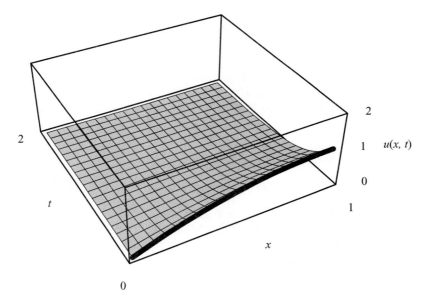

Figure 5.4: The graph of $u(x,t)$ for $0 \le t \le 2$.

$$b_n = \frac{2}{1}\int_0^1 (-x^2 + 2x) \sin\frac{(2n-1)\pi x}{2}\, dx = \frac{32}{(2n-1)^3\pi^3}.$$

Replacing the values for b_n in the Fourier sine series yields the specific solution for $u(x,t)$, which is

$$u(x,t) = \sum_{n=1}^{\infty} \frac{32}{(2n-1)^3\pi^3} e^{-1.14\lambda_n t} \sin\frac{(2n-1)\pi x}{2}.$$

A graph of the solution is shown in Figure (5.4) for the partial sum of $n = 25$ with $0 \leq t \leq 2$. Also, the initial temperature distribution is outlined on the graph by a boldface line. It is interesting to note from Figure (5.4) how fast $e^{-1.14\lambda_n t}$ reduces the initial temperature distribution to the approximate steady-state solution. ∎

EXERCISES 5.3

5.3.1. Separate the following PDEs into appropriate ODEs:

(1) $\dfrac{\partial u(x,t)}{\partial t} = k\dfrac{\partial^2 u(x,t)}{\partial x^2} + u(x,t).$

(2) $\dfrac{\partial u(x,t)}{\partial t} = k\dfrac{\partial^2 u(x,t)}{\partial x^2} - m\dfrac{\partial u(x,t)}{\partial x} + u(x,t).$

(3) $\dfrac{\partial u(x,t)}{\partial t} = \dfrac{\partial}{\partial x}\left[k(x)\dfrac{\partial u(x,t)}{\partial x}\right] + u(x,t).$

5.3.2. Consider the PDE

$$\frac{\partial u(x,t)}{\partial t} = 4\frac{\partial^2 u(x,t)}{\partial x^2}.$$

For each set of BCs and ICs, solve the initial value problem; using your favorite mathematical software, graph the solution for $0 \leq t \leq 5$. Clearly indicate the initial temperature distribution on your graph.

(1) BCs: $\begin{cases} u(0,t) = 0 \\[2mm] \dfrac{\partial u(\pi,t)}{\partial x} = 0 \end{cases}$ and IC: $u(x,0) = x^2 - 2\pi x.$

(2) BCs: $\begin{cases} u(0,t) = 0 \\[2mm] u(2\pi,t) = 0 \end{cases}$ and IC: $u(x,0) = \sin x.$

(3) BCs: $\begin{cases} \dfrac{\partial u(0,t)}{\partial x} = 0 \\[2mm] u\left(\dfrac{3\pi}{2},t\right) = 0 \end{cases}$ and IC: $u(x,0) = \cos x.$

(4) BCs: $\begin{cases} \dfrac{\partial u(0,t)}{\partial x} = 0 \\[2mm] \dfrac{\partial u(\pi,t)}{\partial x} = 0 \end{cases}$ and IC: $u(x,0) = x^3 - \dfrac{3\pi}{2}x^2.$

5.3.3. Consider the PDE

$$\frac{\partial u(x,t)}{\partial t} = \frac{\partial^2 u(x,t)}{\partial x^2}.$$

(a) Solve the initial value problem subject to

BCs: $\begin{cases} u(-\pi, t) = u(\pi, t)) \\[2mm] \dfrac{\partial u(-\pi, t)}{\partial x} = \dfrac{\partial u(\pi, t)}{\partial x} \end{cases}$ and   IC: $u(x,0) = 7 \sin 3x - 2 \cos 2x$.

(b) In five sentences or less give a physical description of this problem.

5.3.4. Consider the following information:

(a) heat conduction in a one-dimensional rod of length $\dfrac{3\pi}{2}$ m,

(b) perfect thermal insulation on the lateral sides,

(c) right end is held at a constant temperature of 0 degrees,

(d) left end has perfect thermal insulation,

(e) thermal diffusivity of 1.7 cm^2/sec (thermal diffusivity of silver),

(f) no internal heat source.

(1) Set up the mathematical model.

(2) Give a series solution of your mathematical model.

(3) Check to see that this solution satisfies the equation, the boundary conditions, and the initial conditions.

(4) Using your favorite mathematical software, graph the solution. Then write a short paragraph discussing the surface at the following times: $t = 1$ sec, $t = 5$ secs, and $t = 10$ secs.

5.3.5. Let a metallic rod 20 cm long be heated to an initial temperature which is modeled by $\left(x \sin \left(\dfrac{7\pi x}{20} \right) \right)^{\circ}$ C. Suppose that at $t = 0$ the ends of the rod are plunged into an ice bath of $0°$C, and thereafter maintain this temperature. Also, suppose no heat is allowed to escape from the lateral surface of the rod. *Note:* You must model and explain your choice of boundary conditions for this experiment.

(1) Solve the initial value problem for a rod made of copper, cast iron, and asbestos. The thermal diffusivity of these materials may be found in Appendix E.

(2) Using your favorite mathematical software graph the solution and the initial condition, then determine the approximate solution, for the following three cases:

(a) Use only the first term in the series for $u(x, t)$ to find the approximate temperature at $x = 7$ cm when $t = 12$ secs for each material.

(b) Use the first three terms in the series for $u(x, t)$ to find the approximate temperature at $x = 7$ cm when $t = 12$ secs for each material. Compare this answer with the previous answer.

(c) Use the first 25 terms in the series for $u(x,t)$ to find the approximate temperature at $x = 7$ cm when $t = 12$ secs for each material. Compare this answer with the previous two answers.

(3) Use the first five terms of the series to determine the amount of time, t, it takes for each material to reach a temperature of $0°C$ at the center of the rod. Compare the amount of time it takes for each material, and write a short essay on the reasons why the time is different or the same.

5.3.6. Consider a uniform one-dimensional rod of length L, without an internal heat source, which is not laterally insulated. (Heat can flow in and out across the lateral boundary.) By experimentation, you discover that heat is flowing across the lateral boundary at a rate proportional to the difference between the temperature $u(x,t)$ and the surrounding medium that is kept at $0°C$.

(1) Given that the convection constant of proportionality is greater than 0 and the ends of the rod are not insulated and held at $0°C$, set up the mathematical model.

(2) Suppose the initial temperature of the rod is x when $0 \leq x < \dfrac{L}{2}$ and $x - L$ when $\dfrac{L}{2} \leq x \leq L$. Give a series solution of your mathematical model.

(3) Check to see if this solution satisfies the equation, the BCs, and the ICs.

5.3.7. Consider a uniform one-dimensional rod of length L that has no internal heat source and is not laterally insulated. (Heat can flow in and out across the lateral boundary.) By experimentation, you discover that heat is flowing across the lateral boundary at a rate proportional to the difference between the temperature $u(x,t)$ and the surrounding medium that is kept at u_0 degrees.

(1) Given the convection constant of proportionality is greater than 0 and the ends of the rod are not insulated and held at u_0 degrees, set up the mathematical model.

(2) Solve the steady-state temperature distribution.

(3) Suppose after steady-state is reached, the rod is suddenly plunged into a $0°C$ bath. State the mathematical model governing this situation and give a series solution of your mathematical model.

(4) Check to see if this solution satisfies the equation, the BCs, and the ICs.

5.3.8. Consider a uniform one-dimensional rod of length L, that has no internal heat source and is not laterally insulated. (Heat can flow in and out across the lateral boundary). By experimentation, you discover that heat is flowing across the lateral boundary at a rate proportional to the difference between the temperature $u(x,t)$ and the surrounding medium that is kept at u_0 degrees.

(1) Given the convection constant of proportionality is greater than 0 and the ends of the rod are perfectly insulated, set up the mathematical model.

(2) Solve the steady-state temperature distribution.

(3) Suppose after steady-state is reached, the rod is suddenly plunged into a $0°C$ bath. State a the mathematical model governing this situation and give a series solution of your mathematical model.

(4) Check to see if this solution satisfies the equation, the BCs, and the ICs.

5.3.9. Consider a thin metallic rod made of cast iron (thermal diffusivity of 0.12) of 3 m length. Suppose the rod does not have lateral insulation and the temperature of the medium in which the rod is held is maintained at $0°C$. Given that the boundaries are also held at $0°C$ and the initial temperature distribution is $\frac{1}{9}x^3 + \frac{2}{3}x^2 - 3x$, and there is no source term, complete the following:

(1) Set up the mathematical model.

(2) Solve the initial value problem for your model and graph the solution for time, $t = 3$. Also, plot the initial temperature distribution on the graph.

(3) Solve the steady-state problem for your model.

(4) For what time t does the initial value problem approximate the steady-state solution?

5.3.10. Consider a thin metallic ring made of copper (thermal diffusivity of 1.14). Suppose the ring has lateral insulation. Determine the solution to the initial value problem if the initial temperature distribution is x^2 and there is no source term.

5.3.11. Consider a thin metallic ring made of aluminum (thermal diffusivity of 0.86). Suppose the ring does not have lateral insulation and the temperature of the medium the ring is held in is maintained at $0°C$. Determine the solution to the initial value problem if the initial temperature distribution is an arbitrary function of x and there is no source term.

5.3.12. Consider

$$\frac{\partial u(x,t)}{\partial t} = 0.245\frac{\partial^2 u(x,t)}{\partial x^2} + x^2 - 8x,$$

subject to

BCs: $\begin{cases} u(0,t) = 0 \\ u(8,t) = 0 \end{cases}$ and IC: $u(x,0) = e^{-x}\sin x - \frac{e^{-8}\sin 8}{8}x.$

(1) Give a physical interpretation of this problem.

(2) For very large time find the following:

(a) the solution.

(b) the heat energy generated per unit time inside the entire rod,

(c) the heat energy flowing out of the rod per unit time at each end, and

(d) the relationship between parts (b) and (c).

(3) Suppose that after Part (5.3.3) has completed, the heat source is turned off.

(a) Find the equations that describe the mathematical model at this time.

(b) Using the equations that you just found, solve the heat flow problem for any time t.

(c) Using your favorite mathematical software, graph the solution for $0 \le t \le 5$.

5.3.13. Diffusion through a Membrane: Fick's Law can be used to derive an analogue of Ohm's Law for a membrane of thickness, L, with different chemical concentrations on each side of the membrane. If the medium is isotropic (diffusion occurs the same regardless of the direction of the measurement), then we get

$$\frac{\partial c}{\partial t} = D \frac{\partial^2 c}{\partial x^2},$$

subject to

$c(0, t) = C_l$, the chemical concentration on the left of the membrane

and

$c(L, t) = C_r$, the chemical concentration on the right of the membrane.

(1) Find the time-dependent solution.

(2) Show that as we let $t \to \infty$, we get the same steady-state solution as in Chapter **??** Section 2.5.[1]

5.4 SEPARATION OF VARIABLES: THE WAVE EQUATION IN A ONE-DIMENSIONAL STRING

Consider the wave equation for a perfectly elastic vibrating string with no external forces

$$\frac{\partial^2 u}{\partial t^2} = c^2 \frac{\partial^2 u}{\partial x^2}, \tag{5.28}$$

subject to the BCs

$$\begin{cases} u(0, t) &= 0 \\ u(L, t) &= 0, \end{cases} \tag{5.29}$$

---

[1] Adapted from James Keener and James Sneyd, *Mathematical Physiology*, ©1998 by Springer-Verlag, New York, pp. 36-38. Reprinted by permission.

and with ICs

$$
\begin{cases}
u(x,0) &= f(x) \\[2mm]
\dfrac{\partial u(x,0)}{\partial t} &= g(x).
\end{cases}
\tag{5.30}
$$

As in the previous section, we want a solution for Equation (5.28) subject to the BCs, Equation (5.29), for any time t and $0 \le x \le L$. To obtain the solution, we must assure ourselves that Equations (5.28 and 5.29) are linear and homogeneous using the techniques developed in Section 2. If Equations (5.28 and 5.29) are linear and homogeneous, we again assume the solution function $u(x,t)$ can be written as the product of two functions, one of time $G(t)$ and one of space $\varphi(x)$. In other words,

$$
u(x,t) = G(t)\varphi(x).
$$

We then find the appropriate second partial derivatives,

$$
\frac{\partial^2 u}{\partial t^2} = G''(t)\varphi(x)
$$

and

$$
\frac{\partial^2 u}{\partial x^2} = G(t)\varphi''(x).
$$

Now substituting into Equation (5.28), we obtain

$$
G''(t)\varphi(x) = c^2 G(t)\varphi''(x).
\tag{5.31}
$$

Equation (5.31) is now ready for separation of variables, where one side of the equation is in t and the other side of the equation is in x. Again, we'll set the equation equal to a separation constant, as we did in the previous section, resulting in

$$
\frac{G''(t)}{c^2 G(t)} = \frac{\varphi''(x)}{\varphi(x)} = -\lambda.
\tag{5.32}
$$

Using Equation (5.32), the equation in time is

$$
G''(t) = -\lambda c^2 G(t),
\tag{5.33}
$$

and the equation in space is

$$
\varphi''(x) = -\lambda\varphi(x).
\tag{5.34}
$$

As in the previous section, we have separated our PDE into two ODEs, one ODE in time and one ODE in space. Now let's separate the BCs, Equation(5.29).

Using our assumption for $u(x,t) = G(t)\varphi(x)$, we find that

$$u(0,t) = G(t)\varphi(0) = 0.$$

If $G(t) = 0$ for any time t, then our solution is trivial. Also, a BC usually refers to a spatial variable; therefore, it makes sense to assume that

$$\varphi(0) = 0.$$

Similarly,

$$u(L,t) = G(t)\varphi(L) = 0,$$

implying that

$$\varphi(L) = 0.$$

Thus, our BCs, Equation (5.29), have become

$$\begin{cases} \varphi(0) & = & 0 \\ \varphi(L) & = & 0. \end{cases} \tag{5.35}$$

We have now completely separated Equation (5.28) and BCs, Equation (5.29), into the boundary value problem

$$\varphi''(x) = -\lambda\varphi(x),$$

$$\begin{cases} \varphi(0) & = & 0 \\ \varphi(L) & = & 0, \end{cases} \tag{5.36}$$

and a time problem

$$G''(t) = -\lambda c^2 G(t).$$

Note: The initial conditions will be applied after a general solution to the problem is found. Thus, they do not need to be separated.

5.4.1 Spatial Problem Solution

The spatial problem, Equation (5.36), is identical to that of the previous section and has solution

$$\left. \begin{array}{l} \lambda_n = \left(\dfrac{n\pi}{L}\right)^2 \\[4mm] \varphi_n(x) = h_n \sin\left(\dfrac{n\pi x}{L}\right) \end{array} \right\}, n = 1, 2, 3, \ldots$$

5.4.2   Time Problem Solution

From the spatial problem, we found that $\lambda_n = \left(\dfrac{n\pi}{L}\right)^2$, $n = 1, 2, 3, \ldots$ This changes Equation (5.33) into

$$G''(t) = -\lambda_n c^2 G(t) = -\left(\frac{n\pi}{L}\right)^2 c^2 G(t).$$

You have seen this problem before, in your ODE course, so we know this has the solution

$$G_n(t) = a_n \cos\left(\frac{n\pi ct}{L}\right) + b_n \sin\left(\frac{n\pi ct}{L}\right), \; n = 1, 2, 3, \ldots$$

5.4.3   The Complete Solution

As in the heat equation example, we have the solution to the time ODE and to the spatial ODE. From our initial assumption

$$u(x, t) = G(t)\varphi(x),$$

we find that for $n = 1, 2, 3, \ldots$,

$$u(x, t) = \left[a_n \cos\left(\frac{n\pi ct}{L}\right) + b_n \sin\left(\frac{n\pi ct}{L}\right)\right] h_n \sin\left(\frac{n\pi x}{L}\right).$$

Since a_n, b_n, and h_n are arbitrary constants and $h_n a_n$ and $h_n b_n$ are the products, we will let $A_n = h_n a_n$ and $B_n = h_n b_n$. Thus, we have for $n = 1, 2, 3, \ldots$,

$$u(x, t) = \left[A_n \cos\left(\frac{n\pi ct}{L}\right) + B_n \sin\left(\frac{n\pi ct}{L}\right)\right] \sin\left(\frac{n\pi x}{L}\right).$$

Again the question remains, which n gives the solution we really want? We know that for $n = 1$, we have

$$u(x, t) = \left[A_1 \cos\left(\frac{\pi ct}{L}\right) + B_1 \sin\left(\frac{\pi ct}{L}\right)\right] \sin\left(\frac{\pi x}{L}\right).$$

When $n = 2$, we obtain

$$u(x, t) = \left[A_2 \cos\left(\frac{2\pi ct}{L}\right) + B_2 \sin\left(\frac{2\pi ct}{L}\right)\right] \sin\left(\frac{2\pi x}{L}\right),$$

which is different from the case $n = 1$. When $n = m$, we have

$$u(x, t) = \left[A_m \cos\left(\frac{m\pi ct}{L}\right) + B_m \sin\left(\frac{m\pi ct}{L}\right)\right] \sin\left(\frac{m\pi x}{L}\right),$$

which is different from the previous two cases. Again, we have an infinite number of independent solutions for $u(x, t)$. By the principle of superposition, we know that any sum of independent solutions is also a solution. Therefore, using summation notation, we have

$$u(x,t) = \sum_{n=1}^{\infty} \left[A_n \cos\left(\frac{n\pi ct}{L}\right) + B_n \sin\left(\frac{n\pi ct}{L}\right) \right] \sin\left(\frac{n\pi x}{L}\right),$$

which we can identify as similar to a Fourier sine series.

To get a specific solution that satisfies our problem, we now use the ICs. To satisfy the first IC in Equation (5.30), which is the initial displacement of the string, we set the following:

$$u(x,0) = f(x) = \sum_{n=1}^{\infty} A_n \sin\left(\frac{n\pi x}{L}\right).$$

Thus $f(x)$ is set equal to a Fourier sine series. Now to find A_n, we use the orthogonality of the Sine function, resulting in

$$A_n = \frac{\int_0^L f(x) \sin\left(\frac{n\pi x}{L}\right) dx}{\int_0^L \sin^2\left(\frac{n\pi x}{L}\right) dx} = \frac{2}{L} \int_0^L f(x) \sin\left(\frac{n\pi x}{L}\right) dx.$$

To satisfy the second IC in Equation (5.30), which is the initial velocity of the string, we set

$$\frac{\partial u(x,0)}{\partial t} = g(x) = \sum_{n=1}^{\infty} B_n \left(\frac{n\pi c}{L}\right) \sin\left(\frac{n\pi x}{L}\right).$$

Just like $f(x)$, $g(x)$ is set equal to a Fourier sine series. Now, to find B_n, we again use the orthogonality of the sine function, resulting in

$$B_n \left(\frac{n\pi c}{L}\right) = \frac{2}{L} \int_0^L g(x) \sin\left(\frac{n\pi x}{L}\right) dx.$$

Solving for B_n, we find that

$$B_n = \frac{2}{n\pi c} \int_0^L g(x) \sin\left(\frac{n\pi x}{L}\right) dx.$$

This completes the solution of Equation (5.28) subject to the BCs, Equation (5.29), with ICs, Equation (5.30). We provide the following example for the wave equation.

EXAMPLE 5.4. Consider

$$\frac{\partial^2 u}{\partial t^2} = 4\frac{\partial^2 u}{\partial x^2}, \tag{5.37}$$

subject to

$$\begin{cases} \dfrac{\partial u(0,t)}{\partial x} = 0 \\[2ex] u(1,t) = 0 \end{cases} \tag{5.38}$$

and

$$\begin{cases} u(x,0) = x^3 + 2x^2 - 3 \\[2ex] \dfrac{\partial u(x,0)}{\partial t} = x^2 - 1. \end{cases} \tag{5.39}$$

Find a time-dependent solution for $u(x,t)$. Since the PDE in Equation (5.37) and BCs, Equation (5.38), are linear and homogeneous, separation of variables technique can be applied. Assuming $u(x,t) = G(t)\varphi(x)$, we have the spatial equation

$$\varphi''(x) = -\lambda \varphi(x), \tag{5.40}$$

subject to

$$\begin{cases} \varphi'(0) = 0 \\[2ex] \varphi(1) = 0 \end{cases} \tag{5.41}$$

and a time equation,

$$G''(t) = -4\lambda G(t). \tag{5.42}$$

Working with the spatial equation first, it can be determined that for $\lambda < 0$ and $\lambda = 0$ there is only the trivial solution. For $\lambda > 0$, we have

$$\varphi(x) = c_1 \cos \sqrt{\lambda}x + c_2 \sin \sqrt{\lambda}x. \tag{5.43}$$

Applying the first BC in Equation (5.41), we find $c_2 = 0$. Applying the second BC yields eigenvalues,

$$\lambda_n = \left[\frac{(2n-1)\pi}{2} \right]^2, \; n = 1, 2, 3, \ldots,$$

with corresponding eigenfunctions,

$$\varphi_n(x) = c_n \cos \frac{(2n-1)\pi x}{2}.$$

Substituting λ_n into Equation (5.42) yields

$$G_n''(t) = -4\lambda_n G_n(t),$$

which has as a solution

$$G_n(t) = a_n \cos\left[(2n-1)\pi t\right] + b_n \sin\left[(2n-1)\pi t\right].$$

Thus, the general solution for $u(x,t)$ is

$$u(x,t) = \sum_{n=1}^{\infty} \left\{A_n \cos\left[(2n-1)\pi t\right] + B_n \sin\left[(2n-1)\pi t\right]\right\} \cos\frac{(2n-1)\pi x}{2},$$

where $A_n = a_n c_n$ and $B_n = b_n c_n$. Applying the IC, $u(x,0) = x^3 + 2x^2 - 3$, yields

$$u(x,0) = x^3 + 2x^2 - 3 = \sum_{n=1}^{\infty} A_n \cos\frac{(2n-1)\pi x}{2}.$$

Using orthogonality, we determine A_n, which is

$$A_n = 2\int_0^1 \left(x^3 + 2x^2 - 3\right)\cos\frac{(2n-1)\pi x}{2}\,dx$$

$$= \frac{-2\left(96 + 80\pi(-1)^{n+1}(1-2n)\right)}{(2n-1)^4\pi^4}. \tag{5.44}$$

Repeating the process with the second IC, $\dfrac{\partial u(x,0)}{\partial t} = x^2 - 1$, yields

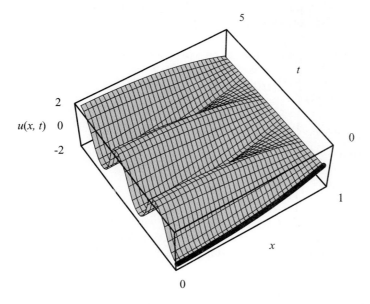

Figure 5.5: The graph of $u(x,t)$ for $0 \le t \le 5$.

$$B_n = \frac{2}{(2n-1)\pi} \int_0^1 \left(x^2 - 1\right) \cos \frac{(2n-1)\pi x}{2} \, dx$$

$$= \frac{-32(-1)^{n+1}}{(2n-1)^4 \pi^4}. \tag{5.45}$$

Therefore, the specific solution for $u(x,t)$ is

$$u(x,t) = \sum_{n=1}^{\infty} \{A_n \cos\left[(2n-1)\pi t\right] + B_n \sin\left[(2n-1)\pi t\right]\} \cos \frac{(2n-1)\pi x}{2}$$

where A_n and B_n are given by Equations (5.44 and 5.45) respectively. Figure (5.5) is the graph of $u(x,t)$ for the partial sum of $n = 15$ with $0 \le t \le 5$. Please look carefully at Figure (5.5). The boundary at $x = 1$ is fixed at 0. Whereas, the boundary at $x = 0$ is freely oscillating. This is exactly what the BCs indicated. Also, the initial displacement of the string is indicated by a boldface line. ■

The one-dimensional problems are useful for developing Fourier Series Solutions to a PDE. However, more interesting problems arise in the multidimensional spatial variable PDEs, which are discussed in the next section.

EXERCISES 5.4

5.4.1. Determine if the following PDEs are separable. If so, separate the PDEs into appropriate ODEs. If not, explain why.

(1) $\dfrac{\partial^2 u(x,t)}{\partial t^2} = c^2 \dfrac{\partial^2 u(x,t)}{\partial x^2} + u(x,t).$

(2) $\dfrac{\partial^2 u(x,t)}{\partial t^2} = c^2 \dfrac{\partial^2 u(x,t)}{\partial x^2} - m\dfrac{\partial u(x,t)}{\partial x} + u(x,t).$

(3) $c(x)\rho(x)\dfrac{\partial^2 u(x,t)}{\partial t^2} = \dfrac{\partial}{\partial x}\left[T(x)\dfrac{\partial u(x,t)}{\partial x}\right] - \dfrac{\partial u(x,t)}{\partial t} + u(x,t).$

5.4.2. Consider the PDE

$$\frac{\partial^2 u(x,t)}{\partial t^2} = 9\frac{\partial^2 u(x,t)}{\partial x^2}.$$

For each set of BCs and ICs, solve the initial value problem. Using your favorite mathematical software, graph the solution for $0 \le t \le 2$. Clearly indicate the initial displacement on your graph.

(1) BCs: $\begin{cases} u(0,t) = 0 \\ u(\pi,t) = 0 \end{cases}$ and ICs: $\begin{cases} u(x,0) = \pi x - x^2 \\ \dfrac{\partial u(x,0)}{\partial t} = 0. \end{cases}$

(2) BCs: $\begin{cases} u(0,t) = 0 \\ \dfrac{\partial u\left(\frac{\pi}{2},t\right)}{\partial x} = 0 \end{cases}$ and ICs: $\begin{cases} u(x,0) = \sin x \\ \dfrac{\partial u(x,0)}{\partial t} = 0. \end{cases}$

(3) BCs: $\begin{cases} \dfrac{\partial u(0,t)}{\partial x} = 0 \\ \dfrac{\partial u(\pi,t)}{\partial x} = 0 \end{cases}$ and ICs: $\begin{cases} u(x,0) = 0 \\ \dfrac{\partial u(x,0)}{\partial t} = \cos x. \end{cases}$

(4) BCs: $\begin{cases} \dfrac{\partial u(0,t)}{\partial x} = 0 \\ u(\pi,t) = 0 \end{cases}$ and IC: $\begin{cases} u(x,0) = \cos x + 1 \\ \dfrac{\partial u(x,0)}{\partial t} = 3x^3 - 3\pi x^2. \end{cases}$

5.4.3. Consider the PDE

$$\frac{\partial^2 u(x,t)}{\partial t^2} = \frac{\partial^2 u(x,t)}{\partial x^2} - \frac{\partial u(x,t)}{\partial t} + \frac{\partial u(x,t)}{\partial x}.$$

(1) In five sentences or less, give a physical description of this problem.

(2) Solve the initial value problem subject to

BCs: $\begin{cases} u(0,t) = 0 \\ u\left(\dfrac{3\pi}{2},t\right) = 0 \end{cases}$ and ICs: $\begin{cases} u(x,0) = \cos x \\ \dfrac{\partial u(x,0)}{\partial t} = x^2 - \left(\dfrac{3\pi}{2}\right)^2. \end{cases}$

5.4.4. A uniform string with mass density 0.03 lbs/ft, and tension 300 lbs, is fixed at the left end and has a freely moving right end. The string has length 20 ft and is initially at rest, with linear displacement from 0 to 1.

(1) Model this problem mathematically. *Note:* You must model and explain your choice of boundary conditions.

(2) Solve the mathematical model.

(3) Write the solution with all known quantities substituted into it.

(4) Check to see that this solution satisfies the equation, the boundary conditions, and the initial conditions.

(5) Using your favorite mathematical software, graph the solution for $0 \le t \le 5$.

5.4.5. Consider the following information:

(a) a perfectly flexible string of length 2π ft,

(b) tension of 50 lbs/ft,

(c) mass density of 0.02 lbs/ft,

(d) vibrating motion that is entirely vertical,

(e) fixed left end at 0,

(f) free right end,

(g) initial displacement of x for $0 < x < \pi$ and of $2\pi - x$ for $\pi < x < 2\pi$,

(h) initial velocity of 0, and

(i) no gravity effects.

(1) Set up the mathematical model.

(2) Solve your mathematical model for any time t.

(3) Check to see that this solution satisfies the equation, the BCs, and the ICs.

(4) Using your favorite mathematical software, graph the solution. Then write a short paragraph discussing the surface at the following times: $t = 1$ sec, $t = 5$ secs, and $t = 10$ secs.

5.4.6. Consider a slightly damped vibrating string with a restoring force that satisfies

$$\frac{\partial^2 u}{\partial t^2} = c^2 \frac{\partial^2 u}{\partial x^2} - \alpha \frac{\partial u}{\partial t} - \beta u, \text{ where } \alpha \text{ and } \beta \text{ are constants.}$$

(1) Explain why $\alpha > 0$.

(2) Explain the action of the restoring force.

(3) Find a series solution subject to

$$\text{BCs: } \begin{cases} u(0, t) = 0 \\ u(L, t) = 0 \end{cases} \quad \text{and} \quad \text{ICs: } \begin{cases} u(x, 0) = f(x) \\ \dfrac{\partial u(x, 0)}{\partial t} = g(x). \end{cases}$$

Explain your choice of the magnitude of α and β as you solve this problem.

5.4.7. A uniform string with fixed ends is excited by the impact of a rigid plane hammer, which gives it the following initial distribution of velocities:

$$\frac{\partial u(x, 0)}{\partial t} = \begin{cases} 0, 0 \le x \le \pi, \\ 225, \pi < x < 2\pi, \\ 0, 2\pi \le x \le 3\pi. \end{cases}$$

Find the vibrations of the string, if the initial displacement was zero.

5.4.8. A uniform string with fixed ends is excited by the impact of a rigid sharp hammer, which gives it the following initial distribution of velocities:

$$\frac{\partial u(x,0)}{\partial t} = \begin{cases} 0, \, 0 \leq x \leq \dfrac{\pi}{2}. \\ 225 \cos{(x - \pi)}, \, \dfrac{\pi}{2} < x < \dfrac{3\pi}{2}. \\ 0, \, \dfrac{3\pi}{2} \leq x \leq 2\pi. \end{cases}$$

Find the vibrations of the string, if the initial displacement was zero.

5.4.9. Consider longitudinal vibrations of a uniform flexible rod with free ends. Determine the following:

(1) The mathematical model if the initial displacement and velocity are arbitrary functions of x in the longitudinal direction.

(2) Solve the initial value problem.
Note: Consider the possibility of uniform linear motion of the rod for the entire problem.

5.4.10. A uniform string with fixed ends has an initial displacement of $2x$ for $0 \leq x < \pi$ and $3\pi - x$ for $\pi \leq x \leq 3\pi$. The initial velocity is zero and the string is vibrating in a medium that resists the vibrations. (The medium produces a resistance proportional to the velocity.) Suppose the resistance constant of proportionality is 0.01. Find the solution to the initial value problem.

5.4.11. A uniform string with a fixed end at 0 and free end at 2π has an initial displacement of $-x$ for $0 \leq x < \dfrac{3\pi}{2}$ and $3x - 6\pi$ for $\dfrac{3\pi}{2} \leq x \leq 2\pi$. It is known that the initial velocity is zero and the string is vibrating in a medium that resists the vibrations. (The medium produces a resistance proportional to the velocity.) Suppose the resistance constant of proportionality is 0.03; find the solution to the initial value problem.

5.4.12. A uniform string with free ends has an initial displacement of

$$u(x,0) = \begin{cases} -x, \, 0 \leq x \leq 1 \\ \dfrac{3}{2}x - \dfrac{5}{2}, \, 1 \leq x < 3 \\ -2x + 8, \, 3 \leq x \leq 4. \end{cases}$$

The initial velocity is zero, and the string is vibrating in a medium that resists the vibrations. (The medium produces a resistance proportional to the velocity.) Suppose the resistance constant of proportionality is 0.13; find the solution to the initial value problem.

5.4.13. This problem develops the mathematical model for longitudinal vibrations of a gas in a tube. Consider an ideal gas performing small longitudinal vibrations when it is enclosed in a cylindrical tube. We know that plane cross sections, consisting of particles of the gas, are not deformed, and that all the gas particles move parallel to the axis of the cylinder. Determine the boundary value problem for the density ρ, the pressure p, the velocity potential ϕ of the gas particles, the velocity v, and the displacement $u(x, t)$ if the ends of the tube are closed by a rigid impermeable surface.

5.5   THE MULTIDIMENSIONAL SPATIAL PROBLEM

We have covered separation of variables for both heat conduction in a one-dimensional rod and transverse vibrations of a one-dimensional string. These are the simplest examples of PDEs. This section is devoted to building on this knowledge of PDEs by considering the multidimensional problem.

An example of a multidimensional problem is the two-dimensional heat problem in a uniform rectangular plate, with no internal source and perfect lateral insulation. What does perfect lateral insulation on a rectangular plate mean, physically? Consider a sheet of paper the rectangular plate. The perfect lateral insulation would cover the front and back of the sheet of paper.

The equation for the problem just described is

$$\frac{\partial u}{\partial t} = k \left(\frac{\partial^2 u}{\partial x^2} + \frac{\partial^2 u}{\partial y^2} \right), \tag{5.46}$$

subject to any necessary BCs and ICs. It is quite easy to see that the heat equation in a uniform three-dimensional parallelepiped (rectangular box), with no internal source and perfect lateral insulation is

$$\frac{\partial u}{\partial t} = k \left(\frac{\partial^2 u}{\partial x^2} + \frac{\partial^2 u}{\partial y^2} + \frac{\partial^2 u}{\partial z^2} \right). \tag{5.47}$$

Again, this equation would be subject to the appropriate boundary and initial conditions. What would the equation be for a four, or even a five, dimensional parallelepiped? Or, for that matter, a circle, a sphere, or an ellipsoid? These equations become quite messy to write, so mathematicians and physicists developed the operator ∇^2, called Del squared, from the operator ∇, called Del. In calculus III, you learned Del is the vector,

$$\nabla = \begin{bmatrix} \dfrac{\partial}{\partial x} \\[2mm] \dfrac{\partial}{\partial y} \end{bmatrix},$$

for a two-dimensional system. Thus, $\nabla^2 = \nabla \cdot \nabla$ or

$$\nabla^2 = \nabla \cdot \nabla = \begin{bmatrix} \dfrac{\partial}{\partial x} \\[2mm] \dfrac{\partial}{\partial y} \end{bmatrix} \cdot \begin{bmatrix} \dfrac{\partial}{\partial x} \\[2mm] \dfrac{\partial}{\partial y} \end{bmatrix} = \frac{\partial^2}{\partial x^2} + \frac{\partial^2}{\partial y^2},$$

which is the dot product of ∇ with ∇.

The operator ∇^2 handles the dimensions of Equations (5.46 and 5.47). For example, we can rewrite Equation (5.46) as

$$\frac{\partial u}{\partial t} = k\nabla^2 u. \tag{5.48}$$

Here,

$$\nabla^2 = \left(\frac{\partial^2}{\partial x^2} + \frac{\partial^2}{\partial y^2} \right).$$

Equation (5.48) is the compact form of Equation (5.46).

For another example, Equation (5.47) has the same compact form as Equation (5.46), which is Equation (5.48) with the added dimension. The expanded ∇^2 operator becomes

$$\nabla^2 = \left(\frac{\partial^2}{\partial x^2} + \frac{\partial^2}{\partial y^2} + \frac{\partial^2}{\partial z^2} \right).$$

In this section, the ∇^2 operator is not used much since we deal with uniform two- and three-dimensional rectangular-shaped plates for both the heat and the wave equations. However, in later chapters, this operator becomes very important, and you need to get accustomed to it.

Consider the problem of heat conduction in a uniform rectangular plate with no internal source, perfect lateral insulation, a prescribed temperature of $0°$ on the sides $x = 0$ and $x = L$, perfect insulation on the sides $y = 0$ and $y = H$, and an initial temperature distribution that is a function of x and y. This problem is mathematically modeled as

$$\frac{\partial u}{\partial t} = k\nabla^2 u = k\left(\frac{\partial^2 u}{\partial x^2} + \frac{\partial^2 u}{\partial y^2} \right), \tag{5.49}$$

subject to the BCs

$$\begin{cases} u(0, y, t) & = \ 0 \\[2mm] u(L, y, t) & = \ 0 \\[2mm] \dfrac{\partial u(x, 0, t)}{\partial y} & = \ 0 \\[2mm] \dfrac{\partial u(x, H, t)}{\partial y} & = \ 0 \end{cases} \tag{5.50}$$

and IC

$$u(x, y, 0) = f(x, y). \tag{5.51}$$

Equation (5.49) and BCs, Equation (5.50), are linear and homogeneous. Hence, the separation of variables technique can be applied as a solution method for this problem. As in the one-dimensional problem, we assume that

$$u(x, y, t) = G(t)\varphi(x, y),$$

so that Equation (5.49) becomes

$$\frac{1}{kG}\frac{dG}{dt} = \frac{1}{\varphi}\left(\frac{\partial^2\varphi}{\partial x^2} + \frac{\partial^2\varphi}{\partial y^2}\right) = -\lambda,$$

where $-\lambda$ is the separation constant. Also, the BCs are separated and become

$$\begin{cases} \varphi(0, y) & = & 0 \\[2mm] \varphi(L, y) & = & 0 \\[2mm] \dfrac{\partial\varphi(x, 0)}{\partial y} & = & 0 \\[2mm] \dfrac{\partial\varphi(x, H)}{\partial y} & = & 0. \end{cases} \tag{5.52}$$

Thus, we have a time problem,

$$\frac{dG}{dt} = -\lambda k G,$$

and a spatial problem,

$$\frac{1}{\varphi}\left(\frac{\partial^2\varphi}{\partial x^2} + \frac{\partial^2\varphi}{\partial y^2}\right) = -\lambda, \tag{5.53}$$

subject to the BCs, Equation (5.52).

The spatial problem is still a PDE, where the PDE and the BCs are linear and homogeneous. Therefore, we can apply the separation of variables technique. If we assume that $\varphi(x, y) = X(x)Y(y)$, then we can rewrite Equation (5.53) as

$$Y\frac{d^2X}{dx^2} + X\frac{d^2Y}{dy^2} = -\lambda XY. \tag{5.54}$$

Separating Equation (5.54) yields

$$\frac{1}{X}\frac{d^2X}{dx^2} = -\lambda - \frac{1}{Y}\frac{d^2Y}{dy^2} = -\tau,$$

where $-\tau$ is the new separation constant. Again, we must separate the boundary conditions, and they become

$$
\begin{cases}
X(0) &=& 0 \\[2mm]
X(L) &=& 0 \\[2mm]
\dfrac{dY(0)}{dy} &=& 0 \\[2mm]
\dfrac{dY(H)}{dy} &=& 0.
\end{cases}
$$

We now have two spatial problems. One in terms of $X(x)$,

$$\frac{d^2 X}{dx^2} = -\tau X, \tag{5.55}$$

subject to the BCs

$$
\begin{cases}
X(0) &=& 0 \\[2mm]
X(L) &=& 0;
\end{cases}
\tag{5.56}
$$

the other in terms of $Y(y)$,

$$\frac{d^2 Y}{dy^2} = -\left(\lambda - \tau\right) Y, \tag{5.57}$$

subject to the BCs

$$
\begin{cases}
\dfrac{dY(0)}{dy} &=& 0 \\[2mm]
\dfrac{dY(H)}{dy} &=& 0.
\end{cases}
\tag{5.58}
$$

5.5.1   Spatial Problem for $X(x)$

We recognize that Equation (5.55), subject to the BCs, Equation (5.56), has eigenvalues

$$\tau_n = \left(\frac{n\pi}{L}\right)^2, \; n = 1, 2, 3, \ldots \tag{5.59}$$

and eigenfunctions

$$X(x) = b_n \sin \frac{n\pi x}{L}, \; n = 1, 2, 3, \ldots \tag{5.60}$$

Thus, the complete solution for the spatial problem for $X(x)$ is

$$
\left.
\begin{aligned}
\tau_n &= \left(\frac{n\pi}{L}\right)^2 \\[4mm]
X_n(x) &= b_n \sin\left(\frac{n\pi x}{L}\right)
\end{aligned}
\right\}, \; n = 1, 2, 3, \ldots
$$

5.5.2   Spatial Problem for $Y(y)$

Letting $\xi = (\lambda - \tau)$, Equation (5.57) becomes

$$\frac{d^2Y}{dy^2} = -\xi Y, \tag{5.61}$$

subject to the BCs, Equation (5.58); from previous work, we recognize that the eigenvalues are

$$\xi_m = \left(\frac{m\pi}{H}\right)^2, \ m = 0, 1, 2, 3, \ldots \tag{5.62}$$

and the eigenfunctions

$$Y(y) = a_m \cos\frac{m\pi y}{H}, \ m = 0, 1, 2, 3, \ldots \tag{5.63}$$

Therefore, the complete solution for the spatial problem for $Y(y)$ is

$$\left.\begin{array}{l} \xi_m = \left(\dfrac{m\pi}{H}\right)^2 \\[2ex] Y_m(y) = a_m \cos\left(\dfrac{m\pi y}{H}\right) \end{array}\right\} , \ m = 0, 1, 2, 3, \ldots$$

Since our original separation constant is in terms of λ, we must determine λ from Equations (5.59 and 5.62). Thus, the eigenvalues are

$$\lambda_{nm} = \tau_n + \xi_m = \left(\frac{n\pi}{L}\right)^2 + \left(\frac{m\pi}{H}\right)^2, \ n = 1, 2, 3, \ldots, \ m = 0, 1, 2, 3, \ldots$$

Note that λ_{nm} has subscripts of n and m. The reason is that λ_{nm} is made up of τ_n and ξ_m.

5.5.3   Time Problem

We have

$$\frac{dG}{dt} = -\lambda_{nm} k G.$$

From previous work, we know the solution is

$$G_{nm}(t) = c_{nm} e^{-\lambda_{nm} k t}. \tag{5.64}$$

5.5.4   The Complete Solution

We know that $\varphi_{nm}(x, y) = X_n(x)Y_m(y)$;

$$X_n(x) = b_n \sin\frac{n\pi x}{L}, \ n = 1, 2, 3, \ldots;$$

and

$$Y_m(y) = a_m \cos \frac{m\pi y}{H}, \ m = 0, 1, 2, 3, \dots$$

Therefore, we have for $n = 1, 2, 3, \dots$ and $m = 0, 1, 2, 3, \dots$,

$$\varphi(x, y) = C_{nm} \sin \frac{n\pi x}{L} \cos \frac{m\pi y}{H}, \tag{5.65}$$

where

$$C_{nm} = b_n a_m. \tag{5.66}$$

Also $u(x, y, t) = G_{nm}(t) \varphi_{nm}(x, y)$. Thus, using Equations (5.64 and 5.65) we get for $n = 1, 2, 3, \dots$ and $m = 0, 1, 2, 3, \dots$,

$$u(x, y, t) = A_{nm} e^{-\lambda_{nm} kt} \sin \frac{n\pi x}{L} \cos \frac{m\pi y}{H}, \tag{5.67}$$

where

$$A_{nm} = C_{nm} c_{nm}, \tag{5.68}$$

which is an infinite collection of solutions.

In previous sections of this chapter, we used the principle of superposition to give a simple summation of all the solutions that would give the correct answer. Equation (5.67) has a slightly different problem. That is, there are two indices, n and m. Let's consider different possible solutions for $u(x, y, t)$. We have

$$u(x, y, t) = A_{1m} e^{-\lambda_{1m} kt} \sin \frac{\pi x}{L} \cos \frac{m\pi y}{H}, \ m = 0, 1, 2, 3, \dots,$$

which, using the reasoning of previous sections of this chapter, naturally leads to

$$u(x, y, t) = \sin \frac{\pi x}{L} \sum_{m=0}^{\infty} A_{1m} e^{-\lambda_{1m} kt} \cos \frac{m\pi y}{H}. \tag{5.69}$$

We could also have

$$u(x, y, t) = A_{2m} e^{-\lambda_{2m} kt} \sin \frac{2\pi x}{L} \cos \frac{m\pi y}{H}, \ m = 0, 1, 2, 3, \dots,$$

which leads to

$$u(x, y, t) = \sin \frac{2\pi x}{L} \sum_{m=0}^{\infty} A_{2m} e^{-\lambda_{2m} kt} \cos \frac{m\pi y}{H}. \tag{5.70}$$

Using the principle of superposition on both, we have

$$u(x, y, t) = \sin \frac{\pi x}{L} \sum_{m=0}^{\infty} A_{1m} e^{-\lambda_{1m} kt} \cos \frac{m\pi y}{H}$$

$$+ \sin \frac{2\pi x}{L} \sum_{m=0}^{\infty} A_{2m} e^{-\lambda_{1m} kt} \cos \frac{m\pi y}{H}.$$

Expanding this process on the cases $n = 1, 2, 3, \ldots$ gives

$$u(x, y, t) = \sin \frac{\pi x}{L} \sum_{m=0}^{\infty} A_{1m} e^{-\lambda_{1m}kt} \cos \frac{m\pi y}{H}$$

$$+ \sin \frac{2\pi x}{L} \sum_{m=0}^{\infty} A_{2m} e^{-\lambda_{1m}kt} \cos \frac{m\pi y}{H}$$

$$+ \ldots + \sin \frac{n\pi x}{L} \sum_{m=0}^{\infty} A_{nm} e^{-\lambda_{nm}kt} \cos \frac{m\pi y}{H} + \ldots \qquad (5.71)$$

But since equation (5.71) is an infinite sum of an infinite sum, it could be easily written as

$$u(x, y, t) = \sum_{n=1}^{\infty} \sin \frac{n\pi x}{L} \sum_{m=0}^{\infty} A_{nm} e^{-\lambda_{nm}kt} \cos \frac{m\pi y}{H},$$

or in the more usual form of

$$u(x, y, t) = \sum_{n=1}^{\infty} \sum_{m=0}^{\infty} A_{nm} e^{-\lambda_{nm}kt} \cos \frac{m\pi y}{H} \sin \frac{n\pi x}{L} =$$

$$\sum_{n=1}^{\infty} A_{n0} e^{-\lambda_{n0}kt} \sin \frac{n\pi x}{L}$$

$$+ \sum_{n=1}^{\infty} \sum_{m=1}^{\infty} A_{nm} e^{-\lambda_{nm}kt} \cos \frac{m\pi y}{H} \sin \frac{n\pi x}{L}, \qquad (5.72)$$

which is the complete general solution to our problem.

To determine the specific solution, we apply the IC, $u(x, y, 0) = f(x, y)$, to Equation (5.72). In doing so, we get

$$u(x, y, 0) = f(x, y) = \sum_{n=1}^{\infty} A_{n0} \sin \frac{n\pi x}{L} + \sum_{n=1}^{\infty} \sum_{m=1}^{\infty} A_{nm} \cos \frac{m\pi y}{H} \sin \frac{n\pi x}{L}.$$

This leaves us the task of determining the equations for the coefficients A_{nm} and A_{n0}.

Let's first determine the equation for the coefficient A_{nm}. In other words, we consider the case when $n = 1, 2, 3, \ldots$ and $m = 1, 2, 3, \ldots$, which is

$$u(x, y, 0) = f(x, y) = \sum_{n=1}^{\infty} \sum_{m=1}^{\infty} A_{nm} \cos \frac{m\pi y}{H} \sin \frac{n\pi x}{L}. \qquad (5.73)$$

Note: To do this we are actually applying the orthogonality of the sine and cosine functions.

In Equation (5.73), the inner sum actually equals a function in y for each value of n. That is,

$$B_n(y) = \sum_{m=1}^{\infty} A_{nm} \cos \frac{m\pi y}{H}. \tag{5.74}$$

Thus, Equation (5.73) may be rewritten as

$$f(x, y) = \sum_{n=1}^{\infty} B_n(y) \sin \frac{n\pi x}{L}.$$

Since $f(x, y)$ is now set equal to a Fourier sine series, we can use the method that we learned in Chapter 3 to solve for $B_n(y)$. The solution is

$$B_n(y) = \frac{2}{L} \int_0^L f(x, y) \sin \frac{n\pi x}{L} dx.$$

Thus, we have $B_n(y)$ equal to a Fourier cosine series. Next, solving for A_{nm} yields

$$A_{nm} = \frac{2}{H} \int_0^H B_n(y) \cos \frac{m\pi y}{H} dy,$$

and on replacing $B_n(y)$ with its integration equation, we get

$$A_{nm} = \frac{2}{H} \int_0^H \frac{2}{L} \int_0^L f(x, y) \sin \frac{n\pi x}{L} dx \cos \frac{m\pi y}{H} dy =$$

$$\frac{4}{HL} \int_0^H \int_0^L f(x, y) \cos \frac{m\pi y}{H} \sin \frac{n\pi x}{L} dx dy. \tag{5.75}$$

Likewise we arrive at

$$A_{n0} = \frac{2}{LH} \int_0^H \int_0^L f(x, y) \sin \frac{n\pi x}{L} dx dy. \tag{5.76}$$

This completes the answer to the problem at the beginning of this section, giving us

$$u(x, y, t) = \sum_{n=1}^{\infty} A_{n0} e^{-\lambda_{n0}kt} \sin \frac{n\pi x}{L}$$

$$+ \sum_{n=1}^{\infty} \sum_{m=1}^{\infty} A_{nm} e^{-\lambda_{nm}kt} \cos \frac{m\pi y}{H} \sin \frac{n\pi x}{L},$$

where

$$A_{n0} = \frac{2}{LH} \int_0^H \int_0^L f(x, y) \sin \frac{n\pi x}{L} dx dy$$

and

$$A_{nm} = \frac{4}{HL} \int_0^H \int_0^L f(x,y) \cos \frac{m\pi y}{H} \sin \frac{n\pi x}{L} dx dy.$$

This completes the solution of Equation (5.49), subject to the BCs, Equation (5.50), with IC, Equation (5.51). See below for a worked example for a multi-dimensional problem.

EXAMPLE 5.5. Find the time-dependent solution for

$$\frac{\partial^2 u}{\partial t^2} = 9 \left(\frac{\partial^2 u}{\partial x^2} + \frac{\partial^2 u}{\partial y^2} \right) \tag{5.77}$$

subject to

$$\begin{cases} u(0,y,t) &= 0 \\[2mm] u(2,y,t) &= 0 \\[2mm] u(x,0,t) &= 0 \\[2mm] u(x,3,t) &= 0, \end{cases} \tag{5.78}$$

with

$$u(x,y,0) = f(x,y) = \begin{cases} xy; & 0 \le x < 1 \quad 0 \le y < 1.5 \\[2mm] x(3-y); & 0 \le x < 1 \quad 1.5 \le y \le 3 \\[2mm] (2-x)y; & 1 \le x \le 2 \quad 0 \le y < 1.5 \\[2mm] (2-x)(3-y); & 1 \le x \le 2 \quad 1.5 \le y \le 3, \end{cases} \tag{5.79}$$

and

$$\frac{\partial u(x,y,0)}{\partial t} = 0. \tag{5.80}$$

The PDE, Equation (5.77), and BCs, Equation (5.78), are linear and homogeneous. Therefore, separation of variables technique is valid. Letting $u(x,y,t) = G(t)\varphi(x,y)$ and substituting into Equation (5.77) and Equation (5.78) yields the time equation

$$G''(t) = -9\lambda G(t), \tag{5.81}$$

and the spatial equation

$$\left(\frac{\partial^2 \varphi(x,y)}{\partial x^2} + \frac{\partial^2 \varphi(x,y)}{\partial y^2} \right) = -\lambda \varphi(x,y), \tag{5.82}$$

subject to the BCs

$$\begin{cases} \varphi(0,y) & = & 0 \\ \varphi(2,y) & = & 0 \\ \varphi(x,0) & = & 0 \\ \varphi(x,3) & = & 0. \end{cases} \tag{5.83}$$

The spatial equation, Equation (5.81), subject to the BCs in Equation (5.83), is a linear and homogeneous system. Hence, separation of variables technique is again valid. Letting $\varphi(x,y) = X(x)Y(y)$, and substituting into Equation (5.81) and Equation (5.83), yields the spatial equation

$$X''(x) = -\tau X(x), \tag{5.84}$$

subject to the BCs

$$\begin{cases} X(0) & = & 0 \\ X(2) & = & 0. \end{cases} \tag{5.85}$$

We know from Section 5.3, Equation (5.84), subject to the BCs, Equation (5.85), has a nontrivial solution only when $\tau > 0$. For $\tau > 0$, the eigenvalues are

$$\tau_n = \left(\frac{n\pi}{2}\right)^2, \ n = 1, \ 2, \ 3, \dots, \tag{5.86}$$

with corresponding eigenfunctions

$$X_n(x) = a_n \sin \frac{n\pi x}{2}, \ n = 1, \ 2, \ 3, \dots \tag{5.87}$$

The separation of variables technique also yields the spatial equation

$$Y''(y) = -(\lambda - \tau)Y(y), \tag{5.88}$$

subject to the BCs

$$\begin{cases} Y(0) & = & 0 \\ Y(3) & = & 0. \end{cases} \tag{5.89}$$

Letting $\xi = (\lambda - \tau)$ in Equation (5.88) yields

$$Y''(y) = -\xi Y(y). \tag{5.90}$$

Again, we know from Section 5.3, Equation (5.90), subject to the BCs, Equation (5.89), has a nontrivial solution only when $\xi > 0$. For $\xi > 0$, the eigenvalues are

$$\xi_m = \left(\frac{m\pi}{3}\right)^2, \ m = 1, \ 2, \ 3, \dots, \tag{5.91}$$

with corresponding eigenfunctions

$$Y_m(y) = b_m \sin \frac{m\pi y}{3}, \quad m = 1, \ 2, \ 3, \ldots \tag{5.92}$$

Using Equations (5.86 and 5.91) and the relationship $\xi = (\lambda - \tau)$, yields

$$\lambda_{mn} = \left(\frac{n\pi}{2}\right)^2 + \left(\frac{m\pi}{3}\right)^2, \tag{5.93}$$

for $n = 1, \ 2, \ 3, \ \ldots$ and $m = 1, \ 2, \ 3, \ \ldots$ Substituting λ_{mn} into Equation (5.81) gives us the solution to the time equation, which is

$$G_{mn}(t) = c_{mn} \cos\left(3\sqrt{\lambda_{mn}} t\right) + h_{mn} \sin\left(3\sqrt{\lambda_{mn}} t\right). \tag{5.94}$$

We now can put together the total solution for $u(x, y, t)$. Remember $\varphi(x, y) = X(x)Y(y)$, so we have

$$u(x, y, t) = \varphi_{mn}(x, y)G_{mn}(t) = X_n(x)Y_m(y)G_{mn}(t) =$$

$$\sum_{n=1}^{\infty} \sum_{m=1}^{\infty} \left[A_{mn} \cos\left(3\sqrt{\lambda_{mn}} t\right) + B_{mn} \sin\left(3\sqrt{\lambda_{mn}} t\right)\right] \sin \frac{m\pi y}{3} \sin \frac{n\pi x}{2}. \tag{5.95}$$

Applying the IC, $u(x, y, 0) = f(x, y)$, yields

$$u(x, y, 0) = f(x, y) = \sum_{n=1}^{\infty} \sum_{m=1}^{\infty} A_{mn} \sin \frac{m\pi y}{3} \sin \frac{n\pi x}{2}. \tag{5.96}$$

Using orthogonality, we determine A_{mn}, which is

$$A_{mn} = \frac{2}{3} \int_0^3 \int_0^2 f(x, y) \sin \frac{m\pi y}{3} \sin \frac{n\pi x}{2} \, dx \, dy$$

$$= \frac{96}{m^2 n^2 \pi^4} \sin \frac{m\pi}{3} \sin \frac{n\pi}{2}. \tag{5.97}$$

Repeating the process with the second IC, $\dfrac{\partial u(x, y, 0)}{\partial t} = 0$, yields

$$\frac{\partial u(x, y, 0)}{\partial t} = 0 = \sum_{n=1}^{\infty} \sum_{m=1}^{\infty} \left[3\sqrt{\lambda_{mn}} B_{mn}\right] \sin \frac{m\pi y}{3} \sin \frac{n\pi x}{2},$$

which implies $B_{mn} = 0$. Therefore, the complete solution is

$$u(x, y, t) = \sum_{n=1}^{\infty} \sum_{m=1}^{\infty} A_{mn} \cos\left(3\sqrt{\lambda_{mn}} t\right) \sin \frac{m\pi y}{3} \sin \frac{n\pi x}{2}$$

where A_{mn} is given in Equation (5.97). Since we can't present four-dimensional graphs, we graph the solution for the partial sums $n = 25$ and $m = 25$ and several different times, t. Figure (5.6) depicts the surface $u(x, y, t)$ at time $t = 0$. Here, the four faces of the figure indicate the initial conditions. Figure (5.7) shows the surface of $u(x, y, t)$ at time $t = 0.2$. The third graph, Figure (5.8), portrays the surface $u(x, y, t)$ at time $t = 0.4$. Finally, Figure (5.9) depicts $u(x, y, t)$ for time $t = 0.8$. ∎

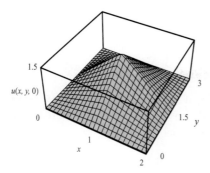

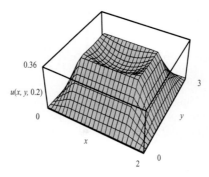

Figure 5.6: The graph of $u(x, y, t)$ for time $t = 0$.

Figure 5.7: The graph of $u(x, y, t)$ for time $t = 0.2$.

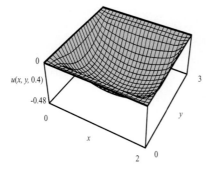

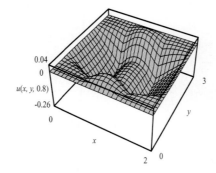

Figure 5.8: The graph of $u(x, y, t)$ for time $t = 0.4$.

Figure 5.9: The graph of $u(x, y, t)$ for time $t = 0.8$.

EXERCISES 5.5

5.5.1. For the following PDE, separate the PDE into its respective ODEs. Also, completely separate the BCs,

$$\frac{\partial^2 u}{\partial t^2} = c^2 \frac{\partial^2 u}{\partial x^2} - b\frac{\partial u}{\partial t} + a\frac{\partial^2 u}{\partial y^2},$$

subject to BCs

$$\begin{cases} u(0, y, t) = 0 \\[2mm] \dfrac{\partial u(L, y, t)}{\partial x} = 0 \\[2mm] \dfrac{\partial u(x, 0, t)}{\partial y} = 0 \\[2mm] u(x, H, t) = 0. \end{cases}$$

5.5.2. For the following PDE, separate the PDE into its respective ODEs.

$$\frac{\partial^2 u}{\partial t^2} = c^2 \left(\frac{\partial^2 u}{\partial x^2} + \frac{\partial^2 u}{\partial y^2} + \frac{\partial^2 u}{\partial z^2} \right) - \left(\frac{\partial u}{\partial x} + \frac{\partial u}{\partial y} \right)$$

5.5.3. Consider a thin rectangular plate of length $L = 2\pi$ meters and width $W = 4\pi$ meters with perfect lateral insulation. Find the temperature distribution in the plate given the following:

(1) The rectangular plate is made of silver (he thermal diffusivity of silver may be found in Appendix E) and is subject to the BCs

$$\begin{cases} u(0, y, t) = 0 \\[2mm] \dfrac{\partial u(L, y, t)}{\partial x} = 0 \\[2mm] u(x, 0, t) = 0 \\[2mm] \dfrac{\partial u(x, W, t)}{\partial y} = 0, \end{cases}$$

with the IC

$$u(x, y, 0) = \begin{cases} \cos(xy) - 1; & 0 \le x < \dfrac{L}{2} \quad 0 \le y < \dfrac{W}{2} \\[3mm] y\cos(x - L); & \dfrac{L}{2} \le x \le L \quad 0 \le y < \dfrac{W}{2} \\[3mm] x\cos(y - W); & 0 \le x < \dfrac{L}{2} \quad \dfrac{W}{2} \le y \le W \\[3mm] \cos((x - L)(y - W)); & \dfrac{L}{2} \le x \le L \quad \dfrac{W}{2} \le y \le W. \end{cases}$$

(2) The rectangular plate is made of granite (the thermal diffusivity of granite may be found in Appendix E) and is subject to the BCs

$$\begin{cases} \dfrac{\partial u(0, y, t)}{\partial x} = 0 \\[2mm] u(L, y, t) = 0 \\[2mm] u(x, 0, t) = 0 \\[2mm] \dfrac{\partial u(x, W, t)}{\partial y} = 0, \end{cases}$$

with the IC

$$u(x,y,0) = \begin{cases} y; & 0 \le x < \dfrac{L}{3} \quad 0 \le y < \dfrac{W}{4} \\[2ex] (y-W); & 0 \le x < \dfrac{L}{3} \quad \dfrac{W}{4} \le y \le W \\[2ex] y(x-L); & \dfrac{L}{3} \le x \le L \quad 0 \le y < \dfrac{W}{4} \\[2ex] (x-L)(y-W); & \dfrac{L}{3} \le x \le L \quad \dfrac{W}{4} \le y \le W. \end{cases}$$

(3) The rectangular plate is made of brick (the thermal diffusivity of brick may be found in Appendix E) and is subject to the BCs

$$\begin{cases} \dfrac{\partial u(0,y,t)}{\partial x} = 0 \\[2ex] \dfrac{\partial u(L,y,t)}{\partial x} = 0 \\[2ex] \dfrac{\partial u(x,0,t)}{\partial y} = 0 \\[2ex] \dfrac{\partial u(x,W,t)}{\partial y} = 0, \end{cases}$$

with IC

$$u(x,y,0) = \cos(3x) \ \ 0 \le x \le L \ \ 0 \le y \le W.$$

5.5.4. Consider a thin rectangular plate of length $L = \dfrac{\pi}{2}$ m and width $W = \pi$ m, which offers no resistance to bending. Find the time-dependent solution given the following conditions:

(1) The plat is subject to the BCs

$$\begin{cases} u(0,y,t) = 0 \\[1.5ex] u(L,y,t) = 0 \\[1.5ex] u(x,0,t) = 0 \\[1.5ex] u(x,W,t) = 0, \end{cases}$$

with the ICs

$$u(x,y,0) = 0,$$

and

$$
\frac{\partial u(x,y,0)}{\partial t} =
\begin{cases}
xy; & 0 \le x < \dfrac{L}{2} \quad 0 \le y < \dfrac{W}{2} \\[2ex]
x(y - W); & 0 \le x < \dfrac{L}{2} \quad \dfrac{W}{2} \le y \le W \\[2ex]
(L - x)y; & \dfrac{L}{2} \le x \le L \quad 0 \le y < \dfrac{W}{2} \\[2ex]
(L - x)(y - W); & \dfrac{L}{2} \le x \le L \quad \dfrac{W}{2} \le y \le W.
\end{cases}
$$

(2) The plate is subject to the BCs

$$
\begin{cases}
\dfrac{\partial u(0,y,t)}{\partial x} = 0 \\[2ex]
\dfrac{\partial u(L,y,t)}{\partial x} = 0 \\[2ex]
\dfrac{\partial u(x,0,t)}{\partial y} = 0 \\[2ex]
u(x, W, t) = 0,
\end{cases}
$$

with the ICs

$$
u(x,y,0) =
\begin{cases}
x^2 y; & 0 \le x < \dfrac{L}{2} \quad 0 \le y < \dfrac{W}{2} \\[2ex]
x^2(W - y); & 0 \le x < 1 \quad \dfrac{W}{2} \le y \le W \\[2ex]
(x^2 - 2Lx)y; & \dfrac{L}{2} \le x \le L \quad 0 \le y < \dfrac{W}{2} \\[2ex]
(x^2 - 2Lx)(W - y); & \dfrac{L}{2} \le x \le L \quad \dfrac{W}{2} \le y \le W,
\end{cases}
$$

and

$$
\frac{\partial u(x,y,0)}{\partial t} = 0.
$$

(3) the plate is subject to the BCs

$$\begin{cases} u(0, y, t) = 0 \\[2mm] \dfrac{\partial u(L, y, t)}{\partial x} = 0 \\[2mm] \dfrac{\partial u(x, 0, t)}{\partial y} = 0 \\[2mm] u(x, W, t) = 0, \end{cases}$$

with the ICs

$$u(x, y, 0) = \begin{cases} y \sin x; & 0 \le x \le L \quad 0 \le y < \dfrac{W}{2} \\[3mm] (y - W) \sin x; & 0 \le x \le L \quad \dfrac{W}{2} \le y \le W, \end{cases}$$

and

$$\frac{\partial u(x, y, 0)}{\partial t} = \begin{cases} x(\cos y + 1); & 0 \le x < \dfrac{L}{2} \quad 0 \le y \le W \\[3mm] (x - L)(\cos y + 1); & \dfrac{L}{2} \le x \le L \quad 0 \le y \le W. \end{cases}$$

5.5.5. Consider a situation with the following characteristics:

(a) vertical vibrations in a rectangular membrane,

(b) membrane length of 3π m,

(c) membrane width of π m,

(d) initial velocity of $x \sin y$,

(e) initial displacement of 0,

(f) tension of 4 kg/m,

(g) mass density of 0.3 kgs,

(h) fixed on two sides meeting at the point $(0, 0)$, and

(i) frictionless moving sides meeting at the point $(3\pi, \pi)$.

(1) Set up the mathematical model.

(2) Solve your mathematical model for any time t.

(3) Using your favorite mathematical software, graph the solution. Then write a short paragraph discussing the surface at the following times: $t = 1$ sec, $t = 5$ secs, and $t = 10$ secs.

5.5.6. A granite brick of length $x = L$, width $y = W$, and height $z = H$ is initially at the uniform temperature of $f(x, y, z)$ degrees C. Suppose that at time $t = 0$ the sides $z = 0$ and $z = H$ are cooled to $0°$C while the other sides are perfectly insulated, and thereafter maintained in that fashion. Find the temperature distribution in the brick at any time t. Assume $L = 2\pi$ ft, $W = \pi$ ft, $H = \dfrac{\pi}{2}$ ft, and $f(x, y, z) = 1 - \cos z - \sin z$. *Note:* The thermal diffusivity of granite may be found in Appendix E. Graph the solution for several different times t.

5.5.7. Consider a thin vibrating rectangular uniform membrane of length $\dfrac{3\pi}{2}$ m and width $\dfrac{\pi}{2}$ m. Suppose the sides $x = 0$ and $y = 0$ are fixed and that the other two sides are free. Given an initial velocity of zero and initial displacement of $f(x, y) = (\sin x)(\sin y)$, determine the time-dependent solution and graph the solution for several different times t.

5.5.8. Consider a perfect laterally insulated thin sheet of glass Pyrex with length 3π ft and width 2π ft. Suppose the sides $x = 0$ and $x = 3\pi$ ft are perfectly insulated and the other two sides are held at zero degrees. Find the solution if the initial temperature distribution in the sheet is given as $f(x, y) = (\cos x)(\sin^2 y)$. *Note:* The thermal diffusivity of glass Pyrex may be found in Appendix E.

5.6   LAPLACE'S EQUATION

Laplace's equation is an extremely important equation in mathematical physics. It naturally arises in electrostatics, steady-state temperature field, magnetostatics, and potential flow of an incompressible liquid. In this section, we derive Laplace's equation from electrostatics. Then we determine the solution for Laplace's equation in the Cartesian coordinate system.

5.6.1   An Electrostatics Derivation of Laplace's Equation

Coulomb's Law—the force of attraction or repulsion between two point charges is directly proportional to the product of the charges and inversely proportional to the square of the distance between them—is the starting point for the study of electrostatics. It directly leads to two important differential equations describing an electric field, **E**. The first is the differential form of Gauss's law, which is

$$\nabla \cdot \mathbf{E} = \rho \qquad (5.98)$$

where ρ is the external charge density. Mathematically, Equation (5.98) is the divergence of the electric field. The second equation is the curl of the electric field as a function of position. It is

$$\nabla \times \mathbf{E} = 0. \qquad (5.99)$$

From Equation (5.99) we know that $\mathbf{E}$ is the gradient of a scalar function, known as the scalar potential, Φ. Thus,

$$\mathbf{E} = -\nabla\Phi. \tag{5.100}$$

Substituting Equation (5.100) into Equation (5.98) yields

$$\nabla \cdot (-\nabla\Phi) = \rho,$$

which may be written as

$$\nabla^2\Phi = -\rho. \tag{5.101}$$

Equation (5.101) is known as Poisson's[2] equation. In regions of space where there is no charge density, ρ, the scalar potential satisfies Laplace's equation,

$$\nabla^2\Phi = 0. \tag{5.102}$$

Since a scalar potential is used to develop Poisson's and Laplace's equations, they are sometimes referred to as the potential equations. (*Note:* In mathematics, we usually use the scalar function u instead of the potential scalar notation of Φ). Thus, Equation (5.102) becomes

$$\nabla^2 u = 0. \tag{5.103}$$

Note: Poisson's and Laplace's equations are time-independent second-order PDEs in two or more spatial dimensions.

Equation (5.103) has the following form in Cartesian coordinates,

$$\nabla^2 u = \frac{\partial^2 u}{\partial x^2} + \frac{\partial^2 u}{\partial y^2} + \frac{\partial^2 u}{\partial z^2} = 0. \tag{5.104}$$

5.6.2 Uniqueness of Solution

We now consider the uniqueness of the solution for Laplace's equation. First, we need to define a harmonic function and state a different form of the maximum principle in $\mathbb{R}^2$, which you first encountered in Chapter 2. For those of you who have already had a complex analysis course, the definition of a harmonic function is quite familiar.

Definition 34. *A **harmonic function** is a function which solves Laplace's equation.*

Theorem 35. *(Maximum Principle) Let Ω be a bounded set in $\mathbb{R}^2$. Let $u(x, y)$ be a harmonic function in Ω, while $u(x, y)$ must be continuous in the union of Ω and the boundary of Ω, denoted $\partial\Omega$. Then, the maximum and minimum values of $u(x, y)$ are attained on the $\partial\Omega$, unless $u(x, y)$ is identically equal to a constant.*

---

[2]Simeon-Denis Poisson (1781–1840) was a French mathematician who once stated that life is good for to things: mathematics and teaching it.

(*Note:* The maximum principle is easily extended to the space $\mathbb{R}^n$). Proof of Theorem 35 may be found in Appendix B. However, a common sense approach gives us the idea behind the maximum principle.

If we consider $u(x, y)$ in Laplace's equation as the steady-state temperature distribution in a plate, then $u(x, y)$ can't be greater at one point in the plate than all the other points of the plate because Fourier's heat conduction axioms say that heat diffuses from high to low. Thus, if one point in the plate is hotter than the rest, then the heat must flow away from that point to the surrounding points, thus, reducing the temperature at the hot point. However, this would mean that the temperature would change with time, a contradiction to the steady-state nature of Laplace's equation. The same reasoning applies to a minimum point.

Having stated the maximum principle, we can state and show uniqueness of Laplace's Equation in $\mathbb{R}^2$ for the Dirichlet problem. The Dirichlet problem is

$$\nabla^2 u = \frac{\partial^2 u}{\partial x^2} + \frac{\partial^2 u}{\partial y^2} = 0 \text{ in } \Omega,$$

subject to

$$u(x, y) = f(x, y) \text{ on } \partial\Omega,$$

where Ω is as given in the previous theorem. The strategy of the proof is to assume there exist two different solutions. Then, show that the solutions must be equal to each other.

Proof. Suppose there exists two solutions to the Dirichlet problem stated earlier, $u_1(x, y)$ and $u_2(x, y)$. Then let $v(x, y) = u_1(x, y) - u_2(x, y)$. Thus, $v(x, y)$ is also a harmonic function in Ω and zero on $\partial\Omega$, because $u_1(x, y) = f(x, y) = u_2(x, y)$ on $\partial\Omega$. Also, $v(x, y)$ is continuous on $\Omega \cap \partial\Omega$ since both $u_1(x, y)$ and $u_2(x, y)$ are continuous and the sum of continuous functions is continuous. Hence, by the maximum principle, $v(x, y)$ must attain its maximum and minimum values on $\partial\Omega$. Thus, $v(x, y) = 0$. Therefore, $v(x, y) = u_1(x, y) - u_2(x, y) = 0$, which implies $u_1(x, y) = u_2(x, y)$. $\qquad\square$

Now that we have shown that the solution to Laplace's equation is unique, we solve Laplace's equation in Cartesian coordinate system.

5.6.3   Laplace's Equation in Cartesian Coordinate System

Consider the problem of steady-state temperature distribution in a rectangular plate with perfect lateral insulation, no source, and length L and height H. Also, suppose the boundary conditions are nonhomogeneous Dirichlet conditions. Figure (5.10) shows the indicated conditions. The mathematical formulation for this problem is the following:

$$\nabla^2 u(x, y) = \nabla^2 u(x, y) = \frac{\partial^2 u(x, y)}{\partial x^2} + \frac{\partial^2 u(x, y)}{\partial y^2} = 0 \qquad (5.105)$$

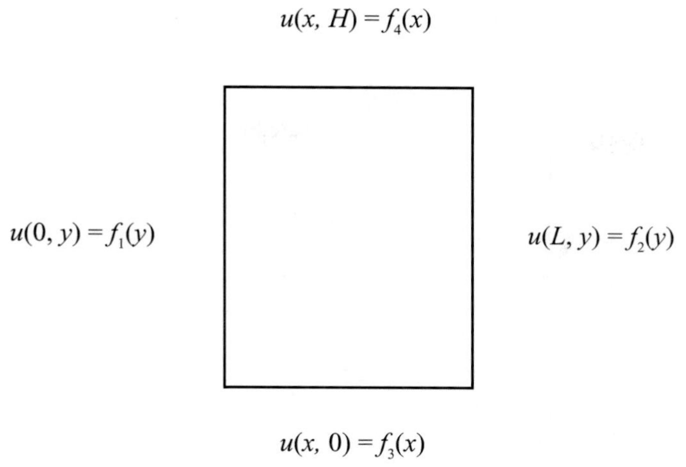

$$u(x, H) = f_4(x)$$

$$u(0, y) = f_1(y) \qquad\qquad\qquad u(L, y) = f_2(y)$$

$$u(x, 0) = f_3(x)$$

Figure 5.10: Heat conduction in a rectangular plate of length L and height H.

subject to the BCs

$$\begin{cases} u(0,y) = f_1(y) \\[2mm] u(L,y) = f_2(y) \\[2mm] u(x,0) = f_3(x) \\[2mm] u(x,H) = f_4(x). \end{cases} \tag{5.106}$$

The only solution technique you currently have to solve a PDE is based on separation of variables, which requires a linear homogeneous PDE and BCs. The BCs, Equation (5.106), are linear, but not homogeneous. Thus, separation of variables does not appear to help us. However, suppose we have four separate solutions to Equation (5.105). Then, by the principle of superposition, the sum of the four solutions would also be a solution. Thus, consider

$$u(x,y) = u_1(x,y) + u_2(x,y) + u_3(x,y) + u_4(x,y). \tag{5.107}$$

We really need to know two things about the four solutions. First, what do the four solutions look like? Second, we know the solutions satisfy Equation (5.105), but how do they satisfy the BCs, Equation (5.106)? The answer to both of these questions is tied together. We let each of the solutions, $u_i(x,y)$, $i = 1, \ldots, 4$, have one nonhomogeneous boundary condition from Equation (5.106) and three homogeneous boundary conditions. Therefore, we have the following solutions, which are graphically illustrated in Figure (5.11):

$\nabla^2 u_1(x,y) = 0$	$\nabla^2 u_2(x,y) = 0$	$\nabla^2 u_3(x,y) = 0$	$\nabla^2 u_4(x,y) = 0$
subject to	subject to	subject to	subject to
$u_1(0,y) = f_1(y)$	$u_2(0,y) = 0$	$u_3(0,y) = 0$	$u_4(0,y) = 0$
$u_1(L,y) = 0$	$u_2(L,y) = f_2(y)$	$u_3(L,y) = 0$	$u_4(L,y) = 0$
$u_1(x,0) = 0$	$u_2(x,0) = 0$	$u_3(x,0) = f_3(x)$	$u_4(x,0) = 0$
$u_1(x,H) = 0$	$u_2(x,H) = 0$	$u_3(x,H) = 0$	$u_4(x,H) = f_4(x)$

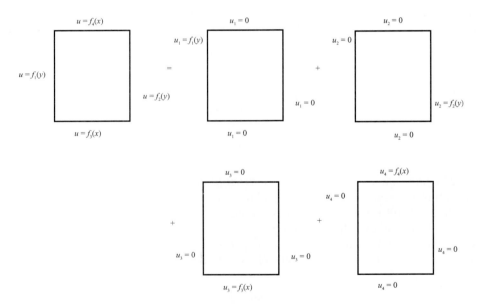

Figure 5.11: Assumption of $u(x,y) = u_1(x,y) + u_2(x,y) + u_3(x,y) + u_4(x,y)$.

From our original supposition, we know the sum of the u_is is a solution to Equation (5.105), and by our assumption about each solution, we know, for instance, at $y = H$ that

$$u_1(x,H) = u_2(x,H) = u_3(x,H) = 0$$

and

$$u_4 = f_4(x).$$

Thus, applying the principle of superposition at $y = H$ yields

$$u(x, H) = u_1(x, H) + u_2(x, H) + u_3(x, H) + u_4(x, H) = f_4(x),$$

the nonhomogeneous condition we require for $u(x, y)$. Similarly, we can show $u(x, 0) = f_3(x)$, $u(L, y) = f_2(y)$, and $u(0, y) = f_1(y)$. Therefore, the original problem may be broken down into four separate problems, each with three homogeneous BCs and one nonhomogeneous BC. Can we use the separation of variables technique on the four separate problems? If we consider each of the four separate problems as a boundary value problem, then we cannot. However, if we consider the variable with the nonhomogeneous condition as a "time-like" variable, then we can use separation of variables.

Since the method of solution is nearly the same for any of the four, the solution for $u_3(x, y)$ is given here. $u_2(x, y)$, and $u_4(x, y)$. Figure (5.12) depicts the rectangular plate $u_3(x, y)$. It demonstrates how a simple shift of a coordinate aids us in the solution. The coordinate shift is possible because the solution of the resulting ODE is invariant under a translation. That is, the translation does not change the solution. The remaining three are left as an exercise.

$$u_3(x, H) = 0$$

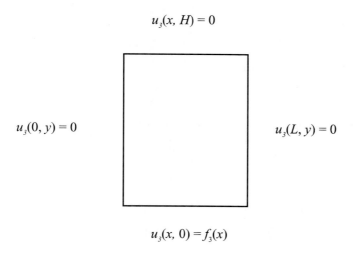

$$u_3(0, y) = 0 \qquad\qquad u_3(L, y) = 0$$

$$u_3(x, 0) = f_3(x)$$

Figure 5.12: Heat conduction in a rectangular plate, $u_3(x, y)$, of length L and height H.

For $u_3(x, y)$, we have

$$\nabla^2 u_3(x, y) = 0 \qquad\qquad (5.108)$$

subject to

$$
\begin{cases}
u_3(0, y) = 0 \\[2mm]
u_3(L, y) = 0 \\[2mm]
u_3(x, 0) = f_3(x) \\[2mm]
u_3(x, H) = 0.
\end{cases}
\tag{5.109}
$$

We assume at $x = 0$ and $x = L$ that we have boundary conditions and at $y = 0$ and $y = H$ that we have "time-like" conditions. Letting $u_3(x, y) = \varphi_3(y)G_3(x)$ and separating in the usual manner we arrive at the two ODEs

$$
\varphi_3''(x) = -\lambda \varphi_3(x)
\tag{5.110}
$$

subject to

$$
\begin{cases}
\varphi_3(0) = 0 \\[2mm]
\varphi_3(L) = 0,
\end{cases}
$$

and

$$
G_3''(y) = \lambda G(y)
\tag{5.111}
$$

subject to

$$
\begin{cases}
G_3(0) = f_3(x) \\[2mm]
G_3(H) = 0.
\end{cases}
$$

The spatial ODE, Equation (5.110), subject to its BCs, has the solution

$$
\left.
\begin{aligned}
\lambda_n &= \left(\frac{n\pi}{L}\right)^2 \\[4mm]
\varphi_{3_n}(x) &= \sin\frac{n\pi x}{L}
\end{aligned}
\right\}
, \; n = 1, 2, 3, \ldots
\tag{5.112}
$$

The second ODE, Equation (5.111) that is now stated as

$$
G_{3_n}''(y) = \lambda_n G_{3_n}(y) = \left(\frac{n\pi}{L}\right)^2 G_{3_n}(y),
\tag{5.113}
$$

subject to

$$
\begin{cases}
G_{3_n}(0) = f_3(x) \\[2mm]
G_{3_n}(H) = 0,
\end{cases}
$$

has the solution

$$G_{3_n}(y) = c_{3_n} \cosh \frac{n\pi y}{L} + d_{3_n} \sinh \frac{n\pi y}{L}. \tag{5.114}$$

Equation (5.114) really does not help us when we want to use the condition $G_{3_n}(H) = 0$. However, a simple shift in the y coordinate may help us. Once we shift the y coordinate, we must make sure it solves Equation (5.113). We replace y with $y - H$ in Equation (5.114). Thus, Equation (5.114) becomes

$$G_{3_n}(y) = c_{3_n} \cosh \frac{n\pi(y - H)}{L} + d_{3_n} \sinh \frac{n\pi(y - H)}{L}. \tag{5.115}$$

As practice, you should show that Equation (5.115) solves Equation (5.113). We may now apply the condition $G_{3_n}(H) = 0$, which yields

$$G_{3_n}(H) = 0 = c_{3_n}.$$

Therefore, we have

$$G_{3_n}(y) = d_{3_n} \sinh \frac{n\pi(y - H)}{L}. \tag{5.116}$$

Combining the solutions for $\varphi_{3_n}(y)$ and $G_{3_n}(y)$, and using the principle of superposition, we arrive at

$$u_3(x, y) = \sum_{n=1}^{\infty} d_{3_n} \sinh \frac{n\pi(y - H)}{L} \sin \frac{n\pi x}{L}.$$

Using the "time-like" BCs, $u_3(x, 0) = f_3(x)$, we may determine the last unknown, d_{3_n}, by using the orthogonality of the Fourier sine series. Thus, we have

$$u_3(x, 0) = f_3(x) = \sum_{n=1}^{\infty} d_{3_n} \sinh \frac{n\pi(-H)}{L} \sin \frac{n\pi x}{L}, \tag{5.117}$$

which implies

$$d_{3_n} = \frac{-2}{L \sinh \frac{n\pi H}{L}} \int_0^L f_3(x) \sin \frac{n\pi x}{L} \, dx. \tag{5.118}$$

The solutions for $u_1(x, y)$, $u_2(x, y)$, and $u_4(x, y)$ are determined similarly, and are left as an exercise. *Note:* For $u_1(x, y)$, a coordinate shift must be determined. You should prove to yourself that the translated solution of the ODE does not change the differential equation. For $u_2(x, y)$ and $u_4(x, y)$, no shift is required. Once all four solutions are found we use the principle of superposition to combine the four solutions, which yields the solution for $u(x, y)$. Also, please remember that Laplace's equation is the primary representative of the elliptic class of linear second-order PDEs.

In Chapter 10, we will again meet Laplace's equation. Only then we will work in polar, cylindrical, and spherical coordinate systems.

EXERCISES 5.6

5.6.1. Show that $u(x, y) = xy$ and $u(x, y) = x^2 + y^2$ are harmonic in $\mathbb{R}^2$.

5.6.2. Find the complete solution of

$$\nabla^2 u(x, y) = \frac{\partial^2 u(x, y)}{\partial x^2} + \frac{\partial^2 u(x, y)}{\partial y^2} = 0,$$

subject to

$$\begin{cases} u(0, y) = f_1(y) \\[2mm] u(L, y) = f_2(y) \\[2mm] u(x, 0) = f_3(x) \\[2mm] u(x, H) = f_4(x). \end{cases}$$

Remember that $u_3(x, y)$ was solved in the text.

5.6.3. Find the complete solution of

$$\nabla^2 u(x, y) = \frac{\partial^2 u(x, y)}{\partial x^2} + \frac{\partial^2 u(x, y)}{\partial y^2} = 0,$$

subject to

$$\begin{cases} \dfrac{\partial u(0, y)}{\partial x} = f_1(y) \\[3mm] \dfrac{\partial u(L, y)}{\partial x} = f_2(y) \\[3mm] \dfrac{\partial u(x, 0)}{\partial y} = f_3(x) \\[3mm] \dfrac{\partial u(x, H)}{\partial y} = f_4(x). \end{cases}$$

5.6.4. Consider Laplace's equation inside a rectangle. Suppose $f_1(y) = f_3(x) = f_4(x) = 0$ and $f_2(y) = y + 1$. Determine the solution.

5.6.5. Given Laplace's equation in Cartesian coordinates

$$\nabla^2 u(x, y) = \frac{\partial^2 u(x, y)}{\partial x^2} + \frac{\partial^2 u(x, y)}{\partial y^2} = 0,$$

find the solution for each of the following BCs:

(1) $u(0, y) = 0$, $\dfrac{\partial u(2, y)}{\partial x} = y$, $u(x, 0) = 0$, and $u(x, 3) = x$.

(2) $u(0, y) = y^2$, $\dfrac{\partial u(\pi, y)}{\partial x} = 0$, $\dfrac{\partial u(x, 0)}{\partial y} = 0$, and $u(x, 2\pi) = \sin x$.

(3) $\dfrac{\partial u(0, y)}{\partial x} = 0$, $u(\dfrac{\pi}{2}, y) = \cos y$, $\dfrac{\partial u(x, 0)}{\partial y} = 0$, and $u(x, 1) = 0$.

(4) $\dfrac{\partial u(0, y)}{\partial x} = 0$, $u(1, y) = \sin y$, $\dfrac{\partial u(x, 0)}{\partial y} = 0$, and $\dfrac{\partial u(x, \pi)}{\partial y} = x^2$.

(5) $u(0, y) = 0$, $u(2, y) - \dfrac{\partial u(2, y)}{\partial x} = 0$, $u(x, 0) = 0$, and $u(x, 3) = 0$.

(6) $u(0, y) - \dfrac{\partial u(0, y)}{\partial x} = 0$, $u(\dfrac{\pi}{2}, y) = \cos y$, $u(x, 0) = 0$, and $\dfrac{\partial u(x, \pi)}{\partial y} = 0$.

(7) $u(0, y) = 0$, $u(\dfrac{\pi}{2}, y) = \sin y$, $u(x, 0) - \dfrac{\partial u(x, 0)}{\partial y} = 0$, and $u(x, 3) = 0$.

(8) $u(0, y) = 0$, $\dfrac{\partial u(\pi, y)}{\partial x} = 0$, $u(x, 0) = 0$, and $u(x, 1) - \dfrac{\partial u(x, 1)}{\partial y} = 0$.

5.6.6. If $u_1(x, y)$ is a solution of Laplace's equation, prove that the partial derivative of $u_1(x, y)$ with respect to one or more of the rectangular coordinates (for example, $\dfrac{\partial u_1}{\partial x}$, $\dfrac{\partial^2 u_1}{\partial x^2}$, and $\dfrac{\partial^2 u_1}{\partial y \partial x}$) are also a solution.

5.6.7. Formulate and solve the general Laplace's equation in a three-dimensional parallelepiped with BCs of the first kind (Dirichlet conditions).

5.6.8. Formulate and solve the general Laplace's equation in a three-dimensional parallelepiped with BCs of the second kind (Neumann conditions).

5.6.9. Formulate and solve the general Laplace's equation in a three-dimensional parallelepiped Dirichlet BCs on the boundaries $x = 0$ and $x = L$ and Neumann BCs on the boundaries $y = 0$ and $y = H$.

5.6.10. Formulate and solve the general Laplace's equation in a three-dimensional parallelepiped Dirichlet BCs on the boundaries $x = 0$ and $y = 0$ and Neumann BCs on the boundaries $x = L$ and $y = H$.

5.6.11. Formulate and solve the general Laplace's equation in a three-dimensional parallelepiped Dirichlet BCs on the boundaries $x = L$ and $y = 0$ and Neumann BCs on the boundaries $x = 0$ and $y = H$.

5.6.12. Consider a region bounded by conducting plates $x = 0$, $y = 0$, and $y = H$. Suppose the plate $x = 0$ is charged to a potential V, the plates $y = 0$ and $y = H$ are earthed, and there are no charges inside the region, find the electrostatic field inside the region.

5.6.13. PROJECT:[3] This project models the fluid in the cochlea surrounding the basilar membrane, which is part of the inner human ear. It is assumed that the fluid in the cochlea is incompressible and inviscid (not thick). If the equations are nondimensionalize we arrive at

$$\frac{\partial^2 u(x,y)}{\partial x^2} + \frac{\partial^2 u(x,y)}{\partial y^2} = 0$$

subject to

$$\frac{\partial u(0,y)}{\partial x} = 1$$

$$u(1,y) = 0$$

$$u(x,0) = \frac{2u(x,0)}{I}$$

$$u(x,H) = 0.$$

where I is called the impedance in the damped harmonic oscillator.

(1) Find the general solution.

(2) Show that after truncating the series at N terms, multiplying by $\cos(m\pi x)$, and integrating from 0 to 1, we obtain the system of linear equations

$$\sum_{n=0}^{N} A_n \alpha_{nm} = f_m,$$

where

$$\alpha_{nm} = 2\cosh(n\pi L) \int_0^1 \frac{\cos(n\pi x)\cos(m\pi x)}{I} \, dx - \frac{1}{2} n\pi \sinh(n\pi H)\delta_{nm}$$

where $\delta_{ij} = 1$ if $i = j$ and 0 otherwise and

$$f_m = L\delta_{m0} - \int_0^1 \frac{x(2-x)\cos(m\pi x)}{I} \, dx.$$

---

[3] Adapted from James Keener and James Sneyd, Mathematical Physiology, ©1998 by Springer-Verlag, New York, pp. 707-711. Reprinted by permission.

Chapter 6

The Calculus of Fourier Series

6.1 INTRODUCTION

Chapter 4 introduced you to Fourier series. You learned that a Fourier Series is a linear combination of orthonormal functions. This linear combination can represent any function in the function space of piecewise smooth functions.

In Chapter 5, you solved PDEs by separation of variables. The separation of variables technique led to a Fourier series solution for the PDE. Also, the Fourier series had a function of time, t, as part of the solution. The natural question to ask is, how do we know the Fourier series solution of the PDE is the actual solution to the PDE? This is a very important question, and it must be answered. There are really two questions here: (1) how and when can a Fourier series be differentiated and integrated?; (2) If a Fourier series can be differentiated and integrated, what is the result?

Before we attempt differentiating or integrating a Fourier series, we must determine if a Fourier series is a function. We'll do that in the next section.

6.2 FOURIER SERIES REPRESENTATION OF A FUNCTION: FOURIER SERIES AS A FUNCTION

What is a function? Here's one definition:

Definition 36. *A function f from a set A, called the domain, to a set B, called the range, is a rule of correspondence that assigns to each x in a certain subset D of A, a uniquely determined element $f(x)$ of B.*

How does this definition apply to a Fourier series? To answer this question, we must identify everything in the definition of a function with a particular object in our study of Fourier series.

First, the sets A and B in the definition are the set of real numbers, $\mathbb{R}$, for Fourier series. *Note:* Both sets A and B can be the same. Second, the subset D of A in the definition refers to the interval $[-L, L]$, a subset of $\mathbb{R}$ for Fourier series. Finally, the very nature of Fourier series is to assign to each element in the interval $[-L, L]$ a unique element in the range. Thus, a Fourier series is a function. Although knowing a Fourier series is a function is important, the really important concept is knowing if the Fourier series is a continuous function.

From Chapter 4, we know the Fourier series representation of a function on the interval $[-L, L]$ converges to the function at all points the function is continuous and to the average value at any point of discontinuity. Consider the following two examples.

EXAMPLE 6.1. Consider

$$f(x) = \begin{cases} \dfrac{-\pi}{2}, & -\pi \leq x < \dfrac{-\pi}{2} \\[2mm] x, & \dfrac{-\pi}{2} \leq x < \dfrac{\pi}{2} \\[2mm] \pi - x, & \dfrac{\pi}{2} \leq x \leq \pi. \end{cases}$$

The function $f(x)$ is a piecewise smooth continuous function. Its Fourier series representation is

$$f(x) = \frac{-\pi}{16} + \sum_{n=1}^{\infty} \left(\frac{-1}{\pi n^2} \left((-1)^n - \cos \frac{n\pi}{2} \right) \right) \cos nx$$

$$+ \sum_{n=1}^{\infty} \left(\left(6 \sin \frac{n\pi}{2} - \pi (-1)^n \right) \frac{1}{2n^2 \pi} \right) \sin nx.$$

The solution is graphed in Figure (6.1) on the interval $[-3\pi, 3\pi]$. It shows the Fourier series representation of $f(x)$ converges to $\dfrac{-\pi}{4}$ at $x = -3\pi$ and $x = 3\pi$. Since Fourier series are periodic with period $2L$ (where L is the length), we know the Fourier series representation of $f(x)$ at $x = -\pi$ or π also converges to $\dfrac{-\pi}{4}$. Also, the Fourier series representation of $f(x)$ at $x = -\pi$ or π converges to $\dfrac{-\pi}{4}$, but the function equals $\dfrac{-\pi}{2}$ and 0 respectively. This means $\dfrac{-\pi}{4} = \dfrac{f(-\pi) + f(\pi)}{2}$, or the average value of the endpoints of the function. Thus, a jump discontinuity exists in the Fourier series representation for $f(x)$. Therefore, the Fourier series representation of $f(x)$ is not continuous and can not be differentiated term-by-term.

Look closely at Figure (6.1). As the Fourier series Representation of $f(x)$ approaches π, from the left or from the right, the graph seems to oscillate. Then the Fourier series representation of $f(x)$ converges to $-\dfrac{\pi}{4}$. The oscillation is known as the Gibbs phenomenon. The Gibbs phenomenon will be discussed at greater length in the last section of this chapter. ∎

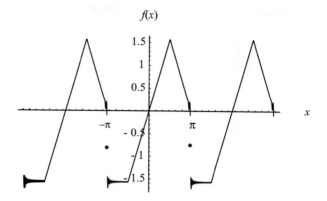

Figure 6.1: The graph of the Fourier series representation of $f(x)$.

EXAMPLE 6.2. Given

$$
g(x) = \begin{cases} \dfrac{\pi}{2}, & -\pi \leq x < \dfrac{-\pi}{2} \\[2mm] -x, & \dfrac{-\pi}{2} \leq x < \dfrac{\pi}{2} \\[2mm] 2x - \dfrac{3\pi}{2}, & \dfrac{\pi}{2} \leq x \leq \pi. \end{cases}
$$

The function $g(x)$ is a piecewise smooth continuous function. Its Fourier series

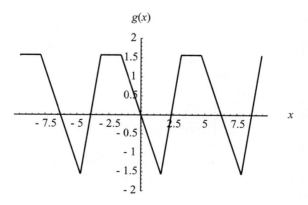

Figure 6.2: The graph of the Fourier series representation of $g(x)$.

representation is

$$g(x) = \frac{3\pi}{16} + \sum_{n=1}^{\infty} \left(\frac{2}{\pi n^2} \left((-1)^n - \cos \frac{n\pi}{2} \right) \right) \cos nx$$

$$+ \sum_{n=1}^{\infty} \left(\frac{-4}{\pi n^2} \sin \frac{n\pi}{2} \right) \sin nx.$$

The solution is graphed in Figure (6.2) on the interval $[-3\pi, 3\pi]$, which shows the Fourier series representation of $g(x)$ converges to $\frac{\pi}{2}$ at $x = -3\pi$ and $x = 3\pi$. As in the last example, since the Fourier series is periodic with period $2L$, we know that the Fourier series converges to $\frac{\pi}{2}$ at $x = -\pi$ or π . Thus, the Fourier series representation for $g(x)$ is continuous at the endpoints. And since there are no other discontinuities in the function, the Fourier series representation of $g(x)$ is continuous on the interval $[-\pi, \pi]$. ∎

Knowing when a particular Fourier series is a continuous function, without being forced to graph the Fourier series, is very important. At the end of Chapter 4, three theorems were given to clarify when a particular Fourier series is continuous. These definitions are restated below:

Theorem 37. *If $f(x)$ is a piecewise smooth continuous function on the interval $[-L, L]$ and $f(-L) = f(L)$, then the Fourier series of the function $f(x)$ is continuous on $[-L, L]$.*

Theorem 38. *If $f(x)$ is a piecewise smooth continuous function on the interval $[0, L]$, then the Fourier cosine series representation of the function $f(x)$ is continuous on $[0, L]$.*

Theorem 39. *If $f(x)$ is a piecewise smooth continuous function on the interval $[0, L]$ and $f(0) = f(L) = 0$, then the Fourier sine series representation of the function $f(x)$ is continuous on $[0, L]$.*

Notice in all three theorems the function $f(x)$ is continuous and piecewise smooth on the interval. Both characteristics are required. A function may be continuous on the interval $[-L, L]$ but not be piecewise smooth. A good example is the $\tan x$ on the interval $[-\frac{\pi}{2}, \frac{\pi}{2}]$. At $-\frac{\pi}{2}$ and $\frac{\pi}{2}$ the $\tan x$ goes to $-\infty$ and ∞, respectively. Likewise, if a function is piecewise smooth and not continuous, the Fourier series representation will have jump discontinuities. Thus, the Fourier series as a function will not be continuous. Also note, for two of the theorems, specific values for the function at the end points are given. The endpoint values are very important because they insure the Fourier series representation of the function in Theorems (37 and 39) converges to the value at the endpoints of the function.

Why is continuity of a Fourier series so important? The answer is quite straight-forward. Since you cannot differentiate a function on an interval if the function has a jump discontinuity, you cannot even consider differentiation of a Fourier series on an interval unless the Fourier series is continuous. This is the first step toward differentiation of a Fourier series; actual differentiation of a Fourier series is the second step, which we go to now.

6.3 DIFFERENTIATION OF FOURIER SERIES

When solving a linear second-order ODE, you were taught to plug the solution back into the ODE to see if you actually got the correct solution. For example, consider

$$y''(x) - 3y'(x) + 2y(x) = 0, \tag{6.1}$$

where the solution is

$$y(x) = c_1 e^{2x} + c_2 e^x. \tag{6.2}$$

Using Equation (6.2), you found

$$y'(x) = 2c_1 e^{2x} + c_2 e^x \tag{6.3}$$

and

$$y''(x) = 4c_1 e^{2x} + c_2 e^x. \tag{6.4}$$

Replacing $y'(x)$ and $y''(x)$ in Equation (6.1) yields

$$4c_1 e^{2x} + c_2 e^{-x} - 3\left(2c_1 e^{2x} + c_2 e^x\right) + 2\left(c_1 e^{2x} + c_2 e^x\right) = 0. \tag{6.5}$$

The obvious conclusion is that you found the correct solution.

Does a similar procedure prove that the solution to a linear second-order PDE is the correct solution? Consider the PDE:

$$\frac{\partial u}{\partial t} = \frac{\partial^2 u}{\partial x^2}$$

subject to

$$u(-L, t) = u(L, t)$$

$$\frac{\partial u(-L, t)}{\partial t} = \frac{\partial u(L, t)}{\partial t}$$

and

$$u(x, 0) = f(x).$$

The solution is

$$u(x, t) = a_0 + \sum_{n=1}^{\infty} \left[a_n e^{-\left(\frac{n\pi}{L}\right)^2 t} \cos\frac{n\pi x}{L} + b_n e^{-\left(\frac{n\pi}{L}\right)^2 t} \sin\frac{n\pi x}{L} \right], \tag{6.6}$$

where the three Fourier coefficients are given by

$$a_0 = \frac{1}{2L} \int_{-L}^{L} f(x) \, dx,$$

$$a_n = \frac{1}{L} \int_{-L}^{L} f(x) \cos \frac{n\pi x}{L} \, dx,$$

and

$$b_n = \frac{1}{L} \int_{-L}^{L} f(x) \sin \frac{n\pi x}{L} \, dx.$$

Note: The Fourier coefficients do not depend on the variable x.

Following the same procedure, which exists for second-order linear ODEs, and noting the Fourier series solution in Equation (6.6) is continuous in the spatial variable by Definition (37) and is assumed to be continuous for the time variable, we must find

$$\frac{\partial u}{\partial t} = \frac{\partial}{\partial t} \left[a_0 + \sum_{n=1}^{\infty} \left[a_n e^{-\left(\frac{n\pi}{L}\right)^2 t} \cos \frac{n\pi x}{L} + b_n e^{-\left(\frac{n\pi}{L}\right)^2 t} \sin \frac{n\pi x}{L} \right] \right]$$

$$= \frac{\partial}{\partial t} \left[a_0 \right] + \frac{\partial}{\partial t} \left[\sum_{n=1}^{\infty} \left[a_n e^{-\left(\frac{n\pi}{L}\right)^2 t} \cos \frac{n\pi x}{L} + b_n e^{-\left(\frac{n\pi}{L}\right)^2 t} \sin \frac{n\pi x}{L} \right] \right], \qquad (6.7)$$

$$\frac{\partial u}{\partial x} = \frac{\partial}{\partial x} \left[a_0 + \sum_{n=1}^{\infty} \left[a_n e^{-\left(\frac{n\pi}{L}\right)^2 t} \cos \frac{n\pi x}{L} + b_n e^{-\left(\frac{n\pi}{L}\right)^2 t} \sin \frac{n\pi x}{L} \right] \right]$$

$$= \frac{\partial}{\partial x} \left[a_0 \right] + \frac{\partial}{\partial x} \left[\sum_{n=1}^{\infty} \left[a_n e^{-\left(\frac{n\pi}{L}\right)^2 t} \cos \frac{n\pi x}{L} + b_n e^{-\left(\frac{n\pi}{L}\right)^2 t} \sin \frac{n\pi x}{L} \right] \right], \qquad (6.8)$$

and

$$\frac{\partial^2 u}{\partial x^2} = \frac{\partial}{\partial x} \left\{ \frac{\partial}{\partial x} \left[a_0 + \sum_{n=1}^{\infty} \left[a_n e^{-\left(\frac{n\pi}{L}\right)^2 t} \cos \frac{n\pi x}{L} + b_n e^{-\left(\frac{n\pi}{L}\right)^2 t} \sin \frac{n\pi x}{L} \right] \right] \right\}$$

$$= \frac{\partial}{\partial x} \left\{ \frac{\partial}{\partial x} \left[a_0 \right] \right\}$$

$$+ \frac{\partial}{\partial x} \left\{ \frac{\partial}{\partial x} \left[\sum_{n=1}^{\infty} \left[a_n e^{-\left(\frac{n\pi}{L}\right)^2 t} \cos \frac{n\pi x}{L} + b_n e^{-\left(\frac{n\pi}{L}\right)^2 t} \sin \frac{n\pi x}{L} \right] \right] \right\}. \qquad (6.9)$$

What we need to perform on the left side of Equations (6.7, 6.8, and 6.9) is quite straightforward. However, the right sides of Equations (6.7, 6.8, and 6.9) present interesting questions. How do you differentiate an infinite series, and what do you get for a solution? Also, an interesting follow-on question is if we can differentiate a Fourier series, which is an infinite series, can we differentiate all infinite series by the same method? This is not a course in general infinite series, but the answer to the last question depends on whether the infinite series is continuous and convergent, as well as if the derivative of the infinite series is uniformly convergent. For Definitions (37, 38, and 39), the Fourier series representation of the function is continuous, converging uniformly to the function. This simple fact makes working with Fourier series much simpler. It means differentiating a Fourier series is possible.

Since differentiation of a Fourier series is possible under certain conditions, what is the end result of differentiation of a Fourier series, and how do we perform the differentiation? We perform the differentiation term-by-term, which must be shown. The end result is hopefully another Fourier series, which is also the Fourier series representation of the derivative of the original function.

To prove these statements, we consider a general Fourier series representation of a function, $u(x, t)$, where all the Fourier coefficients depend on the variable t:

$$u(x,t) = a_0(t) + \sum_{n=1}^{\infty} \left[a_n(t) \cos \frac{n\pi x}{L} + b_n(t) \sin \frac{n\pi x}{L} \right]. \tag{6.10}$$

Also, we assume that the function $u(x, t)$ is continuous for both time and space variables and $u(-L, t) = u(L, t)$. These conditions insure the Fourier series representation of $u(x, t)$ is a continuous function on the interval $[-L, L]$. Next, we assume that $u(x, t)$ is continuously differentiable with respect to both variables. Finally, we hope that the derivative with respect to either the spatial variable, x, or the time variable, t, of a Fourier series is another Fourier series, which converges to the derivative of the original function, $u(x, t)$. Then, we assume that the derivative with respect to either variable of $u(x, t)$ is a piecewise smooth function. We find the derivative with respect to time first, then we state the resulting theorem. Next we find the derivative with respect to the spatial variable, and we state the resulting theorem.

We have

$$u(x,t) = a_0(t) + \sum_{n=1}^{\infty} \left[a_n(t) \cos \frac{n\pi x}{L} + b_n(t) \sin \frac{n\pi x}{L} \right], \tag{6.11}$$

where $u(x, t)$ is continuously differentiable with respect to t, and the Fourier series representation of $u(x, t)$ is continuous. Also, we know that the derivative of $u(x, t)$, $\dfrac{\partial u(x, t)}{\partial t}$, is piecewise smooth, and thus, $\dfrac{\partial u(x, t)}{\partial t}$ may be written as a Fourier series, which is

$$\frac{\partial u(x,t)}{\partial t} = A_0(t) + \sum_{n=1}^{\infty} \left[A_n(t) \cos \frac{n\pi x}{L} + B_n(t) \sin \frac{n\pi x}{L} \right]. \tag{6.12}$$

Note: The coefficients of Equation (6.12) are different from the coefficients in Equation (6.11).

It suffices to show term-by-term differentiation of a Fourier series with respect to the parameter t is possible if we can show that the Fourier coefficients of Equation (6.12) are derived from the Fourier coefficients of Equation (6.11).

The equations for the Fourier coefficients corresponding to Equation (6.11) are

$$a_0(t) = \frac{1}{2L} \int_{-L}^{L} u(x,t) \, dx, \tag{6.13}$$

$$a_n(t) = \frac{1}{L} \int_{-L}^{L} u(x,t) \cos \frac{n\pi x}{L} \, dx, \tag{6.14}$$

and

$$b_n(t) = \frac{1}{L} \int_{-L}^{L} u(x,t) \sin \frac{n\pi x}{L} \, dx. \tag{6.15}$$

The equations for the Fourier coefficients corresponding to Equation (6.12) are

$$A_0(t) = \frac{1}{2L} \int_{-L}^{L} \frac{\partial u(x,t)}{\partial t} \, dx, \tag{6.16}$$

$$A_n(t) = \frac{1}{L} \int_{-L}^{L} \frac{\partial u(x,t)}{\partial t} \cos \frac{n\pi x}{L} \, dx, \tag{6.17}$$

and

$$B_n(t) = \frac{1}{L} \int_{-L}^{L} \frac{\partial u(x,t)}{\partial t} \sin \frac{n\pi x}{L} \, dx. \tag{6.18}$$

The form of the integrals in Equations (6.16, 6.17, and 6.18) should remind you of Leibniz's theorem in Chapter 2, which we restate here.

Theorem 40. *Suppose $f(x,t)$ and the partial derivative $\dfrac{\partial f(x,t)}{\partial t}$ are continuous in some region of the xt-plane where $a \leq x \leq b$, then*

$$\frac{d}{dt}\left[\int_a^b f(x,t) \, dx\right] = \int_a^b \frac{\partial f(x,t)}{\partial t} \, dx.$$

Thus, if we further assume that $\dfrac{\partial u(x,t)}{\partial t}$ is continuous in some region of the xt-plane where $-L \leq x \leq L$, then Equations (6.16, 6.17, and 6.18) become

$$A_0(t) = \frac{d}{dt}\left[\frac{1}{2L} \int_{-L}^{L} u(x,t) \, dx\right], \tag{6.19}$$

$$A_n(t) = \frac{d}{dt}\left[\frac{1}{L}\int_{-L}^{L} u(x,t)\cos\frac{n\pi x}{L}\,dx\right], \tag{6.20}$$

and

$$B_n(t) = \frac{d}{dt}\left[\frac{1}{L}\int_{-L}^{L} u(x,t)\sin\frac{n\pi x}{L}\,dx\right]. \tag{6.21}$$

Using Equations (6.13, 6.14, 6.15), and substituting equivalent terms, we find that Equations (6.19, 6.20, and 6.21), respectively, become

$$A_0(t) = \frac{d}{dt}\left[\frac{1}{2L}\int_{-L}^{L} u(x,t)\,dx\right] = \frac{d}{dt}[a_0(t)] = a_0'(t),$$

$$A_n(t) = \frac{d}{dt}\left[\frac{1}{L}\int_{-L}^{L} u(x,t)\cos\frac{n\pi x}{L}\,dx\right] = \frac{d}{dt}[a_n(t)] = a_n'(t),$$

and

$$B_n(t) = \frac{d}{dt}\left[\frac{1}{L}\int_{-L}^{L} u(x,t)\sin\frac{n\pi x}{L}\,dx\right] = \frac{d}{dt}[b_n(t)] = b_n'(t).$$

Thus, the Fourier coefficients of Equation (6.12) are derived from the Fourier coefficients of Equation (6.11). Therefore, term-by-term differentiation with respect to a parameter t is valid, and we have proved the following theorem:

Theorem 41. *If $u(x,t)$ and $\dfrac{\partial u(x,t)}{\partial t}$ are continuous functions, then the Fourier series representation of $u(x,t)$ on the interval $[-L, L]$,*

$$u(x,t) = a_0(t) + \sum_{n=1}^{\infty}\left[a_n(t)\cos\frac{n\pi x}{L} + b_n(t)\sin\frac{n\pi x}{L}\right],$$

may be differentiated term-by-term with respect to the parameter t, and the result is the Fourier series representation of the derivative of $u(x,t)$ with respect to t,

$$\frac{\partial u(x,t)}{\partial t} = a_0'(t) + \sum_{n=1}^{\infty}\left[a_n'(t)\cos\frac{n\pi x}{L} + b_n'(t)\sin\frac{n\pi x}{L}\right].$$

Next, we show that term-by-term differentiation of a Fourier series with respect to the variable x is valid. Again, we have

$$u(x,t) = a_0(t) + \sum_{n=1}^{\infty}\left[a_n(t)\cos\frac{n\pi x}{L} + b_n(t)\sin\frac{n\pi x}{L}\right], \tag{6.22}$$

where $u(x,t)$ is continuously differentiable with respect to x, $u(-L,t) = u(L,t)$. Therefore, the Fourier series representation of $u(x,t)$ is continuous. Also, we assume the derivative of $u(x,t)$, $\dfrac{\partial u(x,t)}{\partial x}$, is piecewise smooth, and thus, $\dfrac{\partial u(x,t)}{\partial x}$ may be written as a Fourier series, which is

$$\frac{\partial u(x,t)}{\partial x} = \alpha_0(t) + \sum_{n=1}^{\infty} \left[\alpha_n(t) \cos \frac{n\pi x}{L} + \beta_n(t) \sin \frac{n\pi x}{L} \right]. \tag{6.23}$$

Again, it suffices to show that the Fourier coefficients of Equation (6.23) may be derived from the Fourier coefficients of Equation (6.22). The equations for the Fourier coefficients corresponding to Equation (6.22) are

$$a_0(t) = \frac{1}{2L} \int_{-L}^{L} u(x,t) \, dx, \tag{6.24}$$

$$a_n(t) = \frac{1}{L} \int_{-L}^{L} u(x,t) \cos \frac{n\pi x}{L} \, dx, \tag{6.25}$$

and

$$b_n(t) = \frac{1}{L} \int_{-L}^{L} u(x,t) \sin \frac{n\pi x}{L} \, dx. \tag{6.26}$$

The equations for the Fourier coefficients corresponding to Equation (6.23) are

$$\alpha_0(t) = \frac{1}{2L} \int_{-L}^{L} \frac{\partial u(x,t)}{\partial x} \, dx, \tag{6.27}$$

$$\alpha_n(t) = \frac{1}{L} \int_{-L}^{L} \frac{\partial u(x,t)}{\partial x} \cos \frac{n\pi x}{L} \, dx, \tag{6.28}$$

and

$$\beta_n(t) = \frac{1}{L} \int_{-L}^{L} \frac{\partial u(x,t)}{\partial x} \sin \frac{n\pi x}{L} \, dx. \tag{6.29}$$

Integrating Equation (6.27) yields

$$\alpha_0(t) = \frac{1}{2L} \int_{-L}^{L} \frac{\partial u(x,t)}{\partial x} \, dx = \frac{1}{2L} u(x,t) \Big|_{-L}^{L}$$

$$= \frac{1}{2L} \left[u(L,t) - u(-L,t) \right] = 0,$$

since $u(L,t) = u(-L,t)$.

The Fourier coefficient $\alpha_0(t) = 0$ is not unexpected. Remember, a function of t is considered a constant when finding the derivative with respect to the variable x.

For Equation (6.28),

$$\alpha_n(t) = \frac{1}{L} \int_{-L}^{L} \frac{\partial u(x,t)}{\partial x} \cos \frac{n\pi x}{L} \, dx,$$

we employ the method of integration by parts with $dv = \dfrac{\partial u(x,t)}{\partial x} \, dx$ and $w = \cos \dfrac{n\pi x}{L}$. This yields

$$\alpha_n(t) = \frac{1}{L} \left[\left(u(x,t) \cos \frac{n\pi x}{L} \right)_{-L}^{L} + \frac{n\pi}{L} \int_{-L}^{L} u(x,t) \sin \frac{n\pi x}{L} \, dx \right]$$

$$= \frac{1}{L} \left[u(L,t) \cos n\pi - u(-L,t) \cos n\pi + \frac{n\pi}{L} \int_{-L}^{L} u(x,t) \sin \frac{n\pi x}{L} \, dx \right]$$

$$= \frac{1}{L} \left[(u(L,t) - u(-L,t)) \cos n\pi + \frac{n\pi}{L} \int_{-L}^{L} u(x,t) \sin \frac{n\pi x}{L} \, dx \right]$$

$$= \frac{1}{L} \left[\frac{n\pi}{L} \int_{-L}^{L} u(x,t) \sin \frac{n\pi x}{L} \, dx \right]. \tag{6.30}$$

Rewriting Equation (6.30) yields

$$\alpha_n(t) = \frac{n\pi}{L} \left[\frac{1}{L} \int_{-L}^{L} u(x,t) \sin \frac{n\pi x}{L} \, dx \right]. \tag{6.31}$$

Using Equation (6.26) and substituting equivalent terms, Equation (6.31) becomes

$$\alpha_n(t) = \frac{n\pi}{L} \left[\frac{1}{L} \int_{-L}^{L} u(x,t) \sin \frac{n\pi x}{L} \, dx \right] = \frac{n\pi}{L} b_n(t).$$

Thus, $\alpha_n(t)$, in Equation (6.23), depends on the coefficient $b_n(t)$, in Equation (6.22). Again, this is not unexpected since $\dfrac{d}{dx} \left[\sin \dfrac{n\pi x}{L} \right] = \dfrac{n\pi}{L} \cos \dfrac{n\pi x}{L}$.

Similarly, we find the Fourier coefficient $\beta_n(t)$ depends on the coefficient $a_n(t)$. Thus, we have proved term-by-term differentiation of a Fourier series representation of a continuous function with respect to the variable x. We state our results in the following theorem:

Theorem 42. *If $u(x,t)$ is a continuous function on the interval $[-L, L]$ with $u(-L,t) = u(L,t)$, and $\dfrac{\partial u(x,t)}{\partial x}$ is a piecewise smooth function, then the Fourier series representation of $u(x,t)$ is continuous and it can be differentiated term-by-term.*

There are two more theorems stated subsequently, with proofs left as exercises. These theorems develop term-by-term differentiation of Fourier Cosine and Fourier sine series representation of a function $u(x, t)$. As you may have guessed, knowing these theorems is very important in the chapters that follow.

Theorem 43. *If $u(x, t)$ is a continuous function on the interval $[0, L]$, and $\dfrac{\partial u(x, t)}{\partial x}$ is a piecewise smooth function, then the Fourier cosine series representation of $u(x, t)$ is continuous and it can be differentiated term-by-term.*

Theorem 44. *If $u(x, t)$ is a continuous function on the interval $[0, L]$ with $u(0, t) = u(L, t) = 0$, and $\dfrac{\partial u(x, t)}{\partial x}$ is a piecewise smooth function, then the Fourier sine series representation of $u(x, t)$ is continuous and it can be differentiated term-by-term.*

We have covered all cases where Fourier series may be differentiated term-by-term. If the function $u(x, t)$ in Theorems (42, 43, and 44) is a function of one variable, that is, $f(x)$, the theorems still hold. Next, we turn our attention to term-by-term integration of a Fourier series.

EXERCISES 6.3

6.3.1. Show that the Fourier coefficient $B_n(t)$ in Equation (6.29) depends on the coefficient $a_n(t)$ in Equation (6.25).

6.3.2. Prove Theorem 43.

6.3.3. Prove Theorem 44.

6.3.4. Given

$$\frac{\partial^2 u}{\partial t^2} = c^2 \frac{\partial^2 u}{\partial x^2},$$

subject to

$$u(0, t) = 0 \text{ and } u(L, t) = 0$$

and

$$u(x, 0) = f(x) \text{ and } \frac{\partial u(x, 0)}{\partial t} = g(x),$$

the solution for $u(x, t)$ is

$$u(x, t) = \sum_{n=1}^{\infty} \left[a_n \cos \frac{cn\pi t}{L} + b_n \sin \frac{cn\pi t}{L} \right] \sin \frac{n\pi x}{L}.$$

Prove, using term-by-term differentiation, that the solution satisfies the PDE and BCs.

6.3.5. Given

$$\frac{\partial u}{\partial t} = \frac{\partial^2 u}{\partial x^2}$$

subject to

$$\frac{\partial u(0,t)}{\partial x} = 0 \text{ and } \frac{\partial u(L,t)}{\partial x} = 0$$

and

$$u(x,0) = f(x),$$

the solution for $u(x,t)$ is

$$u(x,t) = a_0 + \sum_{n=1}^{\infty} a_n e^{-\left(\frac{n\pi}{L}\right)^2 t} \cos \frac{n\pi x}{L}.$$

Prove, using term-by-term differentiation, that the solution satisfies the PDE and BCs. State reasons for all differentiations.

6.3.6. Given

$$f(x) = \begin{cases} x, \ 0 \le x < \pi \\ \\ 2\pi - x, \ \pi \le x \le 2\pi, \end{cases}$$

find the Fourier sine series representation of $f(x)$. Then, determine the Fourier cosine series representation of $f'(x)$ using term-by-term differentiation.

6.3.7. Consider the Fourier sine series

$$f(x) \approx \sum_{n=1}^{\infty} b_n \sin \frac{n\pi x}{L},$$

where $0 \le x \le L$ and $f(x)$ is a continuous function, $\approx$ means approximately, and $f'(x)$ is piecewise smooth. Suppose $f(0) = \alpha \neq 0$ and $f(L) = \beta \neq 0$, and show that

$$f'(x) \approx \frac{1}{L} [\beta - \alpha] + \sum_{n=1}^{\infty} \left[\frac{n\pi}{L} b_n + \frac{2}{L} \left((-1)^n \beta - \alpha\right) \right] \cos \frac{n\pi x}{L}.$$

6.3.8. Consider

$$\cosh x = a_0 + \sum_{n=1}^{\infty} a_n \cos \frac{n\pi x}{L}, \ 0 \le x \le L.$$

Differentiating both sides of the equation with respect to x yields

$$\sinh x = \sum_{n=1}^{\infty} \frac{-n\pi}{L} a_n \sin \frac{n\pi x}{L}, \ 0 \le x \le L.$$

Differentiating again yields

$$\cosh x = \sum_{n=1}^{\infty} -\left(\frac{n\pi}{L}\right)^2 a_n \cos \frac{n\pi x}{L}, \; 0 \le x \le L.$$

Since the Fourier cosine series representation of a function on an interval is unique, we must have

$$a_0 + \sum_{n=1}^{\infty} a_n \cos \frac{n\pi x}{L} = \sum_{n=1}^{\infty} -\left(\frac{n\pi}{L}\right)^2 a_n \cos \frac{n\pi x}{L}.$$

This implies that

$$a_0 = 0 \text{ and } a_n = 0.$$

This conclusion is clearly false. Determine the error in the logic.

6.3.9. Consider

$$\sinh x = a_0 + \sum_{n=1}^{\infty} a_n \cos \frac{n\pi x}{L}, \; 0 \le x \le L.$$

Differentiating both sides of the equation with respect to x yields

$$\cosh x = \sum_{n=1}^{\infty} \frac{-n\pi}{L} a_n \sin \frac{n\pi x}{L}, \; 0 \le x \le L.$$

Differentiating again yields

$$\sinh x = \sum_{n=1}^{\infty} -\left(\frac{n\pi}{L}\right)^2 a_n \cos \frac{n\pi x}{L}, \; 0 \le x \le L.$$

Since the Fourier cosine series representation of a function on an interval is unique, we must have

$$a_0 + \sum_{n=1}^{\infty} a_n \cos \frac{n\pi x}{L} = \sum_{n=1}^{\infty} -\left(\frac{n\pi}{L}\right)^2 a_n \cos \frac{n\pi x}{L}.$$

This implies that

$$a_0 = 0 \text{ and } a_n = 0.$$

This conclusion is clearly false. Determine the error in the logic.

6.3.10. Given

$$\frac{\partial u}{\partial t} = k_0 \frac{\partial^2 u}{\partial x^2},$$

subject to

$$u(0,t) = 0 \text{ and } \frac{\partial u(L,t)}{\partial x} = 0$$

and

$$u(x,0) = f(x)$$

the solution for $u(x,t)$ is

$$u(x,t) = \sum_{n=1}^{\infty} b_n e^{-\lambda_n k_0 t} \sin \frac{(2n-1)\pi x}{2L},$$

$$\lambda_n = \left[\frac{(2n-1)\pi}{2L} \right]^2, \; n = 1, 2, 3, \ldots$$

Prove, using term-by-term differentiation, that the solution satisfies the PDE and BCs. Briefly discuss any assumptions you make.

6.3.11. Given

$$\frac{\partial^2 u}{\partial t^2} = 16 \frac{\partial^2 u}{\partial x^2},$$

subject to

$$u(0,t) = 0 \text{ and } u(L,t) = 0$$

and

$$u(x,0) = f(x) \text{ and } \frac{\partial u(x,0)}{\partial t} = g(x),$$

the solution for $u(x,t)$ is

$$u(x,t) = \sum_{n=1}^{\infty} \left(a_n \cos \frac{4n\pi t}{L} + b_n \sin \frac{4n\pi t}{L} \right) \sin \frac{n\pi x}{L},$$

$$\lambda_n = \left[\frac{n\pi}{L} \right]^2, \; n = 1, 2, 3, \ldots$$

Prove, using term-by-term differentiation, that the solution satisfies the PDE and BCs. Briefly discuss any assumptions you make.

6.3.12. Given

$$\frac{\partial u}{\partial t} = k_0 \frac{\partial^2 u}{\partial x^2},$$

subject to

$$u(-L, t) = u(L, t) \text{ and } \frac{\partial u(-L, t)}{\partial x} = \frac{\partial u(L, t)}{\partial x}$$

and

$$u(x, 0) = f(x),$$

the solution for $u(x, t)$ is

$$u(x, t) = a_0 + \sum_{n=1}^{\infty} e^{-\lambda_n k_0 t} \left(a_n \cos \frac{n\pi x}{L} + b_n \sin \frac{n\pi x}{L} \right),$$

$$\lambda_n = \left[\frac{n\pi}{L} \right]^2, \, n = 1, 2, 3, \ldots$$

Prove, using term-by-term differentiation, that the solution satisfies the PDE and BCs. Briefly discuss any assumptions you make.

6.4   INTEGRATION OF FOURIER SERIES

When expressing a piecewise smooth function, $f(x)$, on the interval $[-L, L]$, we use the orthogonality of the sine and cosine functions to determine the Fourier coefficients. This requires multiplying the equation

$$f(x) = a_0 + \sum_{n=1}^{\infty} \left[a_n \cos \frac{n\pi x}{L} + b_n \sin \frac{n\pi x}{L} \right]$$

by either the $\sin \frac{m\pi x}{L}$ or $\cos \frac{m\pi x}{L}$ and integrating the equation from $-L$ to L. Thus, for instance, when finding b_n we have

$$\int_{-L}^{L} f(x) \sin \frac{m\pi x}{L} \, dx = a_0 \int_{-L}^{L} \sin \frac{m\pi x}{L} \, dx$$

$$+ \int_{-L}^{L} \sum_{n=1}^{\infty} \left[a_n \cos \frac{n\pi x}{L} + b_n \sin \frac{n\pi x}{L} \right] \sin \frac{m\pi x}{L} \, dx.$$

Since $\sin \frac{m\pi x}{L}$ is an odd function,

$$a_0 \int_{-L}^{L} \sin \frac{m\pi x}{L} \, dx = 0.$$

Therefore, we have remaining

$$\int_{-L}^{L} f(x) \sin \frac{m\pi x}{L} \, dx = \int_{-L}^{L} \sum_{n=1}^{\infty} \left[a_n \cos \frac{n\pi x}{L} + b_n \sin \frac{n\pi x}{L} \right] \sin \frac{m\pi x}{L} \, dx. \quad (6.32)$$

In Equation (6.32), once we know the function $f(x)$, the left integral is feasible. However, to perform the integration of the infinite series in Chapter 5, you were asked to believe the integral of the infinite sum is equal to the infinite sum of the integrals, and the result was a formula for the term b_n. This would tend to mean term-by-term integration of a Fourier series representation of any piecewise smooth functions on the interval $[-L, L]$ is possible. We confirm this suspicion with a theorem.

Theorem 45. *If $f(x)$ is a piecewise smooth function on the interval $[-L, L]$, then the Fourier series representation of $f(x)$ can always be integrated term-by-term. The result is an infinite series, not necessarily a Fourier series, which converges to the integral of the function $f(x)$ for $-L \leq x \leq L$. Note: It does not matter if the function $f(x)$ has jump discontinuities.*

The proof of Theorem (45) can be found in the book titled *Infinite Series*, by Earl D. Rainville, published in 1967.

Two examples are given showing term-by-term integration of Fourier series. These examples demonstrate the unique features of Theorem (45). Remember, you are guaranteed an infinite series that converges to the integral of the function the Fourier series represents.

EXAMPLE 6.3. Consider the function $f(x) = x$ on the interval $[-2, 2]$. Since $f(x)$ is an odd function, the Fourier series representation of $f(x) = x$ is the same as the Fourier sine series representation of $f(x) = x$ on the interval $[0, 2]$. Therefore,

$$x = \sum_{n=1}^{\infty} \frac{-4(-1)^n}{n\pi} \sin \frac{n\pi x}{2}. \tag{6.33}$$

Integrating Equation (6.33) will yield an infinite series for the function $\dfrac{x^2}{2}$. That is,

$$\int_0^x \tau \, d\tau = \int_0^x \sum_{n=1}^{\infty} \frac{-4(-1)^n}{n\pi} \sin \frac{n\pi \tau}{2} \, d\tau =$$

$$\sum_{n=1}^{\infty} \frac{-4(-1)^n}{n\pi} \int_0^x \sin \frac{n\pi \tau}{2} \, d\tau. \tag{6.34}$$

Notice the integrals in Equation (6.34) are from 0 to x. We want the integrals this way because we want a function after integration. Performing the integration yields

$$\frac{x^2}{2} = \sum_{n=1}^{\infty} \frac{-4(-1)^n}{n\pi} \left[\frac{-2}{n\pi} \cos \frac{n\pi \tau}{2} \right]_0^x =$$

$$\sum_{n=1}^{\infty} \frac{-4(-1)^n}{n\pi} \left[\frac{-2}{n\pi} \left(\cos \frac{n\pi x}{2} - 1 \right) \right]. \tag{6.35}$$

Rewriting Equation (6.35) yields

$$\frac{x^2}{2} = \sum_{n=1}^{\infty} \frac{8(-1)^n}{(n\pi)^2} \left(\cos \frac{n\pi x}{2} - 1 \right) = \sum_{n=1}^{\infty} \left(\frac{8(-1)^n}{(n\pi)^2} \cos \frac{n\pi x}{2} - \frac{8(-1)^n}{(n\pi)^2} \right)$$

$$= \sum_{n=1}^{\infty} \frac{8(-1)^n}{(n\pi)^2} \cos \frac{n\pi x}{2} - \sum_{n=1}^{\infty} \frac{8(-1)^n}{(n\pi)^2}$$

or

$$x^2 = \sum_{n=1}^{\infty} \frac{16(-1)^n}{(n\pi)^2} \cos \frac{n\pi x}{2} - \sum_{n=1}^{\infty} \frac{16(-1)^n}{(n\pi)^2}$$

$$= -\sum_{n=1}^{\infty} \frac{16(-1)^n}{(n\pi)^2} + \sum_{n=1}^{\infty} \frac{16(-1)^n}{(n\pi)^2} \cos \frac{n\pi x}{2}. \qquad \blacksquare \qquad (6.36)$$

We know the infinite series in Equation (6.36) converges to x^2. But, as expected, the infinite series in Equation (6.36) is not a Fourier series. Sometimes, we really need a Fourier series solution, and with a little algebraic manipulation we can transform the infinite series, like that in Equation (6.36), into a Fourier series.

EXAMPLE 6.4. Starting with the infinite series in Equation (6.36),

$$x^2 = -\sum_{n=1}^{\infty} \frac{16(-1)^n}{(n\pi)^2} + \sum_{n=1}^{\infty} \frac{16(-1)^n}{(n\pi)^2} \cos \frac{n\pi x}{2}. \qquad (6.37)$$

We notice the second term on the right has a cosine function. The cosine function indicates the possibility of transforming it into a Fourier cosine series. Thus, we set up the standard Fourier cosine series for $g(x) = x^2$,

$$g(x) = x^2 = a_0 + \sum_{n=1}^{\infty} a_n \cos \frac{n\pi x}{2}. \qquad (6.38)$$

Comparing similar terms, we find the term $-\sum_{n=1}^{\infty} \frac{16(-1)^n}{(n\pi)^2}$, in Equation (6.37), must be the a_0 term in Equation (6.38), and the term $\frac{16(-1)^n}{(n\pi)^2}$, in Equation (6.37), must be the a_n term in Equation (6.38). From Equation (6.38), we determine a_n and a_0. We have

$$a_n = \frac{2}{L} \int_0^L g(x) \cos \frac{n\pi x}{L} \, dx = \int_0^2 x^2 \cos \frac{n\pi x}{L} \, dx =$$

$$\frac{16(-1)^n}{(n\pi)^2} \qquad (6.39)$$

and

$$a_0 = \frac{1}{2} \int_0^2 x^2 \, dx = \frac{4}{3}. \tag{6.40}$$

Thus, Equation (6.38) becomes

$$g(x) = x^2 = \frac{4}{3} + \sum_{n=1}^{\infty} \frac{16(-1)^n}{(n\pi)^2} \cos \frac{n\pi x}{2}. \tag{6.41}$$

The right side of Equation (6.41) is almost what we have in Equation (6.37). The a_n term in Equation (6.41) is identical to the term $\dfrac{16(-1)^n}{(n\pi)^2}$, in Equation (6.37). To complete the process, we must show

$$\frac{4}{3} = -\sum_{n=1}^{\infty} \frac{16(-1)^n}{(n\pi)^2}.$$

Consider Equation (6.41). Evaluated at $x = 0$,

$$g(0) = 0^2 = \frac{4}{3} + \sum_{n=1}^{\infty} \frac{16(-1)^n}{(n\pi)^2} \cos \frac{n\pi(0)}{2} = \frac{4}{3} + \sum_{n=1}^{\infty} \frac{16(-1)^n}{(n\pi)^2},$$

which implies

$$\frac{4}{3} = -\sum_{n=1}^{\infty} \frac{16}{\pi^2} \frac{(-1)^n}{n^2}. \tag{6.42}$$

Therefore, we rewrite the infinite series in Equation (6.37) as

$$x^2 = \frac{4}{3} + \sum_{n=1}^{\infty} \frac{16(-1)^n}{(n\pi)^2} \cos \frac{n\pi x}{2}, \tag{6.43}$$

which is the Fourier cosine series of the function $f(x) = x^2$. ∎

Suppose we would like to integrate Equation (6.43) and express the solution as the Fourier sine series representation of x^3. This is done in the next example.

EXAMPLE 6.5. Consider

$$x^2 = \frac{4}{3} + \sum_{n=1}^{\infty} \frac{16(-1)^n}{(n\pi)^2} \cos \frac{n\pi x}{2} \tag{6.44}$$

on the interval $[0, 2]$. Integrate Equation (6.44) and express the solution as a Fourier sine series representation of x^3.

Solution: Integrate Equation (6.44) from 0 to x. That is,

$$\int_0^x \tau^2 \, d\tau = \frac{4}{3} \int_0^x \, d\tau + \int_0^x \sum_{n=1}^{\infty} \frac{16(-1)^n}{(n\pi)^2} \cos \frac{n\pi\tau}{2} \, d\tau$$

$$= \frac{4}{3} \int_0^x \, d\tau + \sum_{n=1}^{\infty} \frac{16(-1)^n}{(n\pi)^2} \int_0^x \cos \frac{n\pi\tau}{2} \, d\tau.$$

Performing the indicated integrations yields

$$\frac{\tau^3}{3} \bigg|_0^x = \frac{4}{3} \tau \bigg|_0^x + \sum_{n=1}^{\infty} \frac{16(-1)^n}{(n\pi)^2} \left[\frac{2}{n\pi} \sin \frac{n\pi\tau}{2} \right]_0^x.$$

This becomes

$$\frac{x^3}{3} = \frac{4}{3} x + \sum_{n=1}^{\infty} \frac{16(-1)^n}{(n\pi)^2} \left[\frac{2}{n\pi} \sin \frac{n\pi x}{2} \right]$$

$$= \frac{4}{3} x + \sum_{n=1}^{\infty} \frac{32(-1)^n}{(n\pi)^3} \sin \frac{n\pi x}{2}. \tag{6.45}$$

However, Equation (6.45) does not look like the Fourier sine series of $f(x) = x^3$. Therefore, first we multiply Equation (6.45) by 3, which yields

$$x^3 = 4x + \sum_{n=1}^{\infty} \frac{96(-1)^n}{(n\pi)^3} \sin \frac{n\pi x}{2}. \tag{6.46}$$

Second, the right side of Equation (6.46) must be completely expressed as a Fourier sine series on the interval $[0, 2]$. Therefore, we must express the $4x$ term in Equation (6.46) as a Fourier sine series on the interval $[0, 2]$. We have

$$4x = \sum_{n=1}^{\infty} b_n \sin \frac{n\pi x}{2},$$

where

$$b_n = \int_0^2 4x \sin \frac{n\pi x}{2} \, dx = \frac{-16(-1)^n}{n\pi}.$$

Thus,

$$4x = \sum_{n=1}^{\infty} \frac{-16(-1)^n}{n\pi} \sin \frac{n\pi x}{2}. \tag{6.47}$$

Combining Equations (6.46 and 6.47) yields

$$x^3 = \sum_{n=1}^{\infty} \frac{-16\,(-1)^n}{n\pi} \sin \frac{n\pi x}{2} + \sum_{n=1}^{\infty} \frac{96(-1)^n}{(n\pi)^3} \sin \frac{n\pi x}{2}$$

$$= \sum_{n=1}^{\infty} \left[\frac{96(-1)^n}{(n\pi)^3} - \frac{16\,(-1)^n}{n\pi} \right] \sin \frac{n\pi x}{2}.$$

This is the correct Fourier sine series for $f(x) = x^3$. ∎

As Examples (6.3 and 6.5) show, term-by-term integration of a Fourier series produces an infinite series. The infinite series, with a little manipulation, may be written in the more useful Fourier series form.

Determining the value of an infinite series is one use of integrating Fourier series. The following example clearly shows this aspect.

EXAMPLE 6.6. Consider the Fourier sine series of the function $f(x) = 1$ on the interval $0 \le x \le L$,

$$1 = \sum_{n=1}^{\infty} b_n \sin \frac{n\pi x}{L}, \tag{6.48}$$

where

$$b_n = \frac{2}{L} \int_0^L \sin \frac{n\pi x}{L}\, dx = \frac{2}{n\pi}(1 - (-1)^n).$$

Integrating both sides of Equation (6.48),

$$\int_0^x ds = \sum_{n=1}^{\infty} \frac{2}{n\pi}(1 - (-1)^n) \int_0^x \sin \frac{n\pi s}{L}\, ds,$$

yields

$$x = \sum_{n=1}^{\infty} \frac{2}{n\pi}(1 - (-1)^n) \left[\frac{-L}{n\pi} \cos \frac{n\pi s}{L} \right]_0^x$$

$$= \sum_{n=1}^{\infty} \frac{-2L}{(n\pi)^2}(1 - (-1)^n) \left[\cos \frac{n\pi x}{L} - 1 \right]$$

$$= \sum_{n=1}^{\infty} \frac{2L}{(n\pi)^2}(1 - (-1)^n) - \sum_{n=1}^{\infty} \frac{2L}{(n\pi)^2}(1 - (-1)^n) \cos \frac{n\pi x}{L}. \tag{6.49}$$

Equation (6.49) has the form of the Fourier cosine series for the function $f(x) = x$
where $a_0 = \sum\limits_{n=1}^{\infty} \dfrac{2L}{(n\pi)^2}(1 - (-1)^n)$. Finding a_0 in the usual way, we have

$$a_0 = \frac{1}{L}\int_0^L x\,dx = \frac{L}{2}.$$

Therefore, the infinite series $\sum\limits_{n=1}^{\infty}\dfrac{2L}{(n\pi)^2}(1 - (-1)^n)$ converges to $\dfrac{L}{2}$, written as

$$\frac{L}{2} = \sum_{n=1}^{\infty}\frac{2L}{(n\pi)^2}(1 - (-1)^n)$$

for any positive number L.   ■

   Another use of term-by-term integration of Fourier series is determining the Fourier coefficients, which is left as an exercise. This concludes our discussion about integrating Fourier series. We move on to the Gibbs phenomenon.

EXERCISES 6.4

6.4.1. Term-by-term integration has other advantages. Using term-by-term integration, determine the Fourier coefficients b_n by integrating both sides of the following equation twice:

$$\sinh x = \sum_{n=1}^{\infty} b_n \sin\frac{nx}{2},\ 0 \le x \le 2\pi.$$

6.4.2. Using term-by-term integration, determine the Fourier coefficients a_0 and a_n by integrating both sides of the following equation twice:

$$e^{-x} = a_0 + \sum_{n=1}^{\infty} a_n \cos n\pi x,\ 0 \le x \le 1.$$

6.4.3. Using term-by-term integration, determine the Fourier coefficients b_n by integrating both sides of the following equation twice:

$$\cosh x = \sum_{n=1}^{\infty} b_n \sin\frac{n\pi x}{2},\ 0 \le x \le 2.$$

6.4.4. Show that

$$-\frac{\pi^2}{12} = \sum_{n=1}^{\infty}\frac{(-1)^n}{n^2}.$$

Hint: Find the Fourier sine series for the function $f(x) = x$, then integrate the Fourier sine series to get the Fourier cosine series of $f(x) = x^2$ where $L = \pi$.

6.4.5. Show that

$$\frac{1}{4} = \sum_{n=1}^{\infty} \frac{6\left((2 - n^2\pi^2)(-1)^n - 2\right)}{(n\pi)^4}.$$

Hint: Find the Fourier sine series for the function $f(x) = x^2$, then integrate the Fourier sine series to get the Fourier cosine series of $f(x) = x^3$ where $L = \pi$.

6.5   FOURIER SERIES AND THE GIBBS PHENOMENON

In our study of PDEs, the stated problems and examples always had matching BCs and ICs. That is, the ICs satisfied the BCs of the problem. The question should arise, what happens if the ICs do not satisfy the BCs of the problem? Also, in Chapter 4 and again in Chapter 6, we had examples of Fourier series that did not converge to the function at the endpoints. In fact, the Fourier series seemed to oscillate rapidly around any jump discontinuity the function had. For example, consider the Figures (4.18 and 4.43) in Chapter 4 and Figure (6.1) in Chapter 6. We indicated that the oscillations were known as the Gibbs phenomenon, and we stated that the Fourier series would converge to the average of the jump discontinuity of the function. In this section, we examine the Gibbs phenomenon.

An examination of the Gibbs phenomenon requires a more in-depth explanation of the convergence of Fourier series. For example, consider the graph function $f(x) = x^2$ given in Figure (6.3) on the interval $[-L, L]$. We know the Fourier series

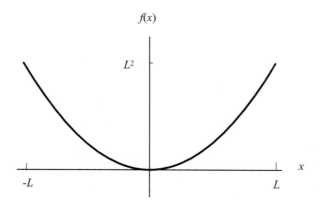

Figure 6.3: The graph of $f(x) = x^2$.

representation of $f(x)$ is

$$x^2 = \frac{L^2}{3} + \sum_{n=1}^{\infty} \frac{4L^2(-1)^n}{(n\pi)^2} \cos\left(\frac{n\pi x}{L}\right). \tag{6.50}$$

Now, consider the following sequence of partial sums of the Fourier series given in Equation (6.50) with their corresponding Figures(6.4–6.9):

$$S_1(x) = \frac{L^2}{3} + \sum_{n=1}^{1} \frac{4L^2(-1)^n}{(n\pi)^2} \cos\left(\frac{n\pi x}{L}\right) = \frac{L^2}{3} - \frac{4L^2}{(n\pi)^2} \cos\left(\frac{\pi x}{L}\right).$$

$$S_2(x) = \frac{L^2}{3} + \sum_{n=1}^{2} \frac{4L^2(-1)^n}{(n\pi)^2} \cos\left(\frac{n\pi x}{L}\right)$$

$$= \frac{L^2}{3} - \frac{4L^2}{(n\pi)^2} \cos\left(\frac{\pi x}{L}\right) + \frac{4L^2}{(n\pi)^2} \cos\left(\frac{2\pi x}{L}\right).$$

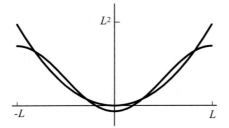

Figure 6.4: The graph of $f(x) = x^2$ and the partial sum $S_1(x)$.

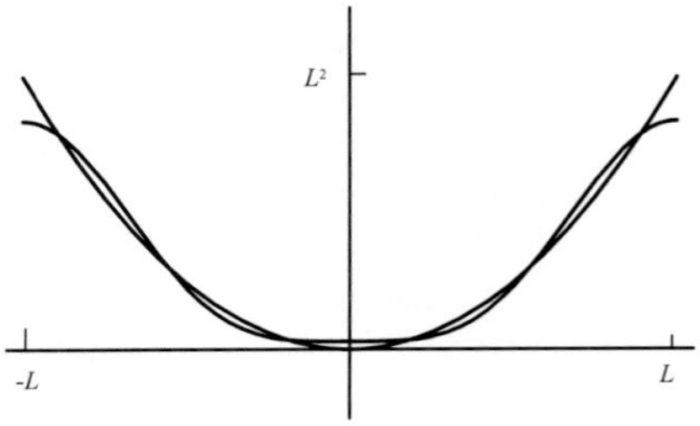

Figure 6.5: The graph of $f(x) = x^2$ and the partial sum $S_2(x)$.

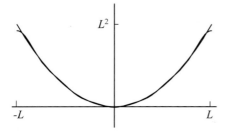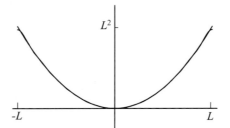

Figure 6.6: The graph of $f(x) = x^2$ and the partial sum $S_5(x)$.

Figure 6.7: The graph of $f(x) = x^2$ and the partial sum $S_{10}(x)$.

$$S_5(x) = \frac{L^2}{3} + \sum_{n=1}^{5} \frac{4L^2(-1)^n}{(n\pi)^2} \cos\left(\frac{n\pi x}{L}\right).$$

$$S_{10}(x) = \frac{L^2}{3} + \sum_{n=1}^{10} \frac{4L^2(-1)^n}{(n\pi)^2} \cos\left(\frac{n\pi x}{L}\right).$$

$$S_{15}(x) = \frac{L^2}{3} + \sum_{n=1}^{15} \frac{4L^2(-1)^n}{(n\pi)^2} \cos\left(\frac{n\pi x}{L}\right),$$

$$S_{100}(x) = \frac{L^2}{3} + \sum_{n=1}^{100} \frac{4L^2(-1)^n}{(n\pi)^2} \cos\left(\frac{n\pi x}{L}\right).$$

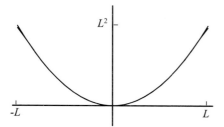

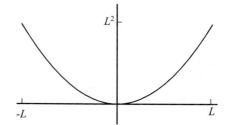

Figure 6.8: The graph of $f(x) = x^2$ and the partial sum $S_{15}(x)$.

Figure 6.9: The graph of $f(x) = x^2$ and the partial sum $S_{100}(x)$.

It is easy to see that the sequence of partial sums of the Fourier series representation of the function x^2 converge uniformly to the function x^2. In fact, by partial sum $S_{100}(x)$ the graph of the partial sum and the graph of the function are almost indistinguishable, which means convergence is relatively fast. Note, each partial sum, S_n, $n = 1, 2, 3, \ldots$, is a trigonometric polynomial. This fact is very important and the following theorem brings it to light.

Theorem 46. *(Weierstrass approximation theorem) If the function $f(x)$ is continuous and has period $2L$, then it can be uniformly approximated by trigonometric polynomials.*

The proof of the Weierstrass[1] approximation theorem may be found in *The Elements of Real Analysis* by Robert G. Bartle.

Remember, trigonometric polynomials are functions of x, and when we discuss uniform approximation by trigonometric polynomials, we are actually considering a sequence of trigonometric polynomial functions uniformly converging to the function $f(x)$ when $f(x)$ is a continuous function.

What happens when the function $f(x)$ is not continuous on the period $2L$, but has a finite number of jump discontinuities? This question is answered in part by the following example.

EXAMPLE 6.7. Consider the Fourier sine series representation of the function

$$f(x) = 5$$

---

[1]Karl Weierstrass, (1815–1897) was the leading mathematical analysts in the second half of the nineteenth century in Berlin

on the interval $0 \leq x \leq L$. Please note the period of this function is still $2L$. Remember, you must do the odd periodic extension shown in Figure (6.10) to generate the Fourier sine series. The Fourier sine series is

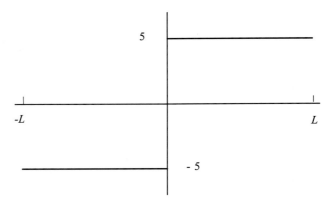

Figure 6.10: The odd periodic extension of $f(x) = 5$ on $[-L, L]$.

$$f(x) = 5 = \sum_{n=1}^{\infty} \frac{10\left[1 - (-1)^n\right]}{n\pi} \sin \frac{n\pi x}{4}.$$

Because of the term $[1 - (-1)^n]$, we recognize that for even natural numbers, $n = 2n$, $S_{2n}(x) = 0$. Therefore, the following sequence of partial sums and graphs–Figures (6.11–6.16)–are for odd n.

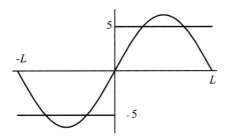

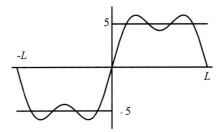

Figure 6.11: The graph of the odd periodic extension of $f(x) = 5$ and the first partial sum $S_1(x)$.

Figure 6.12: The graph of the odd periodic extension of $f(x) = 5$ and the first partial sum $S_3(x)$.

$$S_1(x) = \sum_{n=1}^{1} \frac{10\left[1 - (-1)^n\right]}{n\pi} \sin \frac{n\pi x}{4} = \frac{20}{\pi} \sin \frac{\pi x}{4}.$$

$$S_3(x) = \sum_{n=1}^{3} \frac{10\left[1 - (-1)^n\right]}{n\pi} \sin \frac{n\pi x}{4} = \frac{20}{\pi} \sin \frac{\pi x}{4} + \frac{20}{3\pi} \sin \frac{3\pi x}{4}.$$

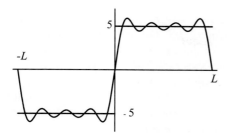

Figure 6.13: The graph of the odd periodic extension of $f(x) = 5$ and the first partial sum $S_7(x)$.

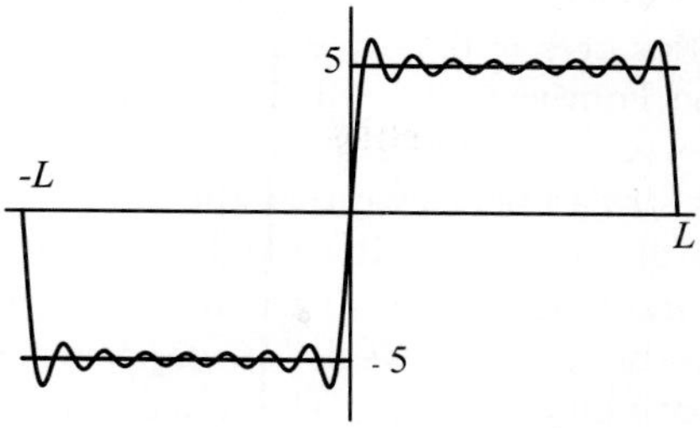

Figure 6.14: The graph of the odd periodic extension of $f(x) = 5$ and the first partial sum $S_{15}(x)$.

$$S_7(x) = \sum_{n=1}^{7} \frac{10\left[1 - (-1)^n\right]}{n\pi} \sin \frac{n\pi x}{4},$$

$$S_{15}(x) = \sum_{n=1}^{15} \frac{10\left[1 - (-1)^n\right]}{n\pi} \sin \frac{n\pi x}{4},$$

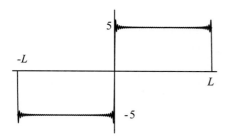
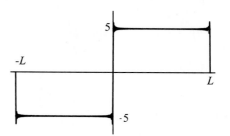

Figure 6.15: The graph of the odd periodic extension of $f(x) = 5$ and the first partial sum $S_{99}(x)$.

Figure 6.16: The graph of the odd periodic extension of $f(x) = 5$ and the first partial sum $S_{201}(x)$.

$$S_{99}(x) = \sum_{n=1}^{99} \frac{10\left[1 - (-1)^n\right]}{n\pi} \sin \frac{n\pi x}{4},$$

$$S_{201}(x) = \sum_{n=1}^{201} \frac{10\left[1 - (-1)^n\right]}{n\pi} \sin \frac{n\pi x}{4}.$$

By the partial sum, $S_{201}(x)$, we see that the Fourier sine series representation of the function $f(x) = 5$ is converging to 5 at points away from the jump discontinuities at $x = 0$, $x = \pm L$. At the points $x = 0$, $x = \pm L$, we notice an overshoot of the Fourier series. This overshoot is known as the Gibbs[2] phenomenon. As $n \longrightarrow \infty$, we could expect the overshoots to disappear. However, in reality they don't. It can be shown that the Fourier series converges to the average value at any jump discontinuity. It should be noted that in higher mathematics, we can show that the Fourier series actually has weak convergence to the endpoints of a jump discontinuity. This weak convergence is from the left and from the right at a jump discontinuity. However, the concept of weak convergence is beyond the scope of this course.

Does the Gibbs phenomenon occur when we solve a PDE? Consider the following example.

EXAMPLE 6.8. Suppose we are given the heat problem

$$\frac{\partial u}{\partial t} = \frac{\partial^2 u}{\partial x^2},$$

subject to

$$u(0, t) = 0$$

$$u(2, t) = 0$$

and

$$u(x, 0) = x^2 + x - 1.$$

We see that the IC has the value of -1 at $x = 0$ and 5 at $x = 2$. From our previous work in Chapter 5, we know that the general solution is

$$u(x, t) = \sum_{n=1}^{\infty} b_n e^{-\left(\frac{n\pi}{2}\right)^2 t} \sin \frac{n\pi x}{2}.$$

Applying the IC yields

$$u(x, 0) = x^2 + x - 1 = \sum_{n=1}^{\infty} b_n \sin \frac{n\pi x}{2}.$$

Thus, we find that b_n is given by

$$b_n = \frac{2}{L} \int_0^L f(x) \sin \frac{n\pi x}{L} \, dx = \frac{2}{2} \int_0^L (x^2 + x - 1) \sin \frac{n\pi x}{2} \, dx$$

$$= \int_0^2 (x^2 + x - 1) \sin \frac{n\pi x}{2} \, dx = \frac{16 \left((-1)^n - 1\right)}{n^3 \pi^3} - \frac{2 \left(1 + 7 (-1)^n\right)}{n\pi}.$$

---

[2] Josiah Gibbs (1839-1903), an American physicist, brought the overshooting phenomenon to the attention of the scientific community in a paper. However, it was first noticed by the mathematician Henry Wilbraham and was rediscovered by the British during World War II while developing radar.

Therefore, the Fourier sine series representation of $x^2 + x - 1$ is

$$x^2 + x - 1 = \sum_{n=1}^{\infty} \frac{16\left((-1)^n - 1\right)}{n^3 \pi^3} - \frac{2\left(1 + 7\left(-1\right)^n\right)}{n\pi} \sin\frac{n\pi x}{2}.$$

Figure (6.17) shows the graph of the polynomial $f(x) = x^2 + x - 1$ on the interval $[0, 2]$. Figure (6.18) is the graph of the Fourier sine series representation of the polynomial $f(x) = x^2 + x - 1$ on the interval $[0, 2]$. Since $f(0) \neq f(2) \neq 0$, the Fourier sine series representation of $f(x) = x^2 + x - 1$ will converge at $f(0)$ and $f(2)$ to 2. Therefore, the Fourier sine series representation of the IC $f(x)$ exhibits the Gibbs phenomenon at the endpoints. Knowing b_n, we can state the specific

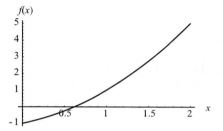

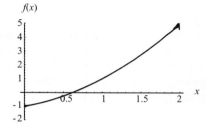

Figure 6.17: The graph of $f(x) = x^2 + x - 1$.

Figure 6.18: The Fourier sine series representation of $f(x) = x^2 + x - 1$.

solution of the problem. It is

$$u(x,t) = \sum_{n=1}^{\infty} \left[\frac{16\left((-1)^n - 1\right)}{n^3 \pi^3} - \frac{2\left(1 + 5\left(-1\right)^n\right)}{n\pi} \right] e^{-\left(\frac{n\pi}{2}\right)^2 t} \sin\frac{n\pi x}{2}.$$

Figure (6.19) shows the graph of $u(x,t)$ for time $0 \leq t \leq 0.5$. Please note how the

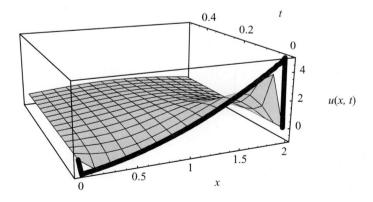

Figure 6.19: $u(x,t)$ for $0 \leq t \leq 0.5$.

boundaries are fixed at 0. Also, the surface of the function has the curve $x^2 + x - 1$ for $0 < x < 2$ subject to the decreasing exponent. The function $f(x) = x^2 + x - 1$ is shown as a bold line where it was produced by a Fourier series partial sum of $n = 500$. Note how the IC does not quite follow the curve of $u(x, t)$ at either end at time $t = 0$. ■

EXERCISES 6.5

6.5.1. Consider the PDE

$$\frac{\partial u(x, t)}{\partial t} = 1.14 \frac{\partial^2 u(x, t)}{\partial x^2}.$$

For each set of BCs and ICs, solve the initial value problem; using your favorite mathematical software, graph the solution for $0 \le t \le 2$. Clearly indicate the initial temperature distribution on your graph and determine if the Fourier series expansion of the initial temperature distribution is accurately represented on the graph of $u(x, t)$.

(1) BCs: $\begin{cases} u(0, t) = 0 \\ \\ u(2\pi, t) = 0 \end{cases}$ and IC: $u(x, 0) = x - 4$.

(2) BCs: $\begin{cases} u(0, t) = 0 \\ \\ \dfrac{\partial u(\pi, t)}{\partial x} = 0 \end{cases}$ and IC: $u(x, 0) = x^3 + 2x - 1$.

(3) BCs: $\begin{cases} \dfrac{\partial u(0, t)}{\partial x} = 0 \\ \\ u\left(\dfrac{3}{2}, t\right) = 0 \end{cases}$ and IC: $u(x, 0) = x^2 - 1$.

(4) BCs: $\begin{cases} \dfrac{\partial u(0, t)}{\partial x} = 0 \\ \\ \dfrac{\partial u(\pi, t)}{\partial x} = 0 \end{cases}$ and IC: $u(x, 0) = x^2 + 2x + 4$.

6.5.2. Consider the PDE

$$\frac{\partial^2 u(x, t)}{\partial t^2} = 16 \frac{\partial^2 u(x, t)}{\partial x^2}.$$

For each set of BCs and ICs, solve the initial value problem. Using your favorite mathematical software, graph the solution for $0 \le t \le 1$. Clearly indicate the initial displacement on your graph and determine if the Fourier series expansion of the initial temperature distribution is accurately represented on the graph of $u(x, t)$.

(1) BCs: $\begin{cases} u(0,t) = 0 \\ \\ u(\pi, t) = 0 \end{cases}$ and ICs: $\begin{cases} u(x,0) = x^2 - 3x + 5 \\ \\ \dfrac{\partial u(x,0)}{\partial t} = 0. \end{cases}$

(2) BCs: $\begin{cases} u(0,t) = 0 \\ \\ \dfrac{\partial u\left(\frac{\pi}{2}, t\right)}{\partial x} = 0 \end{cases}$ and ICs: $\begin{cases} u(x,0) = \cos x \\ \\ \dfrac{\partial u(x,0)}{\partial t} = 0. \end{cases}$

(3) BCs: $\begin{cases} \dfrac{\partial u(0,t)}{\partial x} = 0 \\ \\ u(3,t) = 0 \end{cases}$ and ICs: $\begin{cases} u(x,0) = 0 \\ \\ \dfrac{\partial u(x,0)}{\partial t} = 3\sin(5x). \end{cases}$

(4) BCs: $\begin{cases} \dfrac{\partial u(0,t)}{\partial x} = 0 \\ \\ \dfrac{\partial u(1,t)}{\partial x} = 0 \end{cases}$ and IC: $\begin{cases} u(x,0) = \begin{cases} -x; & 0 \le x < .5 \\ x - 1; & .5 \le x \le 1 \end{cases} \\ \\ \dfrac{\partial u(x,0)}{\partial t} = 0. \end{cases}$

6.5.3. Let a metallic rod 45 cm long be heated to an initial temperature which is modeled by $(x \sin 2x + 3)^0$C. Suppose at $t = 0$, the ends of the rod are plunged into an ice bath of 0^0C and thereafter maintain this temperature. Also, suppose no heat is allowed to escape from the lateral surface of the rod. *Note:* You must model and explain your choice of boundary conditions for this experiment.

(1) Solve the initial value problem if the rod is made of gold. *Note:* The thermal diffusivity of these materials may be found in Appendix E.

(2) Using your favorite mathematical software, graph the solution for each material, then determine the approximate solution for the following three cases:

   (a) Use only the first term in the series for $u(x,t)$ to find the approximate temperature at $x = 15$ cm when $t = 30$ secs.

   (b) Use the first three terms in the series for $u(x,t)$ to find the approximate temperature at $x = 15$ cm when $t = 30$ secs. Compare this answer with the previous answer.

   (c) Use the first 25 terms in the series for $u(x,t)$ to find the approximate temperature at $x = 15$ cm when $t = 30$ secs. Compare this answer with the previous two answers.

   (d) Explain, in a short essay, the differences in the above three approximations.

(3) Determine the amount of time required for the temperature of the rod to reach zero degrees at $x = 15$ cm.

6.5.4. Solve Exercise 6.5.3 if the rod is made of steel.

6.5.5. Solve Exercise 6.5.3 if the rod is made of tungsten.

6.5.6. Consider the following information:

(a) a perfectly flexible string of length 2π ft,

(b) tension of 50 lbs/ft,

(c) mass density of 0.02 lbs/ft,

(d) vibrating motion which is entirely vertical,

(e) fixed left end at 0,

(f) free right end,

(g) initial displacement of $2x - 5$,

(h) initial velocity of 0, and

(i) no gravity effects.

(1) Set up the mathematical model.

(2) Solve your mathematical model for any time t.

(3) Check that this solution satisfies the equation, the BCs, and the ICs.

(4) Using your favorite mathematical software, graph the solution. Then write a short paragraph discussing the surface at the following times: $t = 1$ sec, $t = 5$ secs, and $t = 10$ secs.

Chapter 7

Separation of Variables: The Nonhomogeneous Problem

7.1 INTRODUCTION

In Chapter 5, we discussed the technique of separation of variables for the linear homogeneous problem using homogeneous Dirichlet and Neumann boundary conditions. We summarize this solution method with the following:

1. Determine that the PDE and BCs are linear and homogeneous.

2. Assume that the solution is a product of two functions. One function is possibly of time, while the other function is composed of spatial variables.

3. Substitute this product into the PDE and move all functions of the time variable to one side, while placing all other components on the opposite side of the equation.

4. Set both sides equal to the same separation constant, thus breaking the original equation into two new equations and separating the variables.

5. Repeat steps (2), (3), and (4) on equations with more than one spatial variable until you have an ODE for each spatial variable.

6. Substitute the product form of the solution into the BCs to obtain BCs for the spatial ODEs.

7. Solve the ODE boundary value problem(s) to obtain the eigenvalues and eigenfunctions.

8. Solve the time ODE using the previously found eigenvalues.

9. Use the principle of superposition to get the general solution as an infinite series.

223

10. Use the ICs and orthogonality to determine the constants for the specific solution.

In this chapter, we discuss the separation of variables technique when applied to various linear nonhomogeneous problems. The nonhomogeneous problem is a very realistic mathematical model of experimental physical phenomena. For instance, we can use it to model a forcing function for the wave equation and an internal heat source for the heat equation. Also, we can model homogeneous heat and wave equations whose solutions are not always dependent on homogeneous boundary conditions. These examples only scratch the surface of a host of scenarios that we could also model. I've selected a few of these, and the following are the ones we will consider—remember all equations are linear:

- nonhomogeneous PDEs with homogeneous BCs,

- homogeneous PDEs with nonhomogeneous constant BCs,

- homogeneous PDEs with nonhomogeneous variable BCs, and

- nonhomogeneous PDEs with nonhomogeneous (constant or variable) BCs.

In the next section, we start our discussion of the nonhomogeneous problem by solving the problem of the nonhomogeneous PDEs with homogeneous BCs.

7.2 NONHOMOGENEOUS PDES WITH HOMOGENEOUS BCS

A typical problem of this type is solving for the temperature distribution in a uniform rod of length L, with perfect lateral insulation and some type of constant or nonconstant internal heat source. The mathematical model is

$$\frac{\partial u}{\partial t} = k\frac{\partial^2 u}{\partial x^2} + Q(x,t). \tag{7.1}$$

There are a host of different BCs from which we could choose, but for purposes of this problem we chose homogeneous Dirichlet BCs modeled as

$$\begin{cases} u(0,t) = 0 \\ \\ u(L,t) = 0. \end{cases} \tag{7.2}$$

Also, there must be an IC, and we model it as

$$u(x,0) = f(x). \tag{7.3}$$

The general technique for solving these types of problems is discovered by remembering how we solved a linear second-order nonhomogeneous ODE: First, solve the homogeneous part. Second, solve the particular part. However, solving the homogeneous part of a PDE by separation of variables requires the BCs to be homogeneous. This means we must develop some new and interesting strategies. The problem that I chose to solve in this section does not demonstrate these strategies, but it sets the stage necessary to develop them.

Step 1: Solve the related homogeneous problem by changing Equation (7.1) to

$$\frac{\partial u}{\partial t} = k \frac{\partial^2 u}{\partial x^2}. \tag{7.4}$$

Remembering our techniques from Chapter 5, we assume the solution $u(x,t)$ is of the form $u(x,t) = G(t)\phi(x)$. Performing the necessary differentiations and replacing $u(x,t)$ in Equation (7.4), we obtain

$$G'(t)\phi(x) = kG(t)\phi''(x).$$

Separating variables and setting the resulting equation equal to a separation constant, we obtain

$$\frac{G'(t)}{kG(t)} = \frac{\phi''(x)}{\phi(x)} = -\lambda.$$

This can be written as two separate equations,

$$\frac{G'(t)}{kG(t)} = -\lambda$$

and

$$\frac{\phi''(x)}{\phi(x)} = -\lambda.$$

Now, remembering that the BCs must be separated, we get

$$\begin{cases} \phi(0) = 0 \\[2mm] \phi(L) = 0. \end{cases}$$

We now have two simple ODEs: a spatial ODE with BCs and a time ODE. Up to now, this is familiar, but here is where we enter new territory. Basically, we throw away the time ODE and only solve the spatial ODE for its eigenvalues and eigenfunctions.

The eigenvalues and eigenfunctions that form the solution to the spatial ODE can be found in Chapter 5. They are

$$\left. \begin{array}{l} \text{eigenvalues: } \lambda_n = \left(\dfrac{n\pi}{L}\right)^2 \\[4mm] \text{eigenfunctions: } \phi_n(x) = c_n \sin\left(\dfrac{n\pi x}{L}\right) \end{array} \right\} \quad n = 1, 2, 3, \dots . \tag{7.5}$$

Step 2: Assume that the general solution for Equation (7.1) should be a Fourier series. In particular, we assume that it will be a Fourier series expanded in the eigenfunctions, Equation (7.5), multiplied by some function of time, t. Therefore, assume the general solution for Equation (7.1) is

$$u(x, t) = \sum_{n=1}^{\infty} c_n(t) \sin\left(\frac{n\pi x}{L}\right). \tag{7.6}$$

Notice in Equation (7.6) the former constant, c_n, is now required to be a function of time, t.

Step 3: Start working on the particular part by expanding $Q(x, t)$ from Equation (7.1) in a Fourier series using the eigenvalues and eigenfunctions that were found in Step 1. This means,

$$Q(x, t) = \sum_{n=1}^{\infty} d_n \sin\left(\frac{n\pi x}{L}\right). \tag{7.7}$$

Since we are already using the name c_n in Equation (7.6), we selected the constants d_n for Equation (7.7).

We must determine the constants d_n. From Chapter 4, we know

$$d_n = \frac{2}{L} \int_0^L Q(x, t) \sin\left(\frac{n\pi x}{L}\right) \, dx.$$

From Chapter 5, we know that d_n must in fact be a function of time, t, and only constant with respect to x. Thus, if we assume that we can perform the integral and let $a_n(t) = d_n$, we can write Equation (7.7) as

$$Q(x, t) = \sum_{n=1}^{\infty} a_n(t) \sin\left(\frac{n\pi x}{L}\right). \tag{7.8}$$

What we just did in Step 3 is called eigenfunction expansion. This means we expanded the nonhomogeneous term, $Q(x, t)$, from Equation (7.1) in the eigenfunctions of the related homogeneous problem, Equation (7.4).

Step 4: Using Equation (7.6), we find $\frac{\partial u}{\partial t}$ and $\frac{\partial^2 u}{\partial x^2}$.

In this step, we must satisfy all the conditions required to do term-by-term differentiation of both the Fourier sine series and the Fourier cosine series. Since we assume that time is a continuous function with no sharp bends, we can say

$$\frac{\partial u}{\partial t} = \frac{\partial}{\partial t}\left[\sum_{n=1}^{\infty} c_n(t) \sin\left(\frac{n\pi x}{L}\right)\right] = \sum_{n=1}^{\infty} \frac{d}{dt}\left(c_n(t)\right) \sin\left(\frac{n\pi x}{L}\right)$$

$$= \sum_{n=1}^{\infty} c_n'(t) \sin\left(\frac{n\pi x}{L}\right). \tag{7.9}$$

Since the BCs are homogeneous and we assume the rod to be continuous, we have the Fourier sine series as a continuous function. Thus, if we differentiate the Fourier sine series with respect to x, we have

$$\frac{\partial u}{\partial x} = \frac{\partial}{\partial x}\left[\sum_{n=1}^{\infty} c_n(t)\sin\left(\frac{n\pi x}{L}\right)\right] = \sum_{n=1}^{\infty} c_n(t)\frac{d}{dx}\left(\sin\left(\frac{n\pi x}{L}\right)\right)$$

$$= \sum_{n=1}^{\infty} \frac{n\pi}{L}c_n(t)\cos\left(\frac{n\pi x}{L}\right).$$

Since the BCs are homogeneous and we assume the heat flow to be continuous, we have the Fourier cosine series as a continuous function. Thus, we can differentiate with respect to x the Fourier cosine series. This yields

$$\frac{\partial^2 u}{\partial x^2} = \frac{\partial}{\partial x}\left[\sum_{n=1}^{\infty} \frac{n\pi}{L}c_n(t)\cos\left(\frac{n\pi x}{L}\right)\right]$$

$$= \sum_{n=1}^{\infty} \frac{n\pi}{L}c_n(t)\frac{d}{dx}\left(\cos\left(\frac{n\pi x}{L}\right)\right)$$

$$= -\sum_{n=1}^{\infty} \left(\frac{n\pi}{L}\right)^2 c_n(t)\sin\left(\frac{n\pi x}{L}\right). \tag{7.10}$$

Step 5: Solve Equation (7.1) by replacing the terms of the equation by what we have determined they equal in Steps 1 through 4. That is,

$$\frac{\partial u}{\partial t} = k\frac{\partial^2 u}{\partial x^2} + Q(x,t)$$

becomes

$$\sum_{n=1}^{\infty} c_n'(t)\sin\left(\frac{n\pi x}{L}\right) = -k\sum_{n=1}^{\infty} \left(\frac{n\pi}{L}\right)^2 c_n(t)\sin\left(\frac{n\pi x}{L}\right)$$

$$+ \sum_{n=1}^{\infty} a_n(t)\sin\left(\frac{n\pi x}{L}\right).$$

We now have the Fourier sine series expansions over the same interval, $[0, L]$, in all three terms of the equation. This means that the Fourier coefficients must be equal. Therefore, we only have to solve the ODE

$$c_n'(t) = -k\left(\frac{n\pi}{L}\right)^2 c_n(t) + a_n(t).$$

Since $\left(\dfrac{n\pi}{L}\right)^2 = \lambda_n$, we may write

$$c_n'(t) = -k\lambda_n c_n(t) + a_n(t)$$

or

$$c_n'(t) + k\lambda_n c_n(t) = a_n(t). \tag{7.11}$$

Equation (7.11) is a first-order ODE. To develop the solution for $c_n(t)$, we use the integrating factor

$$e^{k\lambda_n t}.$$

Doing so yields a solution for $c_n(t)$:

$$c_n(t) = e^{-k\lambda_n t}\int^t e^{k\lambda_n \tau} a_n(\tau)\, d\tau + e^{-k\lambda_n t} b_n.$$

Note that the integral does not have a lower limit. We assume that when both sides of the equation are integrated, we will get constants of integration on both sides of the equation, which we combine and call b_n.

Step 6: Write the solution to Equation (7.1) and apply the IC, Equation (7.3). We have

$$u(x,t) = \sum_{n=1}^{\infty}\left[e^{-k\lambda_n t}\int^t e^{k\lambda_n \tau} a_n(\tau)\, d\tau + e^{-k\lambda_n t} b_n\right]\sin\left(\frac{n\pi x}{L}\right). \tag{7.12}$$

Applying the IC, we find a general form of $u(x,0)$ as

$$u(x,0) = f(x) = \sum_{n=1}^{\infty} A_n(0) + b_n \sin\left(\frac{n\pi x}{L}\right),$$

where

$$A_n(0) = \int^t e^{k\lambda_n \tau} a_n(\tau)\, d\tau \bigg|_{t=0}.$$

This implies

$$b_n = -A_n(0) + \frac{2}{L}\int_0^L f(x)\sin\left(\frac{n\pi x}{L}\right)\, dx. \tag{7.13}$$

Therefore, the complete solution to Equation (7.1), subject to the BCs, Equations (7.2), and IC, Equation (7.3), is

$$u(x,t) = \sum_{n=1}^{\infty}\left[e^{-k\lambda_n t}\int^t e^{k\lambda_n \tau} a_n(\tau)\, d\tau + e^{-k\lambda_n t} b_n\right]\sin\left(\frac{n\pi x}{L}\right),$$

where the constant b_n is given by Equation (7.13).

The following example illustrates this method. Also, see the graphical representation of the solution provided in Figure (7.1).

EXAMPLE 7.1. Consider

$$\frac{\partial^2 u}{\partial t^2} = \frac{\partial^2 u}{\partial x^2} + Q(x),$$

(7.14)

where

$$Q(x) = \begin{cases} x, & 0 \le x < \pi \\ 2\pi - x, & \pi \le x \le 2\pi, \end{cases}$$

subject to

$$\begin{cases} u(0, t) = 0 \\ u(2\pi, t) = 0 \end{cases}$$

(7.15)

and

$$\begin{cases} u(x, 0) = x^2 - 2\pi x \\ \dfrac{\partial u(x, 0)}{\partial t} = 0. \end{cases}$$

(7.16)

Find the time-dependent solution for $u(x, t)$. For this problem, the BCs are linear and homogeneous. However, the PDE in Equation (7.14) is not homogeneous. Therefore, we need to follow the method developed in this section.

Step 1: Solve the related homogeneous problem. We basically ignore the term $Q(x)$, and we derive the eigenvalues and corresponding eigenfunctions for

$$\frac{\partial^2 u}{\partial t^2} = \frac{\partial^2 u}{\partial x^2},$$

subject to the BCs, Equation (7.15). Remember, we are only solving for eigenvalues and corresponding eigenfunctions of the spatial problem after we separate the variables. Thus, we must solve

$$\varphi''(x) = -\lambda \varphi(x).$$

(7.17)

Equation (7.17), subject to the BCs indicated in Equation (7.15), has eigenvalues and corresponding eigenfunctions of

$$\left.\begin{aligned} \lambda_n &= \left(\frac{n}{2}\right)^2 \\ \varphi_n(x) &= b_n \sin \frac{nx}{2} \end{aligned}\right\}, \quad n = 1, 2, 3, \dots$$

Step 2: Assume the general solution for Equation (7.14) has the form

$$u(x, t) = \sum_{n=1}^{\infty} b_n(t) \sin\left(\frac{nx}{2}\right).$$

(7.18)

Step 3: Expand the term $Q(x)$ in a Fourier series using the eigenvalues and corresponding eigenfunctions found in Step 1. That is,

$$Q(x) = \sum_{n=1}^{\infty} a_n \sin\left(\frac{nx}{2}\right). \tag{7.19}$$

In Equation (7.19), the a_n term is not a function of time, t. The reason for this is that the function $Q(x)$ is only a function of x. Solving for a_n yields

$$\begin{aligned}
a_n &= \frac{1}{\pi}\int_0^{2\pi} Q(x)\sin\left(\frac{nx}{2}\right)\,dx \\
&= \frac{1}{\pi}\left\{\int_0^{\pi} x\sin\left(\frac{nx}{2}\right)\,dx + \int_0^{2\pi}(2\pi - x)\sin\left(\frac{nx}{2}\right)\,dx\right\} \\
&= \frac{8\sin\left(\dfrac{n\pi}{2}\right)}{\pi n^2}.
\end{aligned}$$

Therefore,

$$Q(x) = \sum_{n=1}^{\infty} \frac{8\sin\left(\dfrac{n\pi}{2}\right)}{\pi n^2}\sin\left(\frac{nx}{2}\right). \tag{7.20}$$

Step 4: Using Equation (7.18) find $\dfrac{\partial^2 u}{\partial t^2}$ and $\dfrac{\partial^2 u}{\partial x^2}$. Remember we must be able to justify term-by-term differentiation. For the derivatives with respect to time, t, we expect time to be continuous. Thus,

$$\frac{\partial^2 u}{\partial t^2} = \sum_{n=1}^{\infty} b_n''(t)\sin\left(\frac{nx}{2}\right). \tag{7.21}$$

Since the string is continuous and the BCs, Equation (7.15), are homogeneous, we know that we can perform term-by-term differentiation with respect to the spatial variable, x. It is

$$\frac{\partial^2 u}{\partial x^2} = \sum_{n=1}^{\infty} -\left(\frac{n}{2}\right)^2 b_n(t)\sin\left(\frac{nx}{2}\right). \tag{7.22}$$

Step 5: Solve Equation (7.14) by using Equations (7.20, 7.21, and 7.22). We have

$$\sum_{n=1}^{\infty} b_n''(t)\sin\left(\frac{nx}{2}\right) = \sum_{n=1}^{\infty} -\left(\frac{n}{2}\right)^2 b_n(t)\sin\left(\frac{nx}{2}\right)$$

$$+ \sum_{n=1}^{\infty} \frac{8\sin\left(\dfrac{n\pi}{2}\right)}{\pi n^2}\sin\left(\frac{nx}{2}\right).$$

Since the Fourier series are all in the same eigenvalues and corresponding eigenfunctions over the same interval, we know the Fourier coefficients must be equal. Therefore,

$$b_n''(t) = -\left(\frac{n}{2}\right)^2 b_n(t) + \frac{8\sin\left(\frac{n\pi}{2}\right)}{\pi n^2}$$

or, written in a more familiar form,

$$b_n''(t) + \left(\frac{n}{2}\right)^2 b_n(t) = \frac{8\sin\left(\frac{n\pi}{2}\right)}{\pi n^2}.$$

This is a second-order linear nonhomogeneous constant coefficient ODE, with a solution of

$$b_n(t) = a_n \cos\left(\frac{nt}{2}\right) + d_n \sin\left(\frac{nt}{2}\right) + \frac{32\sin\left(\frac{n\pi}{2}\right)}{\pi n^4}.$$

Step 6: Write the solution for $u(x,t)$ and apply the ICs (Equation (7.16)). We have

$$u(x,t) = \sum_{n=1}^{\infty} \left[a_n \cos\left(\frac{nt}{2}\right) + d_n \sin\left(\frac{nt}{2}\right) + \frac{32\sin\left(\frac{n\pi}{2}\right)}{\pi n^4} \right] \sin\left(\frac{nx}{2}\right).$$

Applying the first IC, $u(x,0) = x^2 - 2\pi x$, yields

$$u(x,0) = x^2 - 2\pi x = \sum_{n=1}^{\infty} \left[a_n + \frac{32\sin\left(\frac{n\pi}{2}\right)}{\pi n^4} \right] \sin\left(\frac{nx}{2}\right).$$

Thus,

$$a_n = \frac{-32\sin\left(\frac{n\pi}{2}\right)}{\pi n^4} + \frac{1}{\pi} \int_0^{2\pi} \left(x^2 - 2\pi x\right) \sin\left(\frac{nx}{2}\right) \, dx$$

$$= \frac{-32\sin\left(\frac{n\pi}{2}\right)}{\pi n^4} + \frac{16\left((-1)^n - 1\right)}{n^3 \pi}. \tag{7.23}$$

Applying the second IC, $\dfrac{\partial u(x,0)}{\partial t} = 0$, yields

$$\frac{\partial u(x,0)}{\partial t} = 0 = \sum_{n=1}^{\infty} \left[\left(\frac{n}{2}\right) d_n \right] \sin\left(\frac{nx}{2}\right),$$

which indicates $d_n = 0$ for all n. Therefore, the complete solution is

$$u(x,t) = \sum_{n=1}^{\infty} \left[a_n \cos\left(\frac{nt}{2}\right) + \frac{32\sin\left(\frac{n\pi}{2}\right)}{\pi n^4} \right] \sin\left(\frac{nx}{2}\right)$$

where a_n is given by Equation (7.23).

In Figure (7.1), the solution is shown for $0 \leq t \leq 20$. Note, the boundaries at $x = 0$ and $x = 2\pi$ remain fixed at 0, the initial displacement is graphically illustrated by a bold-faced curve, and the repeating curve caused by the ICs and damping function $Q(x)$. ■

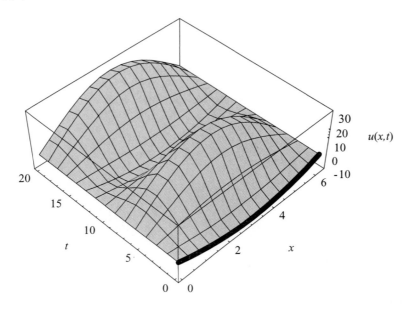

Figure 7.1: $u(x, t)$ **for** $0 \leq t \leq 20$.

The general method and problem we have solved in this section, nonhomogeneous PDEs with homogeneous BCs, is the basis for the other nonhomogeneous problems we will solve later in this chapter. So, it is well worth your time to fully understand the method. My suggestion is to work through the problem by picking a value for k, such as 1, and some simple function for $Q(x, t)$, such as xt. After solving the problem with values for k and $Q(x, t)$, you will notice that we only add a couple of steps to the solution method of separation of variables, from the chapter introduction. At the end of this chapter, we summarize the steps to follow when solving the nonhomogeneous problem by the technique of separation of variables.

EXERCISES 7.2

7.2.1. Consider the PDE

$$\frac{\partial u}{\partial t} = k \frac{\partial^2 u}{\partial x^2} + Q(x, t).$$

Find the time-dependent solution for $u(x, t)$ subject to the IC

$$u(x, 0) = f(x),$$

and the following sets of BCs:

(1) $\dfrac{\partial u(0, t)}{\partial x} = 0$ and $\dfrac{\partial u(L, t)}{\partial x} = 0$.

(2) $u(0, t) = 0$ and $\dfrac{\partial u(L, t)}{\partial x} = 0$.

(3) $\dfrac{\partial u(0, t)}{\partial x} = 0$ and $u(L, t) = 0$.

7.2.2. Consider the PDE

$$\frac{\partial u}{\partial t} = \frac{\partial^2 u}{\partial x^2} + x e^{-t},$$

subject to

$$\text{BCs:} \begin{cases} u(0, t) = 0 \\[2mm] \dfrac{\partial u(\pi, t)}{\partial x} = 0 \end{cases} \quad \text{and} \quad \text{IC:} u(x, 0) = x.$$

Find the time-dependent solution for $u(x, t)$, and using your favorite mathematical software, plot your solution. By inspection, determine if the graph of your solution is reasonable for the problem.

7.2.3. Consider the PDE

$$\frac{\partial u}{\partial t} = \frac{\partial^2 u}{\partial x^2} + x \cos t,$$

subject to

$$\text{BCs:} \begin{cases} \dfrac{\partial u(0, t)}{\partial x} = 0 \\[2mm] u(2\pi, t) = 0 \end{cases} \quad \text{and} \quad \text{IC: } u(x, 0) = \begin{cases} x, & 0 \le x \le \pi \\[1mm] 2\pi - x & \pi < x \le 2\pi. \end{cases}$$

Find the time-dependent solution for $u(x, t)$, and using your favorite mathematical software, plot your solution. By inspection, determine if the graph of your solution is reasonable for the problem.

7.2.4. Consider the PDE

$$\frac{\partial^2 u}{\partial t^2} = c^2 \frac{\partial^2 u}{\partial x^2} + Q(x, t).$$

Find the time-dependent solution for $u(x, t)$ subject to the ICs

$$u(x, 0) = f(x) \quad \text{and} \quad \frac{\partial u(x, 0)}{\partial t} = g(x),$$

and the following sets of BCs:

(1) $u(0,t) = 0$ and $\dfrac{\partial u(L,t)}{\partial x} = 0$.

(2) $\dfrac{\partial u(0,t)}{\partial x} = 0$ and $\dfrac{\partial u(L,t)}{\partial x} = 0$.

(3) $\dfrac{\partial u(0,t)}{\partial x} = 0$ and $u(L,t) = 0$.

7.2.5. Consider the PDE

$$\frac{\partial^2 u}{\partial t^2} = \frac{\partial^2 u}{\partial x^2} + x\sin t,$$

subject to the BCs

$$\frac{\partial u(0,t)}{\partial x} = 0 \text{ and } \frac{\partial u(2,t)}{\partial x} = 0$$

and to the ICs

$$u(x,0) = \begin{cases} -x, & 0 \le x \le 1.5 \\ \\ 5x - 9 & 1.5 < x \le 2 \end{cases} \text{ and } \frac{\partial u(x,0)}{\partial t} = 0.$$

Find the time-dependent solution for $u(x,t)$, and using your favorite mathematical software, plot your solution. By inspection, determine if the graph of your solution is reasonable for the problem.

7.2.6. Consider the PDE

$$\frac{\partial u}{\partial t} = 16\frac{\partial^2 u}{\partial y^2} + y\sin t,$$

subject to the BCs

$$\frac{\partial u(0,t)}{\partial y} = 0 \text{ and } u(2\pi,t) = 0$$

and IC

$$u(y,0) = y.$$

Find the time-dependent solution for $u(y,t)$, and using your favorite mathematical software, plot your solution. By inspection, determine if the graph of your solution is reasonable for the problem.

7.2.7. Consider the PDE

$$\frac{\partial u}{\partial t} = \frac{\partial^2 u}{\partial x^2} + \frac{\partial^2 u}{\partial y^2} + (x+y)e^{-t},$$

subject to the BCs

$$u(0,y,t) = 0, \quad \frac{\partial u(\pi,y,t)}{\partial x} = 0, \quad \frac{\partial u(x,0,t)}{\partial x} = 0, \text{ and } u(x,2\pi,t) = 0$$

and IC

$$u(x, y, 0) = x + y.$$

Find the time-dependent solution for $u(x, y, t)$, and using your favorite mathematical software, plot your solution for several different times t. By inspection, determine if the graph of your solution is reasonable for the problem.

7.2.8. Consider the PDE

$$\frac{\partial^2 u}{\partial t^2} = \frac{\partial^2 u}{\partial x^2} + \frac{\partial^2 u}{\partial y^2} + \sin t \, \cos xy,$$

subject to the BCs

$$\frac{\partial u(\pi, y, t)}{\partial x} = 0, \quad u\left(\frac{\pi}{2}, y, t\right) = 0$$

$$\frac{\partial u(x, 0, t)}{\partial x} = 0, \quad u(x, 2\pi, t) = 0$$

and the ICs

$$u(x, y, 0) = \sin xy \text{ and } \frac{\partial u(x, y, 0)}{\partial t} = 0.$$

Find the time-dependent solution for $u(x, y, t)$, and using your favorite mathematical software, plot your solution for several different times t. By inspection, determine if the graph of your solution is reasonable for the problem.

7.2.9. Consider the PDE

$$\frac{\partial^2 u}{\partial t^2} = 4\frac{\partial^2 u}{\partial x^2} - 2\frac{\partial u}{\partial t} + g, \text{ where } g \text{ is gravity,}$$

subject to the BCs

$$u(0, t) = 0 \text{ and } u(2\pi, t) = 0$$

and the ICs

$$u(x, 0) = \begin{cases} \dfrac{6x}{\pi}, & 0 \le x \le \dfrac{\pi}{2} \\[2ex] \dfrac{2(2\pi - x)}{\pi} & \dfrac{\pi}{2} < x \le 2\pi \end{cases} \quad \text{and} \quad \frac{\partial u(x, 0)}{\partial t} = 0.$$

Find the time-dependent solution for $u(x, t)$, and using your favorite mathematical software, plot your solution. By inspection, determine if the graph of your solution is reasonable for the problem. This problem reflects vibrations which arise from a gravity field in a medium with resistance proportional to the velocity; the ends of the string are fixed at the same height.

7.2.10. Consider the PDE

$$\frac{\partial^2 u}{\partial t^2} = 9\frac{\partial^2 u}{\partial x^2} + g, \text{ where } g \text{ is gravity,}$$

subject to the BCs

$$u(0, t) = 0 \text{ and } \frac{\partial u(\pi, t)}{\partial x} = 0$$

and to the ICs

$$u(x, 0) = 0 \text{ and } \frac{\partial u(x, 0)}{\partial t} = 100.$$

Find the time-dependent solution for $u(x, t)$, and using your favorite mathematical software, plot your solution. By inspection, determine if the graph of your solution is reasonable for the problem. This problem reflects a vertical flexible rod of length π (with its left end rigidly attached to a freely falling lift, while its right end is free) that, having attained a velocity of 100 ft/sec, stops instantaneously.

7.2.11. Given a one-dimensional rod of length π m. A heater moves along the surface of the rod with constant velocity of 1 cm/min. Thus, a convective heat exchange takes place. The flow of heat from the heater to the rod is $6e^{-\beta t}$ watts where β is the convection coefficient. Suppose the ends of the rod are held at $0°$C, and the initial temperature distribution is $3\sin(5x)$.

(1) Mathematically model this problem.

(2) Suppose $\beta = .12$, and solve the time-dependent problem.

(3) Graph the time-dependent problem.

(4) Solve the steady-state problem.

7.3   HOMOGENEOUS PDE WITH NONHOMOGENEOUS BCS

In this section, we cover two different types of nonhomogeneous BCs. First, we consider constant, nonzero BCs. Second, we examine BCs that contain a variable. Although the methods are similar for each, the first is straightforward while the second requires some intuition.

7.3.1   Homogeneous PDE—Nonhomogeneous Constant BCs

Consider a perfect laterally insulated uniform rod of length L, with thermal conductivity k, and some type of initial condition. Also, suppose at $x = 0$, the end of the rod is in a bath held at a constant temperature of T_1; at the end $x = L$, the rod has some type of insulation, which constricts the flow of heat to a constant value of A_2. The mathematical model for this problem is

$$\frac{\partial u}{\partial t} = k \frac{\partial^2 u}{\partial x^2}, \tag{7.24}$$

subject to the BCs

$$u(0, t) = T_1$$

$$\frac{\partial u(L, t)}{\partial x} = A_2, \tag{7.25}$$

with IC

$$u(x, 0) = f(x). \tag{7.26}$$

In the last section, we said the nonhomogeneous PDE with homogeneous BCs problem was the basis for all the other nonhomogeneous problems we will solve in this chapter. However, how is a homogeneous PDE with nonhomogeneous BCs turned into the basic nonhomogeneous problem? The answer is by creating a new function. So what does this new function look like? To answer this question, we must consider all the things that the new function must do.

The new function must have homogeneous BCs. It must replace the function $u(x, t)$ with no loss of information. Finally, the new function must turn the non-homogeneous BCs into a source term that makes the PDE nonhomogeneous. The question that still remains is how? To answer this question, consider what we have learned so far in the solution techniques for PDEs, particularly PDEs that involve temperature distribution in a one-dimensional rod. So far, two techniques have been studied: separation of variables and steady-state. We know that we cannot use the separation of variables technique yet. So, by default, that leaves us with the steady-state technique. Let's investigate this method to determine what it gives us.

Basically, in the steady-state problem, we assume that $u(x, t) = u_{ss}(x)$. This assumption implies that $\frac{\partial u_{ss}}{\partial t} = 0$ and $\frac{\partial^2 u_{ss}}{\partial x^2} = \frac{d^2 u_{ss}}{dx^2}$. The general steady-state solution for Equation (7.24) is

$$u_{ss}(x) = c_1 x + c_2. \tag{7.27}$$

Applying the BCs to Equation (7.27) yields

$$u_{ss}(x) = A_2 x + T_1. \tag{7.28}$$

We now want to create a new function, using Equation (7.28), that has homogeneous BCs. Suppose we create a function $v(x, t)$ such that

$$v(x, t) = u(x, t) - u_{ss}(x). \tag{7.29}$$

Then

$$v(0, t) = u(0, t) - u_{ss}(0) = T_1 - T_1 = 0$$

and

$$\frac{\partial v(L,t)}{\partial x} = \frac{\partial u(L,t)}{\partial x} - \frac{\partial u_{ss}(L)}{\partial x} = A_2 - A_2 = 0.$$

Therefore, $v(x,t)$ is a function that has homogeneous BCs. Ideally, we'd like to replace $u(x,t)$ with $v(x,t)$ in Equation (7.24). To do this, we must determine $u(x,t)$, $\frac{\partial u}{\partial t}$, and $\frac{\partial^2 u}{\partial x^2}$. From Equation (7.29), we have

$$u(x,t) = v(x,t) + u_{ss}(x). \tag{7.30}$$

Using Equation (7.30) and Equation (7.28), we find that

$$\frac{\partial u}{\partial t} = \frac{\partial v}{\partial t} + \frac{\partial u_{ss}(x)}{\partial t} = \frac{\partial v}{\partial t} + \frac{\partial (A_2 x + T_1)}{\partial t} = \frac{\partial v}{\partial t} \tag{7.31}$$

and

$$\frac{\partial^2 u}{\partial x^2} = \frac{\partial^2 v}{\partial x^2} + \frac{\partial^2 u_{ss}(x)}{\partial x^2} = \frac{\partial^2 v}{\partial x^2} + \frac{\partial^2 (A_2 x + T_1)}{\partial x^2} = \frac{\partial^2 v}{\partial x^2}. \tag{7.32}$$

Thus, $u(x,t)$ and $v(x,t)$ are interchangeable in Equation (7.24). However, instead of ending up with a nonhomogeneous PDE with homogeneous BCs, we have a homogeneous PDE with homogeneous BCs, as summarized below:

$$\frac{\partial v}{\partial t} = k \frac{\partial^2 v}{\partial x^2}, \tag{7.33}$$

subject to

$$v(0,t) = 0$$

$$\frac{\partial v(L,t)}{\partial x} = 0. \tag{7.34}$$

We know from our studies in Chapter 5 that Equation (7.33), subject to Equation (7.34), has the solution

$$v(x,t) = \sum_{n=1}^{\infty} b_n e^{-k\lambda_n t} \sin\left[\frac{(2n-1)\pi x}{2L}\right], \tag{7.35}$$

$$\lambda_n = \left[\frac{(2n-1)\pi}{2L}\right]^2, \, n = 1, 2, 3, \ldots$$

However, Equation (7.35) only gives us the solution for $v(x,t)$, not $u(x,t)$. Equation (7.30) and Equation (7.35) determine the general solution for $u(x,t)$. Thus, the general solution is

$$u(x,t) = v(x,t) + u_{ss}(x)$$

$$= \sum_{n=1}^{\infty} b_n e^{-k\lambda_n t} \sin\left[\frac{(2n-1)\pi x}{2L}\right] + A_2 x + T_1. \tag{7.36}$$

To avoid any confusion as to the contents within the summation sign, this may be rewritten as

$$u(x,t) = A_2 x + T_1 + \sum_{n=1}^{\infty} b_n e^{-k\lambda_n t} \sin\left[\frac{(2n-1)\pi x}{2L}\right]. \tag{7.37}$$

Applying the IC

$$u(x,0) = f(x),$$

we can determine the specific solution

$$u(x,0) = f(x) = A_2 x + T_1 + \sum_{n=1}^{\infty} b_n \sin\left[\frac{(2n-1)\pi x}{2L}\right],$$

which can be written as

$$f(x) - A_2 x - T_1 = \sum_{n=1}^{\infty} b_n \sin\left[\frac{(2n-1)\pi x}{2L}\right].$$

Using orthogonality, we find

$$b_n = \frac{2}{L} \int_0^L \left[f(x) - A_2 x - T_1\right] \sin\left[\frac{(2n-1)\pi x}{2L}\right] \, dx. \tag{7.38}$$

Thus, the specific solution to the problem in Equation (7.24), subject to the BCs, Equation (7.25), and the IC, Equation (7.26), is

$$u(x,t) = A_2 x + T_1 + \sum_{n=1}^{\infty} b_n e^{-k\lambda_n t} \sin\left[\frac{(2n-1)\pi x}{2L}\right]$$

where b_n is given by Equation (7.38).

The following example illustrates this method. Also, there is a graphic representation of the solution.

EXAMPLE 7.2. Consider

$$\frac{\partial u}{\partial t} = 3\frac{\partial^2 u}{\partial x^2}, \tag{7.39}$$

subject to

$$u(0,t) = -1$$

$$u(\pi,t) = 1 \tag{7.40}$$

and

$$u(x,0) = -\cos 7x. \tag{7.41}$$

Find the time-dependent solution for $u(x, t)$.

Since the PDE is linear and homogeneous, and the BCs are linear but not homogeneous, we must determine the steady-state solution, $u_{ss}(x)$. The general steady state solution is

$$u_{ss}(x) = c_1 x + c_2.$$

Applying the first boundary condition, $u(0) = -1$, yields $c_2 = -1$. Applying the second boundary condition, $u(\pi) = 1$, yields $c_1 = \dfrac{2}{\pi}$. Therefore, the specific steady-state solution is

$$u_{ss}(x) = \frac{2}{\pi} x - 1. \tag{7.42}$$

Letting the function $v(x, t) = u(x, t) - u_{ss}(x)$, we find $v(0, t) = 0$ and $v(\pi, t) = 0$. Also,

$$\frac{\partial v}{\partial t} = \frac{\partial u}{\partial t}$$

$$\frac{\partial^2 v}{\partial x^2} = \frac{\partial^2 u}{\partial x^2}.$$

Thus, we have linear homogeneous PDE and BCs in terms of $v(x, t)$. We can now apply the separation of variables technique. We should recognize from Chapter 5 that

$$\frac{\partial v}{\partial t} = 3 \frac{\partial^2 v}{\partial x^2},$$

subject to

$$v(0, t) = 0$$

$$v(\pi, t) = 0,$$

has, as a general solution,

$$v(x, t) = \sum_{n=1}^{\infty} b_n e^{-3n^2 t} \sin nx.$$

Therefore,

$$u(x, t) = \frac{2}{\pi} x - 1 + \sum_{n=1}^{\infty} b_n e^{-3n^2 t} \sin nx.$$

Applying the IC, Equation (7.41), yields

$$u(x, 0) = -\cos 7x = \frac{2}{\pi} x - 1 + \sum_{n=1}^{\infty} b_n \sin nx,$$

which can be written as

$$-\cos 7x - \frac{2}{\pi}x + 1 = \sum_{n=1}^{\infty} b_n \sin nx.$$

Using orthogonality of the Fourier sine series provides the value for b_n, which is

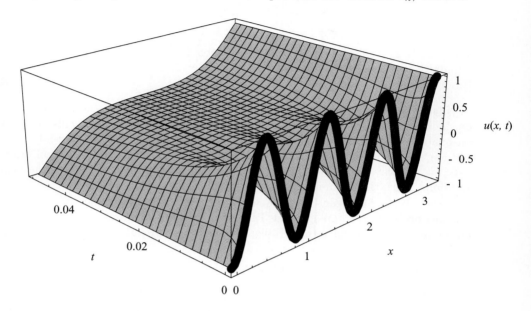

Figure 7.2: $u(x,t)$ **for** $0 \le t \le 0.05$.

$$b_n = \frac{2}{\pi}\int_0^{\pi}\left(1 - \frac{2}{\pi}x - \cos 7x\right)\sin nx \; dx = \frac{98(n-(-1)^n)}{\pi n\left(n^2 - 49\right)}.$$

Note: When $n = 7$, $b_n = 0$ by orthogonality. Therefore, the complete solution for $u(x,t)$ is

$$u(x,t) = \frac{2}{\pi}x - 1 - \frac{49}{3\pi}\left(\frac{e^{-3t}\sin x}{4} + \frac{e^{-12t}\sin 2x}{15} + \frac{e^{-27t}\sin 3x}{5}\right)$$

$$- \frac{49}{\pi}\left(\frac{e^{-48t}\sin 4x}{22} + \frac{e^{-75t}\sin 5x}{10} + \frac{5e^{-108t}\sin 6x}{39}\right) +$$

$$\sum_{n=8}^{\infty}\frac{98(n-(-1)^n)}{\pi n\left(n^2 - 49\right)}e^{-3n^2 t}\sin nx;$$

or it may be written as

$$u(x,t) = \frac{2}{\pi}x - 1 + \sum_{n=1}^{6} \frac{98(n - (-1)^n)}{\pi n \left(n^2 - 49\right)} e^{-3n^2 t} \sin nx$$

$$+ \sum_{n=8}^{\infty} \frac{98(n - (-1)^n)}{\pi n \left(n^2 - 49\right)} e^{-3n^2 t} \sin nx.$$

In Figure (7.2), the solution for $u(x,t)$ is shown for $0 \le t \le 0.05$. Please note the boundaries are at -1 when $x = 0$ and 1 when $x = \pi$. Also, the initial temperature distribution is indicated by a bold-faced curve. ∎

Now, a question that should come to your mind is: The method works great for the heat equation, but how does it apply to the wave equation? This is a very reasonable question considering the wave equation does not have a steady state solution. But, the wave equation does have an equilibrium, and the equilibrium state for the wave equation is equivalent to the steady-state solution of the heat equation. Thus, this technique will work with the wave equation.

Although the homogeneous PDE with constant nonhomogeneous BCs is an interesting problem, the homogeneous PDE with variable nonhomogeneous BCs is far more interesting. We solve this problem in the next subsection.

7.3.2   Homogeneous PDE—Nonhomogeneous Variable BCs

Consider the temperature distribution in a one-dimensional uniform rod with perfect lateral insulation. Suppose that the boundary, $x = 0$, is placed in perfect thermal contact with a medium that changes temperature with time t; at the boundary $x = L$, the flow of heat is described by a function that changes with time t. Also, suppose we know the initial temperature distribution in the rod.

The mathematical model that describes this physical situation is

$$\frac{\partial u}{\partial t} = k \frac{\partial^2 u}{\partial x^2}, \tag{7.43}$$

subject to

$$u(0,t) = \alpha(t)$$

$$\frac{\partial u(L,t)}{\partial x} = \beta(t) \tag{7.44}$$

and

$$u(x,0) = f(x). \tag{7.45}$$

As in the previous subsection, our goal is to determine a function that has homogeneous boundary conditions and can model the temperature distribution in the rod. The approach we will take is called construction. As the name suggests, we will construct a function $v(x,t)$ such that $v(x,t) = u(x,t) - r(x,t)$, with $v(0,t) = 0$ and $\dfrac{\partial v(L,t)}{\partial x} = 0$. This, in turn, requires $r(x,t)$ to have the following properties:

$$r(0,t) = \alpha(t)$$

$$\frac{\partial r(L,t)}{\partial x} = \beta(t). \tag{7.46}$$

Using Equation (7.46) as a starting point, we realize that integrating both sides of $\dfrac{\partial r(x,t)}{\partial x} = \beta(t)$ with respect to x yields $r(x,t) = x\beta(t) + c(t)$. (*Note:* As far as the variable x is concerned the function $c(t)$ is a constant.) Furthermore, applying the first BC, $r(0,t) = \alpha(t)$, we find $\alpha(t) = c$. Thus, it would seem reasonable to assume that $r(x,t) = x\beta(t) + \alpha(t)$. This representation of $r(x,t)$ yields the necessary properties described in Equation (7.46). However, there are an infinite number of functions $r(x,t)$ that may be generated or picked. For instance, $r_1(x,t) = \alpha(t)\left[x\cos L + \cos x\right] - \dfrac{L}{\pi}\beta(t)\sin\left(\dfrac{\pi x}{L}\right)$ will also be a continuous function that satisfies the necessary properties in Equation (7.46). Although $r_1(x,t)$ will work, it is a much more complicated equation than $r(x,t) = x\beta(t) + \alpha(t)$, which we will now use. We may now construct $v(x,t)$ and replace $u(x,t)$ in Equations (7.43 and 7.44) to yield

$$v(x,t) = u(x,t) - r(x,t) = u(x,t) - (x\beta(t) + \alpha(t)). \tag{7.47}$$

This implies that

$$\frac{\partial v}{\partial t} = \frac{\partial u}{\partial t} - \frac{\partial r}{\partial t} = \frac{\partial u}{\partial t} - (x\beta'(t) + \alpha'(t)).$$

Solving for $\dfrac{\partial u}{\partial t}$ yields

$$\frac{\partial u}{\partial t} = \frac{\partial v}{\partial t} + (x\beta'(t) + \alpha'(t)),$$

while $\dfrac{\partial^2 u}{\partial x^2}$ yields

$$\frac{\partial^2 v}{\partial x^2} = \frac{\partial^2 u}{\partial x^2} - \frac{\partial^2 r}{\partial x^2} = \frac{\partial^2 u}{\partial x^2}.$$

Therefore, replacing $u(x,t)$ in Equation (7.43) with an equivalent equation in $v(x,t)$ gives

$$\frac{\partial v}{\partial t} + (x\beta'(t) + \alpha'(t)) = k\frac{\partial^2 v}{\partial x^2}.$$

Thus, we now solve the nonhomogeneous PDE

$$\frac{\partial v}{\partial t} = k\frac{\partial^2 v}{\partial x^2} - \left(x\beta'(t) + \alpha'(t)\right), \tag{7.48}$$

subject to the homogeneous BCs

$$v(0,t) = 0$$

$$\frac{\partial v(L,t)}{\partial x} = 0. \tag{7.49}$$

To do so, we apply what we learned in section two of this chapter to Equation (7.48), subject to BCs, Equation (7.49).

Step 1: Solve the related homogeneous problem,

$$\frac{\partial v}{\partial t} = k\frac{\partial^2 v}{\partial x^2},$$

for eigenvalues and eigenfunctions. From Chapter 4, we know the eigenvalues and eigenfunctions are

$$\left.\begin{array}{l} \lambda_n = \left[\dfrac{(2n-1)\pi}{2L}\right]^2 \\[2em] \varphi_n(x) = c_n \sin\left[\sqrt{\lambda_n}x\right] \end{array}\right\} n = 1, 2, 3, \ldots$$

Step 2: Assume the solution for $v(x,t)$ is a Fourier series in the eigenfunctions just found. We have

$$v(x,t) = \sum_{n=1}^{\infty} c_n(t) \sin\left[\sqrt{\lambda_n}x\right], \tag{7.50}$$

where $\lambda_n = \left[\dfrac{(2n-1)\pi}{2L}\right]^2$, $n = 1, 2, 3, \ldots$ Remember c_n is a function of time t.

Step 3: Expand the term $(x\beta'(t) + \alpha'(t))$ in a Fourier series using the eigenvalues and eigenfunctions found in Step 1. This yields

$$(x\beta'(t) + \alpha'(t)) = \sum_{n=1}^{\infty} d_n \sin\left[\sqrt{\lambda_n}x\right],$$

where

$$d_n = \frac{2}{L}\int_0^L (x\beta'(t) + \alpha'(t))\sin\left[\sqrt{\lambda_n}x\right]\, dx = a_n(t).$$

Therefore, we have

$$(x\beta'(t) + \alpha'(t)) = \sum_{n=1}^{\infty} a_n(t) \sin\left[\sqrt{\lambda_n}x\right]. \tag{7.51}$$

Step 4: Find $\dfrac{\partial v}{\partial t}$ and $\dfrac{\partial^2 v}{\partial x^2}$. Remember, this requires us to satisfy all of the conditions necessary to do term-by-term differentiation of a Fourier series. Assuming all the conditions were met, we obtain

$$\frac{\partial v}{\partial t} = \sum_{n=1}^{\infty} c_n'(t) \sin\left[\sqrt{\lambda_n}\, x\right] \tag{7.52}$$

and

$$\frac{\partial^2 v}{\partial x^2} = -\sum_{n=1}^{\infty} \lambda_n c_n(t) \sin\left[\sqrt{\lambda_n}\, x\right]. \tag{7.53}$$

Step 5: Solve Equation (7.49) by replacing the terms of the equation with what we determined they equal in Steps 1 through 4. Thus,

$$\frac{\partial v}{\partial t} = k\frac{\partial^2 v}{\partial x^2} - (x\beta'(t) + \alpha'(t))$$

becomes

$$\sum_{n=1}^{\infty} c_n'(t) \sin\left[\sqrt{\lambda_n}\, x\right] = -k\sum_{n=1}^{\infty} \lambda_n c_n(t) \sin\left[\sqrt{\lambda_n}\, x\right] - \sum_{n=1}^{\infty} a_n(t) \sin\left[\sqrt{\lambda_n}\, x\right].$$

We now have the Fourier series expansions with respect to the same eigenvalues over the interval $[0, L]$, in all three terms of the equation. This means that the Fourier coefficients must be equal. Extracting the coefficients yields the ODE

$$c_n'(t) = -k\lambda_n c_n(t) - a_n(t)$$

or

$$c_n'(t) + k\lambda_n c_n(t) = -a_n(t). \tag{7.54}$$

Equation (7.54) is a first-order ODE. We develop the solution by using the integrating factor

$$e^{k\lambda_n t}.$$

Thus, the formal solution for $c_n(t)$ is

$$c_n(t) = -e^{-k\lambda_n t} \int^{t} e^{k\lambda_n \tau} a_n(\tau)\, d\tau + b_n e^{-k\lambda_n t}.$$

Note that the integral does not have a lower limit. We assume that when both sides of the equation are integrated, we get constants of integration on both sides of the equation, which we combine and call b_n.

Step 6: Write the solution to Equation (7.49). We have

$$v(x, t) = \sum_{n=1}^{\infty} \left[-e^{-k\lambda_n t} \int^{t} e^{k\lambda_n \tau} a_n(\tau)\, d\tau + b_n e^{-k\lambda_n t} \right] \sin\left[\sqrt{\lambda_n}\, x\right].$$

We cannot apply the initial condition since this is the solution to $v(x,t)$, and we want the solution to $u(x,t)$. But, from Equation (7.47) we know

$$u(x,t) = v(x,t) + r(x,t) = v(x,t) + x\beta(t) + \alpha(t).$$

Substituting, we get

$$u(x,t) = x\beta(t) + \alpha(t) +$$

$$\sum_{n=1}^{\infty} \left[-e^{-k\lambda_n t} \int^t e^{k\lambda_n \tau} a_n(\tau) \, d\tau + b_n e^{-k\lambda_n t} \right] \sin\left[\sqrt{\lambda_n} x\right]. \tag{7.55}$$

We now apply the IC, Equation (7.45), to Equation (7.55) yielding

$$u(x,0) = f(x) = x\beta(0) + \alpha(0) + \sum_{n=1}^{\infty} A_n(0) + b_n \sin\left[\sqrt{\lambda_n} x\right],$$

where $A_n(0) = \int^t e^{k\lambda_n \tau} a_n(\tau) \, d\tau|_{t=0}.$

Rewriting this equation yields

$$f(x) - (x\beta(0) + \alpha(0)) = \sum_{n=1}^{\infty} A_n(0) + b_n \sin\left[\sqrt{\lambda_n} x\right].$$

Using the orthogonality of the sine function, we see that

$$b_n = -A_n(0) + \frac{2}{L} \int_0^L [f(x) - (x\beta(0) + \alpha(0))] \sin\left[\sqrt{\lambda_n} x\right] \, dx. \tag{7.56}$$

So now the complete solution to Equation (7.43), subject to BCs, Equations (7.44), and an IC, Equation (7.45), is

$$u(x,t) = x\beta(t) + \alpha(t) +$$

$$\sum_{n=1}^{\infty} \left[-e^{-k\lambda_n t} \int^t e^{k\lambda_n \tau} a_n(\tau) \, d\tau + e^{-k\lambda_n t} b_n \right] \sin\left[\sqrt{\lambda_n} x\right],$$

where b_n is given by Equation (7.56).

The following example illustrates this method, and it provides a graphic representation of the solution.

EXAMPLE 7.3. Find the time-dependent solution for $u(x,t)$ if

$$\frac{\partial u}{\partial t} = \frac{\partial^2 u}{\partial x^2}, \tag{7.57}$$

subject to

$$u(0, t) = \alpha(t) = e^{-t}$$

$$u\left(\frac{\pi}{2}, t\right) = \beta(t) = t \tag{7.58}$$

and

$$u(x, 0) = \cos 3x. \tag{7.59}$$

The PDE in Equation (7.57) is linear and homogeneous. However, the BCs, Equation (7.58), are linear but not homogeneous. Thus, to use the separation of variables technique, we must develop a new PDE in terms of $v(x, t)$, where

$$v(x, t) = u(x, t) - r(x, t). \tag{7.60}$$

This means the function $r(x, t)$ must equal e^{-t} when $x = 0$. It must equal t when $x = \frac{\pi}{2}$. The BCs in Equation (7.58) suggest that $r(x, t) = \alpha(t) + \frac{x}{L}(\beta(t) - \alpha(t)) = e^{-t} + \frac{2x}{\pi}(t - e^{-t})$. (*Note:* Although $r(x, t)$ is one of the easier continuous functions to work with, there are an infinite number which may be found.) Thus, $r(0, t) = e^{-t}$ and $r\left(\frac{\pi}{2}, t\right) = t$.

Since $v(x, t) = u(x, t) - r(x, t)$, we have $v(0, t) = 0 = v\left(\frac{\pi}{2}, t\right)$. We must now replace $u(x, t)$ in Equation (7.57) with $v(x, t)$. Using Equation (7.60) yields

$$u(x, t) = v(x, t) + r(x, t) = v(x, t) + e^{-t} + \frac{2x}{\pi}(t - e^{-t}).$$

Thus,

$$\frac{\partial u}{\partial t} = \frac{\partial v}{\partial t} + \frac{\partial r}{\partial t} = \frac{\partial v}{\partial t} - e^{-t} + \frac{2x}{\pi}(1 + e^{-t}),$$

and

$$\frac{\partial^2 u}{\partial x^2} = \frac{\partial^2 v}{\partial x^2} + \frac{\partial^2 r}{\partial x^2} = \frac{\partial^2 v}{\partial x^2},$$

since $\frac{\partial^2 r}{\partial x^2} = 0$.

We can now rewrite Equation (7.57), subject to the BCs in Equation (7.58), in terms of $v(x, t)$. We find that

$$\frac{\partial v}{\partial t} - e^{-t} + \frac{2x}{\pi}(1 + e^{-t}) = \frac{\partial^2 v}{\partial x^2}$$

or, written in the usual form

$$\frac{\partial v}{\partial t} = \frac{\partial^2 v}{\partial x^2} + e^{-t} - \frac{2x}{\pi}(1 + e^{-t}), \tag{7.61}$$

subject to

$$v(0, t) = 0$$

$$v\left(\frac{\pi}{2}, t\right) = 0,$$

(7.62)

where the source term $Q(x, t)$ equals $e^{-t} - \frac{2x}{\pi}(1 + e^{-t})$. We now have a linear PDE with linear and homogeneous BCs.

Step 1: Solve the related homogeneous problem, which is

$$\frac{\partial v}{\partial t} = \frac{\partial^2 v}{\partial x^2},$$

subject to

$$v(0, t) = 0$$

$$v\left(\frac{\pi}{2}, t\right) = 0.$$

Solving the related homogeneous problem for eigenvalues and eigenfunctions yields

$$\left. \begin{array}{l} \lambda_n = 4n^2 \\[2mm] \varphi_n(x) = \sin 2nx \end{array} \right\}, \quad n = 1, 2, 3, \ldots$$

Step 2: Assume the solution for $v(x, t)$ is a Fourier series in the eigenfunctions described earlier. We have

$$v(x, t) = \sum_{n=1}^{\infty} b_n(t) \sin 2nx.$$

(7.63)

Step 3: Expand $Q(x, t)$ in a Fourier series using the eigenfunctions found in Step 1. This yields

$$e^{-t} - \frac{2x}{\pi}(1 + e^{-t}) = \sum_{n=1}^{\infty} a_n(t) \sin 2nx.$$

Remember to use orthogonality when solving for $a_n(t)$. Thus, we have

$$a_n(t) = \frac{4}{\pi} \int_0^{\frac{\pi}{2}} \left(e^{-t} - \frac{2x}{\pi}(1 + e^{-t}) \right) \sin 2nx \ dx$$

$$= \frac{2}{n\pi} \left(e^{-t} + (-1)^n \right).$$

Therefore,

$$e^{-t} - \frac{2x}{\pi}(1 + e^{-t}) = \sum_{n=1}^{\infty} a_n(t) \sin 2nx$$

$$= \sum_{n=1}^{\infty} \frac{2}{n\pi}\left(e^{-t} + (-1)^n\right) \sin 2nx. \tag{7.64}$$

Step 4: Using Equation (7.63), find $\dfrac{\partial v}{\partial t}$ and $\dfrac{\partial^2 v}{\partial x^2}$. Remember, this requires us to satisfy all of the conditions necessary to do term-by-term differentiation of a Fourier series. Assuming that all the conditions are met, we have

$$\frac{\partial v}{\partial t} = \sum_{n=1}^{\infty} b'_n(t) \sin 2nx \tag{7.65}$$

and

$$\frac{\partial^2 v}{\partial x^2} = \sum_{n=1}^{\infty} -4n^2 b_n(t) \sin 2nx. \tag{7.66}$$

Step 5: Solve Equation (7.61) using Equations (7.64, 7.65, and 7.66). Thus,

$$\frac{\partial v}{\partial t} = \frac{\partial^2 v}{\partial x^2} + e^{-t} - \frac{2x}{\pi}(1 + e^{-t})$$

becomes

$$\sum_{n=1}^{\infty'} b'_n(t) \sin 2nx = \sum_{n=1}^{\infty} -4n^2 b_n(t) \sin 2nx$$

$$+ \sum_{n=1}^{\infty} \frac{2}{n\pi}\left(e^{-t} + (-1)^n\right) \sin 2nx. \tag{7.67}$$

In Equation (7.67), all the Fourier series expansions are over the same interval $\left[0, \dfrac{\pi}{2}\right]$ and are expanded in the same eigenfunctions. Since Fourier series representation of a function is unique, we may solve the ODE formed by the Fourier coefficients, which is

$$b'_n(t) = -4n^2 b_n(t) + \frac{2}{n\pi}\left(e^{-t} + (-1)^n\right). \tag{7.68}$$

The ODE in Equation (7.68) can be solved by methods reviewed in Appendix C. The solution is

$$b_n(t) = \left[\frac{4n^2 e^{-t} + (4n^2 - 1)(-1)^n}{n^2(4n^2 - 1)\pi} + d_n e^{-4n^2 t}\right].$$

Therefore,

$$v(x,t) = \sum_{n=1}^{\infty} \left[\frac{4n^2 e^{-t} + (4n^2 - 1)(-1)^n}{n^2(4n^2 - 1)\pi} + d_n e^{-4n^2 t} \right] \sin 2nx.$$

Thus, the general solution for $u(x,t)$ is

$$u(x,t) = r(x,t) + v(x,t)$$

$$= e^{-t} + \frac{2x}{\pi}(t - e^{-t})$$

$$+ \sum_{n=1}^{\infty} \left[\frac{4n^2 e^{-t} + (4n^2 - 1)(-1)^n}{n^2(4n^2 - 1)\pi} + d_n e^{-4n^2 t} \right] \sin 2nx. \qquad (7.69)$$

Equation (7.69) indicates the general solution for $u(x,t)$. We now find the specific solution for $u(x,t)$ by applying the IC (Equation (7.59)), $u(x,0) = \cos 3x$. Thus,

$$u(x,0) = \cos 3x$$

$$= 1 - \frac{2x}{\pi} + \sum_{n=1}^{\infty} \left[\frac{4n^2 e^{-t} + (4n^2 - 1)(-1)^n}{n^2(4n^2 - 1)\pi} + d_n e^{-4n^2 t} \right] \sin 2nx,$$

or

$$\cos 3x - 1 + \frac{2x}{\pi} = \sum_{n=1}^{\infty} \left[\frac{4n^2 + (4n^2 - 1)(-1)^n}{n^2(4n^2 - 1)\pi} + d_n \right] \sin 2nx. \qquad (7.70)$$

Using orthogonality, the constants d_n in Equation (7.70) are

$$d_n = -\frac{4n^2 + (4n^2 - 1)(-1)^n}{n^2(4n^2 - 1)\pi}$$

$$+ \frac{4}{\pi} \int_0^{\frac{\pi}{2}} \left(\cos 3x - 1 + \frac{2x}{\pi} \right) \sin 2nx \; dx$$

$$= -\frac{4n^2 + (4n^2 - 1)(-1)^n}{n^2(4n^2 - 1)\pi} + \frac{18}{n(4n^2 - 9)\pi}. \qquad (7.71)$$

Therefore, the complete solution for $u(x,t)$ is

$$u(x,t) = e^{-t} + \frac{2x}{\pi}(t - e^{-t})$$

$$+ \sum_{n=1}^{\infty} \left[\frac{4n^2 e^{-t} + (4n^2 - 1)(-1)^n}{n^2(4n^2 - 1)\pi} + d_n e^{-4n^2 t} \right] \sin 2nx,$$

where d_n is given in Equation (7.71).

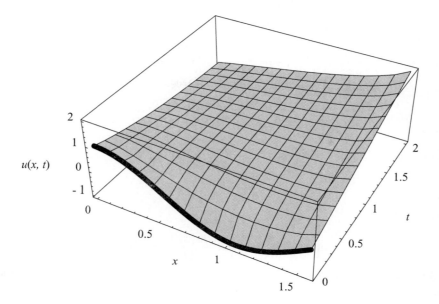

$u(x, t)$

x

t

Figure 7.3: The graph of the solution for $u(x,t)$ for $0 \leq t \leq 2$.

A graph of the solution is given in Figure (7.3) for $0 \leq t \leq 2$, and the initial temperature distribution is indicated by a bold-faced curve. Note how the BCs affect the surface as time progresses. ■

In the previous two sections, we solved the problems of the nonhomogeneous PDE with homogeneous BCs and the homogeneous PDE with nonhomogeneous BCs. Both of these problem types are useful in solving physical models. More importantly, they help us solve the more general problem of a nonhomogeneous PDE with nonhomogeneous BCs. The nonhomogeneous PDE with nonhomogeneous BCs applies to many more physical models. We address this general problem in the next section.

EXERCISES 7.3

7.3.1. Consider the heat equation

$$\frac{\partial u}{\partial t} = k \frac{\partial^2 u}{\partial x^2},$$

subject to

$$u(x,0) = f(x).$$

Solve for a time-dependent solution given the following nonhomogeneous boundary conditions:

(1) $\dfrac{\partial u(0,t)}{\partial x} = \alpha$ and $\dfrac{\partial u(L,t)}{\partial x} = \beta$.

(2) $u(0,t) = \alpha$ and $\dfrac{\partial u(L,t)}{\partial x} = \beta$.

(3) $\dfrac{\partial u(0,t)}{\partial x} = \alpha(t)$ and $\dfrac{\partial u(L,t)}{\partial x} = \beta(t)$.

(4) $\dfrac{\partial u(0,t)}{\partial x} = \alpha(t)$ and $u(L,t) = \beta$.

7.3.2. Consider the wave equation

$$\frac{\partial^2 u}{\partial t^2} = c^2 \frac{\partial^2 u}{\partial x^2},$$

subject to

$$u(x,0) = f(x) \text{ and } \frac{\partial u(x,0)}{\partial t} = g(x).$$

Solve for a time-dependent solution given the following nonhomogeneous boundary conditions:

(1) $u(0,t) = \alpha$ and $u(L,t) = \beta$.

(2) $\dfrac{\partial u(0,t)}{\partial x} = \alpha$ and $u(L,t) = \beta$.

(3) $u(0,t) = \alpha(t)$ and $u(L,t) = \beta(t)$.

(4) $u(0,t) = \alpha$ and $\dfrac{\partial u(L,t)}{\partial x} = \beta(t)$.

7.3.3. Liquid is input at time $t = 0$ by the amount of 3 drops/sec into the right end of a tube of length 2. The left end of the tube is connected to a large tank in which liquid pressure remains invariant. Assuming that, until the change of input at the end $x = 2$, the pressure and input to the tube are constant, find the change of input into the tube for $t > 0$ and the change in pressure in the section $x = 2$ for $t > 0$. The boundary value problem that governs this system is

$$\left. \begin{aligned} -\frac{\partial p}{\partial x} &= \frac{\partial w}{\partial t} - 2w \\[2mm] -\frac{\partial p}{\partial t} &= \frac{\partial w}{\partial x} \end{aligned} \right\} \quad 0 \le x \le 2, \, 0 < t,$$

subject to

$$p(0,t) = 0 \text{ and } w(2,t) = 3 \text{ drops/second}$$

and

$$w(x,0) = 0$$

$$p(x,0) = 0.$$

(*Hint:* Eliminate $p(x, t)$ in the equations previously listed and obtain a second-order wave equation in $w(x, t)$. If you are unsure how to eliminate $p(x, t)$, review Chapter 2.)

7.3.4. Consider the mathematical model

$$\frac{\partial^2 u}{\partial t^2} = 4 \frac{\partial^2 u}{\partial x^2},$$

subject to

$$u(0, t) = 6 \text{ and } u(\pi, t) = 1$$

and

$$u(x, 0) = 6 \cos x + \frac{7}{\pi} x \text{ and } \frac{\partial u(x, 0)}{\partial t} = 0.$$

Find the time-dependent solution for $u(x, t)$, and using your favorite mathematical software, plot your solution. By inspection, determine if the graph of your solution is reasonable for the problem.

7.3.5. Consider the mathematical model

$$\frac{\partial^2 u}{\partial t^2} = c^2 \frac{\partial^2 u}{\partial x^2},$$

subject to

$$u(0, t) = \cos \omega t \text{ and } u(L, t) = 0$$

and

$$u(x, 0) = 0 \text{ and } \frac{\partial u(x, 0)}{\partial t} = 0.$$

Find the time-dependent solution for $u(x, t)$.

7.3.6. Consider the mathematical model

$$\frac{\partial^2 u}{\partial t^2} = 16 \frac{\partial^2 u}{\partial x^2},$$

subject to

$$u(0, t) = 0 \text{ and } u(5, t) = \sin 3t$$

and

$$u(x, 0) = 0 \text{ and } \frac{\partial u(x, 0)}{\partial t} = 0.$$

Find the time-dependent solution for $u(x, t)$, and using your favorite mathematical software, plot your solution. By inspection, determine if the graph of your solution is reasonable for the problem.

7.3.7. Consider the heat equation

$$\frac{\partial u}{\partial t} = \frac{\partial^2 u}{\partial x^2},$$

subject to

$$u(0, t) = 0 \text{ and } \frac{\partial u(L, t)}{\partial x} = \sin \omega t$$

and

$$u(x, 0) = f(x).$$

Find the time-dependent solution for $u(x, t)$.

7.3.8. Consider the heat equation

$$\frac{\partial u}{\partial t} = 1.14 \frac{\partial^2 u}{\partial x^2},$$

subject to

$$\frac{\partial u(0, t)}{\partial x} = \sin \omega t \text{ and } \frac{\partial u(7, t)}{\partial x} = 0$$

and

$$u(x, 0) = \begin{cases} -x, & 0 \le x < 2 \\ -.5x^2 + 4.5x - 9, & 2 \le x \le 5 \\ 1, & 5 < x \le 7. \end{cases}$$

Find the time-dependent solution.

7.3.9. Consider the PDE

$$\frac{\partial^2 v}{\partial x^2} - LC \frac{\partial^2 v}{\partial t^2} - (RC + GL)\frac{\partial v}{\partial t} - GRv = 0,$$

where v is voltage, L is self-inductance, R is resistance, G is leakage conductance, and C is capacitance. Find the voltage if

$$v(0, t) = E \text{ (a constant electromagnetic force)}$$

$$\frac{\partial v(8, t)}{\partial x} = 0 \text{ (perfect insulation)}$$

and

$$v(x, 0) = 0 \text{ and } \frac{\partial v(x, 0)}{\partial t} = 0.$$

7.3.10. Given the PDE

$$\frac{\partial^2 v}{\partial x^2} - LC\frac{\partial^2 v}{\partial t^2} - (RC + GL)\frac{\partial v}{\partial t} - GRv = 0,$$

where v is voltage, L is self-inductance, R is resistance, G is leakage conductance, and C is capacitance. Find the voltage if

$$\frac{\partial v(0, t)}{\partial x} = 0 \text{ (perfect insulation) and } v(\pi, t) = 5\sin\omega_0 t$$

and

$$v(x, 0) = 0 \text{ and } \frac{\partial v(x, 0)}{\partial t} = 0.$$

7.3.11. Consider a rectangular plate made of nickel with length 2 yards and width 1.5 yards. Further, suppose the plate has perfect lateral insulation, and the initial temperature distribution is given by

$$u(x, y, 0) = \sin(xy).$$

At the boundaries $x = 0$ and $y = 0$ the plate is held in a $0°C$ bath, the boundary at $x = 2$ is imperfectly insulated, which is known to vary not only with time, t, but with position, y, and is estimated to follow the equation $ye^{-0.2t}$; and the boundary at $y = 1.5$ is perfectly insulated. Perform the following:

(1) Mathematically model this problem. *Note:* The thermal diffusivity of nickel may be found in Appendix E.

(2) Find a time-dependent solution of your mathematical model.

(3) Graph the solution of your mathematical model for several fixed times, t.

7.3.12. Consider vibrations in a perfectly flexible rectangular membrane of length 2 m and width 3 m. Suppose the tension is 15 kgs, the density is 1 kg/m, the initial position is given by

$$u(x, y, 0) = f(x, y) = \begin{cases} xy; & 0 \le x < 1 \quad 0 \le y < 1.5 \\ x(3 - y); & 0 \le x < 1 \quad 1.5 \le y \le 3 \\ (2 - x)y; & 1 \le x \le 2 \quad 0 \le y < 1.5 \\ (2 - x)(3 - y); & 1 \le x \le 2 \quad 1.5 \le y \le 3, \end{cases}$$

and the initial velocity is zero. Further, suppose that the edge $x = 0$ is rigidly fixed at 0.5 m, the edges $x = 2$ and $y = 3$ are freely moving, and the edge $y = 0$ oscillates to the function $0.5\cos t$. Determine the time-dependent solution. Plot your solution for several different times t.

7.3.13. Given a perfectly flexible rectangular membrane of length 4 ft and width 2 ft, if the tension is 12.5 lbs, the density is 2.5 lbs/ft (light rubber), the initial displacement is $f(x, y) = 0.1(4x - x^2)(2y - y^2)$ feet, the initial velocity is 0, the edge $x = 0$ is fixed at 2, the edge $x = 4$ is fixed at zero, and the edges $y = 0$ and $y = 2$ are free moving:

(1) Find the time-dependent solution.

(2) Using your favorite mathematical software, plot your solution. By inspection, determine if the graph of your solution is reasonable for the problem.

(3) Use only the first term in the series for the vertical vibrations $u(x, y, t)$ to find the approximate solution at $t = 30$ secs, $x = 1$ ft, and $y = 0.5$ ft.

(4) Use the first two terms in the series for the vertical vibrations $u(x, y, t)$ to find the approximate solution at $t = 30$ secs, $x = 1$ ft, and $y = 0.5$ ft. Compare this answer with the previous answer.

7.3.14. Consider

$$\frac{\partial u(x, t)}{\partial t} = 16 \frac{\partial^2 u(x, t)}{\partial x^2} + 9x,$$

subject to

$$\text{BC:} \begin{cases} u(0, t) = 0 \\ \\ u(4, t) = 36 \end{cases} \quad \text{and} \quad \text{IC:} \quad u(x, 0) = 9xe^{x-4}.$$

(1) Give a physical interpretation of this problem.

(2) For very large time find

(a) the solution,

(b) the heat energy generated per unit time inside the entire rod,

(c) the heat energy flowing out of the rod per unit time at each end, and

(d) state the relationship between parts (b) and (c).

(3) After Part (2) is completed, the heat source is turned off. Perform the following:

(a) state the equations that describe the mathematical model at this time and

(b) using the equations that you just stated, solve the heat flow problem for any time t.

7.3.15. A uniform string with mass density 3 lbs/ft and tension 300 lbs is fixed at the left end at 0 and driven at the right end by an oscillator with amplitude and angular frequency of 1. The string has length 200 and is initially at rest with linear displacement from 0 to 1.

(1) Model this problem mathematically.

(2) Solve the mathematical model.

(3) Write the solution with all known quantities substituted into it.

(4) Check that this solution satisfies the equation, the boundary conditions, and the initial conditions.

(5) Using your favorite mathematical software, plot your solution. By inspection, determine if the graph of your solution is reasonable for the problem.

7.3.16. Consider the PDE

$$\frac{\partial^2 z}{\partial t^2} = 9\frac{\partial^2 z}{\partial x^2} - 2\frac{\partial z}{\partial t},$$

subject to

$$\frac{\partial z(0, t)}{\partial x} = 0 \text{ and } z(10, t) = 5$$

and

$$u(x, 0) = 0 \text{ and } \frac{\partial u(x, 0)}{\partial t} = 0.$$

Find the time-dependent solution for $z(x, t)$, and using your favorite mathematical software, plot your solution. By inspection, determine if the graph of your solution is reasonable for the problem.

This problem reflects the input of liquid into the right end of a tube of length 10 ft dropping at $t = 0$ by an amount of 5 gals/sec. The left end of the tube is attached to a large tank in which the liquid pressure remains invariant. If you assume that until the change of input into the right end, the pressure and input were constant, we can find the change in input for $t > 0$ using the equations listed previously, and the change in pressure in the section $x = 10$ for $t > 0$ by the following equation

$$-\frac{\partial p}{\partial x} = \frac{\partial z}{\partial t} + 2z.$$

7.3.17. Consider the PDE

$$\frac{\partial u}{\partial t} = \frac{\partial^2 u}{\partial x^2},$$

subject to

$$u(0, t) = 0 \text{ and } u(5, t) = 2t$$

and

$$u(x, 0) = x^2 + 2x - 4.$$

Find the time-dependent solution for $u(x, t)$.

7.4   NONHOMOGENEOUS PDE AND BCs

Suppose we have transverse vibrations in a perfectly flexible uniform one-dimensional string with a time-dependent forcing function. Also, suppose that the boundary $x = 0$ is fixed at some constant height and, at the boundary $x = L$, a time-dependent force is being applied. Finally, to complete the description of the physical problem, there is an initial displacement and velocity.

The mathematical model which describes this problem is

$$\frac{\partial^2 u}{\partial t^2} = c^2 \frac{\partial^2 u}{\partial x^2} + Q(x,t), \tag{7.72}$$

subject to BCs

$$u(0,t) = A$$

$$\frac{\partial u(L,t)}{\partial x} = B(t) \tag{7.73}$$

and ICs

$$u(x,0) = f(x)$$

$$\frac{\partial u(x,0)}{\partial t} = g(x). \tag{7.74}$$

Equations (7.72–7.74) contain parts of every problem we have discussed in this chapter to this point. We have a nonhomogeneous PDE, and the nonhomogeneous BCs contain both a nonzero constant and a function of time t. Therefore, to solve this problem a solution strategy is mapped out in the following steps:

1. Replace $u(x,t)$ with a function $v(x,t)$ which has homogeneous BCs. This requires constructing a function $r(x,t)$ so that $u(0,t) - r(0,t) = 0 = \dfrac{\partial u(L,t)}{\partial x} - \dfrac{\partial r(L,t)}{\partial x}$.

2. Solve the related homogeneous PDE in $v(x,t)$ for eigenvalues and eigenfunctions.

3. Assume a Fourier series solution for $v(x,t)$, where the coefficient in the Fourier series is a function of time t.

4. Perform an eigenfunction expansion of the time-dependent function, $Q(x,t)$, using the eigenfunctions found in Step 2.

5. Determine the derivatives of the Fourier series solution for $v(x,t)$.

6. Replace the PDE by its equivalent Fourier series representation and determine the ODE, which must be solved.

7. Solve the ODE.

8. Replace the coefficient in the Fourier series representation for $v(x, t)$ with the solution from the ODE.

9. Determine the general solution for $u(x, t)$ from the solution for $v(x, t)$.

10. Determine the constants in the general solution of $u(x, t)$ by using the ICs and orthogonality.

This is our plan. Now, we will carry it out.

Step 1: Replace $u(x, t)$ with $v(x, t)$ where $v(x, t)$ has homogeneous boundary conditions. This requires the construction of a function $r(x, t)$ so that $u(0, t) - r(0, t) = 0 = \dfrac{\partial u(L, t)}{\partial x} - \dfrac{\partial r(L, t)}{\partial x}$.

To construct a function $r(x, t)$ where $r(0, t) = A$ and $\dfrac{\partial r(L, t)}{\partial x} = B(t)$, we simply integrate $\dfrac{\partial r(x, t)}{\partial x} = B(t)$ with respect to x. This gives $r(x, t) = xB(t) + c$. Now, evaluating $r(x, t) = xB(t) + c$ at $x = 0$ implies $c = A$. Therefore, the function we want is $r(x, t) = xB(t) + A$. Remember that

$$v(x, t) = u(x, t) - r(x, t), \tag{7.75}$$

and replacing $r(x, t)$ with $xB(t) + A$ in Equation (7.75) we generate

$$v(x, t) = u(x, t) - r(x, t) = u(x, t) - xB(t) - A. \tag{7.76}$$

This means the BCs

$$v(0, t) = 0$$

$$\dfrac{\partial v(L, t)}{\partial x} = 0 \tag{7.77}$$

are homogeneous. Rewriting Equation (7.76), we get

$$u(x, t) = v(x, t) + xB(t) + A. \tag{7.78}$$

Using Equation (7.78), we can replace $u(x, t)$ in Equation (7.72) with $v(x, t)$. This requires the determination of $\dfrac{\partial^2 u}{\partial t^2}$ and $\dfrac{\partial^2 u}{\partial x^2}$, since

$$\dfrac{\partial^2 u}{\partial t^2} = \dfrac{\partial^2 v}{\partial t^2} + xB''(t) \tag{7.79}$$

and

$$\dfrac{\partial^2 u}{\partial x^2} = \dfrac{\partial^2 v}{\partial x^2}, \tag{7.80}$$

$$\frac{\partial^2 u}{\partial t^2} = c^2 \frac{\partial^2 u}{\partial x^2} + Q(x,t)$$

becomes

$$\frac{\partial^2 v}{\partial t^2} + x B''(t) = c^2 \frac{\partial^2 v}{\partial x^2} + Q(x,t)$$

or

$$\frac{\partial^2 v}{\partial t^2} = c^2 \frac{\partial^2 v}{\partial x^2} + Q(x,t) - x B''(t), \tag{7.81}$$

subject to BCs, Equation (7.77).

Step 2: Solve the related homogeneous problem in terms of $v(x,t)$ for eigenvalues and eigenfunctions.

The related homogeneous problem is

$$\frac{\partial^2 v}{\partial t^2} = c^2 \frac{\partial^2 v}{\partial x^2},$$

subject to BCs, Equation (7.77). After separating variables and applying the BCs to the spatial problem, we conclude that

$$\left.\begin{aligned}
\lambda_n &= \left[\frac{(2n-1)\pi}{2L}\right]^2 \\[2ex]
\varphi_n(x) &= b_n \sin\left[\frac{(2n-1)\pi x}{2L}\right]
\end{aligned}\right\} \quad n = 1,2,3,\ldots \tag{7.82}$$

Step 3: Assume a Fourier series solution for $v(x,t)$, where the coefficient in the Fourier series is a function of time t.

We have

$$v(x,t) = \sum_{n=1}^{\infty} b_n(t) \sin\left[\frac{(2n-1)\pi x}{2L}\right]$$

$$= \sum_{n=1}^{\infty} b_n(t) \sin\left(\sqrt{\lambda_n}\, x\right). \tag{7.83}$$

Step 4: Perform an eigenfunction expansion of the time-dependent function, $Q(x,t)$, using the eigenfunctions found in Step 2.

We must remember the time-dependent forcing function in Equation (7.72) was changed to the time-dependent forcing function $Q(x,t) - x B''(t)$ in Equation (7.81). Therefore, we have

$$Q(x,t) - x B''(t) = \sum_{n=1}^{\infty} d_n \sin\left(\sqrt{\lambda_n}\, x\right),$$

where, after integration, the constant d_n becomes a function of time t. Thus, assuming we can perform the required integration, we have

$$Q(x,t) - xB''(t) = \sum_{n=1}^{\infty} d_n(t) \sin\left(\sqrt{\lambda_n}x\right). \tag{7.84}$$

Step 5: Determine the derivatives of the Fourier series solution for $v(x,t)$.

Remember, we must satisfy all the conditions that allow term-by-term differentiation of a Fourier sine series, which is covered in Chapter 5. Assuming these conditions are satisfied, we have

$$\frac{\partial v(x,t)}{\partial x} = \sum_{n=1}^{\infty} \sqrt{\lambda_n}b_n(t) \cos\left(\sqrt{\lambda_n}x\right).$$

The Fourier cosine series of a continuous function is continuous; thus, term-by-term differentiation is valid, and

$$\frac{\partial^2 v(x,t)}{\partial x^2} = -\sum_{n=1}^{\infty} \lambda_n b_n(t) \sin\left(\sqrt{\lambda_n}x\right). \tag{7.85}$$

Since we always expect time to be continuous, and therefore any function of time to be continuous in both the first and second derivatives, we see that

$$\frac{\partial^2 v(x,t)}{\partial t^2} = \sum_{n=1}^{\infty} b_n''(t) \sin\left(\sqrt{\lambda_n}x\right). \tag{7.86}$$

Step 6: Replace the PDE by its equivalent Fourier series representation and determine the ODE that must be solved.

Using Equations (7.84, 7.85, and 7.86), Equation (7.81)

$$\frac{\partial^2 v}{\partial t^2} = c^2 \frac{\partial^2 v}{\partial x^2} + Q(x,t) - xB''(t)$$

becomes

$$\sum_{n=1}^{\infty} b_n''(t) \sin\left(\sqrt{\lambda_n}x\right) = -c^2 \sum_{n=1}^{\infty} \lambda_n b_n(t) \sin\left(\sqrt{\lambda_n}x\right)$$

$$+ \sum_{n=1}^{\infty} d_n(t) \sin\left(\sqrt{\lambda_n}x\right). \tag{7.87}$$

Equation (7.87) equates one Fourier series to the sum of two Fourier series all expanded in the same eigenfunctions over the same interval. This means the constants must be equal. Extracting coefficients yields

$$b_n''(t) = -c^2 \lambda_n b_n(t) + d_n(t). \tag{7.88}$$

This is a second-order nonhomogeneous ODE.

Step 7: Solve the ODE. We solve a second-order nonhomogeneous ODE by solving the related homogeneous part, solving the particular part, and then summing the two parts together. In general, unless we know the form of the functions $b_n(t)$ and $d_n(t)$ in Equation (7.88), we may not be able to determine a solution. However, if we assume that a function of time, $b_n(t)$, is positive for all time t, and we know c and λ are positive, then the solution to the homogeneous part is

$$b_{n_h}(t) = \alpha_n \cos\left[c\sqrt{\lambda_n}t\right] + \beta_n \sin\left[c\sqrt{\lambda_n}t\right].$$

We will assume the particular part has a solution and call it $b_{n_p}(t)$. Therefore, the solution to Equation (7.88) is

$$b_n(t) = b_{n_h}(t) + b_{n_p}(t) = \alpha_n \cos\left[c\sqrt{\lambda_n}t\right] + \beta_n \sin\left[c\sqrt{\lambda_n}t\right] + b_{n_p}(t). \qquad (7.89)$$

Step 8: Replace the coefficient in the Fourier series representation for $v(x,t)$ with the solution from the ODE.

From Equation (7.83), we have

$$v(x,t) = \sum_{n=1}^{\infty} b_n(t) \sin\left(\sqrt{\lambda_n}x\right) \qquad (7.90)$$

$$= \sum_{n=1}^{\infty} \left[\alpha_n \cos\left[c\sqrt{\lambda_n}t\right] + \beta_n \sin\left[c\sqrt{\lambda_n}t\right] + b_{n_p}(t)\right] \sin\left(\sqrt{\lambda_n}x\right).$$

Step 9: Determine the solution for $u(x,t)$ from the solution for $v(x,t)$. Equation (7.78), which is

$$u(x,t) = v(x,t) + xB(t) + A = xB(t) + A + v(x,t),$$

yields the general form of $u(x,t)$. Replacing $v(x,t)$ in Equation (7.78) then gives the solution for $u(x,t)$. Thus, we have

$$u(x,t) = xB(t) + A \qquad (7.91)$$

$$+ \sum_{n=1}^{\infty} \left[\alpha_n \cos\left[c\sqrt{\lambda_n}t\right] + \beta_n \sin\left[c\sqrt{\lambda_n}t\right] + b_{n_p}(t)\right] \sin\left(\sqrt{\lambda_n}x\right).$$

Step 10: Use the ICs and orthogonality to determine the constants in the general solution of $u(x,t)$.

Equations (7.74) tell us

$$u(x,0) = f(x)$$

$$\frac{\partial u(x,0)}{\partial t} = g(x).$$

Thus, we find that the first IC implies

$$u(x, 0) = f(x) = xB(0) + A + \sum_{n=1}^{\infty} \left[\alpha_n + b_{n_p}(0)\right] \sin\left(\sqrt{\lambda_n}x\right)$$

or

$$f(x) - (xB(0) + A) = \sum_{n=1}^{\infty} \left[\alpha_n + b_{n_p}(0)\right] \sin\left(\sqrt{\lambda_n}x\right).$$

By using orthogonality, the constant within the summation becomes

$$\alpha_n + b_{n_p}(0) = \frac{2}{L}\int_0^L \left[f(x) - (xB(0) + A)\right] \sin\left(\sqrt{\lambda_n}x\right) dx,$$

which means that

$$\alpha_n = \frac{2}{L}\int_0^L \left[f(x) - (xB(0) + A)\right] \sin\left(\sqrt{\lambda_n}x\right) dx - b_{n_p}(0). \tag{7.92}$$

The second IC implies that

$$\frac{\partial u(x, 0)}{\partial t} = g(x) = xB'(0) + \sum_{n=1}^{\infty} \left[\beta_n + b'_{n_p}(0)\right] \sin\left(\sqrt{\lambda_n}x\right).$$

Therefore, in a similar method as described earlier, we have

$$\beta_n = \frac{2}{L}\int_0^L \left[g(x) - xB'(0)\right] \sin\left(\sqrt{\lambda_n}x\right) dx - b'_{n_p}(0). \tag{7.93}$$

We now have the complete solution for $u(x, t)$ from Equation (7.72) subject to BCs, Equation (7.73), and ICs, Equation (7.74). It is

$$u(x, t) = xB(t) + A$$
$$+ \sum_{n=1}^{\infty} \left[\alpha_n \cos\left[c\sqrt{\lambda_n}t\right] + \beta_n \sin\left[c\sqrt{\lambda_n}t\right] + b_{n_p}(t)\right] \sin\left(\sqrt{\lambda_n}x\right),$$

where α_n and β_n are given in Equation (7.92) and Equation (7.93), respectively, and λ_n is given in Equation (7.82).

EXAMPLE 7.4. Consider

$$\frac{\partial u}{\partial t} = \frac{\partial^2 u}{\partial x^2} - xt, \tag{7.94}$$

subject to

$$\frac{\partial u(0, t)}{\partial x} = 0$$
$$\frac{\partial u(\pi, t)}{\partial x} = t \tag{7.95}$$

and

$$u(x,0) = \sin\frac{11x}{2}. \tag{7.96}$$

Solve for the time-dependent solution using the ten steps developed in this section (Nonhomogeneous PDE with Nonhomogeneous BCs).

Step 1: Replace $u(x,t)$ with a function $v(x,t)$.

Let $v(x,t) = u(x,t) - r(x,t)$ where $r(x,t) = \dfrac{x^2 t}{2\pi}$, which implies $u(x,t) = v(x,t) + r(x,t)$. Therefore,

$$\frac{\partial u}{\partial t} = \frac{\partial v}{\partial t} + \frac{x^2}{2\pi} \tag{7.97}$$

and

$$\frac{\partial^2 u}{\partial x^2} = \frac{\partial^2 v}{\partial x^2} + \frac{t}{\pi}. \tag{7.98}$$

Replacing $u(x,t)$ with $v(x,t)$ in Equation (7.94) yields

$$\frac{\partial v}{\partial t} + \frac{x^2}{2\pi} = \frac{\partial^2 v}{\partial x^2} + \frac{t}{\pi} - xt$$

or

$$\frac{\partial v}{\partial t} = \frac{\partial^2 v}{\partial x^2} + \left(\frac{1}{\pi} - x\right)t - \frac{x^2}{2\pi}, \tag{7.99}$$

which has the following BCs

$$\frac{\partial v(0,t)}{\partial x} = 0$$
$$\frac{\partial v(\pi,t)}{\partial x} = 0 \tag{7.100}$$

as corresponding homogeneous BCs. We have completed Step 1, and we can move on to Step 2.

Step 2: Solve the related homogeneous PDE in $v(x,t)$ for eigenvalues and eigenfunctions only.

The related homogeneous PDE in $v(x,t)$ is

$$\frac{\partial v}{\partial t} = \frac{\partial^2 v}{\partial x^2} \tag{7.101}$$

subject to BCs, Equations (7.100). We have solved for the eigenvalues and eigenfunctions several times in the past chapters. The eigenvalues are

$$\lambda_n = n^2, \; n = 0, 1, 2, 3, \ldots \tag{7.102}$$

The related eigenfunctions are

$$\varphi_n(x) = \begin{cases} a_0, \ n = 0 \\ \\ a_n \cos nx, \ n > 0. \end{cases} \tag{7.103}$$

Step 3: Assume a Fourier series solution for $v(x,t)$ where the Fourier coefficients are functions of the variable t.
The Fourier series solution is

$$v(x,t) = a_0(t) + \sum_{n=1}^{\infty} a_n(t) \cos nx. \tag{7.104}$$

Remember, the Fourier coefficient must be a function of time, t. We now move on to Step 4.

Step 4: Do an eigenfunction expansion of the time-dependent forcing function $Q(x,t)$. Using the eigenfunctions stated in Equation (7.103), we have

$$\left(\frac{1}{\pi} - x\right) t - \frac{x^2}{2\pi} = b_0(t) + \sum_{n=1}^{\infty} b_n(t) \cos nx,$$

where

$$b_0(t) = \frac{1}{\pi} \int_0^{\pi} \left(\frac{1}{\pi} - x\right) t - \frac{x^2}{2\pi} \ dx$$

$$= \left(\frac{1}{\pi} - \frac{\pi}{2}\right) t - \frac{\pi}{6},$$

and

$$b_n(t) = \frac{2}{\pi} \int_0^{\pi} \left(\frac{1}{\pi} - x\right) t - \frac{x^2}{2\pi} \cos nx \ dx$$

$$= \frac{2}{\pi n^2} \left[(1 - (-1)^n) t - (-1)^n \right].$$

Thus,

$$\left(\frac{1}{\pi} - x\right) t - \frac{x^2}{2\pi} = \left(\frac{1}{\pi} - \frac{\pi}{2}\right) t - \frac{\pi}{6} +$$

$$\sum_{n=1}^{\infty} \frac{2}{\pi n^2} \left[(1 - (-1)^n) t - (-1)^n \right] \cos nx. \tag{7.105}$$

Step 5: Determine the derivatives of the Fourier series solution for $v(x, t)$. Using Equation (7.104) and assuming the required conditions hold, we have

$$\frac{\partial v(x, t)}{\partial t} = a_0'(t) + \sum_{n=1}^{\infty} a_n'(t) \cos nx \qquad (7.106)$$

and

$$\frac{\partial^2 v(x, t)}{\partial x^2} = \sum_{n=1}^{\infty} -n^2 a_n(t) \cos nx. \qquad (7.107)$$

Step 6: Replacing

$$\frac{\partial v}{\partial t} = \frac{\partial^2 v}{\partial x^2} + \left(\frac{1}{\pi} - x \right) t - \frac{x^2}{2\pi}$$

with its equivalent Fourier series representation yields

$$a_0'(t) + \sum_{n=1}^{\infty} a_n'(t) \cos nx = \sum_{n=1}^{\infty} -n^2 a_n(t) \cos nx + \qquad (7.108)$$

$$\left(\frac{1}{\pi} - \frac{\pi}{2} \right) t - \frac{\pi}{6} +$$

$$\sum_{n=1}^{\infty} \frac{2}{\pi n^2} \left[(1 - (-1)^n) t - (-1)^n \right] \cos nx.$$

From Chapter 4 we know that two Fourier series that are equal must have equal coefficients. Therefore, Equation (7.108) can be reduced to the ODEs

$$a_0'(t) = \left(\frac{1}{\pi} - \frac{\pi}{2} \right) t - \frac{\pi}{6} \qquad (7.109)$$

and

$$a_n'(t) = -n^2 a_n(t) + \frac{2}{\pi n^2} \left[(1 - (-1)^n) t - (-1)^n \right]$$

or

$$a_n'(t) + n^2 a_n(t) = \frac{2}{\pi n^2} \left[(1 - (-1)^n) t - (-1)^n \right]. \qquad (7.110)$$

Step 7: Solve the ODE.
The ODE in Equation (7.109) is a first-order ODE, and it has the solution

$$a_0(t) = \left(\frac{1}{\pi} - \frac{\pi}{2} \right) \frac{t^2}{2} - \frac{\pi t}{6} + d_0. \qquad (7.111)$$

The ODE in Equation (7.110) is a first-order ODE. We arrive at the solution by using an integrating factor. Thus, we have

$$\left[e^{-n^2 t} a_n(t) \right]' = \frac{2}{\pi n^2} \left[(1 - (-1)^n) t - (-1)^n \right] e^{-n^2 t}. \qquad (7.112)$$

Solving for $a_n(t)$ yields

$$a_n(t) = \frac{2}{\pi n^4}\left[\left(1 - (-1)^n\right)\left(t - \frac{1}{n^2}\right) - (-1)^n\right] + d_n e^{-n^2 t}. \tag{7.113}$$

Step 8: Replace the coefficients in the Fourier series representation for $v(x,t)$. Replacing $a_0(t)$ and $a_n(t)$, given in Equations (7.111 and 7.113), in Equation (7.104) yields

$$v(x,t) = \left(\frac{1}{\pi} - \frac{\pi}{2}\right)\frac{t^2}{2} - \frac{\pi t}{6} + d_0 + \tag{7.114}$$

$$\sum_{n=1}^{\infty}\left\{\frac{2}{\pi n^4}\left[\left(1 - (-1)^n\right)\left(t - \frac{1}{n^2}\right) - (-1)^n\right] + d_n e^{-n^2 t}\right\}\cos nx.$$

Step 9: Determine the general solution for $u(x,t)$.
Remembering $u(x,t) = v(x,t) + r(x,t) = r(x,t) + v(x,t)$ implies

$$u(x,t) = \frac{x^2 t}{2\pi} + \left(\frac{1}{\pi} - \frac{\pi}{2}\right)\frac{t^2}{2} - \frac{\pi t}{6} + d_0 + \tag{7.115}$$

$$\sum_{n=1}^{\infty}\left\{\frac{2}{\pi n^4}\left[\left(1 - (-1)^n\right)\left(t - \frac{1}{n^2}\right) - (-1)^n\right] + d_n e^{-n^2 t}\right\}\cos nx.$$

Step 10: Determine the constants in the general solution of $u(x,t)$ by using the ICs and orthogonality.
Equation (7.96) indicates the initial temperature distribution in the one-dimensional rod is $u(x,0) = 0$. Therefore,

$$u(x,0) = \sin\frac{5x}{2} = d_0 +$$

$$\sum_{n=1}^{\infty}\left\{\frac{2}{\pi n^4}\left[\frac{-\left(1 - (-1)^n\right)}{n^2} - (-1)^n\right] + d_n\right\}\cos nx,$$

which implies

$$d_0 = \frac{1}{\pi}\int_0^\pi \sin\frac{11x}{2}\,dx = \frac{2}{11\pi}$$

and

$$d_n = \frac{2}{\pi}\int_0^\pi \sin\frac{11x}{2}\cos nx\,dx + \frac{2}{\pi n^4}\left[\frac{\left(1 - (-1)^n\right)}{n^2} + (-1)^n\right]$$

$$= \frac{-44}{\pi\left(4n^2 - 121\right)} + \frac{2}{\pi n^4}\left[\frac{\left(1 - (-1)^n\right)}{n^2} + (-1)^n\right]. \tag{7.116}$$

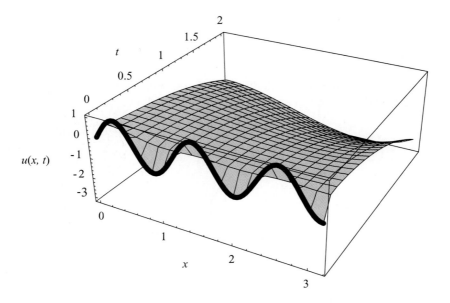

Figure 7.4: The graph of $u(x,t)$ **for** $0 \le t \le 2$.

Thus, our complete solution for $u(x,t)$ is

$$u(x,t) = \frac{x^2 t}{2\pi} + \left(\frac{1}{\pi} - \frac{\pi}{2}\right)\frac{t^2}{2} - \frac{\pi t}{6} + \frac{2}{5\pi} + \tag{7.117}$$

$$\sum_{n=1}^{\infty}\left\{\frac{2}{\pi n^4}\left[(1-(-1)^n)\left(t-\frac{1}{n^2}\right) - (-1)^n\right] + d_n e^{-n^2 t}\right\}\cos nx$$

where d_n is given in Equation (7.116). Figure (7.4) is a graph of the solution presented in Equation (7.117) $0 \le t \le 2$. The bold-faced curve indicates the initial temperature distribution. ∎

For a final example, we solve a multidimensional problem.

EXAMPLE 7.5. Consider

$$\frac{\partial u(x,y,t)}{\partial t} = \frac{\partial^2 u(x,y,t)}{\partial x^2} + \frac{\partial^2 u(x,y,t)}{\partial y^2} + \sin t, \tag{7.118}$$

subject to the BCs

$$u(0,y,t) = 0 \text{ and } \frac{\partial u(\pi,y,t)}{\partial x} = 1 \text{ watt}, \tag{7.119}$$

$$\frac{\partial u(x,0,t)}{\partial y} = 0 \text{ and } u(x,2\pi,t) = x^\circ C, \tag{7.120}$$

and IC

$$u(x, y, 0) = 0°C \tag{7.121}$$

A possible physical model is temperature distribution in a perfect laterally insulated rectangle of length 2π and width π. The initial temperature is 0°C. At the start of the experiment the boundary $y = 0$ is perfectly insulated, the boundary $x = \pi$ is imperfectly insulated, the boundary at $x = 0$ is held in a 0°C bath, and the boundary at $y = 2\pi$ has a temperature that varies as x does from 0 to π. Also, there is a source/sink term depending on time, t.

Solution: First, we must construct a function $r(x, y, t)$ so that r is a continuous function and it has the same BCs as $u(x, y, t)$. One possible function is

$$r(x, y, t) = x. \tag{7.122}$$

Since Equation (7.122) yields $r(0, y, t) = 0$, $\dfrac{\partial r(x, y, t)}{\partial x} = 1$, $\dfrac{\partial r(x, y, t)}{\partial y} = 0$, and $r(x, 2\pi, t) = x$. Thus, $v(x, y, t)$ becomes

$$v(x, y, t) = u(x, y, t) - r(x, y, t)$$

and

$$\frac{\partial v(x, y, t)}{\partial t} = \frac{\partial u(x, y, t)}{\partial t} - \frac{\partial r(x, y, t)}{\partial t} = \frac{\partial u(x, y, t)}{\partial t}, \tag{7.123}$$

$$\frac{\partial^2 v(x, y, t)}{\partial x^2} = \frac{\partial^2 u(x, y, t)}{\partial x^2} - \frac{\partial^2 r(x, y, t)}{\partial x^2} = \frac{\partial^2 u(x, y, t)}{\partial x^2} \tag{7.124}$$

and

$$\frac{\partial^2 v(x, y, t)}{\partial y^2} = \frac{\partial^2 u(x, y, t)}{\partial y^2} - \frac{\partial^2 r(x, y, t)}{\partial y^2} = \frac{\partial^2 u(x, y, t)}{\partial y^2}. \tag{7.125}$$

Rewriting Equation (7.118) in terms of Equations (7.123, 7.124, and 7.125) yields

$$\frac{\partial v(x, y, t)}{\partial t} = \frac{\partial^2 v(x, y, t)}{\partial x^2} + \frac{\partial^2 v(x, y, t)}{\partial y^2} + \sin t. \tag{7.126}$$

subject to the BCs

$$v(0, y, t) = 0 \text{ and } \frac{\partial v(\pi, y, t)}{\partial x} = 0 \tag{7.127}$$

and

$$\frac{\partial v(x, 0, t)}{\partial y} = 0 \text{ and } v(x, 2\pi, t) = 0. \tag{7.128}$$

Now, we solve the homogeneous problem.

$$\frac{\partial v(x,y,t)}{\partial t} = \frac{\partial^2 v(x,y,t)}{\partial x^2} + \frac{\partial^2 v(x,y,t)}{\partial y^2}, \tag{7.129}$$

subject to the BCs Equations (7.127 and 7.128).
Applying separation of variables yields the spatial equations

$$X''(x) = -\tau X(x) \tag{7.130}$$

and

$$Y''(y) = -\xi Y(y) \tag{7.131}$$

where $\xi = \lambda - \tau$, which implies $\lambda = \xi + \tau$, subject to the BCs

$$X(0) = 0 \text{ and } X'(\pi) = 0 \tag{7.132}$$

and

$$Y'(0) = 0 \text{ and } Y(2\pi) = 0, \tag{7.133}$$

respectively. We know that the solution to Equations (7.130 and 7.131) subject to Equations (7.132 and 7.133) are

$$X(x) = \sin\left[\frac{(2n-1)x}{2}\right] \text{ and } Y(y) = \cos\left[\frac{(2m-1)y}{4}\right], \tag{7.134}$$

where

$$\tau = \left[\frac{(2n-1)}{2}\right]^2, \, n = 1, 2, 3, \ldots$$

and

$$\xi = \left[\frac{(2m-1)}{2}\right]^2, \, m = 1, 2, 3, \ldots$$

Therefore,

$$\lambda^2 = \tau^2 + \xi^2 = \left[\frac{(2n-1)}{2}\right]^2 + \left[\frac{(2m-1)}{2}\right]^2, \, n, \, m = 1, 2, 3, \ldots$$

Thus, we write the solution for $v(x, y, t)$, which is

$$v(x,y,t) = \sum_{n=1}^{\infty}\sum_{m=1}^{\infty} a_{nm}(t) \cos\left[\frac{(2m-1)y}{4}\right] \sin\left[\frac{(2n-1)x}{2}\right]. \tag{7.135}$$

Second, we must expand the new source term of Equation (7.126) in terms of the eigenvalues and eigenfunctions found for the homogeneous problem. We have

$$\sin t = \sum_{n=1}^{\infty}\sum_{m=1}^{\infty} b_{nm}(t) \cos\left[\frac{(2m-1)y}{4}\right] \sin\left[\frac{(2n-1)x}{2}\right],$$

where

$$b_{nm}(t) = \frac{2}{\pi^2} \int_0^{2\pi} \int_0^{\pi} \sin t \sin\left[\frac{(2n-1)x}{2}\right] \cos\left[\frac{(2m-1)y}{4}\right] dx dy$$

$$= \frac{16(-1)^{m+1} \sin t}{\pi^2 (2m-1)(2n-1)}$$

$$= b_{nm} \sin t,$$

where $b_{nm} = \dfrac{16(-1)^{m+1}}{\pi^2 (2m-1)(2n-1)}$.

Third, we must find $\dfrac{\partial v(x,y,t)}{\partial t}$, $\dfrac{\partial^2 v(x,y,t)}{\partial x^2}$, and $\dfrac{\partial^2 v(x,y,t)}{\partial y^2}$, and they are

$$\frac{\partial v(x,y,t)}{\partial t} = \sum_{n=1}^{\infty} \sum_{m=1}^{\infty} a'_{nm}(t) \cos\left[\frac{(2m-1)y}{4}\right] \sin\left[\frac{(2n-1)x}{2}\right],$$

$$\frac{\partial^2 v(x,y,t)}{\partial x^2} = \sum_{n=1}^{\infty} \sum_{m=1}^{\infty} -a_{nm}(t) \left[\frac{(2n-1)}{2}\right]^2 \cos\left[\frac{(2m-1)y}{4}\right] \sin\left[\frac{(2n-1)x}{2}\right],$$

and

$$\frac{\partial^2 v(x,y,t)}{\partial y^2} = \sum_{n=1}^{\infty} \sum_{m=1}^{\infty} -a_{nm}(t) \left[\frac{(2m-1)}{4}\right]^2 \cos\left[\frac{(2m-1)y}{4}\right] \sin\left[\frac{(2n-1)x}{2}\right].$$

Next, we put it all the pieces together and solve the resulting ODE. We have

$$\sum_{n=1}^{\infty} \sum_{m=1}^{\infty} a'_{nm}(t) \cos\left[\frac{(2m-1)y}{4}\right] \sin\left[\frac{(2n-1)x}{2}\right]$$

$$= \sum_{n=1}^{\infty} \sum_{m=1}^{\infty} -a_{nm}(t) \left[\frac{(2n-1)}{2}\right]^2 \cos\left[\frac{(2m-1)y}{4}\right] \sin\left[\frac{(2n-1)x}{2}\right]$$

$$+ \sum_{n=1}^{\infty} \sum_{m=1}^{\infty} -a_{nm}(t) \left[\frac{(2m-1)}{4}\right]^2 \cos\left[\frac{(2m-1)y}{4}\right] \sin\left[\frac{(2n-1)x}{2}\right]$$

$$+ b_{nm} \sin t,$$

which yields the ODE

$$a'_{nm}(t) = -a_{nm}(t) \left[\frac{(2n-1)}{2}\right]^2 - a_{nm}(t) \left[\frac{(2m-1)}{4}\right]^2 + b_{nm} \sin t \qquad (7.136)$$

Equation (7.136) may be written as

$$a'_{nm}(t) = -a_{nm}(t)\lambda^2 + b_{nm} \sin t.$$

solving the first-order ODE yields

$$a_{nm}(t) = b_{nm}\left\{\frac{\lambda^2 \sin t - \cos t}{\lambda^2 + 1}\right\} + C_{nm}e^{-\lambda^2 t}.$$

Therefore,

$$v(x, y, t) = \sum_{n=1}^{\infty}\sum_{m=1}^{\infty} a_{nm}(t) \cos\left[\frac{(2m-1)y}{4}\right]\sin\left[\frac{(2n-1)x}{2}\right]$$

becomes

$$v(x, y, t) = \sum_{n=1}^{\infty}\sum_{m=1}^{\infty}\left[\left(b_{nm}\left\{\frac{\lambda^2 \sin t - \cos t}{\lambda^2 + 1}\right\}\right.\right.$$
$$+\quad \left.\left. C_{nm}e^{-\lambda^2 t}\right)\cos\left[\frac{(2m-1)y}{4}\right]\sin\left[\frac{(2n-1)x}{2}\right]\right].$$

We now may state the general solution for $u(x, y, t)$, which is

$$u(x, y, t) = v(x, y, t) + r(x, y, t) = r(x, y, t) + v(x, y, t) = x + v(x, y, t).$$

Applying the IC $u(x, y, 0) = 0$, we have

$$u(x, y, 0) = 0 \quad = \quad x + v(x, y, 0)$$

$$= \quad x + \sum_{n=1}^{\infty}\sum_{m=1}^{\infty}\left[\left(\left\{\frac{-b_{nm}}{\lambda^2 + 1}\right\}\right.\right.$$
$$+\quad \left.\left. C_{nm}\right)\cos\left[\frac{(2m-1)y}{4}\right]\sin\left[\frac{(2n-1)x}{2}\right]\right]$$

or

$$-x = \sum_{n=1}^{\infty}\sum_{m=1}^{\infty}\left(\left\{\frac{-b_{nm}}{\lambda^2 + 1}\right\} + C_{nm}\right)\cos\left[\frac{(2m-1)y}{4}\right]\sin\left[\frac{(2n-1)x}{2}\right],$$

which means

$$C_{nm} = \frac{2}{\pi^2}\int_0^{2\pi}\int_0^{\pi}(-x)\cos\left[\frac{(2m-1)y}{4}\right]\sin\left[\frac{(2n-1)x}{2}\right]dxdy + \frac{b_{nm}}{\lambda^2 + 1}.$$

Thus,

$$C_{nm} = \frac{32(-1)^{m+1}(-1)^n}{(2m-1)(2n-1)^2\pi^2} + \frac{b_{nm}}{\lambda^2 + 1}. \tag{7.137}$$

Therefore, the solution for $u(x, y, t)$ is

$$u(x, y, t) = x \ + \ \sum_{n=1}^{\infty} \sum_{m=1}^{\infty} \left[b_{nm} \left\{ \frac{\lambda^2 \sin t - \cos t}{\lambda^2 + 1} \right\} \right.$$

$$\left. + \ C_{nm} e^{-\lambda^2 t} \cos \left[\frac{(2m-1)y}{4} \right] \sin \left[\frac{(2n-1)x}{2} \right] \right],$$

where C_{nm} is given by Equation (7.137).

In the next section, we summarize the solution technique, called the eigenfunction expansion technique for solving a nonhomogeneous PDE problem by separation of variables.

EXERCISES 7.4

7.4.1. Consider the heat equation

$$\frac{\partial u}{\partial t} = k \frac{\partial^2 u}{\partial x^2} + Q(x, t),$$

subject to

$$u(x, 0) = f(x).$$

Solve for a time-dependent solution given the following nonhomogeneous boundary conditions:

(1) $\dfrac{\partial u(0, t)}{\partial x} = A$ and $\dfrac{\partial u(L, t)}{\partial x} = B.$

(2) $u(0, t) = A$ and $\dfrac{\partial u(L, t)}{\partial x} = B.$

(3) $u(0, t) = A(t)$ and $u(L, t) = B(t).$

(4) $u(0, t) = A$ and $\dfrac{\partial u(L, t)}{\partial x} = B(t).$

7.4.2. Consider the wave equation

$$\frac{\partial^2 u}{\partial t^2} = c^2 \frac{\partial^2 u}{\partial x^2} + Q(x, t),$$

subject to

$$u(x, 0) = f(x) \quad \text{and} \quad \frac{\partial u(x, 0)}{\partial t} = g(x).$$

Solve for a time-dependent solution given the following nonhomogeneous boundary conditions:

(1) $u(0,t) = A$ and $u(L,t) = B$.

(2) $\dfrac{\partial u(0,t)}{\partial x} = A$ and $u(L,t) = B$.

(3) $u(0,t) = A(t)$ and $u(L,t) = B(t)$.

(4) $u(0,t) = A(t)$ and $\dfrac{\partial u(L,t)}{\partial x} = B(t)$.

7.4.3. Consider the following information:

(a) heat conduction in a one-dimensional rod of length $\dfrac{3\pi}{2}$ m,

(b) perfect thermal insulation on the lateral sides,

(c) initial condition of $x + \cos x$ for $0 < x < \dfrac{3\pi}{4}$ and of $x - \sin x$ for $\dfrac{3\pi}{4} < x < \dfrac{3\pi}{2}$,

(d) left boundary is time-dependent and is $\cos t$,

(e) right end is held at a constant temperature of $\dfrac{3\pi + 2}{2}$ degrees,

(f) a uniform rod made of pure iron, and

(g) no source.

(1) Set up the mathematical model. *Note:* The thermal diffusivity of pure iron may be found in Appendix E.

(2) Give a series solution of your mathematical model.

(3) Using your favorite mathematical software, plot your solution. By inspection, determine if the graph of your solution is reasonable for the problem.

(4) Check that this solution satisfies the equation, the BCs, and the ICs.

7.4.4. Consider the following information:

(a) a perfectly flexible string of length 2π ft,

(b) tension of 50 lbs,

(c) mass density of 0.02 lbs/ft,

(d) vibrating motion which is entirely vertical,

(e) fixed left end at 1,

(f) fixed right end at 1,

(g) initial displacement of $x+1$ for $0 < x < \pi$ and of $2\pi - x + 1$ for $\pi < x < 2\pi$,

(h) initial velocity of 0, and

(i) with gravitational effects of 32 ft/sec/sec at the start of the experiment that decreases exponentially as time increases.

(1) Set up the mathematical model.

(2) Solve your mathematical model for any time t.

(3) Using your favorite mathematical software, plot your solution. By inspection, determine if the graph of your solution is reasonable for the problem.

(4) Check that this solution satisfies the equation, the BCs, and ICs.

7.4.5. Consider a uniform one-dimensional rod, without a source, which is not laterally insulated. (Heat can flow in and out across the lateral surface.) By experimentation, you have discovered that heat is flowing across the lateral boundary at a rate proportional to the difference between the temperature $u(x,t)$ and the surrounding medium that is kept at $10x$ degrees where $0 < x < 2\pi$ ft.

(1) Given that the convection constant of proportionality is .071 and the end at $x = 0$ is fixed at one degree, the end $x = 2\pi$ is fixed at $e^{2\pi}$ degrees, and the rod is composed of zinc, set up the mathematical model. *Note:* The thermal diffusivity of zinc may be found in Appendix E.

(2) Suppose the initial temperature of the rod is e^x. Give a series solution of your mathematical model.

(3) Using your favorite mathematical software, plot your solution. By inspection, determine if the graph of your solution is reasonable for the problem.

(4) Check whether this solution satisfies the equation, the BCs, and ICs.

7.4.6. Consider a vibrating string with friction and a time-periodic forcing function

$$\frac{\partial^2 u}{\partial t^2} = c^2 \frac{\partial^2 u}{\partial x^2} - \beta \frac{\partial u}{\partial t} + x \cos \omega t,$$

subject to

$$u(0,t) = 0 \text{ and } u(\pi, t) = \pi$$

and

$$u(x,0) = x \text{ and } \frac{\partial u(x,0)}{\partial t} = 0.$$

(1) Solve the initial value problem if β is small $(0 < \beta < 2c)$.

(2) Suppose that the length of the string was arbitrary, that is, L. What would the bounds be on β in this case?

(3) Suppose $\beta > 2c$. What would happen in this case?

7.4.7. Solve the initial-value problem

$$\frac{\partial u(x,y,t)}{\partial t} = \frac{\partial^2 u(x,y,t)}{\partial x^2} + \frac{\partial^2 u(x,y,t)}{\partial y^2} + t,$$

subject to the BCs

$$u(0, y, t) = y \text{ and } \frac{\partial u(\pi, y, t)}{\partial x} = 0,$$

and

$$\frac{\partial u(x, 0, t)}{\partial y} = 1 \text{ and } u(x, 2\pi, t) = 0,$$

and IC

$$u(x, y, 0) = 0$$

7.4.8. Solve the wave equation in a rectangular sheet with Dirichlet BCs and a time-periodic forcing function, $\cos t$. Assume that the sheet is initially at rest with displacement

$$u(x, y, 0) = f(x, y) = \begin{cases} xy; & 0 \le x < 1 \quad 0 \le y < 1.5 \\ x(3 - y); & 0 \le x < 1 \quad 1.5 \le y \le 3 \\ (2 - x)y; & 1 \le x \le 2 \quad 0 \le y < 1.5 \\ (2 - x)(3 - y); & 1 \le x \le 2 \quad 1.5 \le y \le 3, \end{cases}$$

7.4.9. Solve the wave equation in a rectangular sheet with Neumann BCs and a time-periodic forcing function, $\sin t$. Assume that the sheet is initially at rest with displacement

$$u(x, y, 0) = f(x, y) = \begin{cases} xy; & 0 \le x < 1 \quad 0 \le y < 1.5 \\ x(3 - y); & 0 \le x < 1 \quad 1.5 \le y \le 3 \\ (2 - x)y; & 1 \le x \le 2 \quad 0 \le y < 1.5 \\ (2 - x)(3 - y); & 1 \le x \le 2 \quad 1.5 \le y \le 3, \end{cases}$$

7.4.10. Consider heat conduction on a 3 ft one-dimensional uniform oak rod with no lateral or boundary insulation, hence the lateral and boundary surfaces are in direct contact with a time/position-dependent medium modeled as $\cos(xt)$, where x is in inches and t is in seconds. The initial temperature distribution of the rod is given as 1.

  (1) State the mathematical model. *Note:* The thermal diffusivity of oak may be found in Appendix E.

  (2) Solve the mathematical model.

(3) Using your favorite mathematical software, plot your solution. By inspection, determine if the graph of your solution is reasonable for the problem.

(4) Check whether the solution satisfies the equation, the boundary conditions, and the initial conditions.

7.4.11. The space shuttle heat shields are shaped as rectangular boxes with curved surfaces matching the curvature of the body of the shuttle. The greater majority of the heat shields are surrounded by other heat shields that are fitted tightly together. One surface is against the body of the shuttle, while the opposite side faces space.

The typical heat shield may be modeled adequately by a rectangular box with no curvature. The model is assumed to have perfect insulation on its four sides, while the side against the shuttle is assumed to be held at 14°C. The temperature of the side open to space at time of re-entry is proportional to the time in seconds, required for re-entry. On re-entry into earth's atmosphere this function is modeled by

$$f(t) = \frac{1}{h+1}\left(e^{\left(\frac{t}{26}\right)} - 1\right).$$

Here h is calculated using distance in meters from the front of, and from the center of, the bottom of the shuttle. *Note:* h decreases to 0 as you approach the front of the shuttle.

For this problem, we will use the thermal diffusivity of asbestos, which may be found in Appendix E. The heat shields are sized by height $= z = 7.5$ cm, width $= y = 15$ cm, and length $= x = 50$ cm. Initial temperature distribution is

$$u(x, y, z, 0) = -e^{\left(\frac{z}{14}\right)} + 15.$$

Note: The side facing space is oriented in the positive z direction.

(1) Construct the mathematical model.

(2) Solve the mathematical model.

(3) Write the solution in a completed form.

(4) Suppose $h = .019$. Determine the temperature distribution on this heat shield for a 3.5 min re-entry.

7.5   SUMMARY

In this chapter, the problems solved by the eigenfunction expansion technique were all linear PDEs with linear BCs. It is important to realize that this method only works on linear PDEs with linear BCs.

We summarize our eigenfunction expansion technique as follows:

1. Determine that the PDE and BCs are linear.

2. If BCs are not homogeneous, construct a function that, when replacing the function in the PDE, will have homogeneous BCs.

3. Determine the related homogeneous PDE and BCs and solve for the eigenvalues and eigenfunctions.

   (a) Assume the solution is a product of two functions. One function is possibly of time, while the other function is composed of spatial variables.

   (b) Substitute this product into the PDE and move all functions of time to one side while placing all other components on the opposite side of the equation.

   (c) Set both sides equal to the same separation constant, breaking the original equation into two new equations, thus separating the variables.

   (d) Repeat steps (a), (b), and (c) on equations with more than one spatial variable until you have an ODE for each spatial variable.

   (e) Substitute the product form of the solution into the BCs to obtain BCs for the ODEs.

   (f) Solve the ODE boundary value problem(s) to obtain the eigenvalues and eigenfunctions.

4. Assume the solution is a Fourier series expanded in the related homogeneous eigenfunctions with time-dependent coefficients.

5. Determine a Fourier series expansion in the eigenfunctions found in Step 2 for any nonhomogeneous terms in the PDE.

6. Differentiate, term-by-term, the Fourier series from Step 4. Remember, all conditions necessary to do term-by-term differentiating must be satisfied.

7. Replace the PDE with its Fourier series equivalent.

8. Equate the coefficients to obtain and solve an ODE for the constant coefficients.

9. If a constructed function was used, obtain the solution for the original function of the original PDE.

10. Use the ICs and orthogonality to determine the constants for the specific solution.

Chapter 8

The Sturm–Liouville Eigenvalue Problem

8.1 INTRODUCTION

The Sturm–Liouville eigenvalue problem is a second-order homogeneous ODE with variable coefficients subject to homogeneous boundary conditions. You should remember second-order ODEs produce eigenvalues and eigenfunctions for the solution space where the ODE is defined. The study of this class of ODEs is important in itself. However, applying this class of ODEs to PDEs is our primary interest.

Consider the following examples:

EXAMPLE 8.1. Given the heat equation

$$\frac{\partial u}{\partial t} = k\frac{\partial^2 u}{\partial x^2},$$

subject to

$$u(0, t) = 0$$

$$u(L, t) = 0$$

and

$$u(x, 0) = f(x),$$

we know the separation of variables technique applies. Letting $u(x, t) = \varphi(x)G(t)$, taking the appropriate derivatives and separating the PDE yields the time equation

$$G'(t) = -\lambda k G(t)$$

and the spatial eigenvalue problem

$$\varphi''(x) = -\lambda\varphi(x), \tag{8.1}$$

subject to

$$\varphi(0) = 0$$

$$\varphi(L) = 0. \qquad \blacksquare \qquad (8.2)$$

EXAMPLE 8.2. Given the wave equation

$$\frac{\partial^2 u}{\partial t^2} = c^2 \frac{\partial^2 u}{\partial x^2},$$

subject to

$$\frac{\partial u(0,t)}{\partial x} = 0$$

$$\frac{\partial u(L,t)}{\partial x} = 0$$

and

$$u(x,0) = f(x)$$

$$\frac{\partial u(x,0)}{\partial t} = g(x),$$

we again apply the separation of variables technique resulting in the time equation

$$G''(t) = -\lambda c^2 G(t),$$

and the spatial equation

$$\varphi''(x) = -\lambda \varphi(x), \qquad (8.3)$$

subject to

$$\varphi'(0) = 0$$

$$\varphi'(L) = 0. \qquad \blacksquare \qquad (8.4)$$

We are not concerned with the solution for these examples. What we are concerned with is the spatial problems, which resulted from using the separation of variables technique. In particular, we notice the spatial equations, Equation (8.1) and Equation (8.3), are identical. However, the BCs, Equation (8.2) and Equation (8.4), are different.

Suppose we rewrote the BCs in the following way:

$$\alpha_1 \varphi(0) + \alpha_2 \varphi'(0) = 0$$

$$\alpha_3 \varphi(L) + \alpha_4 \varphi'(L) = 0. \qquad (8.5)$$

Then, we need α_2 and α_4 to equal 0, and α_1 and α_3 to equal 1 for Equation (8.2) to be true; for Equation (8.4) to be true, we need α_1 and α_3 to equal 0, and α_2 and α_4 to equal 1. Thus, Equation (8.5) is really a generalized expression of the BCs in Equations (8.2 and 8.4). Therefore, we may restate the spatial eigenvalue problem given by Equation (8.1), subject to the BCs, Equations (8.2 and 8.4), as the general spatial problem

$$\varphi''(x) = -\lambda\varphi(x), \tag{8.6}$$

subject to

$$\alpha_1\varphi(0) + \alpha_2\varphi'(0) = 0$$
$$\alpha_3\varphi(L) + \alpha_4\varphi'(L) = 0. \tag{8.7}$$

This chapter develops known facts about the generalized spatial problem. In the next section, we develop the complete formulation of the generalized spatial problem, define its characteristics in theorems, and prove several of those theorems. The section titled, "Rayleigh Quotient," proves another theorem and discusses some uses of the Rayleigh quotient. Next, we solve a general heat flow/wave motion problem. We conclude this chapter by considering PDE problems involving Robin's boundary conditions.

8.2 DEFINITION OF THE STURM–LIOUVILLE EIGENVALUE PROBLEM

The generalized spatial problem is important for two major reasons. First, we can study an entire class of ODEs and develop a solution method for that class. Second, it allows us to go beyond the problems of heat in a one-dimensional uniform rod, or wave motion in a one-dimensional uniform horizontal string. For instance, consider the following example.

EXAMPLE 8.3. Suppose we have heat flow in a nonuniform one-dimensional rod of length L with a source term dependent on the temperature distribution multiplied by a proportionality spatial function, and subject to homogeneous Robin's boundary conditions on both ends, and an appropriate initial condition. The mathematical model for these conditions is

$$c(x)\rho(x)\frac{\partial u}{\partial t} = \frac{\partial}{\partial x}\left[K_0(x)\frac{\partial u}{\partial x}\right] + q(x)u(x,t), \tag{8.8}$$

subject to

$$-K_0(0)\frac{\partial u(0,t)}{\partial x} = -h_1 u(0,t)$$
$$K_0(L)\frac{\partial u(L,t)}{\partial x} = -h_2 u(L,t) \tag{8.9}$$

and

$$u(x,0) = f(x).$$

Since the PDE and BCs are homogeneous, the separation of variables technique applies. Letting $u(x,t) = \varphi(x)G(t)$, taking the appropriate derivatives, and rewriting Equation (8.8) yields

$$c(x)\rho(x)\varphi(x)G'(t) = \frac{d}{dx}\left[K_0(x)\varphi'(x)\right]G(t) + q(x)\varphi(x)G(t). \qquad (8.10)$$

Equation (8.10) separates into a time equation,

$$G'(t) = -\lambda G(t),$$

and a spatial equation,

$$\frac{d}{dx}\left[K_0(x)\varphi'(x)\right] + q(x)\varphi(x) = -\lambda c(x)\rho(x)\varphi(x).$$

The spatial equation may be written as

$$\frac{d}{dx}\left[K_0(x)\varphi'(x)\right] + q(x)\varphi(x) + \lambda c(x)\rho(x)\varphi(x) = 0. \qquad (8.11)$$

Applying $u(x,t) = \varphi(x)G(t)$ to the BCs yields

$$-K_0(0)\varphi'(0) = -h_1\varphi(0)$$

$$K_0(L)\varphi'(L) = -h_2\varphi(L)$$

after separation. The BCs also may be rewritten as

$$h_1\varphi(0) - K_0(0)\varphi'(0) = 0$$
$$\qquad\qquad\qquad\qquad\qquad\qquad\qquad (8.12)$$
$$h_2\varphi(L) + K_0(L)\varphi'(L) = 0.$$

Thus, we have the spatial equation, Equation (8.11), subject to the BCs, Equation (8.12).   ■

At this point, we usually solve the spatial problem for eigenvalues and eigenfunctions. However, the spatial problem described by Equations (8.11 and 8.12) is not our usual spatial problem. Therefore, we must determine if a solution exists, and if we can find it.

Equations (8.11 and 8.12) may look somewhat messy, but, by making the following substitutions, Equations (8.11 and 8.12) become more compact. First, let $K_0(x) = p(x)$. Second, let $c(x)\rho(x) = \sigma(x)$. Finally, since h_1, $K_0(0)$, h_2, and $K_0(L)$ are all constants, we let them equal α_1, α_2, α_3, and α_4, respectively. These substitutions yield the spatial problem

$$\frac{d}{dx}\left[p(x)\varphi'(x)\right] + q(x)\varphi(x) + \lambda\sigma(x)\varphi(x) = 0,\ 0 \le x \le L, \qquad (8.13)$$

subject to

$$\alpha_1 \varphi(0) + \alpha_2 \varphi'(0) = 0$$

$$\alpha_3 \varphi(L) + \alpha_4 \varphi'(L) = 0. \tag{8.14}$$

If $p(x)$, $q(x)$, and $\sigma(x)$ are real continuous functions on the interval $0 \leq x \leq L$; $p(x)$ and $\sigma(x)$ are always greater than 0 on the interval $0 \leq x \leq L$; and α_i, $i = 1$ to 4 are real numbers, then Equations (8.13 and 8.14) are known as the **regular Sturm–Liouville eigenvalue problem** on the interval $0 \leq x \leq L$. A more generalized version of the regular Sturm–Liouville eigenvalue problem is stated on the interval $a \leq x \leq b$. However, for our purposes we generally use the interval $0 \leq x \leq L$.

The fact that Equations (8.13 and 8.14) are regular Sturm–Liouville eigenvalue problems doesn't mean we have a solution. It means we know the class that ODE Equations (8.13 and 8.14) belong to, and there are four very important theorems which apply to any Regular Sturm–Liouville Eigenvalue Problem. These theorems are as follows:

Theorem 47. *There exists an infinite number of eigenvalues, λ_n, $n = 1, 2, 3, \ldots$ Furthermore, the eigenvalues, λ_n, $n = 1, 2, 3, \ldots$, have the following properties*

- *All eigenvalues, λ_n, $n = 1, 2, 3, \ldots$, are real numbers.*

- *There is a smallest eigenvalue, usually denoted λ_1.*

- *There is no largest eigenvalue. This means $\lambda_n \to \infty$ as $n \to \infty$.*

- *For the eigenvalues λ_n and λ_m, we have $\lambda_n \neq \lambda_m$ unless $n = m$. Therefore, the eigenvalues may be ordered in the following manner:*

$$\lambda_1 < \lambda_2 < \lambda_3 < \cdots < \lambda_n < \cdots.$$

- *Corresponding to each eigenvalue, λ_n, $n = 1, 2, 3, \ldots$, there is an unique eigenfunction, denoted $\varphi_n(x)$, $n = 1, 2, 3, \ldots$, which has exactly $n - 1$ zeros on $0 < x < L$.*

We will only prove selected parts of Theorem (47). For a complete set of proofs, I refer the reader to *An Introduction to Partial Differential Equations* by Michael Renardy and Robert C. Rogers. We prove that all eigenvalues are real numbers, and eigenfunctions corresponding to each eigenvalue are unique, after we state and prove the next theorem.

Theorem 48. *The eigenfunctions, $\varphi_n(x)$, $n = 1, 2, 3, \ldots$, corresponding to different eigenvalues, λ_n, $n = 1, 2, 3, \ldots$, are orthogonal relative to the weight function $\sigma(x)$. This means*

$$\int_0^L \varphi_n(x)\varphi_m(x)\sigma(x) \ dx = 0, \ \ whenever \ \lambda_n \neq \lambda_m.$$

It should be noted that eigenfunctions are unique up to an arbitrary constant multiplier.

Proof. Let $\varphi_n(x)$ and $\varphi_m(x)$, with corresponding eigenvalues $\lambda_n \neq \lambda_m$, be solutions to

$$\frac{d}{dx}\left[p(x)\varphi'(x)\right] + q(x)\varphi(x) + \lambda\sigma(x)\varphi(x) = 0, \ 0 \leq x \leq L,$$

subject to

$$\alpha_1\varphi(0) + \alpha_2\varphi'(0) = 0$$

$$\alpha_3\varphi(L) + \alpha_4\varphi'(L) = 0.$$

Thus, we have

$$\frac{d}{dx}\left[p(x)\varphi'_n(x)\right] + q(x)\varphi_n(x) + \lambda_n\sigma(x)\varphi_n(x) = 0, \ 0 \leq x \leq L, \tag{8.15}$$

subject to

$$\begin{aligned}\alpha_1\varphi_n(0) + \alpha_2\varphi'_n(0) &= 0 \\ \alpha_3\varphi_n(L) + \alpha_4\varphi'_n(L) &= 0,\end{aligned} \tag{8.16}$$

and

$$\frac{d}{dx}\left[p(x)\varphi'_m(x)\right] + q(x)\varphi_m(x) + \lambda_m\sigma(x)\varphi_m(x) = 0, \ 0 \leq x \leq L, \tag{8.17}$$

subject to

$$\begin{aligned}\alpha_1\varphi_m(0) + \alpha_2\varphi'_m(0) &= 0 \\ \alpha_3\varphi_m(L) + \alpha_4\varphi'_m(L) &= 0.\end{aligned} \tag{8.18}$$

Multiplying Equation (8.15) by $\varphi_m(x)$ and Equation (8.17) by $\varphi_n(x)$ yields

$$\frac{d}{dx}\left[p(x)\varphi'_n(x)\right]\varphi_m(x) + q(x)\varphi_n(x)\varphi_m(x) + \lambda_n\sigma(x)\varphi_n(x)\varphi_m(x) = 0 \tag{8.19}$$

and

$$\frac{d}{dx}\left[p(x)\varphi'_m(x)\right]\varphi_n(x) + q(x)\varphi_m(x)\varphi_n(x) + \lambda_m\sigma(x)\varphi_m(x)\varphi_n(x) = 0, \tag{8.20}$$

respectively. Subtracting Equation (8.19) from Equation (8.20) yields

$$0 = \frac{d}{dx}\left[p(x)\varphi'_m(x)\right]\varphi_n(x) - \frac{d}{dx}\left[p(x)\varphi'_n(x)\right]\varphi_m(x)$$

$$+ \lambda_m\sigma(x)\varphi_m(x)\varphi_n(x) - \lambda_n\sigma(x)\varphi_n(x)\varphi_m(x). \tag{8.21}$$

Equation (8.21) may be rewritten as

$$(\lambda_n - \lambda_m)\,\sigma(x)\varphi_n(x)\varphi_m(x) = \frac{d}{dx}\left[p(x)\varphi'_m(x)\right]\varphi_n(x)$$
$$-\frac{d}{dx}\left[p(x)\varphi'_n(x)\right]\varphi_m(x). \tag{8.22}$$

Integrating both sides of Equation (8.22) with respect to x from 0 to L yields

$$(\lambda_n - \lambda_m)\int_0^L \sigma(x)\varphi_n(x)\varphi_m(x)\,dx = \int_0^L \frac{d}{dx}\left[p(x)\varphi'_m(x)\right]\varphi_n(x)\,dx$$

$$-\int_0^L \frac{d}{dx}\left[p(x)\varphi'_n(x)\right]\varphi_m(x)\,dx. \tag{8.23}$$

We may integrate the right side of Equation (8.23) by using integration-by-parts. Thus, we have

$$(\lambda_n - \lambda_m)\int_0^L \sigma(x)\varphi_n(x)\varphi_m(x)\,dx$$

$$= p(x)\varphi'_m(x)\varphi_n(x)\big|_0^L - \int_0^L p(x)\varphi'_m(x)\varphi'_n(x)\,dx$$

$$-p(x)\varphi'_n(x)\varphi_m(x)\big|_0^L + \int_0^L p(x)\varphi'_n(x)\varphi'_m(x)\,dx,$$

which becomes

$$(\lambda_n - \lambda_m)\int_0^L \sigma(x)\varphi_n(x)\varphi_m(x)\,dx$$

$$= p(x)\varphi'_m(x)\varphi_n(x)\big|_0^L - p(x)\varphi'_n(x)\varphi_m(x)\big|_0^L. \tag{8.24}$$

Evaluating the right side of Equation (8.24) yields

$$(\lambda_n - \lambda_m)\int_0^L \sigma(x)\varphi_n(x)\varphi_m(x)\,dx$$

$$= p(L)\varphi'_m(L)\varphi_n(L) - p(0)\varphi'_m(0)\varphi_n(0)$$

$$- p(L)\varphi'_n(L)\varphi_m(L) + p(0)\varphi'_n(0)\varphi_m(0),$$

which may be written as

$$(\lambda_n - \lambda_m) \int_0^L \sigma(x)\varphi_n(x)\varphi_m(x) \; dx$$

$$= p(L) \left[\varphi_m'(L)\varphi_n(L) - \varphi_n'(L)\varphi_m(L) \right]$$

$$- p(0) \left[p(0)\varphi_n'(0)\varphi_m(0) - p(0)\varphi_m'(0)\varphi_n(0) \right]. \tag{8.25}$$

Using Equations (8.16 and 8.18) in the forms

$$\alpha_1 \varphi_n(0) = -\alpha_2 \varphi_n'(0)$$

$$\alpha_3 \varphi_n(L) = -\alpha_4 \varphi_n'(L)$$

and

$$\alpha_1 \varphi_m(0) = -\alpha_2 \varphi_m'(0)$$

$$\alpha_3 \varphi_m(L) = -\alpha_4 \varphi_m'(L),$$

we evaluate the right side of Equation (8.25). The evaluation yields

$$(\lambda_n - \lambda_m) \int_0^L \sigma(x)\varphi_n(x)\varphi_m(x) \; dx = 0.$$

Since $\lambda_n \neq \lambda_m$, we must have

$$\int_0^L \sigma(x)\varphi_n(x)\varphi_m(x) \; dx = 0,$$

thus completing the proof. $\qquad\qquad\square$

We can now prove, from Theorem 47, that all eigenvalues are real and that eigenfunctions corresponding to each eigenvalue are unique.

Proof. Suppose the eigenvalues, λ_n, $n = 1, 2, 3, \ldots$, are not real numbers. Hence, they are complex. Thus, each eigenvalue, λ_n, $n = 1, 2, 3, \ldots$, has the form $a + bi$. Let λ be a complex eigenvalue and $\varphi(x)$ be the corresponding eigenfunction. The function, $\varphi(x)$, must be complex since the differential equation defining the eigenfunction is complex. Thus, $\varphi(x)$, with its corresponding eigenvalue λ, is a solution of

$$\frac{d}{dx} \left[p(x)\varphi'(x) \right] + q(x)\varphi(x) + \lambda\sigma(x)\varphi(x) = 0, \; 0 \leq x \leq L, \tag{8.26}$$

subject to

$$\alpha_1 \varphi(0) + \alpha_2 \varphi'(0) = 0$$

$$\alpha_3 \varphi(L) + \alpha_4 \varphi'(L) = 0. \tag{8.27}$$

Please note the functions $p(x)$, $q(x)$, and $\sigma(x)$ are real functions. Since λ is a complex eigenvalue with corresponding eigenfunction $\varphi(x)$, then $\overline{\lambda}$ (complex conjugate) is an eigenvalue with corresponding eigenfunction $\overline{\varphi(x)}$. This statement is developed from a similar theorem about complex roots of quadratic equations. Thus, $\overline{\varphi(x)}$, with its corresponding eigenvalue $\overline{\lambda}$, is a solution of Equations (8.26 and 8.27). Hence, we have

$$\frac{d}{dx}\left[p(x)\overline{\varphi'(x)}\right] + q(x)\overline{\varphi(x)} + \overline{\lambda}\sigma(x)\overline{\varphi(x)} = 0,\ 0 \le x \le L, \tag{8.28}$$

subject to

$$\alpha_1\overline{\varphi(0)} + \alpha_2\overline{\varphi'(0)} = 0$$

$$\alpha_3\overline{\varphi(L)} + \alpha_4\overline{\varphi'(L)} = 0. \tag{8.29}$$

By Theorem (48) we know $\sigma(x)$ and $\overline{\varphi(x)}$ must be orthogonal. Therefore,

$$\left(\lambda - \overline{\lambda}\right) \int_0^L \sigma(x)\varphi(x)\overline{\varphi(x)}\ dx = 0, \tag{8.30}$$

which may be written as

$$\left(\lambda - \overline{\lambda}\right) \int_0^L \sigma(x)\left|\varphi(x)\right|^2 dx = 0$$

since $\varphi(x)\overline{\varphi(x)} = |\varphi(x)|^2 \ge 0$. Also, $\sigma(x)$ is defined to be greater than 0 for all x on the interval. Thus,

$$\int_0^L \sigma(x)\left|\varphi(x)\right|^2 dx \ge 0. \tag{8.31}$$

Equation (8.31) may only equal 0 if $\varphi(x) \equiv 0$ (identically equal to 0), which we know can't be true since we may only have $n - 1$ zeros for $0 < x < L$. Hence, we must have $\left(\lambda - \overline{\lambda}\right) = 0$. Therefore, $\lambda = \overline{\lambda}$, which implies λ is real. $\qquad\square$

Finally, we prove eigenfunctions corresponding to each eigenvalue are unique up to an arbitrary constant multiplier.

Proof. Suppose $\varphi_1(x)$ and $\varphi_2(x)$ are eigenfunctions corresponding to the same eigenvalue, λ. Then, we have

$$\frac{d}{dx}\left[p(x)\varphi_1'(x)\right] + q(x)\varphi_1(x) + \lambda\sigma(x)\varphi_1(x) = 0,\ 0 \le x \le L, \tag{8.32}$$

subject to

$$\alpha_1\varphi_1(0) + \alpha_2\varphi_1'(0) = 0$$

$$\alpha_3\varphi_1(L) + \alpha_4\varphi_1'(L) = 0, \tag{8.33}$$

and

$$\frac{d}{dx}\left[p(x)\varphi_2'(x)\right] + q(x)\varphi_2(x) + \lambda\sigma(x)\varphi_2(x) = 0,\ 0 \le x \le L, \qquad (8.34)$$

subject to

$$\alpha_1\varphi_2(0) + \alpha_2\varphi_2'(0) = 0$$
$$\alpha_3\varphi_2(L) + \alpha_4\varphi_2'(L) = 0. \qquad (8.35)$$

If we multiply Equation (8.32) by $\varphi_2(x)$, then subtract from it Equation (8.34) multiplied by $\varphi_1(x)$, we have, after some algebraic manipulation,

$$\frac{d}{dx}\left[p(x)\varphi_1'(x)\right]\varphi_2(x) - \frac{d}{dx}\left[p(x)\varphi_2'(x)\right]\varphi_1(x) = 0,$$

which may be written as

$$\frac{d}{dx}\left[p(x)\varphi_1'(x)\right]\varphi_2(x) = \frac{d}{dx}\left[p(x)\varphi_2'(x)\right]\varphi_1(x). \qquad (8.36)$$

We may integrate Equation (8.36) with respect to x over the interval $0 \le x \le L$. We have

$$\int_0^L \frac{d}{dx}\left[p(x)\varphi_1'(x)\right]\varphi_2(x)\ dx = \int_0^L \frac{d}{dx}\left[p(x)\varphi_2'(x)\right]\varphi_1(x)\ dx.$$

Performing the integration by parts yields

$$\varphi_2(x)\left[p(x)\varphi_1'(x)\right]\big|_0^L - \int_0^L \left[p(x)\varphi_1'(x)\right]\varphi_2'(x)\ dx$$

$$= \varphi_1(x)\left[p(x)\varphi_2'(x)\right]\big|_0^L - \int_0^L \left[p(x)\varphi_2'(x)\right]\varphi_1'(x)\ dx,$$

which becomes

$$0 = \varphi_2(x)\left[p(x)\varphi_1'(x)\right]\big|_0^L - \varphi_1(x)\left[p(x)\varphi_2'(x)\right]\big|_0^L$$

$$+ \int_0^L p(x)\varphi_2'(x)\varphi_1'(x)\ dx - \int_0^L p(x)\varphi_1'(x)\varphi_2'(x)\ dx,$$

or just

$$0 = \varphi_2(x)\left[p(x)\varphi_1'(x)\right]\big|_0^L - \varphi_1(x)\left[p(x)\varphi_2'(x)\right]\big|_0^L. \qquad (8.37)$$

Equation (8.37) may be written as

$$p(x)\left(\varphi_2(x)\varphi_1'(x) - \varphi_1(x)\varphi_2'(x)\right)\big|_0^L = 0. \qquad (8.38)$$

By definition, $p(x)$ is greater than 0 for all x on the interval, $0 \leq x \leq L$. Therefore, we may divide both sides of Equation (8.38) by $p(x)$. This yields

$$(\varphi_2(x)\varphi_1'(x) - \varphi_1(x)\varphi_2'(x))|_0^L = 0. \tag{8.39}$$

Multiplying both sides of Equation (8.40) by $\dfrac{1}{[\varphi_2(x)]^2}$ yields

$$\left.\frac{(\varphi_2(x)\varphi_1'(x) - \varphi_1(x)\varphi_2'(x))}{[\varphi_2(x)]^2}\right|_0^L = 0. \tag{8.40}$$

Equation (8.40) may now be rewritten as

$$\left.\frac{d}{dx}\left(\frac{\varphi_1(x)}{\varphi_2(x)}\right)\right|_0^L = 0,$$

which implies

$$\varphi_1(x) = c\varphi_2(x).$$

Thus, $\varphi_2(x)$ is a constant multiplier of $\varphi_1(x)$. Therefore, eigenfunctions corresponding to each eigenvalue are unique up to an arbitrary constant multiplier. This completes the proof. $\qquad\square$

Theorem 49. *Any piecewise smooth function may be represented by a Fourier series of the eigenfunctions $\varphi_n(x)$. This means*

$$f(x) \approx \sum_{n=1}^{\infty} a_n\varphi_n(x),$$

where

$$a_n = \frac{\displaystyle\int_0^L f(x)\varphi_n(x)\sigma(x)\,dx}{\displaystyle\int_0^L \varphi_n^2(x)\sigma(x)\,dx}.$$

Furthermore, the infinite series converges to $f(x)$ wherever $f(x)$ is continuous on $0 \leq x \leq L$, and it converges to $\dfrac{f(x^+) + f(x^-)}{2}$ where $f(x)$ has a jump discontinuity. Sometimes the eigenfunctions, $\varphi_n(x)$, $n = 1, 2, 3, \ldots$, are referred to as forming a "complete" set.

The proof of Theorem (49) is beyond the scope of this course. A generalized proof may be found in *Introductory Functional Analysis with Applications, Seventh Edition* by Erwin Kreyszig.

Theorem 50. *The expression*

$$\lambda_n = \frac{-p(x)\varphi_n(x)\varphi_n'(x)\big|_0^L + \int_0^L \left[p(x) \left(\varphi_n'(x)\right)^2 - q(x)\varphi_n^2(x) \right] dx}{\int_0^L \varphi_n^2(x)\sigma(x)\ dx}, \qquad (8.41)$$

called the **Rayleigh quotient**, *relates any eigenvalue,* λ_n, *to its corresponding eigenfunction,* $\varphi_n(x)$, *for* $n = 1, 2, 3, \ldots$

Proof of the Rayleigh quotient is provided in the next section.

A more general expression of the above theorems results if we state the theorems with the bounds $a < x < b$ instead of $0 < x < L$.

EXERCISES 8.2

8.2.1. Consider the eigenvalue problem

$$x^2\varphi''(x) + x\varphi'(x) + \lambda\varphi(x) = 0, \qquad (8.42)$$

subject to

$$\varphi(1) = 0 \text{ and } \varphi(b) = 0, \ 1 < b.$$

(1) Show that multiplying Equation (8.42) by $\dfrac{1}{x}$ and performing some algebraic manipulation allows you to put Equation (8.42) in the regular Sturm–Liouville form.

(2) Show that $\lambda \geq 0$.

8.2.2. Consider the non-Sturm–Liouville differential equation

$$\varphi''(x) + \alpha(x)\varphi'(x) + \gamma(x)\varphi(x) + \lambda\beta(x)\varphi(x) = 0.$$

Multiply this equation by $H(x)$. Then, determine $H(x)$ so that the equation may be reduced to the standard Sturm–Liouville form:

$$\frac{d}{dx}\left[p(x)\varphi'(x)\right] + q(x)\varphi(x) + \lambda\sigma(x)\varphi(x) = 0.$$

8.2.3. Consider the following equations:

Equation Name	Equation
(1) **Bessel's**	$x^2u'' + xu' + \left(x^2 - n^2\right)u = 0,$
(2) **Chebyshev's**	$\left(1 - x^2\right)u'' - xu' + n^2u = 0,$
(3) **Euler's**	$ax^2u'' + bxu' + cu = 0,$

Equation Name	Equation
(4) **Hermite's**	$u'' - 2xu' + 2nu = 0,$
(5) **Legendre's**	$\left(1 - x^2\right) u'' - 2xu' + n\left(n+1\right) u = 0.$

Reduce them to the standard Sturm–Liouville form.

8.2.4. Generate a set of orthogonal polynomials from the sequence 1, x, x^2, x^3, $\ldots$ on the $(-1, 1)$. Are the orthogonal polynomials orthonormal?

8.2.5. Given

$$\varphi''(x) + \lambda\varphi(x) = 0,$$

find the eigenvalues and eigenfunctions and graph the first few (more than 3) of the eigenfunctions for each of the following boundary conditions:

(1) $\varphi'(0) = 0$ and $\varphi(\pi) = 0.$

(2) $\varphi(0) = 0$ and $\varphi'(2\pi) = 0.$

(3) $\varphi(0) + \varphi'(0) = 0$ and $\varphi(1) = 0.$

(4) $\varphi(0) - \varphi'(0) = 0$ and $\varphi'(2) = 0.$

(5) $\varphi(0) = 0$ and $\varphi(\pi) + \varphi'(\pi) = 0.$

8.2.6. Consider the following Legendre Polynomials, $P_n(x)$:

$$P_0 \quad = \quad 1,$$

$$P_1 \quad = \quad x,$$

$$P_2 \quad = \quad \frac{3x^2 - 1}{2},$$

$$P_3 \quad = \quad \frac{5x^3 - 3x}{2}, \text{ and}$$

$$P_4 \quad = \quad \frac{35x^4 - 30x^2 + 3}{8}.$$

(1) Show by direct calculation that they are orthogonal to each other on $-1 < x < 1$.

(2) Show by direct calculation that the integral of the square of each of the Legendre polynomials on $-1 < x < 1$ is equal to

$$\frac{2}{2n + 1},$$

where n is the subscript of the respective Legendre polynomial.

8.2.7. Consider the following Laguerre polynomials, $L_n(x)$:

$$
\begin{aligned}
L_0(x) &= 1, \\
L_1(x) &= 1 - x, \\
L_2(x) &= x^2 - 4x + 2, \text{ and} \\
L_3(x) &= -x^3 + 9x^2 - 18x + 6.
\end{aligned}
$$

(1) Show by direct calculation that they are orthogonal to each other on $0 < x < \infty$ with weight function e^{-x}.

(2) Show by direct calculation that the integral of the square of each of the Laguerre polynomials on $0 < x < \infty$ with weight function e^{-x} is equal to

$$(n!)^2$$

where n is the subscript of the respective Laguerre polynomial.

8.2.8. Consider the following Hermite polynomials, $H_n(x)$:

$$
\begin{aligned}
H_0(x) &= 1, \\
H_1(x) &= 2x, \\
H_2(x) &= 4x^2 - 2, \text{ and} \\
H_3(x) &= 8x^3 - 12x.
\end{aligned}
$$

(1) Show by direct calculation that they are orthogonal to each other on $-\infty < x < \infty$ with weight function e^{-x^2}.

(2) Show by direct calculation that the integral of the square of each of the Hermite polynomials on $-\infty < x < \infty$ with weight function e^{-x^2} is equal to

$$2^n n! \sqrt{\pi},$$

where n is the subscript of the respective Hermite polynomial.

8.2.9. Consider

$$(x\varphi'(x))' + \lambda \left(\frac{1}{x}\right) \varphi(x), \ 1 \le x \le 5,$$

subject to

$$\varphi(1) = 0 \text{ and } \varphi(5) = 0.$$

The general solution is

$$\varphi(x) = a\cos\left(\sqrt{\lambda}\ln x\right) + b\sin\left(\sqrt{\lambda}\ln x\right).$$

Find the eigenvalues and eigenfunctions and verify that the eigenfunctions are orthogonal directly by integration.

8.3 RAYLEIGH QUOTIENT

In the previous section, we stated the Rayleigh quotient without proving it. In this section, we restate and prove the Rayleigh quotient. Also, we discuss several properties of the Rayleigh quotient, and some ways the Rayleigh quotient is used.

Given the regular Sturm–Liouville eigenvalue problem

$$\frac{d}{dx}\left[p(x)\varphi'(x)\right] + q(x)\varphi(x) + \lambda\sigma(x)\varphi(x) = 0,\, 0 \le x \le L, \tag{8.43}$$

subject to

$$\alpha_1\varphi(0) + \alpha_2\varphi'(0) = 0$$
$$\alpha_3\varphi(L) + \alpha_4\varphi'(L) = 0, \tag{8.44}$$

we show each eigenvalue, λ_n, $n = 1, 2, 3, \ldots$, is related to its eigenfunction, $\varphi_n(x)$, $n = 1, 2, 3, \ldots$ Thus, we show

$$\lambda_n = \frac{-p(x)\varphi_n(x)\varphi_n'(x)\big|_0^L + \int_0^L \left[p(x)\left(\varphi'(x)\right)^2 - q(x)\varphi^2(x)\right]\,dx}{\int_0^L \varphi^2(x)\sigma(x)\,dx}.$$

Proof. Let $\varphi_n(x)$ be a solution of Equation (8.43). Then, Equation (8.43) becomes

$$\frac{d}{dx}\left[p(x)\varphi_n'(x)\right] + q(x)\varphi_n(x) + \lambda_n\sigma(x)\varphi_n(x) = 0,\, 0 \le x \le L. \tag{8.45}$$

Multiplying Equation (8.45) by $\varphi_n(x)$ yields

$$\frac{d}{dx}\left[p(x)\varphi_n'(x)\right]\varphi_n(x) + q(x)\varphi_n^2(x) + \lambda_n\sigma(x)\varphi_n^2(x) = 0,$$

which may be written as

$$q(x)\varphi_n^2(x) + \lambda_n\sigma(x)\varphi_n^2(x) = -\frac{d}{dx}\left[p(x)\varphi_n'(x)\right]\varphi_n(x). \tag{8.46}$$

Integrating both sides of Equation (8.46) with respect to x from 0 to L gives us

$$\int_0^L q(x)\varphi_n^2(x) + \lambda_n\sigma(x)\varphi_n^2(x)\,dx = \int_0^L -\frac{d}{dx}\left[p(x)\varphi_n'(x)\right]\varphi_n(x)\,dx. \tag{8.47}$$

Rearranging the left side of Equation (8.47), and noting λ_n is a constant, yields

$$\int_0^L q(x)\varphi_n^2(x)\ dx + \lambda_n \int_0^L \sigma(x)\varphi_n^2(x)\ dx = \int_0^L -\frac{d}{dx}\left[p(x)\varphi_n'(x)\right]\varphi_n(x)\ dx. \quad (8.48)$$

Using integration by parts on the right side of Equation (8.47), Equation (8.48) becomes

$$\int_0^L q(x)\varphi_n^2(x)\ dx + \lambda_n \int_0^L \sigma(x)\varphi_n^2(x)\ dx$$

$$= -p(x)\varphi_n(x)\varphi_n'(x)\big|_0^L + \int_0^L p(x)\left(\varphi'(x)\right)^2 dx. \quad (8.49)$$

Performing some algebra manipulations on Equation (8.49) results in

$$\lambda_n = \frac{-p(x)\varphi_n(x)\varphi_n'(x)\big|_0^L + \int_0^L \left[p(x)\left(\varphi_n'(x)\right)^2 - q(x)\varphi_n^2(x)\right]dx}{\int_0^L \varphi_n^2(x)\sigma(x)\ dx}.$$

Thus, we have shown the Rayleigh quotient relates each eigenvalue λ_n with its corresponding eigenfunction $\varphi_n(x)$. $\qquad\square$

One use of the Rayleigh quotient is to directly prove that λ_n, $n = 1,\ 2,\ 3,\ \ldots$ is greater than or equal to 0 if

$$1) \quad -p(x)\varphi_n(x)\varphi_n'(x)\big|_0^L \geq 0 \quad \text{and}$$

$$2) \quad q(x) \leq 0 \text{ for } 0 \leq x \leq L.$$

A second use of the Rayleigh quotient occurs when the eigenfunctions of the Sturm–Liouville eigenvalue problem cannot be determined. In this case, we can generally find a lower bound for the first eigenvalue. Hopefully, it is 0. Also, we can always find an upper bound for the first eigenvalue. The proof for finding the upper bound for the first eigenvalue is shown using the theory of an applied branch of mathematics called calculus of variations. We will not do the proof here. However, we state the resulting theorem.

Theorem 51. *The minimum value of the Rayleigh quotient over all continuous functions satisfying the boundary conditions, $\alpha_1\varphi(0) + \alpha_2\varphi'(0) = 0$ and $\alpha_3\varphi(L) + \alpha_4\varphi'(L) = 0$, is the lowest eigenvalue, λ_1.*

We will demonstrate the theorem using a simple example.

EXAMPLE 8.4. Given

$$\varphi''(x) + \lambda\varphi(x) = 0, \quad (8.50)$$

subject to

$$\varphi(0) = 0$$

$$\varphi'(1) = 0. \tag{8.51}$$

From previous work, we know $\lambda_1 = \dfrac{\pi^2}{4} \sim 2.467$ and $\varphi_1(x) = \sin\dfrac{\pi x}{2}$. However, the point of this example is finding a continuous function satisfying the boundary conditions, which is not $\varphi_1(x)$. First, we find the lower bound of the eigenvalues. From Equation (8.50), we immediately determine $p(x) = 1$, $\sigma(x) = 1$, and $q(x) = 0$. This, in turn, means the Rayleigh quotient becomes

$$\frac{\varphi_n(x)\varphi_n'(x)\big|_0^1 + \displaystyle\int_0^1 (\varphi_n'(x))^2 \, dx}{\displaystyle\int_0^1 \varphi_n^2(x) \, dx}. \tag{8.52}$$

Using the BCs stated in Equation (8.51), we have $\varphi_n(x)\varphi_n'(x)\big|_0^1 = 0$. Thus, since $(\varphi_n'(x))^2 \geq 0$ and $\varphi_n(x) \neq 0$ for all n, we know Equation (8.52) is greater than or equal to 0. Therefore, the lower bound for the first eigenvalue is 0. To find an upper bound, we must find a continuous function that satisfies the BCs, Equation (8.51). Consider the trial function

$$f_t(x) = x^2 - 2x.$$

We immediately see that $f_t(0) = 0$ and $f_t'(1) = 0$. Thus, $f_t(x)$ is a continuous function which satisfies the BCs, Equation (8.51). Next, using the Rayleigh quotient with $f_t(x)$ replacing $\varphi(x)$, we find an upper bound. Since $f_t(x)f_t'(x)\big|_0^1 = 0$, we have

$$0 \leq \lambda_1 \leq \frac{\displaystyle\int_0^1 (f_t'(x))^2 \, dx}{\displaystyle\int_0^1 f_t^2(x) \, dx} = \frac{\displaystyle\int_0^1 (2x-2)^2 \, dx}{\displaystyle\int_0^1 (x^2-2x)^2 \, dx} = \frac{\displaystyle\int_0^1 4x^2 - 8x + 4 \, dx}{\displaystyle\int_0^1 x^4 - 4x^3 + 4x^2 \, dx}$$

or

$$0 \leq \lambda_1 \leq \frac{\frac{4x^3}{3} - 4x^2 + 4x \Big|_0^1}{\frac{x^5}{5} - x^4 + \frac{4x^3}{3} \Big|_0^1} = 2.5.$$

Therefore, $0 \leq \lambda_1 \leq 2.5$. We note the exact value of $\lambda_1 = \dfrac{\pi^2}{4} \approx 2.467 < 2.5$. ∎

Next, a slightly more complicated example using the more general form of the Rayleigh quotient is considered.

EXAMPLE 8.5. Given

$$(x\varphi'(x))' + \lambda \left(\frac{1}{x}\right)\varphi(x) = 0, \ 1 \le x \le 2, \tag{8.53}$$

subject to

$$\varphi'(1) - \varphi(1) = 0$$
$$\varphi(2) = 0, \tag{8.54}$$

find the lower and upper bound for the first eigenvalue.

Solution: First, determine $p(x) = x$, $\sigma(x) = \frac{1}{x}$, and $q(x) = 0$, and set up the Rayleigh quotient, which is

$$\frac{-x\varphi_n(x)\varphi_n'(x)\big|_1^2 + \int_1^2 x\left(\varphi_n'(x)\right)^2 dx}{\int_1^2 \varphi_n^2(x)\frac{1}{x}\, dx}.$$

Since $q(x) = 0$, $\varphi_n^2(x)\frac{1}{x} > 0$ on $1 \le x \le 2$, $x\left(\varphi_n'(x)\right)^2 \ge 0$ on $1 \le x \le 2$ for all n, and $\int_1^2 \varphi_n^2(x)\frac{1}{x}\, dx > 0$, we must determine if $-x\varphi_n(x)\varphi_n'(x)\big|_1^2 \ge 0$.

We have $-x\varphi_n(x)\varphi_n'(x)\big|_1^2 = -2\varphi_n(2)\varphi_n'(2) + \varphi_n(1)\varphi_n'(1) = \varphi_n(1)\varphi_n'(1)$ by the second BC. Using the first BC we know $\varphi'(1) = \varphi(1)$. This means $\varphi_n(1)\varphi_n'(1)$ becomes $\varphi_n(1)\varphi_n(1) = \varphi_n^2(1) \ge 0$. Thus, 0 is the lower bound for λ_1.

Next, we determine a trial function and find the upper bound. Determining a trial function, while sometimes not easy, is always possible. We let $f_t(x) = Ax^2 + Bx + C$ and apply the BCs to determine A, B, and C. We have from the first BC $f_t'(1) - f_t(1) = 0$ or $2A + B - A - B - C = 0$, which implies $A = C$. Thus, by the first BC we have $f_t(x) = A(x^2 + 1) + Bx$. The second BC tells us $f_t(2) = 0$ or $5A + 2B = 0$, which implies $B = \frac{-5}{2}A$. If we let $A = -2$, then $B = 5$, and our trial function becomes $f_t(x) = -2(x^2+1) + 5x$. Now, we substitute $f_t(x)$ for $\varphi_n(x)$ in the Rayleigh quotient. We have

$$0 \le \lambda_1 \le \frac{-x\left(-2(x^2+1)+5x\right)(-4x+5)\big|_1^2 + \int_1^2 x\left(-4x+5\right)^2 dx}{\int_1^2 \left(-2(x^2+1)+5x\right)^2 \frac{1}{x}\, dx} \approx 17.05.$$

Therefore, $0 \le \lambda_1 \le 17.05$. The large gap between 0 and 17.05 may mean our trial function was not very good. If we had chosen as our trial function $x^3 - 5x + 2$, then our upper bound would become approximately 11.52, which decreases the gap.

That is, $0 \leq \lambda_1 \leq 11.52 \leq 17.05$. A far better approximation occurs with the trial function

$$f_t(x) = \begin{cases} x, & 1 \leq x < 1.5 \\ -3x + 6, & 1.5 \leq x \leq 2, \end{cases}$$

which, when used in the Rayleigh quotient, yields an upper bound of 0.583734. ■

Today, one of the major uses of the Rayleigh quotient is a numerical estimate on the beginning bounds of eigenvalues when solving a variable coefficient PDE numerically with a computer. We now develop a general solution to a variable coefficient PDE.

EXERCISES 8.3

8.3.1. Use the Rayleigh quotient to obtain a reasonably accurate upper bound for the lowest eigenvalue of the following:

(1) $\varphi''(x) - x^2\varphi(x) + \lambda\varphi(x) = 0$ with $\varphi'(0) = 0$ and $\varphi(1) = 0$.

(2) $\varphi''(x) - x\varphi(x) + \lambda\varphi(x) = 0$ with $\varphi'(0) = 0$ and $\varphi'(1) + 2\varphi(1) = 0$.

(3) $\varphi''(x) + \lambda\varphi(x) = 0$ with $\varphi(0) = 0$ and $\varphi'(1) + \varphi(1) = 0$.

8.3.2. Given

$$(x\varphi'(x))' + \lambda\left(\frac{1}{x}\right)\varphi(x) = 0, \, 1 \leq x \leq 2,$$

subject to

$$\varphi'(1) - \varphi(1) = 0 \text{ and } \varphi(2) = 0,$$

show that the trial function

$$f_t(x) = \begin{cases} x, & 1 \leq x < 1.5 \\ -3x + 6, & 1.5 \leq x \leq 2 \end{cases}$$

yields an upper bound on λ_1 of 0.583734.

8.3.3. Consider the eigenvalue problem

$$\varphi''(x) - x^2\varphi(x) + \lambda\varphi(x) = 0,$$

subject to

$$\varphi'(0) = 0 \text{ and } \varphi'(1) = 0.$$

Show that $\lambda > 0$.

8.3.4. For the eigenvalue problem

$$\frac{d^4\varphi}{dx^2} = -\lambda e^x \varphi,$$

subject to the boundary conditions

$$\varphi(0) = 0, \ \varphi(1) = 0,$$

$$\frac{d\varphi(0)}{dx} = 0, \text{ and } \frac{d^2\varphi(1)}{dx^2} = 0.$$

Show that the eigenvalues are less than or equal to 0. Also, answer the problem if $\lambda = 0$?

8.3.5. Determine all negative eigenvalues (if any) for

$$\varphi''(x) + 5\varphi(x) + \lambda\varphi(x) = 0,$$

subject to

$$\varphi(0) = 0 \text{ and } \varphi(\pi) = 0.$$

8.4   THE GENERAL PDE EXAMPLE

Consider a one-dimensional rod of length L with perfect lateral insulation. Suppose the rod is composed of several materials that are not uniformly mixed. Thus, thermal conductivity, mass density, and specific heat are functions of the spatial variable. Also, suppose there is a sink term, which is proportional to the temperature distribution. Determine the temperature distribution for all time in the rod if it is known the boundary conditions are homogeneous Dirichlet conditions, and the initial temperature distribution is $f(x)$.

The problem just stated has the following equations:

$$c(x)\rho(x)\frac{\partial u}{\partial t} = \frac{\partial}{\partial x}\left[K_0(x)\frac{\partial u}{\partial x}\right] - \alpha u(x,t), \tag{8.55}$$

subject to

$$u(0,t) = 0$$
$$u(L,t) = 0 \tag{8.56}$$

and

$$u(x,0) = f(x). \tag{8.57}$$

Since Equation (8.55) and BCs, Equation (8.56), are linear and homogeneous, we know separation of variables is a valid solution method. Thus, letting $u(x,t) = \varphi(x)G(t)$ yields the time equation of

$$G'(t) = -\lambda G(t), \tag{8.58}$$

which we know has a solution

$$G(t) = ae^{-\lambda t} \tag{8.59}$$

and spatial equation

$$[K_0(x)\varphi'(x)]' - \alpha\varphi(x) + \lambda c(x)\rho(x)\varphi(x) = 0, \tag{8.60}$$

with spatial BCs

$$\varphi(0) = 0$$
$$\varphi(L) = 0. \tag{8.61}$$

We know from the description of the problem that $K_0(x)$ and $c(x)\rho(x)$ are positive functions for the length of the rod. Therefore, Equations (8.60 and 8.61) are of the regular Sturm–Liouville type where $K_0(x)$ is $p(x)$, $-\alpha$ is $q(x)$, and $c(x)\rho(x)$ is $\sigma(x)$. Thus, we are guaranteed the existence of eigenvalues and eigenfunctions, although we may not be able to identify them. Setting up the Rayleigh quotient, with the appropriate functions replaced, results in

$$\lambda_n = \frac{-K_0(x)\varphi_n(x)\varphi_n'(x)|_0^L + \int_0^L \left[K_0(x)\left(\varphi_n'(x)\right)^2 - \alpha\varphi_n^2(x) \right] dx}{\int_0^L \varphi_n^2(x)c(x)\rho(x)\ dx}.$$

Applying the BCs, Equation (8.61), tells us λ_n is greater than 0. Thus, we have a lower bound on our first eigenvalue. Using the solution to the time equation, Equation (8.59), and what we know about the solution to the spatial equation, yields the general solution for $u(x,t)$, which is

$$u(x,t) = \sum_{n=1}^{\infty} A_n e^{-\lambda_n t}\varphi_n(x). \tag{8.62}$$

We find the specific solution by applying the initial condition. Thus, we have

$$u(x,0) = f(x) = \sum_{n=1}^{\infty} A_n\varphi_n(x)$$

or simply

$$f(x) = \sum_{n=1}^{\infty} A_n\varphi_n(x). \tag{8.63}$$

We find A_n by applying orthogonality. This means we multiply both sides of Equation (8.63) by $\varphi_m(x)$ and integrate over the interval where the rod is defined. Remember, we must also multiply by $\sigma(x) = c(x)\rho(x)$, which is the weight function.

Thus, we have

$$\int_0^L f(x)c(x)\rho(x)\varphi_m(x)\ dx = \int_0^L \sum_{n=1}^{\infty} A_n\varphi_n(x)c(x)\rho(x)\varphi_m(x)\ dx$$

$$= \sum_{n=1}^{\infty} A_n \int_0^L \varphi_n(x)\varphi_m(x)c(x)\rho(x)\ dx.$$

By orthogonality,

$$\int_0^L \varphi_n(x)\varphi_m(x)c(x)\rho(x)\ dx = \begin{cases} 0,\ n \neq m \\ \int_0^L \varphi_n^2(x)c(x)\rho(x)\ dx,\ n = m. \end{cases} \tag{8.64}$$

Thus, the form of the solution for A_n is

$$A_n = \frac{\int_0^L f(x)\varphi_n(x)c(x)\rho(x)\ dx}{\int_0^L \varphi_n^2(x)c(x)\rho(x)\ dx}. \tag{8.65}$$

Therefore, the solution to Equations (8.55, 8.56, and 8.57) is Equation (8.62) where A_n is given by Equation (8.65).

EXERCISES 8.4

8.4.1. Given

$$p(x)\frac{\partial^2 u}{\partial t^2} = \frac{\partial}{\partial x}\left[\tau(x)\frac{\partial u}{\partial x}\right] - \beta u,$$

subject to

$$\frac{\partial u(0, t)}{\partial x} = 0 \text{ and } \frac{\partial u(L, t)}{\partial x} = 0.$$

and

$$u(x, 0) = f(x) \text{ and } \frac{\partial u(x, 0)}{\partial t} = g(x),$$

determine the general solution to include the Rayleigh quotient for λ_n, the orthogonality relationships, and the equations for the coefficients of the generalized Fourier series. You may call the eigenfunctions $\varphi(x)$.

8.4.2. Consider

$$p(x)\frac{\partial u}{\partial t} = \frac{\partial}{\partial x}\left[\tau(x)\frac{\partial u}{\partial x}\right] - \beta(x)u + Q(x),$$

subject to

$$u(0,t) = T_1 \text{ and } u(L,t) = T_2$$

and

$$u(x,0) = f(x).$$

(1) Let $v(x)$ be a solution of the problem

$$[\tau(x)v']' - \beta(x)v = -Q(x),$$

subject to

$$v(0) = T_1 \text{ and } v(L) = T_2.$$

If $z(x,t) = u(x,t) - v(x)$, find the boundary value problem satisfied by $z(x,t)$.

(2) Consider the same PDE, only now subject to the BCs

$$\frac{\partial u(0,t)}{\partial x} - \alpha_1 u(0,t) = T_1 \text{ and } \frac{\partial u(L,t)}{\partial x} + \alpha_2 u(L,t) = T_2.$$

Find the general solution.

(3) Using the methods of this problem solve

$$\frac{\partial u}{\partial t} = \frac{\partial^2 u}{\partial x^2} - 3,$$

subject to

$$u(0,t) = 2 \text{ and } u(\pi,t) = -5,$$

and

$$u(x,0) = x^2 + 2x - 1.$$

(4) Using the methods of this problem solve

$$\frac{\partial u}{\partial t} = \frac{\partial^2 u}{\partial x^2} - \sin 2x,$$

subject to

$$u(0,t) = 0 \text{ and } \frac{\partial u(\pi,t)}{\partial x} = 1$$

and

$$u(x,0) = \cos x.$$

8.5 PROBLEMS INVOLVING HOMOGENEOUS BCS OF THE THIRD KIND

Boundary conditions of the third kind, or Robin's boundary conditions, were last discussed in Chapter 3. In this section, we thoroughly investigate Robin's boundary conditions. While we discuss homogeneous Robin's BCs, as shown in Chapters 2 and 3, Robin's BCs may be nonhomogeneous. Some problems in the exercise section of this chapter have nonhomogeneous Robin's BCs. We start with a general problem.

Consider a thin metallic rod made of a uniform material of length, L, with perfect lateral insulation and an initial temperature distribution, which is an arbitrary function of x. Also, suppose that both ends are subject to homogeneous Robin's Boundary Conditions. Determine the solution of the initial value problem.

The mathematical model for this physical situation is

$$\frac{\partial u(x,t)}{\partial t} = k\frac{\partial^2 u(x,t)}{\partial x^2}, \tag{8.66}$$

subject to the IC

$$u(x,0) = f(x) \tag{8.67}$$

and BCs

$$K_0\frac{\partial u(0,t)}{\partial x} + h_0 u(0,t) = 0$$

$$\tag{8.68}$$

$$K_L\frac{\partial u(L,t)}{\partial x} + h_L u(L,t) = 0.$$

In Equation (8.68), the thermal diffusivity of both ends, K_0 and K_L, and the coefficient of heat transfer at both ends, h_0 and h_L, may be the same or different. However, since the physical description of the problem does not state specifically that they are the same, we must model them as being different. Normally, K_0 would be modeled as $-K_0$. However, for this theoretical model, we want to look at all possible eigenvalues. If we modeled K_0 as $-K_0$, then we would only get positive eigenvalues, which are generally the normal case. Also, note that the thermal diffusivity of the rod may be different from that of the ends. Thus, the thermal diffusivity of the rod is modeled as k.

The mathematical model is linear and homogeneous for both the heat equation and the BCs. Therefore, separation of variables may be applied. Letting $u(x,t) = G(t)\varphi(x)$, performing the necessary derivatives, and separating the equations yields the time equation

$$G'(t) = -\lambda k G(t) \tag{8.69}$$

and the spatial equation

$$\varphi''(x) = -\lambda\varphi(x), \tag{8.70}$$

subject to the BCs

$$K_0\varphi'(0) + h_0\varphi(0) = 0$$

$$K_L\varphi'(L) + h_L\varphi(L) = 0,$$

(8.71)

which may be written as

$$H_0\varphi(0) = -\varphi'(0)$$

$$H_L\varphi(l) = -\varphi'(L).$$

(8.72)

In Equation (8.72), $H_0 = \dfrac{h_0}{K_0}$ and $H_L = \dfrac{h_L}{K_L}$.

From experience, we know that Equation (8.69) has the solution

$$G_n(t) = C_n e^{-k\lambda_n t},$$

(8.73)

where the λ_n are the eigenvalues found when solving the spatial equation.

When solving the spatial equation, Equation (8.70) subject to the BCs, Equation (8.72), we must remember to consider all three cases for λ. That is, $\lambda < 0$, $\lambda = 0$, and $\lambda > 0$.

Case 1: $\lambda < 0$: Let $\lambda = -s$, where $s > 0$. Then Equation (8.70) becomes

$$\varphi''(x) = s\varphi(x),$$

(8.74)

which has solution

$$\varphi(x) = C_1 \cosh(\sqrt{s}x) + C_2 \sinh(\sqrt{s}x).$$

(8.75)

Applying the first BC of Equation (8.72) to Equation (8.75), we find that

$$C_1 = \frac{-\sqrt{s}C_2}{H_0}.$$

Applying the second BC of Equation (8.72) to Equation (8.75), we get the transcendental equation

$$\tanh(\sqrt{s}L) = \frac{s - H_0}{(H_0 - H_L)\sqrt{s}}.$$

(8.76)

Letting $z = \sqrt{s}L$, we can separate the transcendental function into two functions

$$f_1(z) = \tanh z$$

and

$$f_2(z) = \frac{z^2 - H_0 L}{z(H_0 - H_L)}.$$

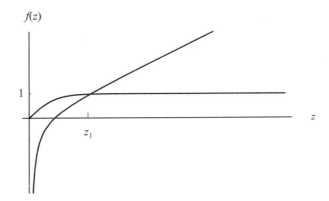

Figure 8.1: Graph of $f_1(z) = \tanh z$ and $f_2(z) = \dfrac{z^2 - H_0 L}{z(H_0 - H_L)}$.

Remember that $s > 0$; then $z > 0$. We graph the two functions and look for intersection points. Figure (8.1) shows one possible intersection point. The intersection point indicates the first eigenvalue. It is

$$\lambda_1 = -\frac{z^2}{L}, \tag{8.77}$$

where the corresponding eigenfunction may be given as

$$\varphi_1(x) = d_1 \left(-\sqrt{-\lambda} \cosh(\sqrt{-\lambda}x) + H_0 \sinh(\sqrt{-\lambda}x) \right). \tag{8.78}$$

Case 2: Let $\lambda = 0$. Then, Equation (8.70) becomes

$$\varphi''(x) = 0, \tag{8.79}$$

which has solution

$$\varphi(x) = C_1 x + C_2. \tag{8.80}$$

Applying the first BC of Equation (8.72) to Equation (8.80), we find that

$$C_1 = -H_0 C_2.$$

Applying the second BC of Equation (8.72) to Equation (8.80), we get

$$H_L C_2 (1 - H_0 L) = -H_0 C_2,$$

which implies $C_2 = 0$, since we must assume that H_L, H_0, and L are not identically 0.

Case 3: Let $\lambda > 0$. Then the solution of Equation (8.70) is

$$\varphi(x) = C_1 \cos\left(\sqrt{\lambda}x\right) + C_2 \sin\left(\sqrt{\lambda}x\right). \tag{8.81}$$

Applying the first BC of Equation (8.72) to Equation (8.81) yields

$$C_1 = -\frac{C_2\sqrt{\lambda}}{H_0}.$$

Applying the second BC of Equation (8.72) to Equation (8.81) gives us the transcendental equation

$$\tan(\sqrt{\lambda}L) = \frac{\sqrt{\lambda}(H_L - H_0)}{\lambda + H_0 H_L}. \tag{8.82}$$

Letting $z = \sqrt{\lambda}L$, we can separate the transcendental function into two functions

$$f_1(z) = \tan z$$

and

$$f_2(z) = \frac{z(H_L - H_0)}{z^2 + H_L H_0 L}.$$

We graph the two functions and look for intersection points. Figure (8.2) gives a possible representation of the intersection points. Also, we know that $C_1 = -\dfrac{C_2\sqrt{\lambda}}{H_0}$.

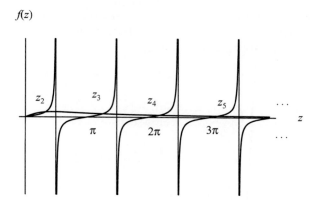

Figure 8.2: Graph of $f_1(z) = \tan z$ and $f_2(z) = \dfrac{z(H_L - H_0)}{z^2 + H_L H_0 L}$.

Thus, we may write the eigenfunction equation as

$$\varphi_n(x) = d_n\left(-\sqrt{\lambda_n}\cos\left(\sqrt{\lambda_n}x\right) + H_0\sin\left(\sqrt{\lambda_n}x\right)\right), \; n = 1, 2, 3, \ldots \tag{8.83}$$

We now have the complete solution to the spatial problem. Therefore, the general solution to Equation (8.66) subject to the BCs, Equation (8.68), may be given by

$$
\begin{aligned}
u(x,t) \;=\; & G_n(t)\varphi_n(x) = \sum_{n=1}^{\infty} D_n e^{-k\lambda_n t}\varphi_n(x) \\[2mm]
=\; & D_1 e^{-k\lambda_1 t}\left(-\sqrt{-\lambda}\cosh(\sqrt{-\lambda}x) + H_0 \sinh(\sqrt{-\lambda}x)\right) \\[2mm]
& + \sum_{n=2}^{\infty} D_n e^{-k\lambda_n t}\left(-\sqrt{\lambda_n}\cos\left(\sqrt{\lambda_n}x\right) + H_0 \sin\left(\sqrt{\lambda_n}x\right)\right),
\end{aligned}
\tag{8.84}
$$

where $D_n = C_n d_n$, $D_1 = C_1 d_1$, and $\lambda_1 = -\dfrac{z^2}{L}$. We may find D_n by using the IC, Equation (8.67), and the orthogonality of the eigenfunctions, and it is

$$
D_n = \frac{\displaystyle\int_0^L \varphi_n(x) f(x)\, dx}{\displaystyle\int_0^L \varphi_n^2(x)\, dx},
\tag{8.85}
$$

where $\varphi_n(x)$ is given by Equations (8.78 and 8.83).

Using Figure (8.2), we may set up a table of the intersection points of the graphs of the two functions $f_1(z) = \tan z$ and $f_2(z) = \dfrac{z(H_L - H_0)}{z^2 + H_L H_0 L}$. The intersection points are the actual points where the two functions are equal. However, we are interested in bounding the intersection points. Bounding the intersection points will yield upper and lower bounds of the eigenvalues. Thus, the following table contains bounding values for each z_n starting with $n = 2$:

Lower Bound		$z_n = \sqrt{\lambda_n}L$		Upper Bound
0	$<$	$z_2 = \sqrt{\lambda_2}L$	$<$	$\dfrac{\pi}{2}$
π	$<$	$z_3 = \sqrt{\lambda_3}L$	$<$	$\dfrac{3\pi}{2}$
2π	$<$	$z_4 = \sqrt{\lambda_4}L$	$<$	$\dfrac{5\pi}{2}$
3π	$<$	$z_5 = \sqrt{\lambda_5}L$	$<$	$\dfrac{7\pi}{2}$
4π	$<$	$z_6 = \sqrt{\lambda_6}L$	$<$	$\dfrac{9\pi}{2}$
$\vdots$	$\vdots$	$\vdots$	$\vdots$	$\vdots$
$(n-2)\pi$	$<$	$z_n = \sqrt{\lambda_n}L$	$<$	$\dfrac{(2n-3)\pi}{2}$

$$\tag{8.86}$$

Using Table (8.86), we may develop upper and lower bounds for each λ_n starting with $n = 2$. In general,

$$\left(\frac{(n-2)\pi}{L}\right)^2 < \lambda_n < \left(\frac{(2n-3)\pi}{2L}\right)^2. \tag{8.87}$$

Again, using Figure (8.2), it is interesting to note that, as $n \to \infty$, $\lambda_n \to \left(\frac{(n-2)\pi}{L}\right)^2$.

We finish this section by presenting an example.

EXAMPLE 8.6. Consider heat conduction in a one-dimensional rod with perfect lateral insulation of length 2π and no internal source. Suppose the boundary at 2π is maintained at the constant temperature of 0^0C, and the boundary at 0 is subject to Newton's law of cooling where it is known to have the relationship $u(0,t) = \dfrac{\partial u(0,t)}{\partial x}$. Also, the initial condition is known to be

$$u(x,0) = f(x) = \begin{cases} x+1 & 0 \le x < \pi \\ \dfrac{-(\pi+1)x}{\pi} + 2(\pi+1) & \pi \le x \le 2\pi. \end{cases} \tag{8.88}$$

The mathematical model for this problem is

$$\frac{\partial u(x,t)}{\partial t} = \frac{\partial^2 u(x,t)}{\partial x^2}, \tag{8.89}$$

subject to

$$u(0,t) = \frac{\partial u(0,t)}{\partial x}$$
$$u(2\pi,t) = 0 \tag{8.90}$$

and the IC, Equation (8.88). The PDE and BCs are linear and homogeneous. Therefore, we may apply the separation of variables technique. If we let $u(x,t) = G(t)\varphi(x)$, and we take the correct partials and substitute into Equation (8.89), we have

$$G'(t)\varphi(x) = G(t)\varphi''(x). \tag{8.91}$$

Separating Equation (8.91) and setting the equations equal to an arbitrary constant yields the time equation

$$G'(t) = -\lambda G(t) \tag{8.92}$$

and the spatial equation

$$\varphi''(x) = -\lambda\varphi(x). \tag{8.93}$$

Applying the same technique to the BCs produces the BCs for the spatial problem, which are

$$\varphi(0) = \varphi'(0)$$

$$\varphi(2\pi) = 0. \tag{8.94}$$

Equations (8.93 and 8.94) are of the regular Sturm–Liouville type where $p(x) = 1 = \sigma(x)$, $q(x) = 0$, $\alpha_1 = 1 = \alpha_3$, $\alpha_2 = -1$, and $\alpha_4 = 0$. Thus, we may use the Rayleigh quotient to find a lower bound for λ. The Rayleigh quotient for this problem is

$$\frac{-\varphi(x)\varphi'(x)\big|_0^{2\pi} + \displaystyle\int_0^{2\pi} [\varphi'(x)]^2\, dx}{\displaystyle\int_0^{2\pi} [\varphi(x)]^2\, dx}.$$

We would like $\lambda \geq 0$. Since $[\varphi'(x)]^2 \geq 0$ and $[\varphi(x)]^2 > 0$, we need only check $-\varphi(x)\varphi'(x)\big|_0^{2\pi}$. Since $\varphi(2\pi) = 0$, then

$$-\varphi(x)\varphi'(x)\big|_0^{2\pi} = \varphi(0)\varphi'(0).$$

Also, we have $\varphi(0) = \varphi'(0)$, which means

$$\varphi(0)\varphi'(0) = [\varphi(0)]^2 > 0.$$

Thus, $\lambda > 0$. Therefore, we need only solve the eigenvalue problem for $\lambda > 0$. We know that $\varphi''(x) = -\lambda\varphi(x)$ has the solution

$$\varphi(x) = c_1 \cos\sqrt{\lambda}x + c_2 \sin\sqrt{\lambda}x. \tag{8.95}$$

The first BC indicates

$$\varphi(0) = c_1 = \varphi'(0) = c_2\sqrt{\lambda},$$

which implies that Equation (8.95) becomes

$$\varphi(x) = c_2\left(\sqrt{\lambda}\cos\sqrt{\lambda}x + \sin\sqrt{\lambda}x\right).$$

Applying the second BC to $\varphi(x)$ yields

$$\varphi(2\pi) = 0 = c_2\left(\sqrt{\lambda}\cos\sqrt{\lambda}2\pi + \sin\sqrt{\lambda}2\pi\right),$$

or

$$0 = \sqrt{\lambda}\cos\sqrt{\lambda}2\pi + \sin\sqrt{\lambda}2\pi. \tag{8.96}$$

Equation (8.96) may be written as

$$-\sqrt{\lambda}\cos\sqrt{\lambda}2\pi = \sin\sqrt{\lambda}2\pi,$$

or in the preferred form

$$-\sqrt{\lambda} = \frac{\sin\sqrt{\lambda}2\pi}{\cos\sqrt{\lambda}2\pi} = \tan\sqrt{\lambda}2\pi. \qquad (8.97)$$

Equation (8.97) is known as a transcendental equation. If we let $z = \sqrt{\lambda}2\pi$, Equation (8.97) may be written as

$$\frac{-z}{2\pi} = \tan z. \qquad (8.98)$$

Our horizontal axis is the z-axis, and we have two functions, $f_1(z) = \tan z$ and $f_2(z) = \dfrac{-z}{2\pi}$. If we graph these two equations, then the intersection points will be the solution of Equation (8.98), which means we can determine λ_n. Figure (8.3) is the graph of $f_1(z)$ and $f_2(z)$.

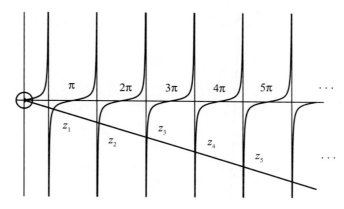

Figure 8.3: Graph of $f_1(z) = \tan z$ and $f_2(z) = \dfrac{-z}{2\pi}$.

Using Figure (8.3), we may set up a table of the intersection points of the graphs of the two functions $f_1(z) = \tan z$ and $f_2(z) = \dfrac{z}{2\pi}$. The intersection points are the actual points that the two functions are equal. However, we are interested in bounding the intersection points. Bounding the intersection points will yield upper and lower bounds of the eigenvalues. Thus, the table of bounding values for each z_n is as follows:

Lower Bound		$z_n = \sqrt{\lambda_n}2\pi$		Upper Bound
$\dfrac{\pi}{2}$	$<$	$z_1 = \sqrt{\lambda_1}2\pi$	$<$	π
$\dfrac{3\pi}{2}$	$<$	$z_2 = \sqrt{\lambda_2}2\pi$	$<$	2π
$\dfrac{5\pi}{2}$	$<$	$z_3 = \sqrt{\lambda_3}2\pi$	$<$	3π

Lower Bound $\qquad z_n = \sqrt{\lambda_n}\,2\pi \qquad$ Upper Bound

$$
\frac{7\pi}{2} \quad < \quad z_4 = \sqrt{\lambda_4}\,2\pi \quad < \quad 4\pi
$$

$$
\frac{9\pi}{2} \quad < \quad z_5 = \sqrt{\lambda_5}\,2\pi \quad < \quad 5\pi \tag{8.99}
$$

$$
\vdots \qquad \vdots \qquad \vdots \qquad \vdots \qquad \vdots
$$

$$
\frac{(2n-1)\pi}{2} \quad < \quad z_n = \sqrt{\lambda_n}\,2\pi \quad < \quad n\pi
$$

Using Table (8.99), we may develop the table for λ_n, which is really what we want. Therefore, Table (8.100) contains upper and lower bounds for each λ_n:

Lower Bound $\qquad \lambda_n \qquad$ Upper Bound

$$
\frac{1}{16} \quad < \quad \lambda_1 \quad < \quad \frac{1}{4}
$$

$$
\frac{9}{16} \quad < \quad \lambda_2 \quad < \quad 1
$$

$$
\frac{25}{16} \quad < \quad \lambda_3 \quad < \quad \frac{9}{4}
$$

$$
\frac{49}{16} \quad < \quad \lambda_4 \quad < \quad 4 \tag{8.100}
$$

$$
\frac{81}{16} \quad < \quad \lambda_5 \quad < \quad \frac{25}{4}
$$

$$
\vdots \qquad \vdots \quad \vdots \qquad \vdots
$$

$$
\frac{(2n-1)^2}{16} \quad < \quad \lambda_n \quad < \quad \frac{n^2}{4}
$$

It is interesting to note that, as $n \to \infty$, $\lambda_n \to \dfrac{(2n-1)^2}{16}$. We have an idea of the bounds on the eigenvalues and of the eigenfunctions, $\varphi_n(x)$. Also, we know the solution to the time problem Equation (8.92), which is

$$
G(t) = d_n e^{-\lambda_n t}.
$$

Therefore, the general solution for $u(x,t)$ is

$$
u(x,t) = \sum_{n=1}^{\infty} a_n e^{-\lambda_n t} \varphi_n(x). \tag{8.101}
$$

The complete solution is Equation (8.101), with the equation for a_n, which is

$$a_n = \frac{\int_0^{2\pi} f(x)\varphi_n(x)\ dx}{\int_0^{2\pi} [\varphi_n(x)]^2\ dx},$$

where $f(x)$ is given by Equation (8.88). ∎

EXERCISES 8.5

8.5.1. Consider the eigenvalue problem

$$\varphi''(x) + \lambda\varphi(x) = 0,$$

subject to

$$\varphi'(0) = 0 \text{ and } \varphi'(1) + 2\varphi(1) = 0.$$

(1) Prove that $\lambda \geq 0$.

(2) Does $\lambda = 0$?

(3) Determine all eigenvalues graphically. Obtain lower and upper bounds. Estimate the large eigenvalues.

8.5.2. Consider the eigenvalue problem

$$\varphi''(x) + \lambda\varphi(x) = 0,$$

subject to

$$\varphi'(0) = 0 \text{ and } \varphi'(1) - 2\varphi(1) = 0.$$

(1) Prove that $\lambda > 0$.

(2) Determine all eigenvalues graphically. Obtain lower and upper bounds. Estimate the large eigenvalues.

8.5.3. Solve

$$\frac{\partial u}{\partial t} = 9\frac{\partial^2 u}{\partial x^2} - 9u,$$

subject to

$$u(0,t) - \frac{\partial u(0,t)}{\partial x} = 0 \text{ and } u(1,t) - 8\frac{\partial u(1,t)}{\partial x} = 0$$

and

$$u(x,0) = f(x).$$

(1) Graphically determine the eigenvalues.

(2) Estimate the large eigenvalues.

(3) Determine the solution.

(4) What happens after a long time?

8.5.4. Solve

$$\frac{\partial u}{\partial t} = 0.4 \frac{\partial^2 u}{\partial x^2},$$

subject to

$$u(0,t) - \frac{\partial u(0,t)}{\partial x} = 0 \text{ and } u(1,t) + \frac{\partial u(1,t)}{\partial x} = 0$$

and

$$u(x,0) = f(x).$$

Note: You may call the relevant eigenfunctions $\varphi_n(x)$ and assume that they are known.

8.5.5. Solve

$$\frac{\partial^2 u}{\partial t^2} = \frac{\partial^2 u}{\partial x^2},$$

subject to

$$\frac{\partial u(0,t)}{\partial x} = u(0,t) \text{ and } \frac{\partial u(400,t)}{\partial x} = -u(400,t)$$

and

$$u(x,0) = 0 \text{ and } \frac{\partial u(x,0)}{\partial t} = 0.01x - 0.000025x^2.$$

(1) Estimate the large eigenvalues.

(2) Estimate λ_1, and using the first term of the series, describe the solution physically.

8.5.6. Consider

$$\frac{\partial^2 u}{\partial t^2} = 3 \frac{\partial^2 u}{\partial x^2} - 3 \frac{\partial u}{\partial t},$$

the equation of a vibrating string in a medium that resists the motion, subject to

$$u(0,t) = \frac{\partial u(0,t)}{\partial x} \text{ and } u(\pi,t) = 4 \frac{\partial u(\pi,t)}{\partial x}$$

and

$$u(x,0) = x^2 + 2x + 1 \text{ and } \frac{\partial u(x,0)}{\partial t} = 2x + 2.$$

(1) Graphically determine the eigenvalues.

(2) Estimate the large eigenvalues.

(3) Determine the solution.

(4) What happens after a long time?

8.5.7. Consider

$$\frac{\partial^2 u}{\partial t^2} = 6\frac{\partial^2 u}{\partial x^2} - 6\frac{\partial u}{\partial t},$$

subject to

$$u(0, t) = -\frac{\partial u(0, t)}{\partial x} \text{ and } u(5, t) = -10\frac{\partial u(5, t)}{\partial x}$$

and

$$u(x, 0) = e^x \text{ and } \frac{\partial u(x, 0)}{\partial t} = \sin x.$$

(1) Graphically determine the eigenvalues.

(2) Estimate the large eigenvalues.

(3) Determine the solution.

(4) What happens after a long time?

8.5.8. Consider

$$\frac{\partial^2 u}{\partial t^2} = \frac{\partial^2 u}{\partial x^2} - \pi\frac{\partial u}{\partial t} - u,$$

subject to

$$u(0, t) = \frac{\partial u(0, t)}{\partial x} \text{ and } u(\pi, t) = 3\frac{\partial u(\pi, t)}{\partial x}$$

and

$$u(x, 0) = x^3 - 3x^2 + 2x + 1 \text{ and } \frac{\partial u(x, 0)}{\partial t} = 3x^2 - 6x + 2.$$

(1) Graphically determine the eigenvalues.

(2) Estimate the large eigenvalues.

(3) Determine the solution.

(4) What happens after a long time?

8.5.9. Consider

$$\frac{\partial^2 u}{\partial t^2} = \frac{\partial^2 u}{\partial x^2} - \frac{\partial u}{\partial t} - \frac{\partial u}{\partial x} - u,$$

subject to

$$u(0,t) = \frac{\partial u(0,t)}{\partial x} \text{ and } u(L,t) = -\frac{\partial u(L,t)}{\partial x}$$

and

$$u(x,0) = f(x) \text{ and } \frac{\partial u(x,0)}{\partial t} = g(x).$$

(1) Graphically determine the eigenvalues.

(2) Estimate the large eigenvalues.

(3) Determine the solution.

(4) What happens after a long time?

8.5.10. Consider the PDE

$$\frac{\partial u(x,t)}{\partial t} = \frac{\partial^2 u(x,t)}{\partial x^2} + \frac{\partial u(x,t)}{\partial x}.$$

Solve the initial value problem subject to the BCs

$$u(0,t) = 0 \text{ and } \frac{\partial u(\pi,t)}{\partial x} = 0$$

and IC

$$u(x,0) = -\sin x.$$

In five sentences or less, give a physical description of this problem.

8.5.11. Consider the PDE

$$\frac{\partial^2 u(x,t)}{\partial t^2} = \frac{\partial^2 u(x,t)}{\partial x^2} + \frac{\partial u(x,t)}{\partial x}.$$

Solve the initial value problem subject to the BCs

$$\frac{\partial u(0,t)}{\partial x} = 0 \text{ and } u\left(\frac{3\pi}{2},t\right) = 0$$

and IC

$$u(x,0) = \cos x \text{ and } \frac{\partial u(x,0)}{\partial t} = x^2 - \left(\frac{3\pi}{2}\right)^2.$$

In five sentences or less, give a physical description of this problem.

8.5.12. Consider the PDE

$$\frac{\partial^2 u(x,t)}{\partial t^2} = \frac{\partial^2 u(x,t)}{\partial x^2} - \frac{\partial u(x,t)}{\partial t} + \frac{\partial u(x,t)}{\partial x}.$$

Solve the initial value problem subject to the BCs

$$u(0, t) = 0 \text{ and } \frac{\partial u(2\pi, t)}{\partial x} = 0$$

and IC

$$u(x, 0) = 1 \text{ and } \frac{\partial u(x, 0)}{\partial t} = x^2.$$

In five sentences or less, give a physical description of this problem.

8.5.13. Consider the PDE

$$\frac{\partial u(x, t)}{\partial t} = 9 \frac{\partial^2 u(x, t)}{\partial x^2} + 4 \frac{\partial u(x, t)}{\partial x}.$$

Solve the initial value problem subject to the BCs

$$\frac{\partial u(0, t)}{\partial t} = 0 \text{ and } u(2\pi, t) - \frac{\partial u(2\pi, t)}{\partial x} = 0$$

and IC

$$u(x, 0) = -\cos x.$$

In five sentences or less, give a physical description of this problem.

Chapter 9

Solution of Linear Homogeneous Variable-Coefficient ODE

9.1 INTRODUCTION

In Appendix C, the general second-order linear homogeneous variable-coefficient ODE is given as follows:

$$S(x)u''(x) + K(x)u'(x) + H(x)u(x) = 0. \tag{9.1}$$

In this chapter, we will refer to Equation (9.1) as a general second-order ODE. Common examples of this type of ODE are

Equation Name	*Equation*
Airy's	$u'' - xu = 0,$
Bessel's	$x^2u'' + xu' + \left(x^2 - n^2\right)u = 0,$
Chebyshev's	$\left(1 - x^2\right)u'' - xu' + n^2u = 0,$
Euler's	$ax^2u'' + bxu' + cu = 0,$
Hermite's	$u'' - 2xu' + 2nu = 0,$
Legendre's	$\left(1 - x^2\right)u'' - 2xu' + n\left(n + 1\right)u = 0.$

$$\tag{9.2}$$

A few examples where these equations arise in problems are astronomy, quantum physics, electromagnetism, propagation of electromagnetic radiation through a coaxial cable, heat distribution in a circular membrane and a sphere, Schrödinger's equation for a harmonic oscillator, and vibrations of a circular drum head. All the Equations (9.2) are known to have series solution except Euler's equation.

This chapter starts with a brief discussion of Equation (9.1). Then, we solve Euler's equation. (Euler's equation does not require power series methods to develop the solution.) Next, we develop the power series solution method. For many students taking this course, this material may be unfamiliar. Therefore, the theorems are presented without proof. However, a reference for the complete proof will be provided for all theorems, most of which may be found in almost any ODE text. Please remember this is a brief overview and IS not meant as a replacement for an in-depth study of this material.

9.2 SOME FACTS ABOUT THE GENERAL SECOND-ORDER ODE

Given the equation

$$S(x)u''(x) + K(x)u'(x) + H(x)u(x) = 0, \tag{9.3}$$

we assume that $S(x)$, $K(x)$, and $H(x)$ are polynomials having no common factors. Then x_0 is said to be an **ordinary point** of Equation (9.3) if $S(x_0) \neq 0$. If x_0 is not an ordinary point, then it is a singular point. For instance, every real number, x_0, is an ordinary point of

$$u'' + xu = 0. \tag{9.4}$$

Whereas, $x_0 = \pm 1$ are singular points of

$$\left(1 - x^2\right) u'' - xu' + n^2 u = 0. \tag{9.5}$$

If x_0 is an ordinary point of Equation (9.3), then Equation (9.3) may be rewritten as

$$u''(x) + \kappa(x)u'(x) + \xi(x)u(x) = 0,$$

where $\kappa(x) = \dfrac{K(x)}{S(x)}$ and $\xi(x) = \dfrac{H(x)}{S(x)}$.

If, for some x_0, $S(x_0) = 0$ in Equation (9.1), then x_0 is either a **regular singular point** or an **irregular singular point**. You can determine if x_0 is a regular singular point if the following holds true:

$$\lim_{x \to x_0} (x - x_0)\frac{K(x)}{S(x)} \text{ is finite} \tag{9.6}$$

and

$$\lim_{x \to x_0} (x - x_0)^2\frac{H(x)}{S(x)} \text{ is finite.} \tag{9.7}$$

For example, consider Euler's equation

$$x^2 u'' + \beta x u' + \chi u = 0.$$

We have

$$S(x) \quad = \quad x^2,$$

$$K(x) \quad = \quad \beta x, \text{ and}$$

$$H(x) \quad = \quad \chi.$$

Hence, for $x_0 = 0$, we have $S(x_0) = 0$;

$$\lim_{x \to x_0} (x - x_0) \frac{K(x)}{S(x)} = \lim_{x \to 0} x \frac{\beta x}{x^2} = \beta, \text{ a finite constant;}$$

and

$$\lim_{x \to x_0} (x - x_0)^2 \frac{H(x)}{S(x)} = \lim_{x \to 0} x^2 \frac{\chi}{x^2} = \chi, \text{ a finite constant.}$$

Therefore, Euler's equation is an example of an equation that has a regular singularity at $x_0 = 0$. Legendre's equation has regular singular points at $x_0 = \pm 1$, which is left as an exercise.

This text examines those variable coefficient ODEs that have only ordinary points or regular singular points, and are relevant to PDEs. ODEs that have irregular singular points are extremely difficult to solve, and their study is left to an advanced course in ODEs.

EXERCISES 9.2

9.2.1. Show that the following equations have ordinary or regular singular points:

    (1) $u'' - xu = 0$.

    (2) $x^2 u'' + xu' + \left(x^2 - n^2\right) u = 0$.

    (3) $\left(1 - x^2\right) u'' - xu' + n^2 u = 0$.

    (4) $u'' - 2xu' + 2nu = 0$.

    (5) $\left(1 - x^2\right) u'' - 2xu' + n(n+1) u = 0$.

9.2.2. Determine if the following equations have an ordinary point, a regular singular point, or an irregular singular point at 0 and 1:

    (1) $xu'' + (1 - x) u' + xu = 0$.

    (2) $x^2 \left(1 - x^2\right) u'' + 5xu' + 8u = 0$.

    (3) $x^5 \left(1 - x^2\right) u'' - 3xu' + u = 0$.

    (4) $2x^4 \left(1 - x^2\right) u'' - \frac{2}{x} u' + 3x^2 u = 0$.

    (5) $u'' + \frac{x}{1+x} u' + 4(1 + x) u = 0$.

    (6) $(x + 2) u'' + 2xu' - \left(1 - x^2\right) u = 0$.

9.3   EULER'S EQUATION

Euler's equation[1] is any equation of the form

$$ax^2u'' + bxu' + cu = 0,$$

where a, b, and c are constants. Since a, b, and c are constants, the usual form of Euler's equation is

$$x^2u'' + \beta xu' + \chi u = 0. \tag{9.8}$$

You can determine the solutions by power series methods. However, sometimes the solution may be determined by a proper guess, which is the way most second order constant-coefficient linear ODEs were originally solved.

For our guess, we'll let $u(x) = x^r$, where $x \neq 0$, since 0 is a singular point of Equation (9.8). Then, $u'(x) = rx^{r-1}$ and $u''(x) = r(r-1)x^{r-2}$. Substituting $u(x)$, $u'(x)$, and $u''(x)$ into Equation (9.8) yields

$$x^2 r(r-1)x^{r-2} + \beta xrx^{r-1} + \chi x^r = 0,$$

which, after some algebraic manipulations, becomes

$$r^2 + (\beta - 1)r + \chi = 0. \tag{9.9}$$

Equation (9.9) looks exactly like the characteristic equations you found in your ODE course for second-order constant-coefficient linear ODEs, and may be solved in the same way. Thus, we have

$$r = \frac{-(\beta - 1) \pm \sqrt{(\beta - 1)^2 - 4\chi}}{2} = -\gamma \pm \delta,$$

where

$$\gamma = \frac{(\beta - 1)}{2} \text{ and}$$

$$\delta = \frac{\sqrt{(\beta - 1)^2 - 4\chi}}{2}.$$

Therefore, the roots of the characteristic equation may be expressed as

$$r_1 = \frac{-(\beta - 1) + \sqrt{(\beta - 1)^2 - 4\chi}}{2} = -\gamma + \delta \tag{9.10}$$

and

$$r_2 = \frac{-(\beta - 1) - \sqrt{(\beta - 1)^2 - 4\chi}}{2} = -\gamma - \delta. \tag{9.11}$$

We have three cases to investigate:

---

[1]Leonhard Euler (1707–1783) was the one of the most famous mathematicians who came from Switzerland. He published over 500 books and papers during his lifetime.

- Case 1: $(\beta - 1)^2 - 4\chi > 0$, produces two real and unequal roots.
- Case 2: $(\beta - 1)^2 - 4\chi = 0$, produces real and equal roots.
- Case 3: $(\beta - 1)^2 - 4\chi < 0$, produces complex roots.

The following examples illustrate the type of solution for each case.

EXAMPLE 9.1. Consider

$$x^2 u'' - xu' - 6u = 0.$$

If we assume a solution $u(x) = x^r$, where $x > 0$, then the characteristic equation becomes

$$r(r - 1) - r - 3 = 0,$$

which is

$$r^2 - 2r - 3 = 0.$$

Thus, we have Case 1: two real and unequal roots, where the roots are -1 and 3. By our assumption, $u(x) = x^r$, we have the two solutions $u_1(x) = x^{-1}4$ and $u_2(x) = x^3$. Therefore, we know that the general solution is

$$u(x) = c_1 x^{-1} + c_2 x^3. \qquad \blacksquare$$

Case 2 is somewhat more interesting. Though we don't cover reduction of order in this text you should be familiar with the process. It is easily found and explained in most ODE texts.

EXAMPLE 9.2. Consider

$$x^2 u'' + 5xu' + 4u = 0.$$

Again, we assume a solution $u(x) = x^r$, where $x > 0$, then the characteristic equation becomes

$$r(r - 1) + 5r + 4 = r^2 + 4r + 4 = (r + 2)^2 = 0.$$

Thus, we have real and equal roots r=-2 with one solution $u_1(x) = x^{-2}$. The second solution may be found by using reduction of order, and it is $u_2(x) = x^{-2} \ln(x)$, which means the general solution is

$$u(x) = x^{-2} \left(c_1 + c_2 \ln(x) \right). \qquad \blacksquare$$

By far, the most interesting case is Case 3. To arrive at the proper solution for Case 3, we must remember some basic mathematics. Since you already know that the solution has complex roots, you probably expect sine and cosine functions to appear. The question is how? Remember that

$$x^r = e^{r \ln x}.$$

From this simple formula, we arrive at the mathematically cleanest solution. We now proceed with the example.

EXAMPLE 9.3. Consider

$$x^2 u'' + 2xu' + u = 0.$$

As before, if we assume a solution $u(x) = x^r$, where $x > 0$, then the characteristic equation becomes

$$r(r - 1) + 2r + 1 = r^2 + r + 1 = 0.$$

Using the quadratic formula, we find that

$$r_1 = \frac{-1 + i\sqrt{3}}{2}$$

$$r_2 = \frac{-1 - i\sqrt{3}}{2}.$$

Thus,

$$
\begin{aligned}
u_1(x) &= x^{r_1} \\[2mm]
&= x^{\frac{-1 + i\sqrt{3}}{2}} \\[2mm]
&= e^{\frac{-1 + i\sqrt{3}}{2} \ln x} \\[2mm]
&= e^{\frac{-1 \ln x}{2}} e^{\frac{i\sqrt{3} \ln x}{2}} \\[2mm]
&= x^{\frac{-1}{2}} e^{\frac{i\sqrt{3} \ln x}{2}}.
\end{aligned}
$$

Similarly,

$$u_2(x) = x^{\frac{-1}{2}} e^{\frac{-i\sqrt{3} \ln x}{2}}.$$

Thus,

$$
\begin{aligned}
u(x) &= c_1 u_1(x) + c_2 u_2(x) \\[2mm]
&= c_1 \left(x^{\frac{-1}{2}} e^{\frac{i\sqrt{3} \ln x}{2}} \right) + c_2 \left(x^{\frac{-1}{2}} e^{\frac{-i\sqrt{3} \ln x}{2}} \right) \\[2mm]
&= x^{\frac{-1}{2}} \left(c_1 \left(e^{\frac{i\sqrt{3} \ln x}{2}} \right) + c_2 \left(e^{\frac{-i\sqrt{3} \ln x}{2}} \right) \right) \\[2mm]
&= x^{\frac{-1}{2}} \left(d_1 \cos\left(\frac{\sqrt{3}}{2} \ln x \right) d_2 \sin\left(\frac{\sqrt{3}}{2} \ln x \right) \right),
\end{aligned}
$$

which completes the solution for Case 3. ∎

Note: In Examples (9.1, 9.2, and 9.3) we assumed $x > 0$. For the complete solution given in Proposition (52) for all $x \neq 0$, we must use absolute values.

Proposition 52. *Given Euler's equation*

$$x^2 u'' + \beta x u' + \chi u = 0. \tag{9.12}$$

We can solve Equation (9.12) in any interval such that $x \neq 0$ by substituting $u(x) = x^r$ and its derivatives into Equation (9.12), solving for the roots r_1 and r_2 of the resulting characteristic equation

$$r^2 + (\beta - 1)r + \chi = 0.$$

If the roots of the characteristic equation are real and unequal, then

$$u(x) = c_1 |x|^{r_1} + c_2 |x|^{r_2} \tag{9.13}$$

is the solution to Euler's equation.
If the roots of the characteristic equation are real and equal, then

$$u(x) = (c_1 + c_2 \ln |x|) |x|^r , \tag{9.14}$$

where $r = \dot{r}_1, r_2$ is the solution to Euler's equation.
If the roots of the characteristic equation are complex, then

$$u(x) = |x|^{-\gamma} [c_1 \cos (\delta \ln |x|) + c_2 \sin (\delta \ln |x|)] , \tag{9.15}$$

where $r = -\gamma \pm i\delta$ is the solution to Euler's equation.

EXERCISES 9.3

9.3.1. In each of the following problems determine the general solution.

(1) $x^2 u'' + 2xu' - u = 0$.

(2) $x^2 u'' + xu' - 9u = 0$.

(3) $x^2 u'' + xu' + 4u = 0$.

(4) $(x + 1)^2 u'' + (x + 1)u' + \dfrac{1}{4} u = 0$.

(5) $x^2 u'' - xu' + u = 0$.

(6) $(x - 1)^2 u'' - 3(x - 1)u' + \dfrac{9}{4} u = 0$.

(7) $x^2 u'' - xu' + 2u = 0$.

9.3.2. Consider the interval $x > 0$. Let $x = e^y$. Knowing that Euler's equation is

$$x^2 u'' + \beta x u' + \chi x u = 0,$$

show that

$$\frac{du}{dx} = \frac{1}{x}\frac{du}{dy} \tag{9.16}$$

and

$$\frac{d^2 u}{dx^2} = \frac{1}{x^2}\frac{d^2 u}{dy^2} - \frac{1}{x^2}\frac{du}{dy}. \tag{9.17}$$

Next, using Equations (9.16 and 9.17), show that Euler's equation becomes

$$\frac{d^2 u}{dy^2} + (\beta - 1)\frac{du}{dy} + \chi u = 0. \tag{9.18}$$

Since Equation (9.18) is now a linear second-order constant-coefficient ODE, find the characteristic equation and show that the roots generate the identical cases as Proposition 52.

9.3.3. In Example (9.2) it was said that a second solution to the real and equal roots case may be found by reduction of order. Given Euler's equation

$$ax^2 u'' + bx u' + cu = 0,$$

and the fact that $r_1 = r_2$ are real roots of the characteristic equation

$$ar(r - 1) + br + c = 0.$$

Thus, $u_1(x) = x^{r_1}$ is one solution of Euler's equation. Use the method of reduction of order to show that $u_2(x) = x^{r_1} \ln x$ is the second solution.

9.4 BRIEF REVIEW OF POWER SERIES

The power series you are most familiar with are Taylor and Maclaurin series. Some common Maclaurin series are

$$e^x \quad = \quad \sum_{n=0}^{\infty} \frac{x^n}{n!}; \qquad -\infty < x < \infty,$$

$$\sin x \quad = \quad \sum_{n=0}^{\infty} \frac{(-1)^n x^{2n+1}}{(2n+1)!}; \qquad -\infty < x < \infty,$$

$$\cos x \quad = \quad \sum_{n=0}^{\infty} \frac{(-1)^n x^{2n}}{(2n)!}; \qquad -\infty < x < \infty,$$

$$\frac{1}{1-x} \quad = \quad \sum_{n=0}^{\infty} x^n; \qquad -1 < x < 1, \text{ and}$$

$$\ln(1+x) \quad = \quad \sum_{n=0}^{\infty} \frac{(-1)^{n+1} x^{2n}}{n} \qquad -1 < x \leq 1.$$

You should have encountered these series in your study of Calculus. However, you may not have encountered a general definition of a power series. It is as follows:

Definition 53. *An infinite series of the form*

$$\sum_{n=0}^{\infty} b_n (x - x_0)^n, \tag{9.19}$$

where x_0 and $b_0, b_1, \ldots, b_n, \ldots$ are constants, is called a power series in $x - x_0$.

Knowing the general definition of a power series only helps us if we know when and how a power series converges. The power series given in Equation (9.19) is said to converge if

$$\lim_{N \to \infty} \sum_{n=0}^{N} b_n (x - x_0)^n$$

exists; otherwise it is said to diverge. If $x = x_0$, the power series must converge. For values of $x \neq x_0$, we may not know whether the series converges. However, if the power series converges, the following theorem applies:

Theorem 54. *If the power series*

$$\sum_{n=0}^{\infty} b_n (x - x_0)^n$$

converges, then either

(1) the power series converges only at $x = x_0$.

(2) the power series converges absolutely and uniformly for all values of x.

(3) there exists a positive number R (called the radius of convergence) such that the power series is absolutely convergent for all $|x - x_0| < R$, and divergent for $|x - x_0| > R$.

The proof of this theorem may be found in the text *Infinite Series* by Earl D. Rainville. Also, note that in the radius of convergence, the power series is often written as

$$f(x) = \sum_{n=0}^{\infty} b_n(x - x_0)^n. \tag{9.20}$$

Since we are developing a power series solution for a general second-order ODE, it would seem useful to know if a power series may be differentiated. This is given in the following theorem. The proof may also be found in the text *Infinite Series* by Earl D. Rainville.

Theorem 55. *If the power series*

$$f(x) = \sum_{n=0}^{\infty} b_n(x - x_0)^n, \ in \ |x - x_0| < R, \tag{9.21}$$

then

$$f'(x) = \sum_{n=1}^{\infty} n b_n(x - x_0)^{n-1}, \ in \ |x - x_0| < R. \tag{9.22}$$

Note: This theorem indicates that within the radius of convergence, we can take as many derivatives as we need. Thus, it is possible to have a power series as a solution to a general second-order ODE. However, there is a difference between the series' given in Equations (9.21 and 9.22). The sums have a different starting point. In Equation (9.21), the series starts at $n = 0$, and in Equation (9.22), the series starts at $n = 1$. This is easily solved by shifting the indices, which is shown in most calculus texts. That is,

Remark 1. *For any integer m, the power series*

$$f(x) = \sum_{n=n_0}^{\infty} b_n(x - x_0)^{n-m}$$

may be rewritten as

$$f(x) = \sum_{n=n_0-m}^{\infty} b_{n+m}(x - x_0)^n.$$

Thus, Equation (9.22) may be rewritten as

$$f'(x) = \sum_{n=0}^{\infty} (n+1)\, b_{n+1}(x - x_0)^n, \text{ in } |x - x_0| < R.$$

In the next section, we show how power series may be used to solve Equation (9.1) at an ordinary point.

EXERCISES 9.4

9.4.1. Find the Maclaurin series expansion of the following functions:

(1) $\ln(1 + x)$.

(2) $\dfrac{1}{1 - 3x}$.

(3) $\dfrac{5}{\sqrt{1 + 2x}}$.

(4) $e^{\frac{-x}{3}}$.

(5) $\sin(4x)$.

9.4.2. Find the Taylor series expansion of the following functions:

(1) e^{2x} in powers of $(x + 1)$.

(2) $\cos x$ in powers of $(x - \pi)$.

(3) $\sin 5x$ in powers of $\left(x + \dfrac{\pi}{2}\right)$.

(4) $\ln(x + 1)$ in powers of $(x + 7)$.

(5) $\dfrac{3}{2x - 1}$ in powers of $(x + 1)$.

9.5   THE POWER SERIES SOLUTION METHOD

The power series solution method is the standard method for solving general second order ODEs. The idea behind solving a variable coefficient second-order ODE at an ordinary point is to

- assume a power series solution with unknown coefficients,

- find all necessary derivatives,

- replace the derivatives in the original equation,

- combine the power series, and

- set the coefficients equal to zero and solve.

As an example, consider the following:

EXAMPLE 9.4. Solve

$$u' - 2u = 0$$

at $x_0 = 0$.

Solution: Let

$$u(x) = \sum_{n=0}^{\infty} b_n x^n.$$

Find $u'(x)$, which is

$$u'(x) = \sum_{n=1}^{\infty} n b_n x^{n-1}.$$

This may be rewritten as

$$u'(x) = \sum_{n=0}^{\infty} (n+1) b_{n+1} x^n.$$

Replace the assumed solution and the derivative in the original equation, and we have

$$\sum_{n=0}^{\infty} (n+1) b_{n+1} x^n - 2 \sum_{n=0}^{\infty} b_n x^n = 0$$

or

$$\sum_{n=0}^{\infty} \left[(n+1) b_{n+1} - 2 b_n \right] x^n = 0.$$

Next, set the coefficient equal to 0, and we get

$$(n+1) b_{n+1} - 2 b_n = 0,$$

which implies that

$$b_{n+1} = \frac{2 b_n}{n+1}. \tag{9.23}$$

Equation (9.23) is called a recurrence relation. This means that all coefficients may be expressed in terms of b_0. For example,

$$b_1 = 2 b_0,$$

$$b_2 = b_1 = 2 b_0,$$

$$b_3 = \frac{2}{3} b_2 = \frac{2}{3} (2 b_0),$$

$$b_4 = \frac{1}{2} b_3 = \frac{1}{3} (2 b_0),$$

$$\vdots$$

Thus, our solution is

$$u(x) = b_0 \left(1 + 2x + 2x^2 + \frac{4}{3}x^3 + \frac{2}{3}x^4 + \dots \right) = b_0 e^{2x},$$

since

$$\left(1 + 2x + 2x^2 + \frac{4}{3}x^3 + \frac{2}{3}x^4 + \dots \right)$$

is the Maclaurin Series for e^{2x}. ∎

Another example is Airy's equation[2].

EXAMPLE 9.5. Find a series solution of Airy's equation

$$u'' - xu = 0 \tag{9.24}$$

in powers of $x - 2$. Since Airy's equation is a second-order ODE, we expect two linearly independent solutions.

Solution: Assume that

$$u(x) = \sum_{n=0}^{\infty} b_n (x - 2)^n. \tag{9.25}$$

Then

$$u''(x) = \sum_{n=2}^{\infty} n(n-1)b_n (x-2)^{n-2},$$

which may be rewritten as

$$u''(x) = \sum_{n=0}^{\infty} (n+2)(n+1)b_{n+2}(x-2)^n. \tag{9.26}$$

Substituting Equations (9.25 and 9.26) for u and u'', respectively, in Equation (9.24) yields

$$\sum_{n=0}^{\infty} (n+2)(n+1)b_{n+2}(x-2)^n - x \sum_{n=0}^{\infty} b_n (x-2)^n = 0.$$

Here, we cannot simply combine the sums. To express Airy's equation in powers of $x - 2$, we must express x in front of the second sum in powers of $x - 2$. This is done

---

[2]G. B. Airy (1801–1892) was the Astronomer Royal of England. He made many contributions to the study of series and integration.

by rewriting x as $2 + (x - 2)$. Thus, we have

$$\sum_{n=0}^{\infty}(n + 2)(n + 1)b_{n+2}(x - 2)^n = [2 + (x - 2)]\sum_{n=0}^{\infty}b_n(x - 2)^n$$

$$= 2\sum_{n=0}^{\infty}b_n(x - 2)^n + \sum_{n=0}^{\infty}b_n(x - 2)^{n+1}. \qquad (9.27)$$

Shifting the index of the last sum in Equation (9.27), we obtain

$$\sum_{n=0}^{\infty}(n + 2)(n + 1)b_{n+2}(x - 2)^n = 2\sum_{n=0}^{\infty}b_n(x - 2)^n + \sum_{n=1}^{\infty}b_{n-1}(x - 2)^n.$$

Equating like powers of $x - 2$ yields

$$(n + 2)(n + 1)b_{n+2} = 2b_n + b_{n-1},$$

which implies

$$b_2 = b_0,$$

$$6b_3 = 2b_1 + b_0,$$

$$12b_4 = 2b_2 + b_1,$$

$$\vdots$$

The general recurrence relation for $n \geq 3$ is

$$b_n = \frac{2b_{n-2} + b_{n-3}}{n(n - 1)}.$$

Thus, we have

$$b_2 = b_0,$$

$$b_3 = \frac{b_1}{3} + \frac{b_0}{6},$$

$$b_4 = \frac{b_2}{6} + \frac{b_1}{12} = \frac{b_0}{6} + \frac{b_1}{12},$$

$$b_5 = \frac{b_3}{10} + \frac{b_2}{20} = \frac{b_1}{30} + \frac{b_0}{15},$$

$$\vdots$$

Thus,

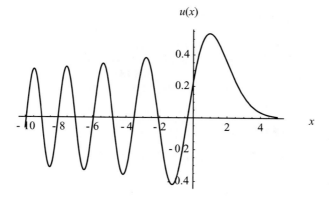

Figure 9.1: Graph of Airy's function.

$$u(x) = b_0 \left(1 + (x-2)^2 + \frac{(x-2)^3}{6} + \frac{(x-2)^4}{6} + \frac{(x-2)^5}{15} + \cdots \right)$$

$$+ b_1 \left((x-2) + \frac{(x-2)^3}{3} + \frac{(x-2)^4}{12} + \frac{(x-2)^5}{30} + \cdots \right)$$

$$= b_0 u_1(x) + b_1 u_2(x),$$

where

$$u_1(x) = \left(1 + (x-2)^2 + \frac{(x-2)^3}{6} + \frac{(x-2)^4}{6} + \frac{(x-2)^5}{15} + \cdots \right)$$

and

$$u_2(x) = \left((x-2) + \frac{(x-2)^3}{3} + \frac{(x-2)^4}{12} + \frac{(x-2)^5}{30} + \cdots \right)$$

and b_0 and b_1 are arbitrary. Figure (9.1) shows the graph of Airy's function from -10 to 5. Notice the oscillatory nature for $x < 0$, whereas for $x > 0$ the function is unbounded. ∎

Example (9.5) demonstrates the following theorem:

Theorem 56. *If x_0 is an ordinary point of*

$$u''(x) + \kappa(x)u'(x) + \xi(x)u(x) = 0, \tag{9.28}$$

then $\kappa(x) = \dfrac{K(x)}{S(x)}$ and $\xi(x) = \dfrac{H(x)}{S(x)}$ have Taylor series expansions at x_0; then, the general solution of Equation (9.28) is given by

$$u(x) = \sum_{n=0}^{\infty} b_n (x - x_0) = c_1 u_1(x) + c_2 u_2(x)$$

where c_1 and c_2 are arbitrary, and $u_1(x)$ and $u_2(x)$ are linearly independent series solutions, which have Taylor series expansions at x_0. Also, the Taylor Series expansions of $u_1(x)$ and $u_2(x)$ have radius of convergence at least as large as the minimum of the Taylor series expansions of $\kappa(x)$ and $\xi(x)$.[3]

Unlike Airy's equation, which may be expanded in powers of any real number, Legendre's equation, given by

$$\left(1 - x^2\right) u'' - 2xu' + n\left(n + 1\right) u = 0, \tag{9.29}$$

has singularities at ± 1. In the next section, we look at Legendre's Equation and its solution.

EXERCISES 9.5

9.5.1. Find a solution to the following ODEs using the Power series method at $x_0 = 0$:

(1) $u' = 5u$.

(2) $u'' - 2u' - 3u = 0$.

(3) $u' - 2u = 0$.

(4) $u'' + 4u' + 4u = 0$.

(5) $u' + 6xu = 0$.

(6) $u'' + 4u = 0$.

(7) $u' - 3u + 2 = 0$.

(8) $u'' + u' + u = 0$.

9.5.2. Find the power series solution for the following ODEs at the indicated value of x_0. That is, find the series solution in powers of $x - x_0$.

(1) $(1 + x^2)u'' + 3xu' + u = 0$, $x_0 = 1$.

(2) $(1 + x^2)u'' + 6xu' + 2u = 0$, $x_0 = -1$.

(3) $(1 + 2x^2)u'' + 3x^2u' - u = 0$, $x_0 = \dfrac{1}{2}$.

(4) $(1 + x^2)u'' - 5xu' + 14u = 0$, $x_0 = 2$.

9.5.3. Find a series solution for Airy's equation in powers of $x - 3$.

9.5.4. Find a series solution for Airy's equation in powers of $x + 1$.

9.5.5. Find a series solution for Airy's equation in powers of $x + 0$.

---

[3] A proof of a more general form of this theorem was completed by Immanuel L. Fuchs (1833–1902) in 1866.

9.5.6. In Hermite's equation,

$$u'' - 2xu' + nu = 0,$$

like Airy's equation, in which every real number x_0 is an ordinary point. Find a series solution to Hermite's equation. The solutions are called Hermite polynomials.

9.5.7. Another way of generating Hermite polynomials is by using Rodrigues' formula, which is

$$H_n(x) = (-1)^n e^{x^2} \frac{d^n e^{-x^2}}{dx^n}.$$

Generate the first five Hermite polynomials using Rodrigues' formula.

9.5.8. Consider $H_0(x) = 1$ and $H_1(x) = 2x$.

(1) Use the recurrence formula $H_{n+1}(x) = 2xH_n(x) - 2nH_{n-1}(x)$ to find the next five Hermite polynomials.

(2) Show that the Hermite polynomials $H_2(x)$ through H_6 found in Part (1) satisfy the recurrence formula $H'_n(x) = 2nH_{n-1}(x)$.

9.5.9. This problem shows how a piecewise smooth function may be expanded in a generalized Fourier series in Hermite polynomials. The weight function for the Hermite polynomials is e^{-x^2}. Also,

$$\int_{-\infty}^{\infty} e^{-x^2} H_n^2(x) \, dx = 2^n n! \sqrt{\pi}.$$

Let

$$f(x) = \sum_{n=0}^{\infty} A_n H_n(x).$$

Show that

$$A_n = \frac{1}{2^n n! \sqrt{\pi}} \int_{-\infty}^{\infty} e^{-x^2} f(x) H_n(x) \, dx.$$

9.5.10. Let

$$f(x) = x.$$

Find the generalized Fourier series in Hermite polynomials. Using your favorite mathematical software graph the function on the interval $(-10, 10)$, and the generalized Fourier series in Hermite polynomials for $n = 10, 25, 50$.

9.5.11. Let

$$f(x) = x^3.$$

Find the generalized Fourier series in Hermite polynomials. Using your favorite mathematical software graph the function on the interval $(-5, 5)$, and the generalized Fourier series in Hermite polynomials for $n = 10, 25, 50$.

9.5.12. Find a power series solution to Weber's equation

$$u'' + \left(n + \frac{1}{2} - \frac{x^2}{4} \right) u = 0,$$

where $n = 0, 1, 2, \ldots$

9.6   LEGENDER'S EQUATION AND LEGENDRE POLYNOMIALS

The solutions to Legendre's equation are called Legendre polynomials, and they are of a class of functions known as *special functions*. Other special functions arise from the solutions of Bessel, Hermite, and Laguerre equations. The solutions to Legendre, Bessel, Hermite, and Laguerre equations form a complete set. Thus, they are used in generalized Fourier series. For instance, Legendre's equation usually occurs when solving a PDE which involves spherical geometry, and Bessel's equation occurs when solving a PDE which involves polar or cylindrical coordinates. Hence, these are very important equations to solve and understand. They will be used extensively in Chapter 10.

We now proceed to solve Legendre's equation

$$\left(1 - x^2 \right) u'' - 2xu' + n \left(n + 1 \right) u = 0. \tag{9.30}$$

Since $S(x) = (1 - x^2)$ is nonzero at 0 we may use the methods we have already developed in the preceeding sections. Here, the radius of convergence is the open interval $(-1, 1)$. Thus, we assume that the solution has the form

$$u(x) = \sum_{m=0}^{\infty} b_m x^m. \tag{9.31}$$

Notice that the index has changed from n to m. We must do this because n already occurs in Equation (9.30). Therefore, the derivatives of Equation (9.31) are

$$u'(x) = \sum_{m=1}^{\infty} mb_m x^{m-1} = \sum_{m=0}^{\infty} (m+1)b_{m+1} x^m$$

and

$$u''(x) = \sum_{m=2}^{\infty} m(m-1)b_m x^{m-2} = \sum_{m=0}^{\infty} (m+2)(m+1)b_{m+2} x^m.$$

Substituting into Equation (9.30), Equation (9.31), and the correct derivatives of Equation (9.31) yields

$$0 = (1 - x^2) \sum_{m=0}^{\infty} (m+2)(m+1)b_{m+2}x^m - 2x \sum_{m=0}^{\infty} (m+1)b_{m+1}x^m$$
$$+ n(n+1) \sum_{m=0}^{\infty} b_m x^m,$$

which becomes

$$0 = \sum_{m=0}^{\infty} (m+2)(m+1)b_{m+2}x^m - \sum_{m=0}^{\infty} (m+2)(m+1)b_{m+2}x^{m+2}$$
$$- 2 \sum_{m=0}^{\infty} (m+1)b_{m+1}x^{m+1} + n(n+1) \sum_{m=0}^{\infty} b_m x^m. \tag{9.32}$$

Expanding the first few terms of Equation (9.32) we obtain

$$0 = \quad 2b_2 \quad + \quad (3)(2)b_3x \quad + \quad (4)(3)b_4x^2 \quad + \quad \cdots$$
$$- \quad 2b_2x^2 \quad + \quad \cdots$$
$$- \quad 2b_1x \quad - \quad (2)(2)b_2x^2 \quad + \quad \cdots$$
$$+ n(n+1)b_0 \quad + \quad n(n+1)b_1x \quad + \quad n(n+1)b_2x^2 \quad + \quad \cdots . \tag{9.33}$$

Since $x \neq 0$ for all values within $(-1, 1)$, we must have

$$2b_2 + n(n+1)b_0 = 0,$$

which implies

$$b_2 = -\frac{n(n+1)}{2}b_0 \tag{9.34}$$

and

$$(3)(2)b_3 + [n(n+1) - 2]\,b_1 = 0$$

or

$$b_3 = -\frac{(n-1)(n+2)}{(3)(2)}b_1. \tag{9.35}$$

The general recursion formula is

$$b_{k+2} = -\frac{(n+k+1)(n-k)}{(k+2)(k+1)}b_k, \text{ for } k = 0, 1, 2, 3, \ldots \tag{9.36}$$

Thus, the coefficients of the infinite series are

$$b_2 = -\frac{(n+1)n}{2}b_0,$$

$$b_3 = -\frac{(n+2)(n-1)}{(3)(2)}b_1,$$

$$b_4 = -\frac{(n+3)(n-2)}{(4)(3)}b_2 = \frac{(n+3)(n+1)n(n-2)}{(4)(3)(2)}b_0, \tag{9.37}$$

$$b_5 = -\frac{(n+4)(n-3)}{(5)(4)}b_3 = \frac{(n+4)(n+2)(n-1)(n-3)}{(5)(4)(3)(2)}b_1,$$

$$\vdots$$

Therefore, one solution to Legendre's equation, Equation (9.30), may be written as

$$u(x) = b_0 \left(1 - \frac{(n+1)n}{2}x^2 + \frac{(n+3)(n+1)n(n-2)}{(4)(3)(2)}x^4 \pm \cdots \right)$$

$$+ b_1 \left(x - \frac{(n+2)(n-1)}{(3)(2)}x^3 + \frac{(n+4)(n+2)(n-1)(n-3)}{(5)(4)(3)(2)} \pm \cdots \right)$$

$$= b_0 u_1(x) + b_1 u_2(x) \tag{9.38}$$

where

$$u_1(x) = 1 - \frac{(n+1)n}{2!}x^2 + \frac{(n+3)(n+1)n(n-2)}{4!}x^4 \pm \cdots$$

and

$$u_2(x) = x - \frac{(n+2)(n-1)}{3!}x^3 + \frac{(n+4)(n+2)(n-1)(n-3)}{5!} \pm \cdots$$

Generally, we will work with Legendre's equation when n is a nonnegative integer. Therefore, whenever $n = k$, either $u_1(x)$ or $u_2(x)$ will be a finite sum while the other is an infinite sum. For example, suppose $n = 2$; we see that in $u_1(x)$ every term after the x^2 term has $n - 2 = 0$ in the product of the numerator. Thus,

$$u_1(x) = 1 - \frac{(n+1)n}{2!}x^2$$

and

$$u_2(x) = x - \frac{(n+2)(n-1)}{3!}x^3 + \frac{(n+4)(n+2)(n-1)(n-3)}{5!} - \cdots$$

So whenever n is an odd integer the sum for $u_2(x)$ is finite; whenever n is an even integer the sum for $u_1(x)$ is finite. These finite polynomials multiplied by constants are called the Legendre polynomials. In most cases, we normalize the Legendre polynomials. That is, for $n = 0$, we let $b_0 = 1$; for $n = 1, 2, 3, \ldots$, we let $b_n = \dfrac{(2n)!}{2^n(n!)^2}$. The normalization allows the polynomials to have a value of 1 when $x = 1$. The formula for the Legendre polynomials is

$$P_n(x) = \sum_{k=0}^{K}(-1)^k \frac{(2n-2k)!}{2^n k!(n-k)!(n-2k)!}x^{n-2k},$$

where $K = \dfrac{n}{2}$ or $\dfrac{n-1}{2}$, whichever is an integer. Thus, a simplified solution to Equation (9.30) is

$$u(x) = c_1 P_n(x).$$

The first five Legendre polynomials are

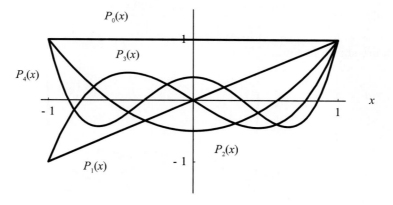

Figure 9.2: The plot of the first five Legendre polynomials, $P_n(x)$.

$$P_0 = 1,$$

$$P_1 = x,$$

$$P_2 = \frac{3x^2 - 1}{2},$$

$$P_3 = \frac{5x^3 - 3x}{2}, \text{ and}$$

$$P_4 = \frac{35x^4 - 30x^2 + 3}{8},$$

which are shown in Figure (9.2).

Another solution to Legendre's equation, Equation (9.30), is Legendre functions of the second kind, denoted $Q_n(x)$. Legendre functions of the second kind are unbounded at ± 1. The first five Legendre functions of the second kind are

$$Q_0 = \frac{1}{2} \ln \left(\frac{1+x}{1-x} \right),$$

$$Q_1 = -1 + \frac{x}{2} \ln \left(\frac{1+x}{1-x} \right),$$

$$Q_2 = -\frac{3x}{2} + \frac{3x^2 - 1}{4} \ln \left(\frac{1+x}{1-x} \right),$$

$$Q_3 = \frac{4 - 15x^2}{6} - \left(\frac{5x^2 - 3x}{4} \right) \ln \left(\frac{1+x}{1-x} \right), \text{ and}$$

$$Q_4 = \frac{55x - 105x^3}{24} + \left(\frac{35x^4 - 30x^2 + 3}{16} \right) \ln \left(\frac{1+x}{1-x} \right),$$

which are shown in Figure (9.3).

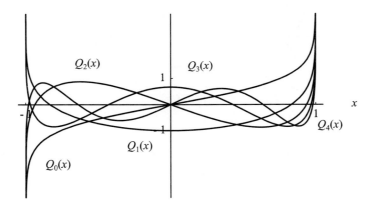

Figure 9.3: The plot of the first five Legendre functions, $Q_n(x)$.

Thus, the complete solution to Legendre's equation, Equation (9.30), is

$$u(x) = a_1 P_n(x) + a_2 Q_n(x).$$

Two other ways of generating the Legendre polynomials are

$$P_n(x) = \frac{1}{2^n n!} \frac{d^n}{dx^n} (x^2 - 1)^n, \tag{9.39}$$

and

$$\frac{1}{\sqrt{1 - 2xt + t^2}} = \sum_{n=0}^{\infty} P_n(x)t^n. \tag{9.40}$$

Equation (9.39) is called the Rodrigues'[4] formula. Equation (9.40) is known as a generating function for Legendre polynomials.

Since Legendre polynomials are used in generalized Fourier series, they must have the same properties as the trigonometric Fourier series. In particular, Legendre polynomials form a complete set and are orthogonal on the interval $-1 < x < 1$, which means

$$\int_{-1}^{1} P_n(x)P_m(x) \, dx = 0, \text{ if } m \neq n$$

and

$$\int_{-1}^{1} [P_n(x)]^2 \, dx = \frac{2}{2n + 1}.$$

These are left for you to show in the exercises. This short introduction to Legendre polynomials completes our coverage of this topic. If you are interested in learning more, I suggest you study the text *Special Functions* by Earl D. Rainville. In the next section, we examine Bessel's equation.

EXERCISES 9.6

9.6.1. Generate the first five Legendre polynomials using the formula

$$P_n(x) = \frac{1}{2^n n!} \frac{d^n}{dx^n} (x^2 - 1)^n.$$

9.6.2. Show that

$$\frac{1}{\sqrt{1 - 2xt + t^2}} = \sum_{n=0}^{\infty} P_n(x)t^n.$$

Hint: Use the binomial theorem on $(1 + \omega)^p$.

9.6.3. Show that Legendre's equation has regular singular points at ± 1.

9.6.4. Prove the recurrence formula

$$P_{n+1}(x) = \frac{2n + 1}{n + 1} x P_n(x) - \frac{n}{n + 1} P_{n-1}(x).$$

Hint: Use the generating function and differentiate with respect to t. Then multiply both sides of the equation by $1 - 2xt + t^2$.

---

[4]Olinde Rodrigues (1794–1851), a French economist and mathematician.

9.6.5. Given $P_1(x) = x$ and $P_2(x) = \dfrac{(3x^2 - 1)}{2}$, find $P_3(x)$, $P_4(x)$, and $P_5(x)$ using the recurrence relation given in the previous problem.

9.6.6. This problem involves a generalized Fourier series in Legendre polynomials. If

$$f(x) = \sum_{n=0}^{\infty} A_n P_n(x), \quad -1 < x < 1,$$

show that

$$A_n = \frac{2n + 1}{n} \int_{-1}^{1} P_n(x) f(x) \, dx.$$

9.6.7. Let

$$f(x) = \begin{cases} x + 1, & -1 < x < 0 \\ \\ 1 - x, & 0 \le x < 1. \end{cases}$$

Expand $f(x)$ in a generalized Fourier series in Legendre polynomials. Use your favorite mathematical software to graph the function, $f(x)$, and the generalized Fourier series in Legendre polynomials for $n = 10$, 25, 50, and 100.

9.6.8. Let

$$g(x) = \begin{cases} -1 - x, & -1 < x \le 0 \\ \\ x - 1, & 0 < x < 1. \end{cases}$$

Expand $g(x)$ in a generalized Fourier series in Legendre polynomials. Use your favorite mathematical software to graph the function, $g(x)$, and the generalized Fourier series in Legendre polynomials for $n = 10$, 25, 50, and 100.

9.6.9. Let

$$h(x) = \sin x, \quad -1 < x < 1.$$

Expand $h(x)$ in a generalized Fourier series in Legendre polynomials. Use your favorite mathematical software to graph the function, $h(x)$, and the generalized Fourier series in Legendre polynomials for $n = 10$, 25, 50, and 100.

9.6.10. Let

$$f(x) = \cos x, \quad -1 < x < 1.$$

Expand $f(x)$ in a generalized Fourier series in Legendre polynomials. Use your favorite mathematical software to graph the function, $f(x)$, and the generalized Fourier series in Legendre polynomials for $n = 10$, 25, 50, and 100.

9.6.11. Let

$$g(x) = \tanh x, \ -1 < x < 1.$$

Expand $g(x)$ in a generalized Fourier series in Legendre polynomials. Use your favorite mathematical software to graph the function, $g(x)$, and the generalized Fourier series in Legendre polynomials for $n = 10, 25, 50,$ and 100.

9.7 METHOD OF FROBENIUS AND BESSEL'S EQUATION

The last equation we will examine is Bessel's equation. Bessel's equation requires a completely new method of solution. That method is known as the method of Frobenius.[5] the method of Frobenius, like the power series method, is another method for solving variable-coefficient ODEs. This method is briefly discussed in this section.

The method of Frobenius is an extremely interesting and important method for an applied mathematician, engineer, and physicist to learn. However, a complete study of this method would require almost a complete chapter by itself. We have neither the space nor the time to present such an in-depth method. If you would like to learn more than just the basics of the method of Frobenius, I would refer you to either *Elementary Differential Equations and Boundary Value Problems* by Boyce and Diprima or *Elementary Differential Equations* by William Trench. Both texts cover this topic extensively and present some very interesting problems. We start our presentation with a short discussion immediately followed by the primary "lengthy" theorem, which we require for Bessel's equation. Then we work through Bessel's equation.

In general, the method of Frobenius is studied at regular singular points, which occur at the origin. In fact, if a regular singular point occurs at a point different than the origin it is easily translated to the origin for convenience. Also, the method of Frobenius requires the use of a Frobenius Series

$$|x|^r \sum_{n=0}^{\infty} b_n x^n, \ b_0 \neq 0. \tag{9.41}$$

Under certain conditions, which we state later, Equation (9.41) always yields one solution. The second solution may or may not exist. The problems we study generally have two solutions. The solutions are either immediately available by knowing that two roots exist or a second solution is "easily" (Remember: Everything is relative) constructed. Since it is easier for the presentation, and it does not effect the method for determining r and all b_n, we only consider $x > 0$. Thus, Equation

---

[5]G. Frobenius (1849–1917), a German mathematician who also made important contributions in group theory.

(9.41) may be written as

$$\sum_{n=0}^{\infty} b_n x^{n+r}, \; b_0 \neq 0. \tag{9.42}$$

As a way to explain the method of Frobenius, we solve a typical problem.

EXAMPLE 9.6. Consider the ODE

$$2x^2 u''(x) + (3x + 2x^2)u'(x) - (1 - x)u(x) = 0. \tag{9.43}$$

First, we determine if there is a regular singular point at $x_0 = 0$. Since $S(x_0) = x_0^2 = 0$ at $x_0 = 0$, we know that Equation (9.43) is a possible candidate. Now, we must determine if Equations (9.6 and 9.7) are finite at $x_0 = 0$. Equation (9.6) gives us

$$\lim_{x \to 0} x \frac{K(x)}{S(x)} = \lim_{x \to 0} x \frac{(3x + 2x^2)}{2x^2} = \frac{3}{2}, \text{ which is a finite constant,}$$

and Equation (9.7) tells us that

$$\lim_{x \to 0} x^2 \frac{H(x)}{S(x)} = \lim_{x \to 0} x^2 \frac{-(1 - x)}{2x^2} = -\frac{1}{2}, \text{ which is also a finite constant.}$$

Thus, $x_0 = 0$ is a regular singular point. Another way to check to see if you have a regular singular point is determining if $xK(x)$ and $x^2 H(x)$ have Maclaurin Series. We know that both $xK(x) = x(3x+2x^2)$ and $x^2 H(x) = -(1-x)x^2$ have Maclaurin series. Thus, at $x_0 = 0$, we have a regular singularity.

Second, assume the solution

$$u(x) = \sum_{n=0}^{\infty} b_n x^{n+r}, \tag{9.44}$$

and find the appropriate derivatives of Equation (9.44), then replace them in Equation (9.43). We have

$$u'(x) = \sum_{n=0}^{\infty} (n + r)b_n x^{n+r-1} \tag{9.45}$$

and

$$u''(x) = \sum_{n=0}^{\infty} (n + r)(n + r - 1)b_n x^{n+r-2}. \tag{9.46}$$

Note: Equations (9.45 and 9.46) seem to not follow the rule for differentiation of a power series. However, when $n = 0$ in Equations (9.44) we have $b_0 x^r$ and the derivatives of this first term are $rb_0 x^{r-1}$ and $r(r - 1)b_0 x^{r-2}$.

Replacing in $u(x)$, $u'(x)$, and $u''(x)$ in Equation (9.43) with Equations (9.44, 9.45, and 9.46) yields

$$0 = 2x^2 \sum_{n=0}^{\infty} (n+r)(n+r-1)b_n x^{n+r-2} + (3x + 2x^2) \sum_{n=0}^{\infty} (n+r)b_n x^{n+r-1}$$
$$- (1-x) \sum_{n=0}^{\infty} b_n x^{n+r},$$

which becomes

$$0 = \sum_{n=0}^{\infty} 2(n+r)(n+r-1)b_n x^{n+r} + \sum_{n=0}^{\infty} 3(n+r)b_n x^{n+r}$$
$$+ \sum_{n=0}^{\infty} 2(n+r)b_n x^{n+r+1} - \sum_{n=0}^{\infty} b_n x^{n+r} + \sum_{n=0}^{\infty} b_n x^{n+r+1}. \tag{9.47}$$

Canceling out the common x^r term in Equation (9.47) and combining like sums yields

$$0 = \sum_{n=0}^{\infty} \left[2(n+r)(n+r-1) + 3(n+r) - 1\right] b_n x^n$$
$$+ \sum_{n=0}^{\infty} \left[2(n+r) + 1\right] b_n x^{n+1}. \tag{9.48}$$

Performing some algebra and writing Equation (9.48) differently yields

$$
\begin{aligned}
0 \;=\;& \left[2r(r-1) + 3r - 1\right] b_0 \\
+\;& \sum_{n=1}^{\infty} (n+r+1)(2n+2r-1)b_n x^n \\
+\;& \sum_{n=1}^{\infty} (2n+2r-1)b_{n-1} x^n.
\end{aligned}
\tag{9.49}
$$

The b_0 term of Equation (9.49) yields the equation

$$\left[2r(r-1) + 3r - 1\right] = 0,$$

called the indicial equation. It tells us that $r = \dfrac{1}{2}, -1$. The second and third lines of Equation (9.49) yield the relationship of the b_n's. We have

$$(n+r+1)(2n+2r-1)b_n = -(2n+2r-1)b_{n-1},$$

which yields

$$b_n = -\frac{1}{(n+r+1)} b_{n-1}. \tag{9.50}$$

Thus,

$$u_1(x) = x^{\frac{1}{2}} \sum_{n=0}^{\infty} b_n \left(\frac{1}{2}\right) x^n$$

where $b_n(\frac{1}{2}) = -\dfrac{2}{2n+3} b_{n-1}(\frac{1}{2})$, and

$$u_2(x) = x^{-1} \sum_{n=0}^{\infty} b_n(-1)x^n,$$

where $b_n(-1) = -\dfrac{1}{n} b_{n-1}(-1)$. Also, $u_1(x)$ and $u_2(x)$ form a fundamental set of solutions. Therefore,

$$u(x) = C_1 u_1(x) + C_2 u_2(x) = C_1 x^{\frac{1}{2}} \sum_{n=0}^{\infty} b_n \left(\frac{1}{2}\right) x^n + C_2 x^{-1} \sum_{n=0}^{\infty} b_n(-1)x^n.$$

One way of finding a general solution is to define $b_0(r) = 1$, then to use the recurrence relation, Equation (9.50). This process yields

$$u(x) = C_1 x^{\frac{1}{2}} \left(1 - \frac{2}{5}x + \frac{4}{35}x^2 - \frac{8}{315}x^3 + \cdots\right)$$

$$+ C_2 x^{-1} \left(1 - x + \frac{1}{2}x^2 - \frac{1}{6}x^3 + \cdots\right). \qquad \blacksquare$$

We now state the general theorem for the method of Frobenius.

Theorem 57. *Given the ODE*

$$S(x)u''(x) + K(x)u'(x) + H(x)u(x) = 0,$$

where $x = 0$ is a regular singular point,

$$\lim_{x \to 0} x \frac{K(x)}{S(x)} \text{ is finite,}$$

and

$$\lim_{x \to 0} x^2 \frac{H(x)}{S(x)} \text{ is finite,}$$

let r_1 and r_2, with $r_1 \geq r_2$, be roots of the indicial equation

$$r(r-1) + k_0 r + h_0 = 0.$$

Then, we have the following three cases:
(1) If $r_1 - r_2$ is not an integer or 0, we have

$$u_1(x) = x^{r_1} \left[1 + \sum_{n=1}^{\infty} b_n(r_1)x^n\right]$$

and

$$u_2(x) = x^{r_2} \left[1 + \sum_{n=1}^{\infty} b_n(r_2)x^n \right].$$

(2) If $r_1 = r_2 = r$, we have

$$u_1(x) = x^r \left[1 + \sum_{n=1}^{\infty} b_n(r)x^n \right]$$

and

$$u_2(x) = u_1(x)\ln x + x^r \sum_{n=1}^{\infty} B_n(r)x^n.$$

(3) If $r_1 - r_2 = N$, a positive integer, we have

$$u_1(x) = x^{r_1} \left[1 + \sum_{n=1}^{\infty} b_n(r_1)x^n \right]$$

and

$$u_2(x) = cu_1(x)\ln x + x^{r_2} \left[1 + \sum_{n=1}^{\infty} B_n(r_2)x^n \right],$$

where c may equal 0.

Proof of this theorem may be found in Appendix 4 of *Advanced Engineering Mathematics, Seventh Edition* by Erwin Kreyszig. We may now consider Bessel's equation.

Bessel's equation arises in the study of heat diffusion in a cylinder or a circular plate and in vibrations of a circular drum head. Bessel's equation is

$$x^2 u'' + xu' + \left(x^2 - n^2 \right), u = 0 \tag{9.51}$$

where, for our purposes, n is a nonnegative integer. Also, $x = 0$ is a regular singular point. We leave you to show this in an exercise. Therefore, letting

$$u(x) = \sum_{n=0}^{\infty} b_n x^{n+r}$$

and substituting $u(x)$ and first and second derivatives of $u(x)$ into Equation (9.51) yields

$$0 = x^2 \sum_{m=0}^{\infty} (m+r)(m+r-1)b_m x^{m+r-2}$$

$$+ x \sum_{m=0}^{\infty} (m+r)b_m x^{m+r-1} + \left(x^2 - n^2 \right) \sum_{m=0}^{\infty} b_m x^{m+r},$$

which becomes, after some algebraic manipulation and canceling the common x^r term,

$$0 = \sum_{m=0}^{\infty} \left[(m+r+n)(m+r-n) \right] b_m x^m$$

$$+ \sum_{m=0}^{\infty} b_m x^{m+2}. \tag{9.52}$$

Equation (9.52) may be rewritten as

$$0 = (r+n)(r-n)b_0 + (1+r+n)(1+r-n)b_1 x$$
$$+ (2+r+n)(2+r-n)b_2 x^2 + \cdots$$
$$+ b_0 x^2 + \cdots$$

Thus, the indicial equation, which comes from the coefficient of the x^0 term, is

$$(r+n)(r-n) = 0, \tag{9.53}$$

which means $r_1 = n \geq 0$ and $r_2 = -n$. Hence $r_1 - r_2 = 2n \geq 0$. The next equation is

$$(1+r+n)(1+r-n)b_1 = 0,$$

which must equal 0 for all choices of nonnegative integers n and values of r. Therefore, b_1 must equal 0. The recurrence relation is given by the equation

$$(2+r+n)(2+r-n)b_2 + b_0 = 0,$$

which can be rewritten as

$$b_m = \frac{-1}{(m+r+n)(m+r-n)} b_{m-2}. \tag{9.54}$$

Since $b_1 = 0$ and n is a nonnegative integer, the recurrence relation indicates $b_3 = b_5 = \cdots = 0$. Thus, m is always even, so we may replace it with $2m$ in Equation (9.54), which becomes

$$b_{2m} = \frac{-1}{(2m+r+n)(2m+r-n)} b_{2m-2}. \tag{9.55}$$

If we make the assumption that $r = n$, Equation (9.55) becomes

$$b_{2m} = \frac{-1}{(2m+2n)2m} b_{2m-2} = \frac{-1}{2^2 m(m+n)} b_{2m-2}, \quad m = 1, 2, 3, \ldots$$

Thus,

$$b_2 = -\frac{b_0}{2^2(1+n)},$$

$$b_4 = \frac{-b_2}{2^2(2)(2+n)} = \frac{b_0}{2^4(2)(2+n)(1+n)},$$

$$b_6 = \frac{-b_4}{2^2(3)(3+n)} = \frac{-b_0}{2^6(3)(2)(3+n)(2+n)(1+n)},$$

$$\vdots$$

The general formula is

$$b_{2m} = \frac{(-1)^m b_0}{2^{2m} m!(1+n)(2+n)(3+n)\cdots(m+n)}.$$

Therefore, if we let $n = 0$, we have

$$u_1(x) = b_0 \left[1 + \sum_{m=1}^{\infty} \frac{(-1)^m x^{2m}}{2^{2m}(m!)^2}\right],$$

which is the Bessel function of the first kind of order zero, and is denoted as $J_0(x)$. Since $n = 0$, we have case (2) of Theorem 57, a double root. Hence, the second solution has the form

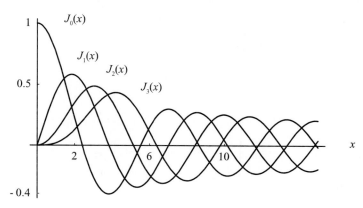

Figure 9.4: Bessel's function of the first kind for $n = 0, 1, 2, 3$.

Figure (9.4) shows the graph of Bessel's function of the first kind for $n = 0, 1, 2, 3$.

$$u_2(x) = u_1(x)\ln x + \sum_{m=1}^{\infty} B_m x^m = J_0(x)\ln x + \sum_{m=1}^{\infty} B_m x^m.$$

B_m can be determined by taking first and second derivatives of $u_2(x)$ and replacing $u_2(x)$, $u_2'(x)$, and $u_2''(x)$ in Equation (9.51). The solution obtained, after some algebraic manipulation, is

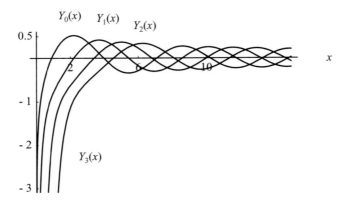

Figure 9.5: Bessel's function of the second kind for $n = 0, 1, 2, 3$.

$$Y_0(x) = \frac{2}{\pi} \left[\left(\gamma + \ln \frac{x}{2} \right) J_0(x) \ln x + \sum_{m=1}^{\infty} \frac{(-1)^{m+1} G_m}{2^{2m}(m!)^2} x^{2m} \right],$$

where

$$G_m = \frac{1}{m} + \frac{1}{m-1} + \cdots + \frac{1}{2} + 1$$

and $\gamma = 0.577215664\ldots$, which is called Euler's constant. $Y_0(x)$ is called Bessel's function of the second kind of order zero. Therefore, the general solution to Equation (9.51) when $n = 0$ is

$$u(x) = C_1 Y_0(x) + C_2 J_0(x).$$

Figure (9.5) shows the graph of Bessel's function of the second kind for $n = 0, 1, 2, 3$.

EXERCISES 9.7

9.7.1. Use the method of Frobenius to find the indicial equation for each of the following ODEs:

(1) $xu'' - 2xu' + u = 0$.

(2) $x^2 u'' + xu' = 0$.

(3) $3xu'' - (x + 2)u' - 2u = 0$.

(4) $x^2u'' + xu = 0$.

(5) $x(u'' - 2u - 1u) = 0$.

(6) $x^2u'' - x^2u' + 3xu = 0$.

9.7.2. Another set of polynomials is Laguerre's polynomials. They can be found by using the method of Frobenius on Laguerre's equation

$$xu'' + (1 - x)u' + nu = 0, \ n = 0, 1, 2, 3, \ldots$$

(1) Show that $x = 0$ is a regular singular point of Laguerre's Equation.

(2) Using the method of Frobenius, develop one fundamental solution of Laguerre's equation.

9.7.3. Use the method of Frobenius to determine one solution to the following Bessel's equation,

$$x^2u'' + xu' + \left(x^2 - \frac{1}{4}\right)u = 0.$$

This is known as the Bessel equation of order one-half.

9.7.4. Show that

$$\int_0^1 x\, [J_0(\lambda x)]^2 \ dx = \frac{1}{\lambda^2} \int_0^\lambda t\, [J_0(t)]^2 \ dt = \frac{1}{2}\left\{[J_0(\lambda)]^2 + [J_0'(\lambda)]^2\right\}.$$

Hint: Consider the ODE $y'' + \dfrac{y'}{t} + y = 0$, where $y = J_0(t)$. Multiply the ODE by $2t^2y'$.

9.7.5. The Hypergeometric equation, sometimes called Gauss hypergeometric equation,

$$x(1 - x)u'' + [c - (a + b + 1)x]u' - abu = 0,$$

is solved by the method of Frobenius.

(1) Show that the hypergeometric equation has a regular singular point at $x = 0$.

(2) Show that the hypergeometric equation has a regular singular point at $x = 1$.

(3) Use the method of Frobenius to solve the hypergeometric equation.

Chapter 10

Classical PDE Problems

10.1 INTRODUCTION

In this chapter, we consider solution techniques for some of the classical problems of partial differential equations. They include Laplace's equation in cylindrical and spherical coordinates, the transverse vibrations of a thin beam, heat conduction in and vibrations of a thin circular membrane, Schrödinger's equation, the Telegrapher's equation, and some interesting problems in diffusion. We start with Laplace's equation.

10.2 LAPLACE'S EQUATION

In Chapter 5, Laplace's equation in a rectangle was discussed. In this chapter, we expand Laplace's equation to other coordinate systems. Again, Laplace's is an extremely important equation in mathematical physics, and it naturally arises in electrostatics, steady-state temperature field, magnetostatics, and potential flow of an incompressible liquid. Laplace's equation has the following forms in Cartesian coordinates,

$$\nabla^2 u = \frac{\partial^2 u}{\partial x^2} + \frac{\partial^2 u}{\partial y^2} + \frac{\partial^2 u}{\partial z^2} = 0; \tag{10.1}$$

in cylindrical coordinates, Figure (10.1) provides a frame of reference,

$$\nabla^2 u = \frac{1}{r} \frac{\partial}{\partial r} \left(r \frac{\partial u}{\partial r} \right) + \frac{1}{r^2} \frac{\partial^2 u}{\partial \theta^2} + \frac{\partial^2 u}{\partial z^2} = 0; \tag{10.2}$$

and in spherical coordinates, Figure(10.2) provides a frame of reference,

$$\nabla^2 u = \frac{1}{r^2} \frac{\partial}{\partial r} \left(r^2 \frac{\partial u}{\partial r} \right) + \frac{1}{r^2 \sin \theta} \frac{\partial}{\partial \theta} \left(\sin \theta \frac{\partial u}{\partial \theta} \right) + \frac{1}{r^2 \sin^2 \theta} \frac{\partial^2 u}{\partial \varphi^2}. \tag{10.3}$$

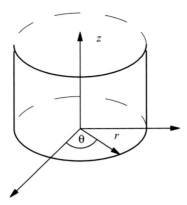

Figure 10.1: Cylindrical coordinate system.

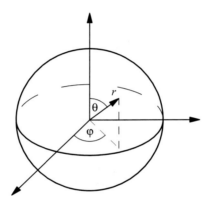

Figure 10.2: Spherical coordinate system.

10.2.1 Laplace's equation in the Polar Coordinate System

We consider steady-state heat flow in a circular plate. Thus, the use of a suitable coordinate system is required. We use the polar coordinate system, which is the two-dimensional version of the cylindrical coordinate system. Therefore, Laplace's equation becomes

$$\nabla^2 u = \frac{1}{r} \frac{\partial}{\partial r} \left(r \frac{\partial u}{\partial r} \right) + \frac{1}{r^2} \frac{\partial^2 u}{\partial \theta^2} = 0. \tag{10.4}$$

We assume the boundary of the circular plate is a function of θ. Thus,

$$u(\alpha, \theta) = f(\theta), \tag{10.5}$$

where α is the radius of the circular plate. Figure (10.3) shows the cylindrical plate.

It appears we only have one boundary condition. However, Equation (10.4) is in polar coordinates, which allows us to make some assumptions.

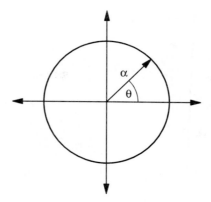

Figure 10.3: Laplace's equation in a circular plate.

When working with a rectangular plate, in Cartesian coordinates, it is natural to assume that the lower left corner is at the origin. Thus, $0 \leq x \leq L$ and $0 \leq y \leq H$. But for a circular plate, in polar coordinates, it is natural to assume the origin is at the center of the plate, and for any value of r, $-\pi \leq \theta \leq \pi$. This implies that heat flow at $-\pi$ and π must be the same. Therefore, $\dfrac{\partial u(r, -\pi)}{\partial \theta} = \dfrac{\partial u(r, \pi)}{\partial \theta}$, and for any value of r, we have $-\pi$ and π as the same point on the circular plate. Thus, we must have $u(r, -\pi) = u(r, \pi)$. These boundary conditions for the variable $u(r, \theta)$ are identical to the conditions we encountered for the one-dimensional circular wire, and they are the natural boundary conditions to assume for the variable θ. Therefore, we may, for practical purposes, restate the problem given by Equations (10.4 and 10.5) as

$$\nabla^2 u = \frac{1}{r} \frac{\partial}{\partial r} \left(r \frac{\partial u}{\partial r} \right) + \frac{1}{r^2} \frac{\partial^2 u}{\partial \theta^2} = 0, \tag{10.6}$$

subject to

$$u(r, -\pi) = u(r, \pi)$$

$$\frac{\partial u(r, -\pi)}{\partial \theta} = \frac{\partial u(r, \pi)}{\partial \theta} \tag{10.7}$$

and IC

$$u(\alpha, \theta) = f(\theta). \tag{10.8}$$

Also, we must note that when $r = 0$, Equation (10.6) would become unbounded. Therefore, it is usual to assume that as $r \to 0$, we have $|u(r, \theta)| < \infty$, which means that Equation (10.6) is bounded as the radius approaches 0.

The single boundary condition, which is nonhomogeneous, immediately tells us that the "time-like" condition is in terms of r. Therefore, we proceed as before, using separation of variables on Equation (10.6). Let $u(r, \theta) = R(r)\Theta(\theta)$. Then, Equation (10.6) becomes

$$\frac{1}{r}\frac{d}{dr}\left(r\frac{dR(r)}{dr}\right)\Theta(\theta) + \frac{1}{r^2}R(r)\frac{d^2\Theta(\theta)}{d\theta^2} = 0, \tag{10.9}$$

which may be written as

$$\frac{-r}{R(r)}\frac{d}{dr}\left(r\frac{dR(r)}{dr}\right) = \frac{1}{\Theta(\theta)}\frac{d^2\Theta(\theta)}{d\theta^2} = -\lambda. \tag{10.10}$$

We now have two separate ODEs–one "time-like" equation

$$r\frac{d}{dr}\left(r\frac{dR(r)}{dr}\right) = \lambda R(r),$$

which may be written as

$$r^2\frac{d^2R(r)}{dr^2} + r\frac{dR(r)}{dr} - \lambda R(r) = 0, \tag{10.11}$$

subject to the condition

$$|R(r)| < \infty \text{ as } r \to 0, \tag{10.12}$$

and the homogeneous equation

$$\frac{d^2\Theta(\theta)}{d\theta^2} = -\lambda\Theta(\theta). \tag{10.13}$$

After separating the BCs, Equation (10.7), the homogeneous equation, Equation (10.13), is subject to the BCs

$$\Theta(-\pi) = \Theta(\pi)$$
$$\Theta'(-\pi) = \Theta'(\pi). \tag{10.14}$$

Equation (10.13), subject to BCs, Equation (10.14), is well known to us and has the solution

$$\lambda_0 = 0, \ \Theta_0(\theta) = a_0$$

$$\left.\begin{array}{l} \lambda_n = n^2 \\ \\ \Theta_n(\theta) = a_n\cos n\theta + b_n\sin n\theta \end{array}\right\}, \ n = 1, 2, 3, \ldots \tag{10.15}$$

The "time-like" equation, Equation (10.11), we studied in Chapter 9 Section 9.3. The form we used there is

$$x^2 y'' + \beta xy' + \chi y = 0,$$

known as the Euler-Cauchy equation.

We have two cases to consider. The first case is $\lambda_n = n^2$ for $n = 1, 2, 3, \ldots$. Therefore, Equation (10.11) becomes

$$r^2 R''(r) + r R'(r) - n^2 R(r) = 0.$$

Letting $R(r) = r^s$, and finding the appropriate derivatives yields the characteristic equation

$$\left[s(s-1) + s - n^2 \right] r^s = 0$$

or $s = \pm n$. Thus, the solution is

$$R(r) = c_1 r^n + c_2 r^{-n}, n = 1, 2, 3, \ldots \tag{10.16}$$

Applying the condition given by Equation (10.12), $|R(r)| < \infty$ as $r \to 0$, implies $c_2 = 0$. Therefore, the solution for Equation (10.16) becomes

$$R(r) = c_1 r^n. \tag{10.17}$$

The second case is $\lambda_0 = 0$. If we follow the same format as for $\lambda_n = n^2$, we would end up with the characteristic equation

$$\left[s(s-1) + s \right] r^s = 0,$$

which yields $s = 0$ or a solution of $R(r) = c_3$. However, we know that a second-order ODE has two solutions. Let us rewrite Equation (10.11) as

$$r \frac{d}{dr} \left(r \frac{dR(r)}{dr} \right) = \lambda R(r) = 0.$$

Thus, we have

$$\frac{d}{dr} \left(r \frac{dR(r)}{dr} \right) = 0,$$

which implies that

$$r \frac{dR(r)}{dr} = c_4, \text{ where } c_4 \text{ is a constant.}$$

This in turn implies that

$$\frac{dR(r)}{dr} = \frac{c_4}{r},$$

which implies that a second solution is

$$R(r) = c_4 \ln |r|.$$

Thus, our complete solution for Equation (10.11) for the case $\lambda_0 = 0$ is

$$R(r) = c_3 + c_4 \ln|r|. \tag{10.18}$$

Applying the condition given by Equation (10.12), $|R(r)| < \infty$ as $r \to 0$ implies that $c_4 = 0$. Therefore, the solution for Equation (10.11) for the case $\lambda_0 = 0$ becomes

$$R(r) = c_3. \tag{10.19}$$

For the general solution to Equation (10.6), subject to the BCs, Equation (10.7), we have $u(r, \theta) = \Theta(\theta)R(r) = a_0 c_3 = A_0$ given by Equations (10.15 and 10.19) for $\lambda = 0$, and $u(r, \theta) = \Theta(\theta)R(r) = (a_n \cos n\theta + b_n \sin n\theta)(c_1 r^n) = A_n r^n \cos n\theta + B_n r^n \sin n\theta$ given by Equations (10.15 and 10.17). Summing all of our solutions, we arrive at

$$u(r, \theta) = A_0 + \sum_{n=1}^{\infty} A_n r^n \cos n\theta + B_n r^n \sin n\theta. \tag{10.20}$$

Applying our one "time-like" condition, Equation (10.8), yields

$$u(\alpha, \theta) = f(\theta) = A_0 + \sum_{n=1}^{\infty} A_n \alpha^n \cos n\theta + B_n \alpha^n \sin n\theta.$$

Using orthogonality we have

$$A_0 = \frac{1}{2\pi} \int_{-\pi}^{\pi} f(\theta) \, d\theta, \tag{10.21}$$

$$A_n = \frac{1}{\pi \alpha^n} \int_{-\pi}^{\pi} f(\theta) \cos n\theta \, d\theta, \tag{10.22}$$

and

$$B_n = \frac{1}{\pi \alpha^n} \int_{-\pi}^{\pi} f(\theta) \sin n\theta \, d\theta. \tag{10.23}$$

We now solve Laplace's equation in spherical coordinates.

10.2.2   Laplace's equation in the Spherical Coordinate System

For a sphere, as shown in Figure (10.4), Laplace's equation in spherical coordinates is given by

$$\nabla^2 u = \frac{1}{r^2} \frac{\partial}{\partial r} \left(r^2 \frac{\partial u}{\partial r} \right) + \frac{1}{r^2 \sin \theta} \frac{\partial}{\partial \theta} \left(\sin \theta \frac{\partial u}{\partial \theta} \right) + \frac{1}{r^2 \sin^2 \theta} \frac{\partial^2 u}{\partial \varphi^2} = 0. \tag{10.24}$$

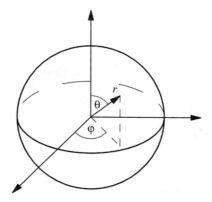

Figure 10.4: The sphere of radius, r.

However, we will consider the problem where $u(r, \theta, \varphi) = u(r, \theta)$. This means we are considering Laplace's equation in spherical coordinates, which is independent of φ. This is not an unrealistic assumption, since many problems in electrostatics are solved in this manner. It is done by setting up the problem in such a way that it is independent of the variable φ. Therefore, Equation (10.24) becomes

$$\nabla^2 u = \frac{1}{r^2} \frac{\partial}{\partial r} \left(r^2 \frac{\partial u}{\partial r} \right) + \frac{1}{r^2 \sin \theta} \frac{\partial}{\partial \theta} \left(\sin \theta \frac{\partial u}{\partial \theta} \right) = 0. \tag{10.25}$$

Like the polar coordinate system, as $r \to 0$, $|u(r, \theta)| < \infty$. Again, this means that the function u is bounded as r approaches 0. Also, we assume the boundary of the sphere is now a function of θ. Thus,

$$u(a, \theta) = f(\theta), \tag{10.26}$$

where a is the radius of the sphere. This is the "time-like" condition, which will be used to find the coefficients of the general Fourier series. Since Equation (10.25) is homogeneous, we let $u(r, \theta) = R(r)\Theta(\theta)$ and proceed with separation of variables. Thus, Equation (10.25) becomes

$$\frac{1}{r^2} \left(r^2 R' \right)' \Theta + \frac{R}{r^2 \sin \theta} \left(\sin \theta \Theta' \right)' = 0,$$

which when separated becomes

$$\frac{1}{R} \left(r^2 R' \right)' = -\frac{1}{\Theta \sin \theta} \left(\sin \theta \Theta' \right)' = -\lambda.$$

Therefore, we have two ODEs,

$$\left(r^2 R' \right)' = -\lambda R \tag{10.27}$$

and

$$\left(\sin \theta \Theta' \right)' = -\lambda \Theta \sin \theta. \tag{10.28}$$

We will work with Equation (10.27) first. After completing the differentiation, it becomes

$$r^2 R'' + 2rR' + \lambda R = 0. \tag{10.29}$$

Equation (10.29) we have seen before; it is Euler's equation. (See Chapter 9.) The usual solution method is to assume the solution has the form $R = r^\rho$, and then, take the appropriate derivatives and determine the characteristic equation. Using our assumed solution, we have

$$r^2 \rho (\rho - 1) r^{\rho-2} + 2r\rho r^{\rho-1} + \lambda r^\rho = \rho (\rho - 1) r^\rho + 2\rho r^\rho + \lambda r^\rho = 0.$$

Dividing by r^ρ, since we assume the radius of the sphere is not 0, we get the characteristic equation

$$\rho^2 + \rho + \lambda = 0,$$

which has solutions

$$\rho = \frac{-1 \pm \sqrt{1 - 4\lambda}}{2}$$

or, written in an easier fashion,

$$\rho_1 = \frac{-1}{2} + \sqrt{\frac{1}{4} - \lambda} \tag{10.30}$$

and

$$\rho_2 = \frac{-1}{2} - \sqrt{\frac{1}{4} - \lambda}. \tag{10.31}$$

If we let Equation (10.30) equal n, then Equation (10.32) becomes $-n - 1$. Also, if we multiply Equation (10.30) by Equation (10.32) we find that the eigenvalues are

$$\lambda_n = n(n+1). \tag{10.32}$$

Thus, the solution to Equation (10.29) is

$$R(r) = C_1 r^n + C_2 r^{-n-1} = C_1 r^n + \frac{C_2}{r^{n+1}}. \tag{10.33}$$

Since $|u(r, \theta)| < \infty$ as $r \to 0$, we must have $C_2 = 0$ in Equation (10.33). Therefore, we have

$$R_n(r) = C_n r^n \tag{10.34}$$

as our final solution for $R(r)$. Next we solve Equation (10.28).

Given

$$(\sin \theta \Theta')' = -\lambda \Theta \sin \theta.$$

We first replace λ by $n(n+1)$, which gives us

$$(\sin\theta\Theta')' + n(n+1)\Theta\sin\theta = 0. \tag{10.35}$$

Next, we let $\omega = \cos\theta$. Then, using the chain rule, we get

$$\frac{d\Theta}{d\theta} = \frac{d\Theta}{d\omega}\frac{d\omega}{d\theta} = -\sin\theta\frac{d\Theta}{d\omega}.$$

Therefore,

$$\sin\theta\Theta' = \sin\theta\frac{d\Theta}{d\theta} = -\sin^2\theta\frac{d\Theta}{d\omega};$$

applying a trigonometric identity followed by substitution yields

$$-\sin^2\theta\frac{d\Theta}{d\omega} = (\cos^2\theta - 1)\frac{d\Theta}{d\omega} = (\omega^2 - 1)\frac{d\Theta}{d\omega}.$$

Thus,

$$(\sin\theta\Theta')' = \frac{d}{d\theta}\left[(\omega^2-1)\frac{d\Theta}{d\omega}\right] = \frac{d}{d\omega}\left[(\omega^2-1)\frac{d\Theta}{d\omega}\right]\frac{d\omega}{d\theta}$$

$$= -\frac{d}{d\omega}\left[(\omega^2-1)\frac{d\Theta}{d\omega}\right]\sin\theta.$$

Hence, Equation (10.35) becomes

$$-\frac{d}{d\omega}\left[(\omega^2-1)\frac{d\Theta}{d\omega}\right]\sin\theta + n(n+1)\Theta\sin\theta = 0,$$

which becomes, after canceling $\sin\theta$,

$$\frac{d}{d\omega}\left[(1-\omega^2)\frac{d\Theta}{d\omega}\right] + n(n+1)\Theta = 0. \tag{10.36}$$

Replacing Θ with v and ω with x in Equation (10.36) yields

$$\left[(1-x^2)v'\right]' + n(n+1)v = 0$$

or

$$(1-x^2)v'' - 2xv' + n(n+1)v = 0. \tag{10.37}$$

Equation (10.37) is Legendre's equation. (See Chapter 9.) We know it has solutions

$$v(x) = A_n P_n(x) + B_n Q_n(x),$$

which may be written as

$$v(\omega) = A_n P_n(\omega) + B_n Q_n(\omega),$$

or, in the preferred form,

$$\Theta(\cos\theta) = A_n P_n(\cos\theta) + B_n Q_n(\cos\theta). \tag{10.38}$$

Since Legendre functions of the second kind are not finite at ± 1, $\cos 0 = 1$, and $\cos\pi = -1$, we have $Q_n(\cos\theta)$ undefined at 0 and π. Therefore, we must have $B_n = 0$. Thus, Equation (10.38) becomes

$$\Theta(\cos\theta) = A_n P_n(\cos\theta). \tag{10.39}$$

Combining Equations (10.34 and 10.39) we have our solution for $u(r,\theta)$, which is

$$u(r,\theta) = R(r)\Theta(\theta) = \sum_{n=0}^{\infty} c_n r^n P_n(\cos\theta). \tag{10.40}$$

Applying the IC, Equation (10.26), we find

$$u(a,\theta) = f(\theta) = \sum_{n=0}^{\infty} c_n a^n P_n(\cos\theta).$$

Thus, the coefficients are found by using orthogonality, and the equation for them is

$$c_n = \frac{2n+1}{2a^n} \int_0^\pi f(\theta) P_n(\cos\theta) \sin\theta \; d\theta.$$

Note:

$$\int_0^\pi P_n^2(\cos\theta) \sin\theta \; d\theta = \frac{2}{2n+1}.$$

EXERCISES 10.2

10.2.1. Given Laplace's equation in polar coordinates

$$\nabla^2 u = \frac{1}{r}\frac{\partial}{\partial r}\left(r\frac{\partial u}{\partial r}\right) + \frac{1}{r^2}\frac{\partial^2 u}{\partial\theta^2} = 0,$$

find the solution for each of the following BCs:

(1) $u(2,\theta) = 5$.

(2) $u(1,\theta) = 3\sin\theta$.

(3) $u(1,\theta) = 2\cos(2\theta)$.

(4) $u(0.5,\theta) = 2\theta - 1$.

(5) $u(\pi, \theta) = 2\cos\theta - \sin\theta$.

(6) $u(3, \theta) = \theta^2 - 2\theta + 1$.

10.2.2. Given Laplace's equation in spherical coordinates, which is independent of the variable φ,

$$\nabla^2 u = \frac{1}{r^2}\frac{\partial}{\partial r}\left(r^2\frac{\partial u}{\partial r}\right) + \frac{1}{r^2 \sin\theta}\frac{\partial}{\partial\theta}\left(\sin\theta\frac{\partial u}{\partial\theta}\right) = 0,$$

find the solution for each of the following BCs:

(1) $u(2, \theta) = 0.5$.

(2) $u(1, \theta) = \sin(3\theta)$.

(3) $u(1, \theta) = 2\cos(\theta)$.

(4) $u(0.5, \theta) = 5\theta + 1$.

(5) $u(\pi, \theta) = 3\sin\theta - 4\cos\theta$.

(6) $u(3, \theta) = \theta^2 - 3\theta + 2$.

10.2.3. Show that the exterior boundary value problem

$$\frac{\partial^2 u(x, y)}{\partial x^2} + \frac{\partial^2 u(x, y)}{\partial y^2} = f(x, y) \qquad \text{in } \Omega,$$

$$u(x, y) = g(x, y) \qquad\qquad \text{on } \partial\Omega,$$

and

$$|u(x, y)| \leq \alpha, \ \alpha \text{ a constant in } \Omega$$

has at most one solution.

10.2.4. Let Ω be a bounded region. Show that the Neumann problem

$$\nabla^2 u + \alpha u = f \text{ in } \Omega \text{ and } \frac{\partial u}{\partial n} = g \text{ on } \partial\Omega$$

has at most one solution if $\alpha < 0$ in Ω.

10.2.5. Let Ω be a bounded region. Show for the Neumann problem

$$\nabla^2 u = f \text{ in } \Omega \text{ and } \frac{\partial u}{\partial \mathbf{n}} = g \text{ on } \partial\Omega.$$

Show that any two solutions differ by a constant.

10.2.6. If we write Laplace's equation in cylindrical coordinates, then assume that it is axially symmetric (no dependence on θ), we arrive at

$$\frac{\partial^2 u(r, z)}{\partial r^2} + \frac{1}{r}\frac{\partial^2 u(r, z)}{\partial z^2} = 0.$$

Use separation of variables where $u(r, z) = R(r)Z(z)$ and show that R and Z satisfy

$$rR''(r) + R'(r) + \lambda r R(r) = 0 \text{ and } Z''(z) - \lambda Z(z) = 0.$$

(1) Identify both of these ODEs.

(2) Pick reasonable BCs and give a series solution.

10.2.7. Determine the three ODEs obtained by separation of variables for Laplace's equation in spherical coordinates.

10.2.8. Solve Laplace's equation in a slot (a vertical strip) given by

$$\frac{\partial^2 u(x,y)}{\partial x^2} + \frac{\partial^2 u(x,y)}{\partial y^2} = 0,$$

subject to

$$u(0,y) \;=\; f_1(y), \quad 0 < y < H;$$

$$u(x,0) \;=\; f_2(x), \qquad 0 < x;$$

and

$$u(x,H) \;=\; f_3(x), \quad 0 < x.$$

$u(x,y)$ must remain bounded as $x \to \infty$. *Hint:* Let $u(x,y) = u_1(x,y) + u_2(x,y)$.

10.2.9. Solve Laplace's equation in a slot given by

$$\frac{\partial^2 u(x,y)}{\partial x^2} + \frac{\partial^2 u(x,y)}{\partial y^2} = 0,$$

subject to

$$u(0,y) \;=\; f_1(y), \quad 0 < y;$$

$$u(L,y) \;=\; f_2(x), \quad 0 < y;$$

and

$$u(x,0) \;=\; f_3(x), \quad 0 < x < L,$$

where $u(x,y)$ must remain bounded as $y \to \infty$. *Hint:* Let $u(x,y) = u_1(x,y) + u_2(x,y)$.

10.2.10. State the complete Laplace's equation in the first quadrant. Write a short essay on how to solve this problem.

10.2.11. Poisson's equation,

$$\nabla^2 u(x,y) = \frac{\partial^2 u(x,y)}{\partial x^2} + \frac{\partial^2 u(x,y)}{\partial y^2} = -f(x,y),$$

is sometimes easy to solve if $f(x,y)$ is a constant or depends on just one variable. Solve Poisson's equation if

(1) $\nabla^2 u(x,y) = -3$.

(2) $\nabla^2 u(x,y) = x$.

(3) $\nabla^2 u(x,y) = y^2$.

10.2.12. Solve Laplace's equation inside a semicircle of radius 1 where $0 < \theta < \pi$, subject to the following BCs

(1) The diameter is a homogenous Dirichlet condition and $u(1,\theta) = \theta^2 + 1$.

(2) The diameter is perfectly insulated and $u(1,\theta) = \cos\theta$.

10.2.13. Solve Laplace's equation inside the wedge $0 < \theta < \dfrac{\pi}{4}$ of radius α, subject to the following BCs

(1) $u(r,0) = 0$, $u\left(r, \dfrac{\pi}{4}\right) = 0$, $u(\alpha,\theta) = \theta$.

(2) $u(r,0) = 0$, $\dfrac{\partial u\left(r, \dfrac{\pi}{4}\right)}{\partial\theta} = 0$, $u(\alpha,\theta) = \sin\theta$.

(3) $\dfrac{\partial u\,(r,0)}{\partial\theta} = 0$, $\dfrac{\partial u\left(r, \dfrac{\pi}{4}\right)}{\partial\theta} = 0$, $u(\alpha,\theta) = \cos\theta$.

10.2.14. Suppose $u(x,y,z)$ satisfies Laplace's equation. Show that the value of $u(x,y,z)$ at any point (x,y,z) is approximately equal to the average of its values at the six surrounding points $(x \pm \delta, y, z)$, $(x, y \pm \delta, z)$, and $(x, y, z \pm \delta)$. *Hint:* Calculate the Taylor series expansion of $u(x+\delta, y, z)$ to the term δ^3, similarly for the other five points.

10.2.15. In the text, we solved the first boundary-value problem for Laplace's equation inside a sphere of radius a, which is independent of φ. Solve the first boundary-value problem for Laplace's equation outside the sphere. *Note:* The first boundary-value problem is a Dirichlet BC.

10.2.16. Complete the following:

(1) Solve Laplace's equation for the general second boundary-value problem inside a sphere of radius a, which is independent of the variable φ. *Note:* The second boundary-value problem is a Neumann BC.

(2) Solve Laplace's equation for the second boundary-value problem inside a sphere of radius a, which is independent of the variable φ, where the BC is

$$\frac{\partial u}{\partial n} = \alpha\cos\theta.$$

10.2.17. Solve Laplace's equation for the second boundary-value problem outside a sphere of radius a, which is independent of the variable φ.

10.2.18. Solve Laplace's equation for the general Sturm–Liouville BCs inside a sphere of radius a, which is independent of the variable φ.

10.2.19. Find the potential ϕ in the interior and exterior of a sphere of radius a if the upper half of the sphere is charged to a potential of ϕ_1 and the lower half of the sphere is charged to a potential ϕ_2.

10.2.20. Find the potential ϕ in the interior and exterior of a sphere of radius 2 when one-half of the surface of the sphere is charged to potential $\phi_0 = 0$ and the other half of the sphere has potential of 0.

10.3 TRANSVERSE VIBRATIONS OF A THIN BEAM

In this section, we investigate transverse vibrations of a thin beam. We start with the derivation of the equation. Then we solve the problem of transverse vibrations of a thin beam, which is simply supported.

10.3.1 Derivation of the Beam Equation

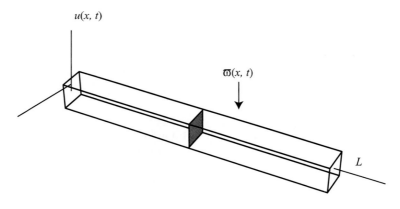

Figure 10.5: An elastic beam.

We derived the wave equation for a tightly stretched perfectly flexible horizontal string in Chapter 3. Those equations also allowed us to model longitudinal vibrations of a uniform flexible rod. (See Problem 4.4.9.) However, transverse vibrations of a thin beam require a new set of equations since, unlike transverse vibrations of a tightly stretched perfectly flexible horizontal string, the thin beam offers resistance to bending.

Consider an elastic beam with uniform rectangular cross section, which lies with the x-axis centered down the beam of length L, as shown in Figure (10.5).

Given a variable load, $\varpi(x, t)$, is on top of the beam, which produces a small downward deflection of the beam, we have for any small section $(x, x + \triangle x)$ of the beam a bending moment, $M(x, t)$. The bending moment is usually given as

$$M(x, t) = -E(x)I(x)C, \tag{10.41}$$

where $E(x)I(x)$ is the **flexural rigidity** and is composed of the Young's modulus of elasticity, $E(x)$, which depends on the material and the **moment of inertia**, $I(x)$. Also, the curvature, C, of the beam under a variable load $\varpi(x, t)$ per unit length x can be found from calculus and is given as

$$C = \frac{\dfrac{\partial^2 u(x, t)}{\partial x^2}}{\left[1 + \left(\dfrac{\partial u(x, t)}{\partial x}\right)^2\right]^{\frac{3}{2}}}. \tag{10.42}$$

When we assume a small slope in the curvature, which is the usual assumption, we have

$$\frac{\partial u(x, t)}{\partial x} \approx 0.$$

Thus, Equation (10.42) is approximated as

$$C = \frac{\partial^2 u(x, t)}{\partial x^2}$$

and Equation (10.41) may be given as

$$M(x, t) = -E(x)I(x)\frac{\partial^2 u(x, t)}{\partial x^2}. \tag{10.43}$$

Newton's law of motion, applied to a small section of the beam, $\triangle x$, is given as the sum of the forces in the u direction equal to mass of the beam times acceleration of the beam in the u direction,

$$\sum F_u = ma_u = m\frac{\partial^2 u(x, t)}{\partial t^2} \quad \text{where } m = \frac{\varpi(x, t)}{g},$$

and the sum of the moments of bending equal to the moments of inertia times the angular acceleration,

$$\sum M(x, t) = I\omega.$$

We assume the angular acceleration, ω, to be 0. We make this assumption because we ignore angular acceleration just as we ignored horizontal vibrations in the derivation of the wave equation for a vibrating string. (See Chapter 3.) Thus,

$$\sum M(x, t) = 0.$$

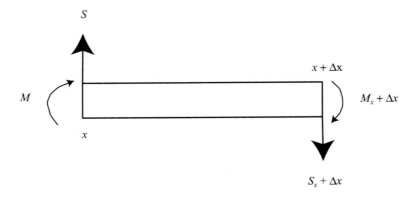

Figure 10.6: A small section $(x, x + \triangle x)$ of an elastic beam.

We also have shear forces acting on the cross section of the small section of the beam, $\triangle x$, shown in Figure (10.6). We denote shear forces as $S(x, t)$. The sum of the forces in the u-direction under a continuous load $\varpi(x, t)$ on the small section of beam, $\triangle x$ is

$$\sum F_u = S(x, t) - \left(S(x, t) + \frac{\partial S(x, t)}{\partial x} \triangle x \right) + \varpi(x, t) \triangle x = 0. \tag{10.44}$$

Dividing Equation (10.44) by $\triangle x$ and taking the limit as $\triangle x \to 0$ yields

$$\frac{\partial S(x, t)}{\partial x} + \varpi(x, t) = m \frac{\partial^2 u(x, t)}{\partial t^2}. \tag{10.45}$$

If we take the sum of the moments counterclockwise about the point x, we arrive at

$$0 = \sum M(x, t) = -M(x, t) + \left(M(x, t) + \frac{\partial M(x, t)}{\partial x} \triangle x \right)$$

$$- \left(S(x, t) + \frac{\partial S(x, t)}{\partial x} \triangle x \right) \triangle x + \varpi(x, t) \triangle x \frac{\triangle x}{2}. \tag{10.46}$$

Again, dividing by $\triangle x$ and taking the limit as $\triangle x \to 0$ yields

$$\frac{\partial M(x, t)}{\partial x} = S(x, t). \tag{10.47}$$

Replacing $M(x, t)$ in Equation (10.47) with its equivalent value found in Equation (10.43) yields

$$\frac{\partial}{\partial x} \left(-E(x) I(x) \frac{\partial^2 u(x, t)}{\partial x^2} \right) = S(x, t). \tag{10.48}$$

Taking the derivative of Equation (10.48) with respect to x and adding $\varpi(x,t)$ to both sides gives us

$$\frac{\partial^2}{\partial x^2}\left(-E(x)I(x)\frac{\partial^2 u(x,t)}{\partial x^2}\right) + \varpi(x,t) = \frac{\partial S(x,t)}{\partial x} + \varpi(x,t) = m\frac{\partial^2 u(x,t)}{\partial t^2},$$

which becomes

$$m(x)\frac{\partial^2 u(x,t)}{\partial t^2} + \frac{\partial^2}{\partial x^2}\left(E(x)I(x)\frac{\partial^2 u(x,t)}{\partial x^2}\right) = \varpi(x,t). \qquad (10.49)$$

If the beam is uniform, then $E(x)$ and $I(x)$ are constant. Also, if the load $\varpi(x,t)$ is constant and does not vary with time, t, Equation (10.49) becomes

$$\frac{\partial^2 u(x,t)}{\partial t^2} + c^2\frac{\partial^4 u(x,t)}{\partial x^4} = g, \qquad (10.50)$$

where $c^2 = \dfrac{EI}{m}$. Since the gravity, g, is small compared with the internal forces of the beam, it may be neglected in most applications. Thus, Equation (10.50) becomes

$$\frac{\partial^2 u(x,t)}{\partial t^2} + c^2\frac{\partial^4 u(x,t)}{\partial x^4} = 0,$$

which is the homogeneous equation for transverse vibrations in a beam.

10.3.2 Transverse Vibrations of a Simply Supported Thin Beam

"Simply supported" means that the ends are fixed to support brackets, which allows the slope of the endpoints to change. Figure (10.7) shows a simply supported beam. Since riveted connections are elastic, a good example of a simply supported beam is a girder made either of wood or steel in the construction of a building or a bridge.

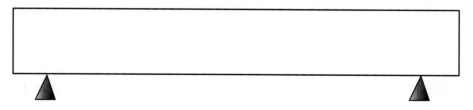

Figure 10.7: Simply supported beam.

We know that the mathematical equation for transverse vibrations of a simply supported thin beam is

$$\frac{\partial^2 u(x,t)}{\partial t^2} + c^2\frac{\partial^4 u(x,t)}{\partial x^4} = 0. \qquad (10.51)$$

Simply supported means that the endpoints are subject to

$$u(0, t) = 0, \qquad\qquad u(L, t) = 0,$$

$$\frac{\partial^2 u(0, t)}{\partial x^2} = 0, \quad \text{and} \quad \frac{\partial^2 u(L, t)}{\partial x^2} = 0, \tag{10.52}$$

and

$$u(x, 0) = f(x) \text{ and } \frac{\partial u(x, 0)}{\partial t} = g(x). \tag{10.53}$$

Equations (10.51 and 10.52) are linear and homogeneous. Therefore separation of variables applies. If we let $u(x, t) = \varphi(x)G(t)$ and replace it in Equation (10.51) we find that Equation (10.51) becomes

$$\varphi(x)G''(t) + c^2 \varphi^{iv}(x)G(t) = 0. \tag{10.54}$$

Separating Equation (10.54) and setting it equal to a separation constant in the usual manner yields

$$\frac{G''(t)}{c^2 G(t)} = -\frac{\varphi^{iv}(x)}{\varphi(x)} = -\lambda,$$

which becomes the time equation

$$G''(t) = -\lambda c^2 G(t) \tag{10.55}$$

and the spatial equation

$$\varphi^{iv}(x) = \lambda \varphi(x). \tag{10.56}$$

Separating the boundary conditions yields

$$\varphi(0) = 0, \qquad\qquad \varphi(L) = 0,$$

$$\varphi''(0) = 0, \quad \text{and} \quad \varphi''(L) = 0. \tag{10.57}$$

Equation (10.56) subject to the BCs given in Equation (10.57) presents a slightly different problem than the equations we got from separation of variables in Chapter 5. Since Equation (10.56) is a fourth-order ODE, we should expect four linearly independent solutions. Also, if λ is negative the time equation will force the destruction of the thin beam because the time equation will produce hyperbolic sine and cosine functions. If $\lambda = 0$, then the spatial equation only has a trivial solution. Thus, if we assume that $\lambda > 0$ and $\varphi(x) = e^{rx}$ is a solution of Equation (10.56), we get the equation

$$r^4 e^{rx} = \lambda e^{rx}.$$

Canceling like terms yields the characteristic equation

$$r^4 - \lambda = 0,$$

which, when factored, becomes

$$(r - (\lambda)^{\frac{1}{4}})(r + (\lambda)^{\frac{1}{4}})(r - i(\lambda)^{\frac{1}{4}})(r + i(\lambda)^{\frac{1}{4}}) = 0.$$

Therefore, we have four roots, $r = \pm\lambda,\ \pm i\lambda$ From Chapter 5 or Appendix C, you should recognize that the solution is

$$\varphi(x) = c_1 e^{(\lambda)^{\frac{1}{4}} x} + c_2 e^{-(\lambda)^{\frac{1}{4}} x} + c_3 e^{i(\lambda)^{\frac{1}{4}} x} + c_4 e^{-i(\lambda)^{\frac{1}{4}} x}. \tag{10.58}$$

An easier form of Equation (10.58) is found by using the noncomplex and complex forms of Euler's equations found in Chapter 5. Thus Equation (10.58) becomes

$$\varphi(x) = C_1 \cosh(\lambda)^{\frac{1}{4}} x + C_2 \sinh(\lambda)^{\frac{1}{4}} x + C_3 \cos(\lambda)^{\frac{1}{4}} x + C_4 \sin(\lambda)^{\frac{1}{4}} x. \tag{10.59}$$

Applying the first BC of Equation (10.57), $\varphi(0) = 0$, yields

$$C_1 = -C_3.$$

Applying the third BC of Equation (10.57), $\varphi''(0) = 0$, yields

$$C_1 = C_3.$$

Thus, $C_1 = C_3 = 0$, and Equation (10.59) becomes

$$\varphi(x) = C_2 \sinh(\lambda)^{\frac{1}{4}} x + C_4 \sin(\lambda)^{\frac{1}{4}} x. \tag{10.60}$$

Applying the second and fourth BCs of Equation (10.57), $\varphi(L) = 0$ and $\varphi''(L) = 0$, yields

$$0 = C_2 \sinh(\lambda)^{\frac{1}{4}} L + C_4 \sin(\lambda)^{\frac{1}{4}} L \tag{10.61}$$

and

$$0 = C_2 \sinh(\lambda)^{\frac{1}{4}} L - C_4 \sin(\lambda)^{\frac{1}{4}} L. \tag{10.62}$$

Adding Equations (10.61 and 10.62) together yields

$$0 = 2C_2 \sinh(\lambda)^{\frac{1}{4}} L,$$

which implies $C_2 = 0$ because the hyperbolic sine function equals 0 only at $x = 0$. Therefore, we are left with

$$0 = C_4 \sin(\lambda)^{\frac{1}{4}} L \text{ and } 0 = -C_4 \sin(\lambda)^{\frac{1}{4}} L,$$

which both yield the same eigenvalue

$$\lambda_n = \left(\frac{n\pi}{L}\right)^4, \ n = 1, 2, 3, \ldots, \tag{10.63}$$

and eigenfunctions

$$\varphi_n(x) = \sin\left(\frac{n\pi x}{L}\right). \tag{10.64}$$

The Time equation, Equation (10.55), becomes

$$G_n''(t) = -\left(\frac{n\pi}{L}\right)^4 c^2 G_n(t),$$

which has the solution

$$G_n(t) = D_1 \cos\left[\left(\frac{\sqrt{c}n\pi}{L}\right)^2 t\right] + D_2 \sin\left[\left(\frac{\sqrt{c}n\pi}{L}\right)^2 t\right]. \tag{10.65}$$

The general solution for $u(x,t)$ becomes

$$u(x,t) = \sum_{n=1}^{\infty} \left\{ A_1 \cos\left[\left(\frac{\sqrt{c}n\pi}{L}\right)^2 t\right] + A_2 \sin\left[\left(\frac{\sqrt{c}n\pi}{L}\right)^2 t\right] \right\} \sin\left(\frac{n\pi x}{L}\right). \tag{10.66}$$

Applying the first IC from Equation (10.53) to Equation (10.66) yields

$$u(x,0) = f(x) = \sum_{n=1}^{\infty} A_1 \sin\left(\frac{n\pi x}{L}\right).$$

Using the orthogonality of the sine function, we arrive at

$$A_1 = \frac{2}{L} \int_0^L f(x) \sin\left(\frac{n\pi x}{L}\right) \, dx. \tag{10.67}$$

Applying the second IC from Equation (10.53) to Equation (10.66) yields

$$\frac{\partial u(x,0)}{\partial t} = g(x) = \sum_{n=1}^{\infty} \left(\frac{\sqrt{c}n\pi}{L}\right)^2 A_2 \sin\left(\frac{n\pi x}{L}\right).$$

Again using the orthogonality of the sine function, we arrive at

$$A_2 = \frac{2L}{(\sqrt{c}n\pi)^2} \int_0^L g(x) \sin\left(\frac{n\pi x}{L}\right) \, dx. \tag{10.68}$$

Therefore, the complete answer for transverse vibrations in a simply supported thin beam is given by Equations (10.66, 10.67, and 10.68).

Other possible boundary conditions are

1. The cantilever beam, Figure (10.8), which has one end rigidly fixed and the other as a free end. An example of this model is an airplane wing. It has the

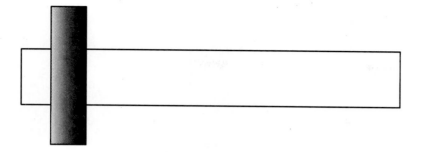

Figure 10.8: The cantilever beam: rigidly fixed on the left end and free on the right end.

following BCs:

$$\text{Rigidly fixed at } x = 0 \text{ end} \begin{cases} u(0,t) &= 0 \\[2mm] \dfrac{\partial u(0,t)}{\partial x} &= 0 \end{cases}$$

$$\text{Free end at } x = L \text{ end} \begin{cases} \dfrac{\partial^2 u(L,t)}{\partial x^2} &= 0 \\[2mm] \dfrac{\partial^3 u(L,t)}{\partial x^3} &= 0. \end{cases}$$

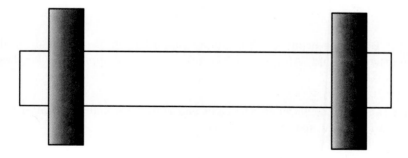

Figure 10.9: Beam rigidly fixed at both ends.

2. Beam rigidly fixed at both ends, Figure (10.9). A very good example of this type of beam is the World War II fighter, Lockheed P-38 Lightning. The P-38 was designed as a twin engine fighter where the engines were attached to long booms, the wings extend from the outside part of the boom, and between the two booms was the central nacelle, which contained the pilot and armament.

Such a rigidly fixed beam has the following BCs:

$$\text{Rigidly fixed at } x = 0 \text{ end} \begin{cases} u(0,t) &= 0 \\ \dfrac{\partial u(0,t)}{\partial x} &= 0 \end{cases}$$

$$\text{Rigidly fixed at } x = L \text{ end} \begin{cases} u(0,t) &= 0 \\ \dfrac{\partial u(0,t)}{\partial x} &= 0. \end{cases}$$

EXERCISES 10.3

10.3.1. Solve the simply supported beam problem of length 2π when $u(x,0) = x(2\pi - x)$ and $\dfrac{\partial u(x,0)}{\partial t} = 0$.

10.3.2. Solve the simply supported beam problem of length π when $u(x,0) = x$ and $\dfrac{\partial u(x,0)}{\partial t} = 0$.

10.3.3. Solve the simply supported beam problem of length 1 when $u(x,0) = 0$ and $\dfrac{\partial u(x,0)}{\partial t} = 1 - x^2$.

10.3.4. State the BCs for a beam which is simply supported on one end and rigidly fixed on the other end. Also, try to determine a physical application for these BCs.

10.3.5. Given a cantilever beam where the right end is free, we know that after separating variables the spatial equation is

$$\varphi(x) = C_1 \cosh(\lambda)^{\frac{1}{4}}x + C_2 \sinh(\lambda)^{\frac{1}{4}}x + C_3 \cos(\lambda)^{\frac{1}{4}}x + C_4 \sin(\lambda)^{\frac{1}{4}}x. \quad (10.69)$$

Show that Equation (10.69) satisfies the cantilever beam BCs if $(\lambda)^{\frac{1}{4}} L$ is a root of $\cosh{(\lambda)^{\frac{1}{4}}} L \cos{(\lambda)^{\frac{1}{4}}} L = -1$.

10.3.6. Given a beam that is rigidly fixed at both ends, we know that after separating variables the spatial equation is Equation (10.69). Show that Equation (10.69) satisfies rigidly fixed beam BCs if $(\lambda)^{\frac{1}{4}} L$ is a root of $\cosh{(\lambda)^{\frac{1}{4}}} L \cos{(\lambda)^{\frac{1}{4}}} L = 1$.

10.4 HEAT CONDUCTION IN A CIRCULAR PLATE

Heat conduction in a circular plate is really no different from heat conduction in a rectangular plate, which has the mathematical equation

$$\frac{\partial u(x,y,t)}{\partial t} = k\nabla^2 u(x,y,t)$$

$$= k\left(\frac{\partial^2 u(x,y,t)}{\partial x^2} + \frac{\partial^2 u(x,y,t)}{\partial y^2}\right), \tag{10.70}$$

subject to the homogeneous Dirichlet BCs

$$u(0,y,t) = u(L,y,t) = u(x,0,t) = u(x,H,t) = 0$$

and

$$u(x,y,t) = f(x,y).$$

If you notice, I used the Laplacian (∇^2), then wrote out the partial derivatives with

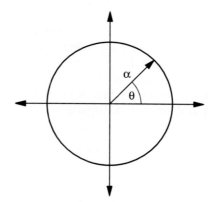

Figure 10.10: Polar coordinate system.

respect to x and y. This makes the transition from a rectangular plate to a circular plate transparent. The only thing that changes is the coordinate system. In a circular plate, the preferred coordinate system is the cylindrical coordinate system, which is called polar form in two dimensions, Figure (10.10). Thus, Equation (10.70) becomes

$$\frac{\partial u(r,\theta,t)}{\partial t} = k\nabla^2 u(r,\theta,t)$$

$$= k\left[\frac{1}{r}\frac{\partial}{\partial r}\left(r\frac{\partial u(r,\theta,t)}{\partial r}\right) + \frac{1}{r^2}\frac{\partial^2 u(r,\theta,t)}{\partial\theta^2}\right], \tag{10.71}$$

subject to

$$u(a, \theta, t) = 0 \qquad (10.72)$$

and

$$u(r, \theta, 0) = g(r, \theta). \qquad (10.73)$$

Just as in Section 10.2.1, it appears that we only have one boundary condition. However, as in Section 10.2.1 the other BCs come from the geometry of the problem, and they are

$$u(r, -\pi, t) = u(r, \pi, t)$$

and

$$\frac{\partial u(r, -\pi, t)}{\partial \theta} = \frac{\partial u(r, \pi, t)}{\partial \theta}, \qquad (10.74)$$

and

$$\lim_{r \to 0} |u(r, \theta, t)| < \infty. \qquad (10.75)$$

Equation (10.75) is not so much a BC as it is a natural condition on temperature at the origin of the plate. If the temperature were allowed to go to infinity the plate would melt down. An unwanted condition in most experiments.

Equation (10.71) and its corresponding BCs are linear and homogeneous. Therefore, we may use separation of variables. We let $u(r, \theta, t) = G(t)\varphi(r, \theta)$. After taking the necessary derivatives and separating Equation (10.71) and the BCs, Equation (10.72, 10.74, and 10.75), we arrive at the time ODE

$$G'(t) = -\lambda k G(t), \qquad (10.76)$$

and the spatial PDE

$$\frac{1}{r} \frac{\partial}{\partial r} \left(r \frac{\partial \varphi(r, \theta)}{\partial r} \right) + \frac{1}{r^2} \frac{\partial^2 \varphi(r, \theta)}{\partial \theta^2} = -\lambda \varphi(r, \theta), \qquad (10.77)$$

subject to

$$\varphi(a, \theta, t) \quad = \quad 0,$$

$$\lim_{r \to 0} |\varphi(r, \theta)| \quad < \quad \infty,$$

$$\varphi(r, -\pi) \quad = \quad \varphi(r, \pi), \qquad (10.78)$$

and

$$\frac{\partial \varphi(r, -\pi)}{\partial \theta} \quad = \quad \frac{\partial \varphi(r, -\pi)}{\partial \theta}.$$

Equation (10.77) is a linear homogeneous PDE with linear homogeneous BCs, Equation (10.78). Thus, we can use separation of variables. Letting $\varphi(r, \theta) = R(r)\Theta(\theta)$, taking the necessary derivatives, and separating the equations, we have

$$\Theta''(\theta) = -\mu\Theta(\theta), \tag{10.79}$$

subject to

$$\Theta(-\pi) = \Theta(\pi)$$

$$\text{and} \tag{10.80}$$

$$\Theta'(-\pi) = \Theta'(\pi),$$

and after some algebraic manipulation,

$$r^2 R''(r) + rR'(r) + (\lambda r^2 - \mu)R(r) = 0, \tag{10.81}$$

subject to

$$R(a) = 0$$

$$\text{and} \tag{10.82}$$

$$\lim_{r \to 0} |R(r)| < 0.$$

We have encountered Equation (10.79) subject to BCs, Equation (10.80), and know its solution, which is

$$\mu_0 = 0, \ \Theta_0(\theta) = A_0$$

$$\left.\begin{array}{c} \mu_n = m^2 \\ \Theta_m(\theta) = A_m \cos m\theta + B_m \sin m\theta \end{array}\right\}, \ m = 1, 2, 3, \ldots \tag{10.83}$$

We must solve Equation (10.81), restated as

$$r^2 R''_m(r) + rR'_m(r) + (\lambda r^2 - m^2)R_m(r) = 0, \tag{10.84}$$

subject to

$$R_m(a) = 0$$

$$\text{and} \tag{10.85}$$

$$\lim_{r \to 0} |R_m(r)| < 0.$$

First, we should recognize Equation (10.84) is a form of Bessel's equation, Equation (9.51),

$$x^2 u'' + xu' + \left(x^2 - n^2\right) u = 0,$$

found in Chapter 8, and we know the general solution, which, for Equation (9.51), is

$$u(x) = C_1 Y_n(x) + C_2 J_n(x).$$

Thus, the general solution to Equation (10.84) is

$$R_m(r) = C_1 Y_m\left(\sqrt{\lambda}r\right) + C_2 J_m\left(\sqrt{\lambda}r\right).$$

Since we know that $\lim_{r \to 0}|R_m(r)| < 0$ and as $r \to 0$, $Y_m\left(\sqrt{\lambda}r\right) \to \infty$, we must have $C_1 = 0$. Therefore, we are left with

$$R_m(r) = C_2 J_m\left(\sqrt{\lambda}r\right). \tag{10.86}$$

Applying the second BC, $R_m(a) = 0$ to Equation (10.86) yields

$$R_m(a) = C_2 J_m\left(\sqrt{\lambda}a\right).$$

For each Bessel function of the first kind of order m, we have an infinite number of 0s. Hence, just like a sine or cosine function, we have an infinite number of possible solutions. Since each Bessel function of the first kind of each order m is a solution to a linear homogeneous ODE, we know the sum of all the solutions is also a solution. Thus, letting y_{mn} represent the 0s of Bessel function of the first kind of each order m, we have

$$\sqrt{\lambda}a = y_{mn},$$

which implies

$$\lambda_{mn} = \left(\frac{y_{mn}}{a}\right)^2.$$

Thus, the general solution to Equation (10.86) is

$$R_m(r) = \sum_{n=1}^{\infty} c_n J_m\left(\sqrt{\lambda_{mn}}r\right),$$

where

$$c_n = \frac{\int_0^a R_m(r) J_m\left(\sqrt{\lambda_{mn}}r\right) r\ dr}{\int_0^a \left[J_m\left(\sqrt{\lambda_{mn}}r\right)\right]^2 r\ dr}.$$

Therefore, the general solution of the PDE given in Equation (10.77) is

$$\varphi(r, \theta) = \sum_{m=0}^{\infty} \sum_{n=1}^{\infty} J_m\left(\sqrt{\lambda_{mn}}r\right) \left[\alpha_{mn} \cos m\theta + \beta_{mn} \sin m\theta\right], \tag{10.87}$$

where $\alpha_{mn} = c_n A_m$ and $\beta_{mn} = c_n B_m$.

We must now solve the time ODE, Equation (10.76),

$$G'_{mn}(t) = -\lambda_{mn} k G_{mn}(t),$$

which is

$$G_{mn}(t) = d_{mn} e^{-\lambda_{mn} kt}. \tag{10.88}$$

We now can give the general solution for $u(r, \theta, t)$, which is

$$u(r, \theta, t) = \sum_{m=0}^{\infty} \sum_{n=1}^{\infty} e^{-\lambda_{mn} kt} J_m\left(\sqrt{\lambda_{mn}} r\right) [D_{mn} \cos m\theta + E_{mn} \sin m\theta], \tag{10.89}$$

where $D_{mn} = d_{mn} \alpha_{mn}$ and $E_{mn} = d_{mn} \beta_{mn}$.

Applying the IC, Equation (10.73),

$$u(r, \theta, 0) = g(r, \theta)$$

yields

$$u(r, \theta, 0) = g(r, \theta) = \sum_{m=0}^{\infty} \sum_{n=1}^{\infty} J_m\left(\sqrt{\lambda_{mn}} r\right) [D_{mn} \cos m\theta + E_{mn} \sin m\theta] \tag{10.90}$$

$$= \sum_{n=1}^{\infty} J_0\left(\sqrt{\lambda_{0n}} r\right) D_{0n} + \sum_{m=1}^{\infty} \sum_{n=1}^{\infty} J_m\left(\sqrt{\lambda_{mn}} r\right) D_{mn} \cos m\theta \tag{10.91}$$

$$+ \sum_{m=1}^{\infty} \sum_{n=1}^{\infty} J_m\left(\sqrt{\lambda_{mn}} r\right) E_{mn} \sin m\theta. \tag{10.92}$$

Thus,

$$D_{0n} = \frac{\int_{-\pi}^{\pi} \int_0^a g(r, \theta) J_0\left(\sqrt{\lambda_{0n}} r\right) r \, dr \, d\theta}{\int_{-\pi}^{\pi} \int_0^a \left[J_0\left(\sqrt{\lambda_{0n}} r\right)\right]^2 r \, dr \, d\theta}; \tag{10.93}$$

$$D_{mn} = \frac{\int_{-\pi}^{\pi} \int_0^a g(r, \theta) J_m\left(\sqrt{\lambda_{mn}} r\right) \cos(m\theta) \, r \, dr \, d\theta}{\int_{-\pi}^{\pi} \int_0^a \left[J_m\left(\sqrt{\lambda_{mn}} r\right) \cos(m\theta)\right]^2 r \, dr \, d\theta}, \quad m > 0; \tag{10.94}$$

and

$$E_{mn} = \frac{\int_{-\pi}^{\pi} \int_0^a g(r, \theta) J_m\left(\sqrt{\lambda_{mn}} r\right) \sin(m\theta) \, r \, dr \, d\theta}{\int_{-\pi}^{\pi} \int_0^a \left[J_m\left(\sqrt{\lambda_{mn}} r\right) \sin(m\theta)\right]^2 r \, dr \, d\theta}, \quad m > 0. \tag{10.95}$$

Therefore, the solution to the circular heat problem is Equation (10.89) where the coefficients are given by Equations (10.93, 10.94, and 10.95).

A similar method is used to solve the wave equation for vibrations of a drumhead.

EXERCISES 10.4

10.4.1. Consider temperature in a circular plate of radius $r = \alpha$ where the initial temperature of the boundary of the plate is a homogeneous Dirichlet condition. Given a non0 initial temperature distribution, briefly explain what occurs when we take $\lim_{t \to \infty} u(r, \theta, t)$.

10.4.2. Solve for temperature distribution in a semicircle plate of radius 1 where $0 < \theta < \pi$ with initial condition of $u(r, \theta, 0) = g(r, \theta)$ and the following BCs:

(1) $u(r, 0, t) = 0$, $u(r, \pi, t) = 0$, $u(1, \theta, t) = 0$.

(2) $\dfrac{\partial u(r, 0, t)}{\partial \theta} = 0$, $\dfrac{\partial u(r, \pi, t)}{\partial \theta} = 0$, $\dfrac{\partial u(1, \theta, t)}{\partial \theta} = 0$.

(3) $\dfrac{\partial u(r, 0, t)}{\partial \theta} = 0$, $\dfrac{\partial u(r, \pi, t)}{\partial \theta} = 0$, $u(1, \theta, t) = 0$.

For each case briefly explain what occurs when we take $\lim_{t \to \infty} u(r, \theta, t)$.

10.4.3. State and solve the general vibrating circular drumhead problem.

10.4.4. Solve the vibrating circular drumhead problem when the radius $r = 1$, the boundary is fixed, the initial velocity is 0, and the initial displacement is $f(r) \sin 2\theta$.

10.4.5. Solve the vibrating circular drumhead problem when the radius $r = \pi$, the boundary is free, the initial velocity is 0, and the initial displacement is $f(r) \cos \theta$.

10.4.6. Consider vertical vibrations of a wedge $0 < \theta < \dfrac{\pi}{4}$ with radius 2. Determine the solution if the BCs are as follows:

(1) $u(r, 0, t) = 0$, $u\left(r, \dfrac{\pi}{4}, t\right) = 0$, $u(2, \theta, t) = 0$.

(2) $\dfrac{\partial u(r, 0, t)}{\partial \theta} = 0$, $\dfrac{\partial u\left(r, \dfrac{\pi}{4}, t\right)}{\partial \theta} = 0$, $\dfrac{\partial u(2, \theta, t)}{\partial \theta} = 0$.

(3) $\dfrac{\partial u(r,0,t)}{\partial \theta} = 0,\ \dfrac{\partial u\left(r,\frac{\pi}{4},t\right)}{\partial \theta} = 0,\ u(2,\theta,t) = 0.$

For all cases, assume the initial velocity is 0 and the initial displacement is a function of the radius and the angle.

10.4.7. State and solve temperature distribution in a circular plate where there is circular symmetry, i.e., $\dfrac{\partial^2 u}{\partial \theta^2} = 0.$

10.4.8. Consider

$$\frac{\partial u(r,\theta,z,t)}{\partial t} = k\nabla^2 u(r,\theta,z,t),$$

subject to

$$u(r,\theta,z,0) = f(r,\theta,z)$$

and the following BCs:

(1) $u(r,\theta,0,t) = 0,\ u(r,\theta,1,t) = 0,\ u(2,\theta,z,t) = 0.$

(2) $u(r,\theta,0,t) = 0,\ \dfrac{\partial u(r,\theta,2,t)}{\partial z} = 0,\ u(0.5,\theta,z,t) = 0.$

(3) $\dfrac{\partial u(r,\theta,0,t)}{\partial z} = 0,\ \dfrac{\partial u(r,\theta,3,t)}{\partial z} = 0,\ u(\pi,\theta,z,t) = 0.$

10.4.9. Consider

$$\nabla^2 u = 0$$

in a cylinder. Find the solution with the following BCs:

(1) $u(r,\theta,0) = 0,\ u(r,\theta,5) = r\sin 2\theta,\ u(1,\theta,z) = 0.$

(2) $\dfrac{\partial u(r,\theta,0)}{\partial z} = 0,\ \dfrac{\partial u(r,\theta,2)}{\partial z} = 0,\ u(\pi,\theta,z) = 0.$

(3) $\dfrac{\partial u(r,\theta,0)}{\partial z} = 0,\ u(r,\theta,\pi) = r\cos\theta,\ u(1,\theta,z) = 0.$

(4) $u(r,\theta,0) = r^2\theta^2,\ \dfrac{\partial u(r,\theta,2)}{\partial z} = 0,\ u(0.5,\theta,z) = 0.$

Does the solution always exist?

10.4.10. Complete the following:

(1) Solve for temperature distribution inside a uniform sphere of radius a, which is independent of the variable φ if the initial temperature distribution inside the sphere is constant throughout the sphere and the BC is a homogeneous Dirichlet condition.

(2) Solve the problem in Part (1) if the radius $a = 5$ feet.

(3) Show that as time, t, goes to infinity and that the solution to Part (1) becomes the solution of Laplace's equation in the sphere of radius a with homogeneous Dirichlet condition.

10.4.11. A uniform hemisphere of radius 1 m has its convex surface temperature kept at $100°$C while its base is kept at a temperature of $0°$C.

(1) Find the temperature distribution inside the hemisphere for all time.

(2) Find the steady-state temperature inside the hemisphere.

10.5 SCHRÖDINGER'S EQUATION

Schrödinger's equation is very important in the study of quantum mechanics, and most of you have heard of it. In this section, we develop the time-independent Schrödinger Equation. Also, in the exercises, several projects are presented that involve nuclear transport. We start our derivation with the wave equation in three-dimensional Cartesian space

$$\frac{\partial^2 u}{\partial t^2} = k^2 \nabla^2 u. \tag{10.96}$$

We are using the constant k^2 instead of the constant c^2 since we want to reserve c^2 for the speed of light. Helmholtz[1] formulated solutions to Equation (10.96) by using separation of variables, where the separation constant is $-\xi^2$. Helmholtz's solution is in the form

$$u = \varphi e^{-i\xi t}, \tag{10.97}$$

where $\xi = 2\pi\nu$ and ν is the frequency of the radiation. The derivation of $\xi = 2\pi\nu$ is far beyond the scope of this text. However, it is easy to show that $G(t) = e^{-i\xi t}$ is a solution of

$$G''(t) = -\xi^2 G(t).$$

The function φ is the solution of the time-independent equation

$$\nabla^2 \varphi = -\varsigma^2 \varphi, \quad \varsigma = \frac{\xi}{k}. \tag{10.98}$$

---

[1]Herman Helmholtz (1821–1894) started his professional life in physiology, then became interested in mathematical physics. His studies were primarily in acoustics and he published results in a work titled *On the Sensations of Tone.*

Since we are interested in the time-independent Schrödinger Equation, we want to focus on Equation (10.98), which can be written as

$$\nabla^2 \varphi + \varsigma^2 \varphi = 0. \tag{10.99}$$

Equation (10.99) is the time-independent wave equation, which is sometimes called the Helmholtz equation. The solutions of Equation (10.99) are called monochromatic waves. From Helmholtz's equation, we want to develop the time-independent Schrödinger Equation. We start by considering Einstein's[2] famous equation

$$E = mc^2. \tag{10.100}$$

Einstein's equation relates energy to mass. From it we see that

$$m = \frac{E}{c^2}. \tag{10.101}$$

The photon is a quantity (quantum) of electromagnetic radiation. We know it has speed and energy where the energy is given by

$$E = h\nu, \tag{10.102}$$

where h is Planck's[3] constant and ν is the frequency of the radiation. This implies that

$$e^{-i\xi t} = e^{-2\pi i \nu t} = e^{\frac{-2\pi i E t}{h}},$$

which implies that the eigenvalues depend on the total energy, E. This is shown later to be exactly the case.

Replacing energy, E, in Equation (10.101) with the formula for energy given in Equation (10.102), we find that

$$m = \frac{h\nu}{c^2} \tag{10.103}$$

Since a photon has speed and energy, it has momentum. Momentum for a single particle of mass (a photon) is given by the equation

$$p = mv, \tag{10.104}$$

where v is the velocity (usually expressed as a vector). Traditionally, the velocity of light is given as c. Thus, the momentum of a photon is

$$p = mc. \tag{10.105}$$

---

[2]Albert Einstein (1879–1955), a brilliant physicist, developed the theory of general relativity, which brought the field of differential geometry back to center stage in the mathematical world.

[3]Max Planck (1858–1947) a German physicist who in 1901 took the first steps toward quantum mechanics.

Using Equation (10.103), we find momentum for a photon is given by

$$p = \frac{h\nu}{c^2}c = \frac{h\nu}{c}. \tag{10.106}$$

In the mid 1920s, de Broglie's hypothesis was presented. It suggested that waves were associated with material particles, with wavelength, Λ, and momentum, p, where

$$\Lambda = \frac{h}{p} = \frac{h}{mc} = \frac{c}{\nu}. \tag{10.107}$$

Thus,

$$p = \frac{h\nu}{c} = \frac{h}{\Lambda}, \tag{10.108}$$

which can be written as

$$p = \frac{h\nu}{c} = \frac{h}{\Lambda} = \frac{h}{2\pi}\frac{2\pi}{\Lambda}. \tag{10.109}$$

Note: $\dfrac{h}{2\pi}$ is called the Dirac-h, denoted $\hbar$. However, we continue the derivation using $\dfrac{h}{2\pi}$. Since k is the velocity of the wave, which is the velocity of light, and therefore c, we have

$$\varsigma = \frac{\xi}{k} = \frac{2\pi\nu}{k} = \frac{2\pi}{\Lambda}.$$

Using Equations (10.105 and 10.107) we have $\varsigma = \dfrac{2\pi mc}{h}$. Replacing ς in Equation (10.99) by $\dfrac{2\pi mv}{h}$ yields

$$\nabla^2\varphi + \left(\frac{2\pi mc}{h}\right)^2\varphi = \nabla^2\varphi + \frac{4\pi^2 m^2 c^2}{h^2}\varphi = 0. \tag{10.110}$$

After rewriting Equation (10.110) slightly, it becomes

$$\nabla^2\varphi + \frac{8\pi^2 m}{h^2}\frac{mc^2}{2}\varphi = 0. \tag{10.111}$$

From physics, we know total that the energy, E, is equal to the sum of the kinetic energy, E_k, and the potential energy, E_p. Also, we know the kinetic energy is equal to $\dfrac{mc^2}{2}$. Thus,

$$E = E_k + E_p = \frac{mc^2}{2} + E_p,$$

which means

$$\frac{mc^2}{2} = E - E_p. \tag{10.112}$$

Note: In Equation (10.112), the potential energy, E_p, is a function of position. Using Equation (10.112), we may replace $\frac{mv^2}{2}$ in Equation (10.111) with $E - E_p$. Thus, we have

$$\nabla^2 \varphi + \frac{8\pi^2 m}{h^2} (E - E_p) \varphi = 0, \tag{10.113}$$

which is the time-independent Schrödinger's equation. In one space dimension, Equation (10.113) may be written as

$$\varphi''(x) + \frac{8\pi^2 m}{h^2} (E - E_p(x)) \varphi(x) = 0. \tag{10.114}$$

Equation (10.114) allows us to employ our knowledge of Sturm–Liouville theory. We can rewrite Equation (10.114) as

$$\varphi''(x) + q(x)\varphi(x) + \lambda\varphi(x) = 0,$$

where $q(x) = -\frac{8\pi^2 m}{h^2} E_p(x)$ and $\lambda = \frac{8\pi^2 m}{h^2} E$. Thus, the eigenvalues are based on the total energy. Also, each eigenvalue, E_n, $n = 1, 2, 3, \ldots$, has a corresponding eigenfunction, φ_n, $n = 1, 2, 3, \ldots$

EXERCISES 10.5

10.5.1. PROJECT: Apply Schrödinger's equation to a harmonic oscillator-quantum harmonic oscillator. Consider a physical body of mass m attached to the bottom of a spring, and the top of the spring is attached to some immovable object, such as a ceiling. We have no external force or damping. Let u represent the displacement of the mass, m, from equilibrium or the at-rest position; then the classical differential equation is given by

$$m\ddot{u} + ku = 0, \tag{10.115}$$

where k is the spring constant and m is the mass. *Note:* ku represents the restoring force. Dividing through by m in Equation (10.115) yields

$$\ddot{u} + \omega_0^2 u = 0$$

where $\omega_0^2 = \frac{k}{m}$, which implies $k = m\omega_0^2$.

(1) Solve the ODE.

(2) Derive the potential, E_p, by integrating the restoring force, ku. *Note:* We must choose the constant of integration so that $E_p = 0$ when the spring–mass system is at equilibrium, $u = 0$.

(3) Replace E_p in Equation (10.114).

(4) Multiply Equation (10.114) by $\dfrac{h}{2\pi m\omega_0}$.

(5) Set $\lambda = \dfrac{4\pi E}{\omega_0 h}$ and obtain the equation

$$\frac{h}{2\pi m\omega_0}\varphi'' + \lambda\varphi - \frac{2\pi m\omega_0}{h}u^2\varphi = 0.$$

(6) Introduce a new independent variable $\varpi = \dfrac{u\sqrt{h}}{\sqrt{2\pi m\omega_0}}$ and let $\varphi(u) = \tilde{\varphi}(\varpi)$. Then find that

$$\frac{d^2\tilde{\varphi}}{d\varpi^2} + \left(\lambda - \varpi^2\right)\tilde{\varphi} = 0. \tag{10.116}$$

(7) We now look for solutions to Equation (10.116) on $(-\infty, \infty)$ by substituting $\tilde{\varphi}(\varpi) = e^{\frac{-\varpi^2}{2}}v(x)$ into Equation (10.116) and dividing out $e^{\frac{-\varpi^2}{2}}$.

(8) Now let $\lambda = 2n + 1$ and our equation becomes identical to Hermite's equation given in Chapter 10.

(9) Solve and graph several Hermite polynomials on the interval $(-5, 5)$.

10.5.2. PROJECT: Consider nuclear transport

(1) The Euler differential equation is one of the few differential equations with variable coefficients that can be solved using a change of variables. The change of variables necessary to transform the Euler differential equation into a linear constant–coefficient differential equation is $|x| = e^z$. With this transformation, find the complete solution to the following equations:

(a) $x^2 y'' - xy' + y = x^5$.

(b) $x^2 y'' - xy' + 2y = 1 + (\ln x)^2$.

(2) You are working in the joint counter-terrorism unit with other federal agencies. While searching for the plutonium brick (which was found using information you supplied about the maximum size of the brick), airport security forces discovered that the same group of terrorists were trying to smuggle a 6.0 cm radius solid sphere of Pu-239 into the U.S. disguised as a child's toy ball. The FBI wants to ship the ball back to a Department of Energy (DOE) lab for evaluation. Since there is a Coast Guard cutter docked at a nearby marina, they plan to send the Pu ball packed in a crate on board ship. An FBI agent goes and gets a big crate

full of small foam packing peanuts. (In the center of which he/she plans to place the Pu sphere for shipping.)

The steady-state neutron diffusion equation with neutron multiplication is:

$$D\nabla^2\varphi - \sigma_a\varphi + \frac{v\sigma_f\varphi}{k} = 0,$$

where

$$\nabla^2\varphi = \frac{1}{r^2}\frac{\partial}{\partial r}\left(r^2\frac{\partial\varphi}{\partial r}\right), \text{ assuming spherical symmetry.}$$

The following nuclear data may be useful:

For the Plutonium Ball: $D = 1.263$ cm, $\sigma_a = 0.0819$ cm^{-1}, $v\varphi_f = 0.214$ cm^{-1} and density $\rho = 15.4$ g/cc.

For the Foam Packing Peanuts: $D \approx 2.1$ cm, $\sigma_a = 10^{-6}$ cm^{-1}, $v\varphi_f = 0$ cm^{-1}, and density $\rho = 1.0$ g/cc.

Notes: The steady-state neutron diffusion (Helmholtz) equation represents a balance of leakage, absorption, and production from fission reactions in the sphere using the fundamental mode eigenvalue. When we apply the "bare sphere" boundary condition that $\varphi(R + 2D) = 0$ at the radius of the sphere (R) plus the nuclear "extrapolation" dimension $(2D)$ to make the physics correct, we are essentially assuming that no neutrons are reflected back into the sphere. In fact, because atmospheric air reflects back a few neutrons (~ 0) that escape the sphere surface, this is a good assumption; therefore, we need not consider solving the diffusion equation in the air immediately surrounding the sphere. However, if we replace the material surrounding the Pu sphere with a hydrocarbon (foam packing peanuts), we could get some very nice reflection of neutrons back into the sphere, and we will have to account for the diffusion of neutrons in the region outside of the sphere.

The multiplication factor (k) of the 6.0 cm radius unreflected ("bare") sphere in open air is $k = 0.8446$. This means that in this case, the steady-state loss rate of neutrons by absorption and leakage from the sphere is greater than the rate neutrons are produced inside the sphere from fission. If we add to the Pu, or change the geometry to ramp up the production from fission reactions to make $k = 1.0$, then a steady state balance of production from fission and loss due to absorption and leakage occur. (Notice how the "k" in the steady-state diffusion equation "adjusts" the production term to satisfy the overall equation. When k exactly equals 1.0, this is known as "criticality," and the mass causing this condition is known as "critical mass"—not a good thing to be standing next to, since radiation coming directly out of a critical mass is lethal.)

(a) What is the critical size (when $k = 1.0$) of a "bare sphere?"

(b) Determine the critical size ($k = 1.0$) for a Pu sphere reflected by an infinite thickness of packing peanuts. (Use $\varphi_1(R) = \varphi_2(R)$, and

$J_1(R) = \dot{J}_2(R)$ (where $J = -D\nabla\varphi \cdot r$) at the interface of the ball and the packing peanuts.)

(c) Based on your answer, should the FBI agent pack the 6.0 cm radius sphere in the crate as planned, and if he/she does, is it safe to stand nearby when he/she packs it? Is using an infinite thickness of packing peanuts a conservative approach? Why or why not? Explain.

(d) Demonstrate that $k = 0.8446$ for the 6.0 cm radius "bare sphere."

10.5.3. PROJECT: It is July 1945, and the future is uncertain. You are designing the Little Boy atomic weapon. Little Boy is a "gun-type" weapon that uses two identical subcritical cylindrical halves. One cylinder of U-235 remains stationary at one end of the device, while the other (identical) subcritical cylinder is fired using a chemical explosive as a projectile down a gun barrel directly into the stationary cylinder. Provided other components are installed correctly in the device, a nuclear yield results. Using the time-dependent neutron diffusion equation with neutron multiplication you find that

$$D\nabla^2\varphi - -\sigma_a\varphi + v\sigma_f\varphi = \frac{1}{V_0}\frac{\partial\varphi}{\partial t},$$

where

$$\nabla^2\varphi = \frac{1}{r}\frac{\partial}{\partial r}\left(r\frac{\partial\varphi}{\partial r} + \frac{\partial^2\varphi}{\partial z^2}\right).$$

For boundary conditions, the origin can be taken to be the center of the cylinder. The flux can be assumed to be 0 at the extrapolated boundaries, and the initial $(t = 0)$ flux profile is given as

$$\varphi\left(\check{R}, z, t\right) = 0 \qquad \left(\check{R} = R + 2D\right) \qquad \text{cylinder extrapolated radius}$$

$$\varphi\left(r, \pm\frac{\check{a}}{2}, t\right) = 0 \qquad \left(\pm\frac{\check{a}}{2} = \pm\left(\frac{a}{2} + 2D\right)\right) \qquad \text{cylinder half height}$$

and

$$\varphi(r, z, 0) = \varphi_0 \qquad \text{initial flux distribution.}$$

Note that $\varphi \geq 0$ everywhere in the cylinder. Also, the following nuclear data for U-235 metal may be useful:

diffusion constant	$D = 1.3175$ cm
mean neutron velocity	$V_0 = 1.4 \times 10^9$ cm/s
absorption probability	$\sigma_a = 0.0722$ cm^{-1}
fission neutron production probability	$v\sigma_f = 0,1687$ cm^{-1}
fission probability	$\sigma_f = 0.0649$ cm^{-1}
metal density	$\rho = 18.0$ g/cc

(1) Solve the time-dependent diffusion equation in cylindrical geometry using separation of variables. (*Hint:* To solve this, begin separation of variables with $\mu = -\lambda$ for the time-separation constant).

(2) Identify the largest (just $\varepsilon = 1\%$ below critical) mass of U-235 that can be safely used for each half cylinder using diffusion theory. Assume that each half cylinder is a square extrapolated cylinder, where twice the extrapolated radius is equal to the total extrapolated height ($2R = a$). (Recall that criticality is achieved when a stable fundamental mode flux is obtained, ($\lambda_1 \to 0$) that is, the fundamental mode flux does not change with time, and other modes will die out with time).

(3) Assuming $\varphi_0 \to 100$ n/cm^2/s, plot, using your favorite mathematical software, the time dependence of neutron flux in the center of the assembled device at $\varphi(0,0,t)$ right after triggering the detonator—right when the two cylindrical halves–each of a size you found in part (b)–meet together to form a single, supercritical cylinder. *Note:* Time should be in the microsecond range.

(4) The heat power released in the supercritical assembly as a function of time is determined by integrating the volumetric fission rate over the cylindrical volume:

$$P(t) \text{ in watts} = 3.204 \times 10^{-11} \int (\sigma_f)(\varphi(r,z,t)) \, dV.$$

The total energy released, in Joules, is

$$E_r = \int P(t) \, dt.$$

Use your favorite mathematical software to compute the energy released after 1 μsec and 2.5 μsec using kiloton TNT equivalents, where 1 kiloton TNT $= 4.18 \times 10^{12}$ Joules.

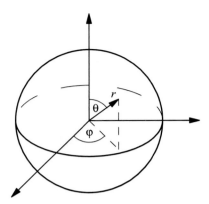

Figure 10.11: A fixed point in three-dimensional space

10.5.4. PROJECT: The time-independent Schrödinger's equation

$$\nabla^2 \varphi + \frac{8\pi^2 m}{h^2} \left(E - E_p\right) \varphi = 0$$

may be developed in spherical coordinates if the potential E_p depends only on a distance r from some fixed point in space. Given the spherical r, θ, and ϕ as shown in Figure (10.11) where

$$x_1 = r \sin \theta \cos \phi, \ x_2 = r \sin \theta \sin \phi, \text{ and } x_3 = r \cos \theta,$$

show that

$$\nabla^2 \varphi = \frac{1}{r^2} \frac{\partial}{\partial r} \left(r^2 \frac{\partial \varphi}{\partial r} \right) + \frac{1}{r^2} \left[\frac{1}{\sin \theta} \frac{\partial}{\partial \theta} \left(\sin \theta \frac{\partial \varphi}{\partial \theta} \right) + \frac{1}{\sin^2 \theta} \frac{\partial^2 \varphi}{\partial \phi^2} \right]. \quad (10.117)$$

(1) Show that Equation (10.117) may also be written in the form

$$\nabla^2 \varphi = \frac{\partial^2 \varphi}{\partial r^2} + \frac{2}{r} \frac{\partial \varphi}{\partial r} + \frac{1}{r^2} \left[\frac{\partial^2 \varphi}{\partial \theta^2} + \cot \theta \frac{\partial \varphi}{\partial \theta} + \frac{1}{\sin^2 \theta} \frac{\partial^2 \varphi}{\partial \phi^2} \right].$$

(2) Using Schrödinger's equation in spherical coordinates, show that separating variables results in

$$R''(r) + \frac{2}{r} R'(r) + \frac{8\pi^2 m}{h^2} \left(E - E_p\right) R(r) - \frac{\lambda}{r} R(r) = 0, \quad (10.118)$$

where λ is the separation constant and

$$\frac{1}{\sin \theta} \frac{\partial}{\partial \theta} \left(\sin \theta \frac{\partial \Psi(\theta, \phi)}{\partial \theta} \right) + \frac{1}{\sin^2 \theta} \frac{\partial \Psi(\theta, \phi)}{\partial \phi^2} + \lambda \Psi(\theta, \phi) = 0. \quad (10.119)$$

(3) Using Equation (10.119), applying separation of variables results in

$$\Theta''(\theta) + \cot\theta\Theta'(\theta) + \left(\lambda - \frac{\tau}{\sin^2\theta}\right)\Theta(\theta) = 0 \qquad (10.120)$$

and

$$\Phi''(\phi) + \Phi(\phi) = 0. \qquad (10.121)$$

(4) Show that $\Phi(\phi) = e^{i\widetilde{m}\phi}$, $\widetilde{m} = \pm 1, \pm 2, \pm 3, \pm\ldots$ is a solution of Equation (10.121). *Note:* $\tau = \left(\widetilde{m}\right)^2$. What conclusions can you make about this solution?

(5) Solve Equation (10.120) where $\tau = \left(\widetilde{m}\right)^2$. Also, set $x = \cos\theta$, $\Theta(\theta) = y(x)$; only consider the case where $\widetilde{m} = 0$ and $\lambda = n(n+1)$, $n = 0, 1, 2, 3, \ldots$ Graph the solution for several values of n. (*Hint:* After replacing $x = \cos\theta$ and $\Theta(\theta) = y(x)$, your new equation should be Legendre's equation; see Chapter 9.)

10.6 THE TELEGRAPHER'S EQUATION

The telegrapher's equation has a long history. It was first developed in 1855 by Lord Kelvin, who studied the Atlantic telegraph cable, which was completed in 1858. It is used today by modern physicists in the analysis of the time-dependent Boltzmann equation in the theory of neutron transport. Another use is in mathematical physiology. Here, the telegrapher's equation models electrical tranmission in a neuron. Once, we develop the standard telegrapher's equation, we will develop the neuron application.

10.6.1 Development of the Telegrapher's Equation

We develop the telegrapher's equation by considering the flow of electricity in a long transmission line. A transmission line is imperfectly insulated so that there is both capacitance and current leakage to the ground. The transmission line runs along the x-axis, and the line will contain the same amount of resistance, inductance, and capacitance for any small length, dx. The following is a list of the essential components, quite similar to the list given in Chapter 3:

- dx is the length of line from p_1 to p_2.

- $v(x,t)$ and $v(x + dx, t)$ are the potential or voltage at any point of the line.

- $i(x,t)$ and $i(x + dx, t)$ are the current along the line.

- R is the resistance per unit length.

- L is the inductance per unit length.

- G is the conductance to ground per unit length of the line.

- C is the capacitance to ground per unit length of line.

We will now develop the telegrapher's equation. The potential at p_2 is equal to the potential at p_1 minus the drop of the potential along the element $p_1 p_2$. Thus, using Kirchoff's first law we have

$$v(x + dx, t) = v(x, t) - R\ dx\ i(x, t) - L\ dx\ \frac{\partial i(x, t)}{\partial t}, \tag{10.122}$$

which may be written as

$$v(x + dx, t) - v(x, t) = -R\ dx\ i(x, t) - L\ dx\ \frac{\partial i(x, t)}{\partial t}. \tag{10.123}$$

Dividing by dx and taking the limit as $dx \to 0$, yields

$$\frac{\partial v(x, t)}{\partial x} = -R\ i(x, t) - L\ \frac{\partial i(x, t)}{\partial t}. \tag{10.124}$$

Similarly, the current at p_2 is equal to the current at p_1 minus the current loss through leakage to the ground and the current loss due to the capacitance. Thus, according to Kirchoff's second law we have

$$i(x + dx, t) = i(x, t) - G\ dx\ v(x, t) - C\ dx\ \frac{\partial v(x, t)}{\partial t}, \tag{10.125}$$

which may be written as

$$i(x + dx, t) - i(x, t) = -G\ dx\ v(x, t) - C\ dx\ \frac{\partial v(x, t)}{\partial t}. \tag{10.126}$$

Dividing by dx and taking the limit as $dx \to 0$, yields

$$\frac{\partial i(x, t)}{\partial x} = -G\ v(x, t) - C\ \frac{\partial v(x, t)}{\partial t}. \tag{10.127}$$

If we differentiate Equation (10.124) with respect to t and multiply it by C, and differentiate Equation (10.127) with respect to x, we have

$$
\begin{aligned}
C\frac{\partial^2 v(x, t)}{\partial t \partial x} &= -RC\ \frac{\partial i(x, t)}{\partial t} - LC\ \frac{\partial^2 i(x, t)}{\partial t^2} \\[2mm]
\frac{\partial^2 i(x, t)}{\partial x^2} &= -G\ \frac{v(x, t)}{\partial x} - C\ \frac{\partial^2 v(x, t)}{\partial x \partial t}.
\end{aligned}
\tag{10.128}
$$

If we subtract the first equation in Equation (10.128) from the second equation in Equation (10.128) and assume the mixed partials are identically equal, we get

$$\frac{\partial^2 i(x, t)}{\partial x^2} = LC\ \frac{\partial^2 i(x, t)}{\partial t^2} + RC\ \frac{\partial i(x, t)}{\partial t} - G\ \frac{v(x, t)}{\partial x}. \tag{10.129}$$

In Equation (10.129) if we substitute Equation (10.124) for $\dfrac{v(x,t)}{\partial x}$, the resulting equation is the telegrapher's equation for $i(x,t)$, which is

$$\frac{\partial^2 i(x,t)}{\partial x^2} = LC\,\frac{\partial^2 i(x,t)}{\partial t^2} + (RC + GL)\frac{\partial i(x,t)}{\partial t} + RGi(x,t). \tag{10.130}$$

Similarly, we can find the telegrapher's equation for $v(x,t)$, which is given in Chapter 3.

If we are working with submarine cables, then it is assumed that $L = G = 0$, and Equation (10.130) becomes

$$\frac{\partial^2 i(x,t)}{\partial x^2} = RC\frac{\partial i(x,t)}{\partial t}, \tag{10.131}$$

which is known as the telegrapher's equation for submarine cables.

10.6.2 Application of the Telegrapher's Equation to a Neuron

So that we are all using the same terminology, the following is a short description of a neuron and its functions. It is directly followed by the development of the telegrapher's equation for a neuron.[4]

A neuron is composed of three parts: the cell body, called soma, which contains the nucleus; the dendrites, which are numerous protoplasmic outgrowths from the soma; and the axon, a single extension longer than the dendrites, which ends in a brushlike filament.

The axon is covered with a fat-containing myelin sheath. The myelin sheath may be looked at as insulation. The axon carries the output of the neuron. The brush-like filaments at the end are called the synapse. The synapse are specialized for the transmission of an electrical signal to other neurons. This transmission to many other neurons is called divergence. The electrical transmission in the axon is active.

The dendrites are the receivers of the neuron. A single neuron may receive input from many other neurons, and this action is called convergence. Electrical current spreading through the dendrites is passive. Thus, the telegrapher's equation may be used to study the electrical current.

This is somewhat different than the normal telegrapher's equation because for a neuron we must consider the intracellular space, the extracellular space, and the cell membrane. In a neuron, the cell is viewed as a membrane surrounding the cytoplasm. The cytoplasm is the transmission line (denoted line), which is considered one-dimensional, and for our purposes the line runs along the x-axis. Wilfrid Rall called this the "core conductance assumption." The line is separated into small pieces of length dx. Along any length, all currents must balance.

---

[4]This subsection is adapted from James Keener and James Sneyd, Mathematical Physiology, ©1998 by Springer-Verlag, New York, pp. 251-256. Reprinted by permission.

The cell membrane acts or takes on the roll of the capacitor, denoted $C = \frac{q}{v}$ where q is the charge across the capacitor and v is the voltage potential. Since we have both intracellular and extracellular space, we must consider voltage and resistance of both. For the intracellular space, the voltage and resistance are denoted as $v_i(x)$ and $v_i(x + dx)$ and R_i dx. For the extracellular space, they are denoted as $v_e(x)$ and $v_e(x + dx)$ and R_e dx. Also, there are two currents to consider; the transmembrane current denoted I_t dx, which is the current running across the membrane, and the axial current denoted I_a, which has components in both the intracellular and extracellular space, denoted I_{a_i} and I_{a_e}, respectively, and considered linear functions of the voltage. There also exists an ionic current, denoted I_I (*Note:* When a neutral atom loses or gains one or more electrons, the loss of electrons results in a positively charged ion and the gain of electrons results in a negatively charged electron–gains and losses may occur during chemical reactions), which occurs on the cell membrane. The different functions are as follows:

- dx–small length of line

- $v_i(x, t)$ and $v_i(x + dx, t)$–intracellular space voltage along the small length dx

- R_i–intracellular space resistance along the small length dx

- $I_{a_i}(x, t)$–intracellular space axial current

- $v_e(x, t)$ and $v_e(x + dx, t)$–extracellular space voltage along the small length dx

- R_e–extracellular space resistance along the small length dx

- $I_{a_e}(x, t)$–extracellular space axial current

- I_t–transmembrane current

- C–cell membrane, which acts like a capacitor

- R_{c_i}–resistance of the cytoplasma in the intracellular space

- A_i–cross-sectional area of the line in the intracellular space

- R_{c_e}–resistance of the cytoplasma in the extracellular space

- A_e–cross-sectional area of the line in the extracellular space, which may be very large in comparison with A_i

- I_I–ionic current on the cell membrane

We start with derivation with the axial currents. The equations are similar to Equation (10.122), and they are

$$
\begin{aligned}
v_i(x + dx) &= v_i(x) - I_{a_i}(x, t) R_i dx \\
v_e(x + dx) &= v_e(x) - I_{a_e}(x, t) R_e dx,
\end{aligned}
\tag{10.132}
$$

which may be written as

$$v_i(x + dx) - v_i(x) = -I_{a_i}(x, t)R_i dx$$
$$v_e(x + dx) - v_e(x) = -I_{a_e}(x, t)R_e dx. \tag{10.133}$$

Dividing the equations in Equation (10.133) by dx and taking the limit as $dx \to 0$ yields

$$\frac{\partial v_i(x, t)}{\partial x} = -I_{a_i}(x, t)R_i$$
$$\frac{\partial v_e(x, t)}{\partial x} \quad -I_{a_e}(x, t)R_e. \tag{10.134}$$

Thus,

$$I_{a_i}(x, t) = -\frac{1}{R_i}\frac{\partial v_i(x, t)}{\partial x}$$
$$I_{a_e}(x, t) = -\frac{1}{R_e}\frac{\partial v_e(x, t)}{\partial x}. \tag{10.135}$$

We use Kirchoff's second law to calculate the change in axial current due to trans-membrane current. We must have

$$I_{a_i}(x, t) - I_{a_i}(x + dx, t) = I_t dx = I_{a_e}(x + dx, t) - I_{a_e}(x, t). \tag{10.136}$$

Dividing Equation (10.136) by dx and taking the limit as $dx \to 0$ yields

$$-\frac{\partial I_{a_i}(x, t)}{\partial x} = I_t = \frac{\partial I_{a_e}(x, t)}{\partial x}. \tag{10.137}$$

We also have that the total axial current $I_{a_T}(x, t) = I_{a_i}(x, t) + I_{a_e}(x, t)$ if no other sources of current exist in the line. Therefore, using the expressions for I_{a_i} and I_{a_e} given in Equation (10.135), we have

$$I_{a_T} = -\frac{1}{R_i}\frac{\partial v_i(x, t)}{\partial x} - \frac{1}{R_e}\frac{\partial v_e(x, t)}{\partial x}$$

or

$$-I_{a_T} = \frac{1}{R_i}\frac{\partial v_i(x, t)}{\partial x} + \frac{1}{R_e}\frac{\partial v_e(x, t)}{\partial x}. \tag{10.138}$$

Adding a factor of $1 = \dfrac{1}{R_e}\dfrac{\partial v_i(x,t)}{\partial x} - \dfrac{1}{R_e}\dfrac{\partial v_i(x,t)}{\partial x}$ to Equation (10.139) yields

$$-I_{a_T} = \frac{1}{R_i}\frac{\partial v_i(x,t)}{\partial x} + \frac{1}{R_e}\frac{\partial v_e(x,t)}{\partial x} + \frac{1}{R_e}\frac{\partial v_i(x,t)}{\partial x} - \frac{1}{R_e}\frac{\partial v_i(x,t)}{\partial x}$$

$$= \frac{1}{R_i}\frac{\partial v_i(x,t)}{\partial x} + \frac{1}{R_e}\frac{\partial v_i(x,t)}{\partial x} + \frac{1}{R_e}\frac{\partial v_e(x,t)}{\partial x} - \frac{1}{R_e}\frac{\partial v_i(x,t)}{\partial x}$$

$$= \left(\frac{1}{R_i} + \frac{1}{R_e}\right)\frac{\partial v_i(x,t)}{\partial x} + \frac{1}{R_e}\left(\frac{\partial v_e(x,t)}{\partial x} - \frac{\partial v_i(x,t)}{\partial x}\right)$$

$$= \left(\frac{R_i + R_e}{R_i R_e}\right)\frac{\partial v_i(x,t)}{\partial x} - \frac{1}{R_e}\left(\frac{\partial v_i(x,t)}{\partial x} - \frac{\partial v_e(x,t)}{\partial x}\right). \tag{10.139}$$

Also, we know that the total voltage is given by $v_T(x,t) = v_i(x,t) - v_e(x,t)$. Therefore, Equation (10.139) may be written as

$$-I_{a_T} = \left(\frac{R_i + R_e}{R_i R_e}\right)\frac{\partial v_i(x,t)}{\partial x} - \frac{1}{R_e}\frac{\partial v_T(x,t)}{\partial x}. \tag{10.140}$$

Equation (10.141) may be rewritten as

$$\frac{-R_e I_{a_T}}{R_i + R_e} = \frac{1}{R_i}\frac{\partial v_i(x,t)}{\partial x} - \frac{1}{R_i + R_e}\frac{\partial v_T(x,t)}{\partial x}$$

or

$$\frac{1}{R_i}\frac{\partial v_i(x,t)}{\partial x} = \frac{1}{R_i + R_e}\frac{\partial v_T(x,t)}{\partial x} - \frac{R_e}{R_i + R_e}I_{a_T}. \tag{10.141}$$

Since $I_{a_i}(x,t) = -\dfrac{1}{R_i}\dfrac{\partial v_i(x,t)}{\partial x}$, Equation (10.141) becomes

$$-I_{a_i}(x,t) = \frac{1}{R_i + R_e}\frac{\partial v_T(x,t)}{\partial x} - \frac{R_e}{R_i + R_e}I_{a_T}. \tag{10.142}$$

If we assume that in Equation (10.142) I_{a_T} is constant, which is a reasonable assumption, and use Equation (10.137), we arrive at

$$I_t = \frac{\partial}{\partial x}\left(\frac{1}{R_i + R_e}\frac{\partial v_T(x,t)}{\partial x}\right). \tag{10.143}$$

The transmembrane current, I_t, is equal to the capacitive current and the ionic current where the capacitive current is given by $C\dfrac{\partial v_T(x,t)}{\partial t}$ and the ionic current is given by I_I. Thus, Equation (10.143) may be written as

$$I_t = p\left(C\frac{\partial v_T(x,t)}{\partial x} + I_I\right) = \frac{\partial}{\partial x}\left(\frac{1}{R_i + R_e}\frac{\partial v_T(x,t)}{\partial x}\right),$$

which becomes

$$p\left(C\frac{\partial v_T(x,t)}{\partial t} + I_I\right) = \frac{\partial}{\partial x}\left(\frac{1}{R_i + R_e}\frac{\partial v_T(x,t)}{\partial x}\right). \tag{10.144}$$

Note: Equation (10.144) is very similar to the telegraph equation for submarine cables, Equation (10.131). Thus, Equation (10.144) is the telegraph or line equation for a neuron.

If we assume that R_i and R_e are constants, then Equation (10.144) becomes

$$C\frac{\partial v_T(x,t)}{\partial t} + I_I = \frac{1}{p\,(R_i + R_e)}\frac{\partial^2 v_T(x,t)}{\partial x^2}. \tag{10.145}$$

Equation (10.145) may be normalized by taking into account certain constraints on the membrane resistivity. Ignoring the extracellular resistance, and then nondimensionalizing the resulting equation. Thus, we get a line equation for a neuron, which is somewhat easier to work with; it is

$$\frac{\partial v}{\partial t} = \frac{\partial^2 v}{\partial x^2} + f(v,t), \tag{10.146}$$

where $f(v,t)$ is a function of voltage and time. (*Note:* This is in the true tradition of mathematical modeling. If you try to keep everything you sometimes end up with an unworkable model, or a model that predicts nothing or the wrong answer quite accurately. It follows the old military axiom: He who protects everything, protects nothing at all.)

Many times $f(v,t)$ can be modeled as $-v$ when we are working with passive activity, such as dendrites. For many other cells, the activity is called passive only if the membrane potential is sufficiently small. Thus, Equation (10.146) becomes

$$\frac{\partial v}{\partial t} = \frac{\partial^2 v}{\partial x^2} - v. \tag{10.147}$$

This concludes our development of the line equation for a neuron.

EXERCISES 10.6

10.6.1. Consider the telegrapher's equation

$$\frac{\partial^2 u(x,t)}{\partial t^2} = \frac{\partial^2 u(x,t)}{\partial x^2} - \alpha\frac{\partial u(x,t)}{\partial t} - \beta u(x,t), \tag{10.148}$$

where α and β are considered nonnegative constants, subject to

$$u(0,t) = 0 \text{ and } u(L,t) = 0 \tag{10.149}$$

or

$$\frac{\partial u(0,t)}{\partial x} = 0 \text{ and } \frac{\partial u(L,t)}{\partial x} = 0 \tag{10.150}$$

and

$$u(x,0) = f(x) \text{ and } \frac{\partial u(x,0)}{\partial t} = g(x). \tag{10.151}$$

(1) Solve the Telegrapher's equation for both sets of BCs, Equations (10.149 and 10.150).

(2) Multiply Equation (10.148) by $2\dfrac{\partial u(x,t)}{\partial t}$ and derive the differential identity

$$\frac{\partial}{\partial x}\left(2\frac{\partial u}{\partial t}\frac{\partial u}{\partial x}\right) - \frac{\partial}{\partial t}\left[\left(\frac{\partial u}{\partial x}\right)^2 + \left(\frac{\partial u}{\partial t}\right)^2 + bu^2\right] - 2a\left(\frac{\partial u}{\partial t}\right)^2 = 0.$$

(3) Prove that if $u(x,t)$ satisfies Equation (10.148) and either BCs, Equation (10.149), or BCs, Equation (10.150), then

$$\int_0^L \left[\left(\frac{\partial u}{\partial x}\right)^2 + \left(\frac{\partial u}{\partial t}\right)^2 + bu^2\right]_{t=T_0} dx \le \int_0^L \left[\left(\frac{\partial u}{\partial x}\right)^2 + \left(\frac{\partial u}{\partial t}\right)^2 + bu^2\right]_{t=0} dx.$$

(4) State and prove the uniqueness theorem for the initial value problem given by Equations (10.148, 10.149, and 10.151).

10.6.2. Given the Telegrapher's equation

$$\frac{1}{c^2}\frac{\partial^2 u(x,t)}{\partial t^2} = \frac{\partial^2 u(x,t)}{\partial x^2} - \alpha\frac{\partial u(x,t)}{\partial t} - \beta u(x,t).$$

Let $u_1(x,t) = u(x,t)$, $u_2(x,t) = \dfrac{\partial u(x,t)}{\partial x}$, and $u_3(x,t) = \dfrac{\partial u(x,t)}{\partial t}$ and show that $u_1(x,t)$, $u_2(x,t)$, and $u_3(x,t)$ satisfy the following system of three equations:

$$\frac{\partial u_1(x,t)}{\partial t} - u_3(x,t) = 0$$

$$\frac{\partial u_2(x,t)}{\partial t} - \frac{\partial u_3(x,t)}{\partial t} = 0$$

$$\frac{\partial u_3(x,t)}{\partial t} - c^2\left(\frac{\partial u_2(x,t)}{\partial x} + \alpha u_3(x,t) + \beta u_1(x,t)\right) = 0.$$

10.6.3. Solve the Telegrapher's equation

$$\frac{\partial^2 u(x,t)}{\partial t^2} = \frac{\partial^2 u(x,t)}{\partial x^2} - \frac{\partial u(x,t)}{\partial t} - u(x,t),$$

subject to

$$u(0,t) = 1 \text{ and } u(L,t) = \cos t$$

and

$$u(x,0) = x^3 - 2x + 1 \text{ and } \frac{\partial u(x,0)}{\partial t} = 0.$$

10.6.4. Solve the Telegrapher's equation

$$\frac{\partial^2 u(x,t)}{\partial t^2} = \frac{\partial^2 u(x,t)}{\partial x^2} - \frac{\partial u(x,t)}{\partial t} - u(x,t),$$

subject to

$$u(0,t) = \sin t \text{ and } \frac{\partial u(L,t)}{\partial x} = \cos t$$

and

$$u(x,0) = x \text{ and } \frac{\partial u(x,0)}{\partial t} = 1.$$

10.6.5. PROJECT:[5] Consider the line equation for a neuron

$$\frac{\partial v}{\partial t} = \frac{\partial^2 v}{\partial x^2} - v, \tag{10.152}$$

subject to the IC

$$v(x,0) = 0. \tag{10.153}$$

(1) Suppose the left boundary at $x = 0$ is a voltage-clamp BC. That is,

$$v(0,t) = v_b,$$

where $v_b = \xi$ volts, and at the right boundary, $x = L$ cm, there is a current injection. Thus, the BC is given as

$$\frac{\partial v(L,t)}{\partial x} = -R_i \sigma_m I(t),$$

where t is given in microseconds, R_i is the intracellular resistance, σ_m is in cm, and $I(t)$ is the current injection.

(2) Another set of BCs would be

$$\frac{\partial v(0,t)}{\partial x} = 0,$$

which means the injected current, $I(t) = 0$, and

$$v(L,t) - \gamma \frac{\partial v(L,t)}{\partial x} = R_s,$$

where the soma acts like a resistance. However, the soma membrane potential is the same at all points. These are two of the three assumptions of the "Rall lumped-soma model." *Note:* The BCs for this part of the problem are simplified.

---

[5]Adapted from James Keener and James Sneyd, Mathematical Physiology, ©1998 by Springer-Verlag, New York, pp. 251-256. Reprinted by permission.

10.7   INTERESTING PROBLEMS IN DIFFUSION

The most famous diffusion example you know is the heat equation, which was derived in Chapter 2. However, the diffusion equation can be used in many areas other than just the heat equation. This section is devoted to problems of this type. Although many of them do involve temperature distribution directly, you really want to stay focused on the primary structure of these problems, which is diffusion. We start and end our discussion with a rather simple example.

EXAMPLE 10.1. Suppose a tungsten rod of length π ft is perfectly insulated on its lateral surface and has an initial temperature distribution of

$$x, \qquad 0 \le x < \frac{\pi}{4};$$

$$\frac{\pi}{4}, \qquad \frac{\pi}{4} \le x < \frac{3\pi}{4};$$

and

$$\pi - x, \qquad \frac{3\pi}{4} \le x \le \pi.$$

Further suppose that the rod has no insulation on its ends and is plunged into a bath, which is held at the temperature of 0^0 F.

(1) Find the temperature distribution of the rod for all time.

(2) Consider the point $x = \frac{\pi}{2}$ and estimate the error made in replacing the series by its partial sum. Then, determine the time required for which the ratio of the sum of all terms, starting with the second, to the first term, is less than $\varepsilon > 0$.

Solution: Part (1), we mathematically model the physical problem. We have

$$\frac{\partial u}{\partial t} = 2.39 \frac{\partial^2 u}{\partial x^2}, \tag{10.154}$$

subject to

$$\begin{aligned} u(0,t) &= 0 \\ u(\pi,t) &= 0 \end{aligned} \tag{10.155}$$

and

$$u(0,t) = \begin{cases} x, & 0 \le x < \frac{\pi}{4} \\ \frac{\pi}{4}, & \frac{\pi}{4} \le x < \frac{3\pi}{4} \\ \pi - x, & \frac{3\pi}{4} \le x \le \pi. \end{cases} \tag{10.156}$$

Equation (10.155) subject to the homogeneous Dirichlet BCs has eigenvalues of

$$\lambda_n = n^2, \ n = 1, 2, 3, \ldots \tag{10.157}$$

Thus, the general solution for $u(x,t)$ is

$$u(x,t) = \sum_{n=1}^{\infty} b_n e^{-2.39n^2 t} \sin nx. \tag{10.158}$$

Using the IC, we solve for a specific solution, which is

$$u(x,t) = \sum_{n=1}^{\infty} \frac{4}{n^2 \pi} \left(\sin \frac{n\pi}{2} \cos \frac{n\pi}{4} \right) e^{-2.39n^2 t} \sin nx. \tag{10.159}$$

Part (2), we solve for $u\left(\dfrac{\pi}{2}, t\right)$, which is

$$u\left(\frac{\pi}{2}, t\right) = \sum_{n=1}^{\infty} \frac{4}{n^2 \pi} \left(\sin \frac{n\pi}{2} \cos \frac{n\pi}{4} \right) e^{-2.39n^2 t} \sin \frac{n\pi}{2}. \tag{10.160}$$

Equation (10.160) is 0 whenever n is even. Thus, we can rewrite Equation (10.160) as

$$u\left(\frac{\pi}{2}, t\right) = \sum_{m=0}^{\infty} \frac{4(-1)^m}{(2m+1)^2 \pi} \left(\sin \frac{(2m+1)\pi}{2} \cos \frac{(2m+1)\pi}{4} \right) e^{-2.39(2m+1)^2 t}. \tag{10.161}$$

The series on the right side of Equation (10.161) satisfies the alternating series test. Therefore, we know that if s is the sum of the series and s_n is the nth partial sum, we have

$$|s - s_n| \leq z_{n+1}$$

where z_{n+1} is the $n+1$ term of the series. Therefore,

$$\left| R_n \left(\frac{\pi}{2}, t \right) \right| = \left| \sum_{m=n+1}^{\infty} \frac{4(-1)^m}{(2m+1)^2 \pi} \left(\sin \frac{(2m+1)\pi}{2} \cos \frac{(2m+1)\pi}{4} \right) e^{-2.39(2m+1)^2 t} \right|$$

$$\leq \frac{4}{(2n+3)^2 \pi} \left(\sin \frac{(2n+3)\pi}{2} \cos \frac{(2n+3)\pi}{4} \right) e^{-2.39(2n+3)^2 t}$$

where $|s - s_n| = \left| R_n \left(\dfrac{\pi}{2}, t \right) \right|$.

We can now estimate the ratio of the sum of all the terms of the series starting with the second term to the first term. This yields

$$\frac{\left| R_0 \left(\dfrac{\pi}{2}, t \right) \right|}{\frac{4}{\pi} \left(\frac{\sqrt{2}}{2} \right) e^{-2.39 t}} \leq \frac{\frac{4}{9\pi} \left(\frac{\sqrt{2}}{2} \right) e^{-2.39(3)^2 t}}{\frac{4}{\pi} \left(\frac{\sqrt{2}}{2} \right) e^{-2.39 t}}$$

$$= \frac{1}{9} e^{-2.39(26) t} \leq \frac{1}{9} e^{-2.39(5^2) t} < \varepsilon.$$

Thus, for $t \geq t_0 = \dfrac{-1}{2.39(5^2)}\ln(9\varepsilon)$, we have the ratio of the sum of all terms starting with the second term to the first term less than $\varepsilon > 0$.   ∎

EXERCISES 10.7

10.7.1. Given a rod made of nickel with perfect lateral insulation and perfect insulation on the end $x = 0$. Suppose the other end of the rod, at $x = 15$ ft, has a convective heat exchange with a medium whose temperature is $0°$F.

    (1) Find the temperature distribution in the rod if the initial temperature distribution is $100°$F.

    (2) Estimate the error made in replacing the sum of the series, representing the point $x = \dfrac{15}{2}$ by its partial sum.

    (3) Determine the time at which a steady-state will occur at $x = \dfrac{15}{2}$ to a degree of accuracy ε.

10.7.2. Given a rod made of lead with no lateral insulation of length 5π m. Suppose the rod is held in a medium of $0°$C, and the ends of the rod are uninsulated..

    (1) Find the temperature distribution of the rod if the initial temperature of the rod is given as

$$f(x) = \begin{cases} x, & 0 \leq x < 3\pi \\ 3\pi, & 3\pi \leq x < 4\pi \\ 5\pi - x, & 4\pi \leq x \leq 5\pi \end{cases}$$

    and the convective constant of proportionality is 0.208.

    (2) Given the same rod, medium, and initial temperature distribution, suppose a heater moves with constant velocity of 5 cm/s along the rod. The flow of heat from the heater to the rod is $Q(t) = 350e^{-0.208t}$ degrees Celsius. Find the temperature distribution in the rod for all time.

10.7.3. PROJECT: Give the mathematical model for the cooling of a uniformly heated rod having the shape of a right circular cone of height L and base radius r. Assume the temperature is constant over the cross sectional area of the rod. Suppose the ends of the rod are perfectly insulated, and there is a lateral heat exchange between the surface of the rod and the medium the rod is in, whose temperature is $0°$C.

    (1) Suppose the height of the right circular cone is 2π m and its radius 0.5 m, and it is made of aluminum. Also, suppose there is perfect thermal

insulation on the lateral surface of the rod and that the ends of the rod are held at $0°C$. Find the temperature distribution if the initial temperature is uniformly $100°C$ throughout the rod.

(2) Suppose the cone is made by rotating the curve $y = 3e^{-x}$, $0 \le x \le 5$ about the x-axis. Suppose this cone is made of copper and is uniformly heated to a temperature of $300°C$. Further, suppose the lateral surface area and the end $x = 5$ is perfectly insulated and the end $x = 0$ is held at $0°C$. Find the temperature distribution in the cone for all time.

10.7.4. Given a parallelepiped made of asbestos with $0 \le x \le \pi$ ft, $0 \le y \le 2\pi$ ft, and $0 \le z \le 3\pi$ ft, find the temperature distribution if the sides of the parallelepiped are maintained at $0°F$ and the initial temperature distribution is $100°F$. Find the time at which a steady-state will occur at the center of the parallelepiped with relative accuracy $\varepsilon > 0$.

10.7.5. Consider a spherical shell made of mild steel with inner radius $r_1 = \pi$ m and outer radius of $r_2 = \dfrac{5\pi}{2}$ m. Suppose the inner and outer surfaces of the sphere have a convective heat exchange with a medium whose temperature is $0°C$, and that the initial temperature of the spherical shell is $r^2 + 1$ for $r_1 < r < r_2$.

10.7.6. Find the temperature distribution of a rod of length 2π m, that has perfect lateral insulation, consisting of two homogeneous materials: the first half of the rod is made of aluminum, the second half of the rod is made of silver. Suppose the left end of the rod is held at $0°C$ and the right end has a heat flow of $\sin t$ watts, where t is time, and the initial temperature of the rod is

$$f(x) = \begin{cases} \sin x, & 0 \le x < \pi \\ \\ 6\left(\cos x - 1\right), & \pi \le x \le 2\pi. \end{cases}$$

10.7.7. A spherical vessel filled with gas moves uniformly for a long time with velocity v_0, and then at time $t = 0$ it is stopped instantaneously and remains stationary. Find the vibrations of the gas in the vessel.

10.7.8. Find the vibrations of a gas in a spherical vessel of radius $r = 3$ produced by small deformations of the wall, beginning at time $t = 0$, if the velocities of the wall are radial and equal

$$6P_n(\cos \theta) \cos(\omega t).$$

Note: $P_n(\cos \theta)$ is a Legendre polynomial.

Chapter 11

Fourier Integrals
and Transform Methods

11.1 INTRODUCTION

In Chapter 3, we discussed d'Alembert's solution to the two-dimensional wave equation. The particular solution developed was for an infinite string. So far this is the only method we discussed for infinite boundaries. In the current chapter, we will develop methods that allow us to solve linear second-order PDEs with semi-infinite and infinite boundaries. These methods involve transforms.

You were first introduced to transforms in your ordinary differential equation course. There you solved linear initial-valued ODEs with Laplace transforms. The method involved transforming an initial-valued ODE into a space where the equation could be solved algebraically. Once the algebraic solution was found, the inverse Laplace transform is applied returning you to the space of the original problem. In ODEs, tables of Laplace transforms were used to simplify the process. Later, in this chapter, we will again visit Laplace transforms. Here, we use them as another solution method for initial-valued PDEs with finite and semi-infinite boundaries. Again tables of Laplace transforms are provided to simplify the process. However, unlike solving ODEs by Laplace transforms, solving PDEs by Laplace transforms does not generate an algebraic equation to be solved but an ODE one to be solved. Before we solve PDEs with Laplace transforms, we will introduce the Fourier integral.

The Fourier integral is a natural extension of Fourier series. By extension, we mean the representation of a piecewise-smooth function with a Fourier integral where the domain of the function is semi-infinite or infinite. This is quite different than Fourier series where the domain is typically $[-L, L]$. However, as in Fourier series representation of a function, there are certain restrictions on a function represented by a Fourier integral. These restrictions are discussed in some detail.

Once the Fourier integral is developed, the Fourier sine and cosine integrals are discussed. The Fourier sine and cosine integrals are the natural extensions of the Fourier sine and cosine series, respectively. Next, we introduce transform solution methods by considering the Laplace transform of PDEs with finite or semi-infinite boundaries. We then move on to a discussion of the Fourier transform. Finally, we use the Fourier and the Fourier sine and cosine transforms to solve PDEs with semi-infinite and infinite boundaries.

11.2   THE FOURIER INTEGRAL

In Chapter 4, we started our discussion of Fourier series. There, we discovered that the Fourier series representation of a piecewise-smooth function on $[-L, L]$ converged to the function wherever the function was continuous and to the average of the left and right values of the function at any jump discontinuity. In Chapter 6, we started with a discussion of Fourier series as the representation of a function and the Fourier series as a function itself. For instance, consider the function $x^3 - 3x^2 - 2x + 3$ on the interval $[-2, 2]$ shown in Figure (11.1). The Fourier series

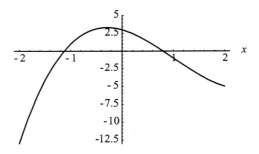

Figure 11.1: Graph of $x^3 - 3x^2 - 2x + 3$ on $[-2, 2]$.

representation of the function $x^3 - 3x^2 - 2x + 3$ is given by

$$-1 + \sum_{n=1}^{\infty} \frac{8(-1)^n}{(n\pi)^3} \left[-6n\pi \cos\left(\frac{n\pi x}{2}\right) + (12 - (n\pi)^2) \sin\left(\frac{n\pi x}{2}\right) \right], \qquad (11.1)$$

which is shown in Figure (11.2). Please note that Figure (11.2) exhibits the Gibbs phenomenon, and that at $x = \pm 2$ the Fourier series converges to -9.

The Fourier series representation of $x^3 - 3x^2 - 2x + 3$ as a function is considered is periodic on $[-\infty, \infty]$. Figure (11.3) shows this periodic nature of the Fourier series function using $x^3 - 3x^2 - 2x + 3$ on $[-2, 2]$ as the original function represented. Thus, we speak of a Fourier series involving "periodic" functions. The question which arises is; what do we do with a nonperiodic function on $[0, \infty]$ or $[-\infty, \infty]$, such as

$$f(x) = \begin{cases} 0, & -2 < x \text{ or } x > 2 \\ x^3 - 3x^2 - 2x + 3, & -2 \le x \le 2. \end{cases} \qquad (11.2)$$

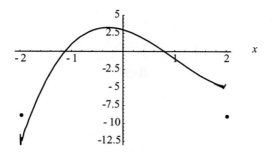

Figure 11.2: Graph of Fourier series representation of $x^3 - 3x^2 - 2x + 3$ on $[-2, 2]$ where $n = 150$.

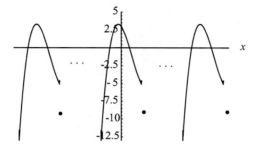

Figure 11.3: Graph of Fourier series function on $[-\infty, \infty]$ where $n = 150$.

Since the Function (11.2) is not periodic, it can't be represented by a Fourier series. However, a solution method is needed, and Fourier integrals were developed to handle exactly these type of cases.

11.2.1 Development of the Fourier Integral

We develop The Fourier integral with an intuitive approach. However, at the conclusion of our approach, we state the Fourier integral theorem. The Fourier integral theorem states all the sufficient conditions necessary for the Fourier integral representation of a function. The proof may be found in *Fourier series and Boundary Value Problems* by Ruel Churchill. We now start with our intuitive approach,

Suppose we have an arbitrary piecewise-smooth function $f(x)$ on the interval $[-L, L]$. Then, we may represent $f(x)$ by the Fourier series

$$f(x) = a_0 + \sum_{n=1}^{\infty} a_n \cos\left(\frac{n\pi x}{L}\right) + b_n \sin\left(\frac{n\pi x}{L}\right), \qquad (11.3)$$

where

$$a_0 = \frac{1}{2L} \int_{-L}^{L} f(x)\, dx,$$

$$a_n = \frac{1}{L} \int_{-L}^{L} f(x) \cos\left(\frac{n\pi x}{L}\right)\, dx,$$

and

$$b_n = \frac{1}{L} \int_{-L}^{L} f(x) \sin\left(\frac{n\pi x}{L}\right)\, dx.$$

If we let $\theta = \dfrac{n\pi}{L}$, and replace a_0, a_n, and b_n in Equation (11.3) with their integral representations, we have

$$f(x) = \frac{1}{2L} \int_{-L}^{L} f(x)\, dx \; + \sum_{n=1}^{\infty} \left\{ \left[\frac{1}{L} \int_{-L}^{L} f(x) \cos\left(\theta x\right)\, dx \right] \cos\left(\theta x\right) \right.$$

$$\left. + \left[\frac{1}{L} \int_{-L}^{L} f(x) \sin\left(\theta x\right)\, dx \right] \sin\left(\theta x\right) \right\}. \qquad (11.4)$$

Now, letting $\Delta\theta = \theta_{n+1} - \theta_n = \dfrac{(n+1)\pi}{L} - \dfrac{n\pi}{L} = \dfrac{\pi}{L}$ and substituting $\Delta\theta$ into Equation (11.4) yields

$$f(x) = \frac{1}{2L} \int_{-L}^{L} f(x)\, dx \; + \sum_{n=1}^{\infty} \left\{ \left[\frac{\Delta\theta}{\pi} \int_{-L}^{L} f(x) \cos\left(\theta x\right)\, dx \right] \cos\left(\theta x\right) \right.$$

$$\left. + \left[\frac{\Delta\theta}{\pi} \int_{-L}^{L} f(x) \sin\left(\theta x\right)\, dx \right] \sin\left(\theta x\right) \right\}$$

$$= \frac{1}{2L} \int_{-L}^{L} f(x)\, dx \; + \frac{1}{\pi} \sum_{n=1}^{\infty} \left\{ \left[\int_{-L}^{L} f(x) \cos\left(\theta x\right)\, dx \right] \Delta\theta \cos\left(\theta x\right) \right.$$

$$\left. + \left[\int_{-L}^{L} f(x) \sin\left(\theta x\right)\, dx \right] \Delta\theta \sin\left(\theta x\right) \right\}$$

$$= \frac{1}{2L} \int_{-L}^{L} f(x)\, dx \; + \frac{1}{\pi} \sum_{n=1}^{\infty} \left\{ \left[\int_{-L}^{L} f(x) \cos\left(\theta x\right)\, dx \right] \cos\left(\theta x\right) \right.$$

$$\left. + \left[\int_{-L}^{L} f(x) \sin\left(\theta x\right)\, dx \right] \sin\left(\theta x\right) \right\} \Delta\theta. \qquad (11.5)$$

Taking the limit of Equation (11.5) as $L \to \infty$ results in some interesting discoveries. First, in Equation (11.5),

$$\frac{1}{2L} \int_{-\infty}^{\infty} f(x)\, dx = 0.$$

Second,

$$\frac{1}{\pi}\sum_{n=1}^{\infty}\left\{\left[\int_{-\infty}^{\infty}f(x)\cos{(\theta x)}\,dx\right]\cos{(\theta x)}+\left[\int_{-\infty}^{\infty}f(x)\sin{(\theta x)}\,dx\right]\sin{(\theta x)}\right\}\Delta\theta$$

looks very similar to

$$\frac{1}{\pi}\sum_{n=1}^{\infty}F(\theta)\Delta\theta \tag{11.6}$$

where

$$F(\theta)=\left[\int_{-\infty}^{\infty}f(x)\cos{(\theta x)}\,dx\right]\cos{(\theta x)}+\left[\int_{-\infty}^{\infty}f(x)\sin{(\theta x)}\,dx\right]\sin{(\theta x)}.$$

Since $L\rightarrow\infty$ implies $\Delta\theta\rightarrow 0$, we could look at Equation (11.6) as

$$\frac{1}{\pi}\lim_{\Delta\theta\rightarrow 0}\sum_{n=1}^{\infty}F(\theta)\Delta\theta. \tag{11.7}$$

Equation (11.7) is very suggestive of the definition of the improper integral

$$\frac{1}{\pi}\int_{0}^{\infty}F(\theta)\,d\theta,$$

which becomes

$$f(x)\;=\;\frac{1}{\pi}\int_{0}^{\infty}\left\{\left[\int_{-\infty}^{\infty}f(x)\cos{(\theta x)}\,dx\right]\cos{(\theta x)}\right.$$
$$\left.+\left[\int_{-\infty}^{\infty}f(x)\sin{(\theta x)}\,dx\right]\sin{(\theta x)}\right\}d\theta. \tag{11.8}$$

Equation (11.8) is called the Fourier integral.

Remember, this was an intuitive approach and left many details out. These details are stated in Fourier's integral theorem.

Theorem 58. *(Fourier's integral theorem) If $f(x)$ is piecewise-smooth on every finite interval on the x-axis, and at every point of jump discontinuity the left and right hand limits exist; furthermore, suppose that $f(x)$ is absolutely integrable, that is,*

$$\int_{-\infty}^{\infty}|f(x)|\,dx$$

exists. Then, at every point x where the one-sided derivatives of $f(x)$ exist, the function $f(x)$ may be represented by the Fourier integral

$$f(x)\;=\;\frac{1}{\pi}\int_{0}^{\infty}\left\{\left[\int_{-\infty}^{\infty}f(\xi)\cos{(\theta\xi)}\,d\xi\right]\cos{(\theta x)}\right.$$
$$\left.+\left[\int_{-\infty}^{\infty}f(\xi)\sin{(\theta\xi)}\,d\xi\right]\sin{(\theta x)}\right\}d\theta. \tag{11.9}$$

Please note, in Theorem (58), that the absolutely integrable function, $f(x)$, eliminates all periodic functions except $f(x) = 0$. Also, a more convienient form of Equation (11.9) is

$$f(x) = \int_0^\infty \{A(\theta)\cos(\theta x) + B(\theta)\sin(\theta x)\}\, d\theta \qquad (11.10)$$

where

$$A(\theta) = \frac{1}{\pi} \int_{-\infty}^\infty f(\xi)\cos(\theta\xi)\, d\xi \qquad (11.11)$$

and

$$B(\theta) = \frac{1}{\pi} \int_{-\infty}^\infty f(\xi)\sin(\theta\xi)\, d\xi \qquad (11.12)$$

We demonstrate Theorem (58) with three examples.

EXAMPLE 11.1. Consider

$$f(x) = \begin{cases} 0, & -1 < x \text{ or } x > 1 \\ 1, & -1 \le x \le 0 \\ 2, & 0 < c \le 1, \end{cases} \qquad (11.13)$$

shown in Figure (11.4). The Fourier integral of the function, $f(x)$, given in Equation

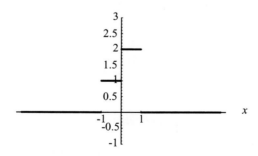

Figure 11.4: Graph of $f(x)$.

(11.13) is

$$f(x) = \int_0^\infty \{A(\theta)\cos(\theta x) + B(\theta)\sin(\theta x)\}\, d\theta,$$

where

$$A(\theta) = \frac{1}{\pi} \int_{-\infty}^{\infty} f(\xi) \cos(\theta\xi) \, d\xi$$

$$= \frac{1}{\pi} \left\{ \int_{-1}^{0} \cos(\theta\xi) \, d\xi + \int_{0}^{1} 2\cos(\theta\xi) \, d\xi \right\}$$

$$= \frac{3\sin(\theta)}{\theta\pi}$$

and

$$B(\theta) = \frac{1}{\pi} \int_{-\infty}^{\infty} f(\xi) \sin(\theta\xi) \, d\xi$$

$$= \frac{1}{\pi} \left\{ \int_{-1}^{0} \sin(\theta\xi) \, dx + \int_{0}^{1} 2\sin(\theta\xi) \, d\xi \right\}$$

$$= \frac{1 - \cos(\theta)}{\theta\pi}.$$

Therefore,

$$f(x) = \int_{0}^{\infty} \frac{3\sin(\theta)}{\theta\pi} \cos(\theta x) + \frac{1 - \cos(\theta)}{\theta\pi} \sin(\theta x) \, d\theta. \tag{11.14}$$

Most Fourier integrals are difficult to integrate. Thus, they are generally left in their integral form. However, many algebraic software packages can evaluate the integral numerically for plotting purposes. Figure (11.5) shows the graph of the Fourier integral representation of $f(x)$. The vertical bars from 0 to 1 at $x = -1$, from 1 to 2 at $x = 0$, and from 2 to 0 at $x = 1$ are a result of the mathematical software being used. ∎

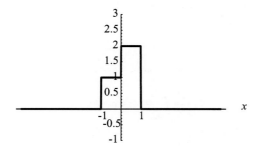

Figure 11.5: Graph of the Fourier integral of $f(x)$

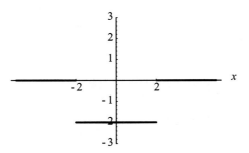

Figure 11.6: Graph of $g(x)$.

EXAMPLE 11.2. Consider

$$g(x) = \begin{cases} 0, & -2 < x \text{ or } x > 2 \\ -2, & -2 \le x \le 2, \end{cases} \tag{11.15}$$

shown in Figure (11.6). The Fourier integral of the function, $g(x)$, given in Equation (11.15) is

$$g(x) = \int_0^\infty \{A(\theta)\cos(\theta x) + B(\theta)\sin(\theta x)\} \, d\theta$$

where

$$\begin{aligned} A(\theta) &= \frac{1}{\pi}\int_{-\infty}^{\infty} g(\xi)\cos(\theta\xi)\,d\xi \\ &= \frac{1}{\pi}\int_{-2}^{2} -2\cos(\theta\xi)\,d\xi \\ &= -\frac{4\sin(2\theta)}{\theta\pi} \end{aligned}$$

and

$$\begin{aligned} B(\theta) &= \frac{1}{\pi}\int_{-\infty}^{\infty} g(\xi)\sin(\theta\xi)\,d\xi \\ &= \frac{1}{\pi}\int_{-2}^{2} -2\sin(\theta\xi)\,d\xi \\ &= 0. \end{aligned}$$

Therefore,

$$g(x) = \int_0^\infty -\frac{4\sin(2\theta)}{\theta\pi}\cos(\theta x)\,d\theta. \tag{11.16}$$

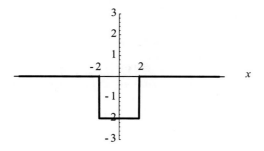

Figure 11.7: Graph of the Fourier integral of $g(x)$.

Figure (11.7) shows the graph of the Fourier integral representation of $g(x)$. The vertical bars from -2 to 0 at $x = \pm 2$ are a result of the mathematical software being used. ∎

EXAMPLE 11.3. Consider the function

$$
h(x) = \begin{cases} 0, & x < -1 \text{ or } x > 1 \\[2mm] -1, & -1 \le x < 0 \\[2mm] 1, & 0 \le x \le 1. \end{cases} \tag{11.17}
$$

Figure (11.8) shows the graph of the Equation (11.17). The Fourier integral of the

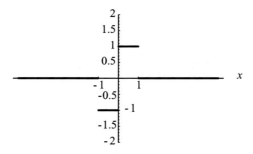

Figure 11.8: Graph of $h(x)$.

function $h(x)$ given in Equation (11.15) is

$$
h(x) = \int_0^\infty \{A(\theta)\cos(\theta x) + B(\theta)\sin(\theta x)\}\, d\theta,
$$

where

$$A(\theta) = \frac{1}{\pi} \int_{-\infty}^{\infty} h(\xi) \cos(\theta\xi) \, d\xi$$

$$= \frac{1}{\pi} \left\{ \int_{-1}^{0} -\cos(\theta\xi) \, d\xi + \int_{0}^{1} \cos(\theta\xi) \, d\xi \right\}$$

$$= 0$$

and

$$B(\theta) = \frac{1}{\pi} \int_{-\infty}^{\infty} h(\xi) \sin(\theta\xi) \, d\xi$$

$$= \frac{1}{\pi} \left\{ \int_{-1}^{0} -\sin(\theta\xi) \, d\xi + \int_{0}^{1} \sin(\theta\xi) \, d\xi \right\}$$

$$= \frac{2}{\pi} \left[\frac{1 - \cos(\theta)}{\theta} \right].$$

Therefore,

$$h(x) = \int_{0}^{\infty} \frac{2}{\pi} \frac{1 - \cos(\theta)}{\theta} \sin(\theta x) \, d\theta. \tag{11.18}$$

Figure (11.9) shows the graph of the Fourier integral representation of $h(x)$. The vertical bars from -1 to 0 at $x = -1$ and 1 to 0 at $x = 1$ are a result of the mathematical software being used. ∎

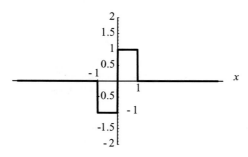

Figure 11.9: Graph of The Fourier integral of $h(x)$.

Fourier's integral,

$$f(x) = \int_{0}^{\infty} \{A(\theta) \cos(\theta x) + B(\theta) \sin(\theta x)\} \, d\theta, \tag{11.19}$$

where

$$A(\theta) = \frac{1}{\pi} \int_{-\infty}^{\infty} f(\xi) \cos(\theta\xi) \, d\xi \tag{11.20}$$

and

$$B(\theta) = \frac{1}{\pi} \int_{-\infty}^{\infty} f(\xi) \sin(\theta\xi)\, d\xi, \tag{11.21}$$

has other convenient forms. For instance,

$$f(x) = \frac{1}{\pi} \int_{0}^{\infty} \int_{-\infty}^{\infty} f(\xi) \cos(\theta(x - \xi))\, d\xi d\theta \tag{11.22}$$

is another form of Fourier's integral.

We develop Equation (11.22) from Equations (11.19, 11.20), and 11.21) in the following manner. First, Equations (11.20 and 11.21) for $A(\theta)$ and $B(\theta)$ are

$$A(\theta) = \frac{1}{\pi} \int_{-\infty}^{\infty} f(\xi) \cos(\theta\xi)\, d\xi \tag{11.23}$$

and

$$B(\theta) = \frac{1}{\pi} \int_{-\infty}^{\infty} f(\xi) \sin(\theta\xi)\, d\xi. \tag{11.24}$$

Next, we place Equations (11.23 and 11.24), $A(\theta)$ and $B(\theta)$, in Equation (11.19). This yields

$$f(x) \;=\; \frac{1}{\pi} \int_{0}^{\infty} \left\{ \left[\int_{-\infty}^{\infty} f(\xi) \cos(\theta\xi)\, du \right] \cos(\theta x) \right.$$
$$\left. + \left[\int_{-\infty}^{\infty} f(\xi) \sin(\theta\xi)\, d\xi \right] \sin(\theta x) \right\} d\theta. \tag{11.25}$$

Equation (11.25) may be rewritten as

$$f(x) \;=\; \frac{1}{\pi} \int_{0}^{\infty} \int_{-\infty}^{\infty} f(\xi) \left\{ \cos(\theta\xi) \cos(\theta x) \right.$$
$$\left. + \sin(\theta\xi) \sin(\theta x) \right\} d\xi d\theta. \tag{11.26}$$

Now, we consider part of the integrand of Equation (11.26), which is

$$\cos(\theta\xi) \cos(\theta x) + \sin(\theta\xi) \sin(\theta x). \tag{11.27}$$

Equation (11.27) reduces to

$$\cos(\theta\xi - \theta x) = \cos(\theta(\xi - x)) = \cos(\theta(x - \xi)) \tag{11.28}$$

by the cosine of the difference of two angles' identity and the fact that the cosine function is an even function. Finally, we place Equation (11.28) in its proper spot in Equation (11.26), which yields

$$f(x) = \frac{1}{\pi} \int_{0}^{\infty} \int_{-\infty}^{\infty} f(\xi) \cos(\theta(x - \xi))\, d\xi d\theta. \tag{11.29}$$

Equation (11.29) also has the form

$$f(x) = \frac{1}{2\pi} \int_{-\infty}^{\infty} \int_{-\infty}^{\infty} f(\xi) \cos\left(\theta\left(x - \xi\right)\right) d\xi d\theta. \tag{11.30}$$

Two other useful forms of Fourier's integral are

$$f(x) = \frac{1}{2\pi} \int_{-\infty}^{\infty} \int_{-\infty}^{\infty} f(\xi) e^{i\theta(x-\xi)} d\xi d\theta \tag{11.31}$$

and

$$f(x) = \frac{1}{2\pi} \int_{-\infty}^{\infty} e^{i\theta x} d\theta \int_{-\infty}^{\infty} f(\xi) e^{-i\theta\xi} d\xi. \tag{11.32}$$

Equations (11.31 and 11.32) are known as the complex form of Fourier's integral, and their derivation is left as exercises.

Since we have considered a Fourier integral as the limiting case of a Fourier series, it is natural to assume that the limiting case for the Fourier sine and cosine series are Fourier sine and cosine integrals. This topic is discussed in the next subsection.

11.2.2   The Fourier Sine and Cosine Integrals

In the previous subsection, we stated Fourier's integral theorem. Also, we gave the preferred form of Fourier's integral, which is

$$f(x) = \int_{0}^{\infty} \{A(\theta)\cos(\theta x) + B(\theta)\sin(\theta x)\} \, d\theta, \tag{11.33}$$

where

$$A(\theta) = \frac{1}{\pi} \int_{-\infty}^{\infty} f(\xi)\cos(\theta\xi) \, d\xi \tag{11.34}$$

and

$$B(\theta) = \frac{1}{\pi} \int_{-\infty}^{\infty} f(\xi)\sin(\theta\xi) \, d\xi. \tag{11.35}$$

In Example (11.2), the function,

$$g(x) = \begin{cases} 0, & -2 < x \text{ or } x > 2 \\ -2, & -2 \le x \le 2 \end{cases} \tag{11.36}$$

was represented by a Fourier integral. However, since $g(x)$ is an even function, $B(\theta)$ in Equation (11.35) equaled 0. Thus, $A(\theta)$ is an integral of $g(x)\cos(\theta x)$, which is the product of two even functions. Therefore, the integral $A(\theta)$ could be written as

$$A(\theta) = \frac{2}{\pi} \int_0^\infty g(\xi) \cos(\theta\xi) \, d\xi$$

$$= \frac{2}{\pi} \int_0^\infty -2 \cos(\theta\xi) \, d\xi$$

$$= \frac{2}{\pi} \int_0^2 -2 \cos(\theta\xi) \, d\xi. \tag{11.37}$$

Using Equation (11.37), we rewrite Equation (11.33) as

$$g(x) = \int_0^\infty A(\theta) \cos(\theta x) \, d\theta. \tag{11.38}$$

Equations (11.38 and 11.37) are an example of the Fourier cosine integral. Formally, the Fourier cosine integral for even functions, $f(x)$, is given as

$$f(x) = \int_0^\infty A(\theta) \cos(\theta x) \, d\theta \tag{11.39}$$

where

$$A(\theta) = \frac{2}{\pi} \int_0^\infty f(\xi) \cos(\theta\xi) \, d\xi. \tag{11.40}$$

Close examination of Example (11.3) indicates that $h(x)$ is an odd function. Thus, $A(\theta) = 0$. This means that Equations (11.33 and 11.35) form the Fourier sine integral. Formally, the Fourier sine integral for odd functions, $f(x)$, is given as

$$f(x) = \int_0^\infty B(\theta) \sin(\theta x) \, d\theta, \tag{11.41}$$

where

$$B(\theta) = \frac{2}{\pi} \int_0^\infty f(\xi) \sin(\theta\xi) \, d\xi. \tag{11.42}$$

In Example (11.4), we examine the Fourier cosine and sine integrals with the function $f(x) = 2e^{-x}$, where $0 \leq x$.

EXAMPLE 11.4. Consider the function

$$f(x) = 2e^{-x} \text{ where } 0 \leq x \tag{11.43}$$

shown in Figure (11.10). First, we find the Fourier cosine integral of $f(x)$. This

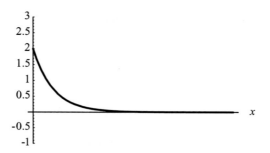

Figure 11.10: Graph of $f(x) = 2e^{-x}$, where $0 \leq x$.

means, we compute

$$
\begin{aligned}
A(\theta) &= \frac{2}{\pi} \int_0^\infty f(\xi) \cos(\theta\xi)\, d\xi \\[2mm]
&= \frac{2}{\pi} \int_0^\infty 2e^{-\xi} \cos(\theta\xi)\, d\xi \\[2mm]
&= \frac{4}{\pi} \lim_{b\to\infty} \int_0^b e^{-\xi} \cos(\theta\xi)\, d\xi \\[2mm]
&= \frac{4}{\pi} \left(\frac{1}{1+\theta^2} + \lim_{b\to\infty} \frac{e^{-b}}{1+\theta^2} \{\theta \sin(b\theta) - \cos(b\theta)\} \right) \\[2mm]
&= \frac{4}{\pi} \frac{1}{1+\theta^2}.
\end{aligned}
\tag{11.44}
$$

Thus, the Fourier cosine integral of $f(x) = 2e^{-x}$ is

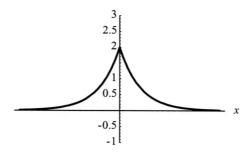

Figure 11.11: Graph of The Fourier cosine integral of $f(x) = 2e^{-x}$, where $0 \leq x$.

$$f(x) = \int_0^\infty A(\theta) \cos(\theta x) \, d\theta$$

$$= \int_0^\infty \frac{4}{\pi} \frac{1}{1+\theta^2} \cos(\theta x) \, d\theta$$

$$= \frac{4}{\pi} \int_0^\infty \frac{1}{1+\theta^2} \cos(\theta x) \, d\theta, \tag{11.45}$$

which is shown in Figure (11.11). Now, we find the Fourier sine integral of $f(x)$. This means, we compute the following:

$$B(\theta) = \frac{2}{\pi} \int_0^\infty f(\xi) \sin(\theta\xi) \, d\xi$$

$$= \frac{2}{\pi} \int_0^\infty 2e^{-\xi} \sin(\theta\xi) \, d\xi$$

$$= \frac{4}{\pi} \lim_{b \to \infty} \int_0^b e^{-\xi} \sin(\theta\xi) \, d\xi$$

$$= \frac{4}{\pi} \left(\frac{\theta}{1+\theta^2} - \lim_{b \to \infty} \frac{e^{-b}}{1+\theta^2} \{ \sin(b\theta) + \theta \cos(b\theta) \} \right)$$

$$= \frac{4}{\pi} \frac{\theta}{1+\theta^2}. \tag{11.46}$$

Thus, the Fourier sine integral of $f(x) = 2e^{-x}$ is

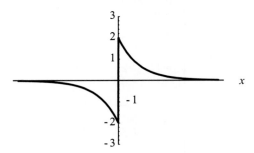

Figure 11.12: Graph of The Fourier sine integral of $f(x) = 2e^{-x}$, where $0 \le x$.

$$f(x) = \int_0^\infty B(\theta) \sin(\theta x) \, d\theta$$

$$= \int_0^\infty \frac{4}{\pi} \frac{\theta}{1+\theta^2} \sin(\theta x) \, d\theta$$

$$= \frac{4}{\pi} \int_0^\infty \frac{\theta}{1+\theta^2} \sin(\theta x) \, d\theta, \tag{11.47}$$

which is shown in Figure (11.12). ■

Another important form of the Fourier cosine integral is

$$f(x) = \frac{2}{\pi} \int_0^\infty \cos(\theta x)\, d\theta \int_0^\infty f(\xi) \cos(\theta \xi)\, d\xi, \tag{11.48}$$

where the variable of integration in the second integral was changed to ξ for convenience. Also, an important form of the Fourier sine integral is

$$f(x) = \frac{2}{\pi} \int_0^\infty \sin(\theta x)\, d\theta \int_0^\infty f(\xi) \sin(\theta \xi)\, d\xi. \tag{11.49}$$

Again, the variable of integration in the second integral was changed for convenience.

Before we proceed with the Fourier transform, our next section discusses Laplace transforms and how they apply to PDEs.

EXERCISES 11.2

11.2.1. Find the Fourier integral representation of the following functions and graph the solution using an algebraic software package:

(1)
$$f(x) = \begin{cases} 0, & |x| > 3 \\ -1, & |x| \le 3. \end{cases}$$

(2)
$$f(x) = \begin{cases} 0, & |x| > 5 \\ 1, & |x| \le 5. \end{cases}$$

(3)
$$f(x) = \begin{cases} -1, & -2 \le x < 0 \\ -2, & 0 \le x < 1 \\ 1, & 1 \le x < 2 \\ 0, & |x| > 2. \end{cases}$$

(4)
$$f(x) = \begin{cases} 1, & -2 \le x < 0 \\ 2, & 0 \le x < 1 \\ -1, & 1 \le x < 2 \\ 0, & |x| > 2. \end{cases}$$

$$(5) \qquad f(x) = \begin{cases} x, & 1 \le x < 0 \\ x^2, & 0 \le x < 1 \\ 0, & |x| > 1. \end{cases}$$

$$(6) \qquad f(x) = \begin{cases} \sin x, & |x| < \pi \\ 0, & |x| > \pi. \end{cases}$$

$$(7) \qquad f(x) = \begin{cases} \cos x, & |x| < \pi \\ 0, & |x| > \pi. \end{cases}$$

$$(8) \qquad f(x) = \begin{cases} 2 - x & -2 \le x < 0 \\ x, & 0 \le x < 1 \\ x - 2, & 1 \le x < 2 \\ 0, & |x| > 2. \end{cases}$$

$$(9) \qquad f(x) = \begin{cases} x^2 - 1, & -3 \le x < 0 \\ x - 1, & 0 \le x < 1 \\ 1 - x^2, & 1 \le x < 3 \\ 0, & |x| > 3. \end{cases}$$

$$(10) \qquad f(x) = \begin{cases} 0, & x < -1 \\ -x - 1, & -1 \le x < 0 \\ -e^{-x}, & 0 \le x. \end{cases}$$

11.2.2. Starting with Fourier's integral, Equation (11.9), show that

$$f(x) = \frac{1}{2\pi} \int_{-\infty}^{\infty} \int_{-\infty}^{\infty} f(\xi) e^{i\theta(x-\xi)} \, d\xi d\theta$$

is another form of Fourier's integral.

11.2.3. Starting with Fourier's integral, Equation (11.9), show that

$$f(x) = \frac{1}{2\pi} \int_{-\infty}^{\infty} e^{i\theta x} \, d\theta \int_{-\infty}^{\infty} f(\xi) e^{-i\theta \xi} \, d\xi$$

is another form of Fourier's integral.

11.2.4. This problem develops the Sine Integral,

$$\text{Si}\,(z) = \int_0^z \frac{\sin\theta}{\theta}\,d\theta.$$

(1) Find the Fourier integral representation of

$$f(x) = \begin{cases} x^2 - 1, & |x| < 1 \\ 0, & |x| > 1. \end{cases}$$

(2) Show that the solution to part (1),

$$\int_0^\infty \frac{\cos(\theta x)\sin\theta}{\theta}\,d\theta = \begin{cases} \frac{\pi}{2}, & 0 \le x < 1 \\ \frac{\pi}{4}, & x = 1 \\ 0, & 1 < x. \end{cases}$$

(3) Evaluate the integral from part (2) at $x = 0$.

(4) Determine that the integral in part (3) is the limit of the sine integral, and using an algebraic software package, graph the sine integral.

11.2.5. Find the Fourier sine and cosine integral representation of the following functions and graph the solution using an algebraic software package:

(1) $$f(x) = \begin{cases} 1, & 0 < x < 1 \\ 0, & 1 < x. \end{cases}$$

(2) $$f(x) = \begin{cases} x^3, & 0 < x < 1 \\ 0, & 1 < x. \end{cases}$$

(3) $$f(x) = \begin{cases} e^x, & 0 < x < 1 \\ 0, & 1 < x. \end{cases}$$

(4) $$f(x) = \begin{cases} 25 - x^2, & 0 < x < 5 \\ 0, & 5 < x. \end{cases}$$

(5)
$$f(x) = \begin{cases} x, & 0 < x < 2 \\ 4 - x, & 2 < x < 4 \\ 0, & 4 < x. \end{cases}$$

(6)
$$f(x) = \begin{cases} \dfrac{1}{x}, & 1 < x < 2 \\ 0, & \text{otherwise.} \end{cases}$$

(7)
$$f(x) = e^{-x} + e^{-2x}, \ 0 < x.$$

(8)
$$f(x) = e^{-x^2}, \ 0 < x.$$

(9)
$$f(x) = \begin{cases} \sin x, & 0 < x < \pi \\ 0, & \text{otherwise.} \end{cases}$$

(10)
$$f(x) = \begin{cases} \cos x, & 0 < x < \pi \\ 0, & \text{otherwise.} \end{cases}$$

11.2.6. Starting with Fourier's cosine integral, Equation (11.39), show that

$$f(x) = \frac{2}{\pi} \int_0^\infty \cos(\theta x)\, d\theta \int_0^\infty f(\xi \cos(\theta \xi)\, d\xi$$

is another form of Fourier's cosine integral.

11.2.7. Starting with Fourier's sine integral, Equation (11.41), show that

$$f(x) = \frac{2}{\pi} \int_0^\infty \sin(\theta x)\, d\theta \int_0^\infty f(\xi) \sin(\theta \xi)\, d\xi.$$

is another form of Fourier's sine integral.

11.3   THE LAPLACE TRANSFORM

When you first encountered Laplace transforms, you learned that they were used for solving ODEs. There you learned to take the transform of the ODE, which turned the ODE into an algebraic problem. You solved the algebraic problem. Then, took the inverse Laplace transform to obtain the answer for the ODE. In this section, we are going to solve PDEs with both finite and semi-infinite spatial boundaries with Laplace transforms. The method is very similar to that for ODEs. However, there are a few minor differences. Instead of getting an algebraic expression after transforming a PDE, we get an ODE, which we must solved.

Before we solve a PDE with Laplace transforms, we'll state several important theorems, list some known properties, and refresh our memories by solving an ODE. (*Note:* The proofs for all the properties may be found in any standard ODE text.) Next, we discuss the error function, which is also known as the probability integral. Finally, we solve PDEs using the Laplace transform.

11.3.1   Laplace transform Solution Method of ODEs

You should remember that the basis of a Laplace transform is an integral equation. In fact, the integral is an improper one. This means that the integral over the unbounded interval is defined as a limit of integrals over finite intervals. For example,

$$\int_0^\infty f(t)\,dt = \lim_{b\to\infty} \int_0^b f(t)\,dt,$$

where b is a positive real number. If the integral exists from 0 to β for each $\beta < b$, and if the $\lim_{b\to\infty}$ exists, then the integral is said to converge to the limiting value. If any part the previous statement is false, then the integral is said to diverge. Given this brief description of an improper integral, we state the primary theorem for Laplace transforms.

Theorem 59. *Suppose that f is a piecewise-smooth function on the interval $0 \le t \le T$ for any $T \in \mathbb{R}^+$ and $|f(t)| \le ke^{\alpha t}$ when $t \ge M$ and where $\alpha \in \mathbb{R}$ and M and $k \in \mathbb{R}^+$. Then, the Laplace transform is given by*

$$\mathcal{L}\{f(t)\} = F(s) = \int_0^\infty e^{-st} f(t)\,dt, \tag{11.50}$$

where $s > \alpha$.

Theorem (59) defines the Laplace transform. However, when we solve initial value problems, it is useful to have the following theorem:

Theorem 60. *Suppose that the functions f, f', f'', ..., f^n are continuous and that f^{n+1} is piecewise-smooth on the interval $0 \le t \le T$ for any $T \in \mathbb{R}^+$. Also, suppose that there exist real constants k, α, and M such that $|f(t)| \le ke^{\alpha t}$, $|f'(t)| \le ke^{\alpha t}$, ..., $|f^n(t)| \le ke^{\alpha t}$ when $t \ge M$. Then, $\mathcal{L}\{f^{n+1}(t)\}$ exists for $s > \alpha$ and is given by*

$$\mathcal{L}\{f^{n+1}(t)\} = s^{n+1}\mathcal{L}\{f(t)\} - s^n f(0) - \cdots - sf^{n-1}(0) - f^n(0). \tag{11.51}$$

The following list contains several more properties of Laplace transforms. You should be familiar with all of them.

Properties of Laplace transforms

$$\mathcal{L}\{af(t)\} = a\mathcal{L}\{f(t)\}.$$

$$\mathcal{L}\{f(t) + g(t)\} = \mathcal{L}\{f(t)\} + \mathcal{L}\{g(t)\}.$$

Properties of Laplace transforms (continued)

$$\mathcal{L}\left\{f'(t)\right\} = sF(s) - f(0).$$

$$\mathcal{L}\left\{f''(t)\right\} = s^2 F(s) - sf(0) - f'(0).$$

$$\mathcal{L}\left\{e^{bt} f(t)\right\} = F(s - b).$$

$$\mathcal{L}\left(\int_0^t f(t')\, dt'\right) = \frac{1}{s} F(s).$$

$$\mathcal{L}\left\{tf(t)\right\} = -\frac{dF(s)}{ds}.$$

$$\mathcal{L}\left\{\frac{1}{t} f(t)\right\} = \int_s^\infty F(s')\, ds'.$$

Theorems (59 and 60) and the other properties of Laplace transforms provide the necessary tools for solving initial-value ODE problems. Consider the following example:

EXAMPLE 11.5. Consider

$$u''(t) + 2u'(t) - 3u(t) = 0, \tag{11.52}$$

subject to

$$u(0) = 1 \text{ and } u'(0) = -2. \tag{11.53}$$

The Laplace transform of Equation (11.52) is

$$\mathcal{L}\left\{u''(t)\right\} + 2\mathcal{L}\left\{u'(t)\right\} - 3\mathcal{L}\left\{u(t)\right\} = 0.$$

Using Theorem (60) and the initial conditions, Equation (11.53), the Laplace transform of Equation (11.52) becomes

$$s^2 \mathcal{L}\left\{u(t)\right\} - su(0) - u'(0) + 2\left[s\mathcal{L}\left\{u(t)\right\} - u(0)\right] - 3\mathcal{L}\left\{u(t)\right\} = 0$$

or

$$\left(s^2 + 2s - 3\right) F(s) - (s + 2)u(0) - u'(0).$$

Solving for $F(s)$ yields

$$F(s) = \frac{s}{s^2 + 2s - 3} = \frac{s}{(s + 3)(s - 1)}.$$

Therefore, taking the inverse Laplace transform we have

$$u(t) = \mathcal{L}^{-1}\left\{\frac{s}{(s + 3)(s - 1)}\right\} = \frac{3e^{-3t} + e^t}{4}$$

where we used the table of Laplace transforms in Appendix F. ∎

It should be noted that transforms come in pairs. Theorem (59) gives a definition of the Laplace transform. The inverse Laplace transform requires knowledge of contour integration. However, the inverse Laplace transform is

$$f(t) = \mathcal{L}^{-1}\{F(s)\} \frac{1}{2\pi i} \int_{\alpha-i\infty}^{\alpha+i\infty} F(s)e^{st}\,ds. \qquad (11.54)$$

This short review is certainly not meant to be complete, just a refresher. For a complete review of this material, I suggest that you read your ODE text. We now move on to an important integral, the error function, which is needed for even an introduction to Laplace transform methods for PDEs.

11.3.2   The Error Function

The error function plays a very important role in many areas of engineering and science. The error function is denoted by erf (x), where

$$\text{erf}\,(x) = \frac{2}{\sqrt{\pi}} \int_0^x e^{-u^2}\,du = 1 - \frac{2}{\sqrt{\pi}} \int_0^\infty e^{-u^2}\,du.$$

Also, there is a complementary error function, denoted erfc (x), where

$$\text{erfc}\,(x) = \frac{2}{\sqrt{\pi}} \int_0^\infty e^{-u^2}\,du = 1 - \text{erf}\,(x).$$

It should be noted that erf $(0) = 0$ and erfc $(0) = 1$. Following is a proof of erfc $(0) = 1$, from which we find that erf $(0) = 0$.

Proof. Let

$$I = \frac{2}{\sqrt{\pi}} \int_0^\infty e^{-x^2}\,dx,$$

and consider I^2. Thus,

$$I^2 = \frac{2}{\sqrt{\pi}} \int_0^\infty e^{-x^2}\,dx \frac{2}{\sqrt{\pi}} \int_0^\infty e^{-y^2}\,dy$$

, where we changed the variable of integration in the second integral. I^2 may also be written as

$$\frac{4}{\pi} \int_0^\infty \int_0^\infty e^{-(x^2+y^2)}\,dx\,dy, \qquad (11.55)$$

transferring to polar coordinates where $r^2 = x^2 + y^2$ and $\theta = \tan^{-1}\frac{y}{x}$. Also, changing to polar coordinates requires a change to the limits of integration. Thus, for r, we have $0 < r < \infty$, and for θ, we have $0 < \theta < \dfrac{\pi}{2}$. Thus, Equation (11.55) becomes

$$\frac{4}{\pi} \int_0^{\frac{\pi}{2}} \int_0^{\infty} e^{-r^2} r \, dr \, d\theta = \frac{4}{\pi} \int_0^{\frac{\pi}{2}} \left[-\frac{1}{2} e^{-r^2} \right]_0^{\infty} d\theta$$

$$= \frac{2}{\pi} \int_0^{\frac{\pi}{2}} d\theta$$

$$= \left[\frac{2}{\pi} \theta \right]_0^{\frac{\pi}{2}} = 1.$$

$\square$

Figure (11.13) shows the graph of erf (x) and erfc (x).

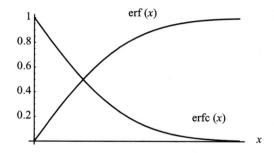

Figure 11.13: The Graph of erf (x) and erfc (x).

In Figure (11.13), please note that as $x \to \infty$, erf $(x) \to 1$ and erfc $(x) \to 0$. Also, in Appendix F, there are several useful Laplace transforms of the complementary error function. These transforms will be needed as you solve the problems at the end of this section and chapter. Next, we look at Laplace transforms and PDEs.

11.3.3   Laplace Transform Solution Method of PDEs

For this section, we work with PDEs that are functions of two variables, x and t. We immediately note that a Laplace transform is designed to work with one variable. However, we learned in Chapter 6, The Calculus of Fourier series, that we could take the derivative of a Fourier series with respect to a parameter t. We will use that same type of technique here. Only now the parameter is x. Thus, using Theorem (59), the definition of the Laplace transform of a function of two variables, $u(x,t)$ becomes

$$\mathcal{L}\left\{u(x,t)\right\} = U(x,s) = \int_0^{\infty} e^{-st} u(x,t) \, dt. \tag{11.56}$$

Definition (11.56) is demonstrated by two examples. Remember, that an integral over an unbounded interval is called an improper integral, and it must be defined as the limit of integrals over finite intervals.

EXAMPLE 11.6. Find the Laplace transform of $u(x, t) = e^{-3t} \sin(\pi x)$. By Definition (11.56), we have

$$\mathcal{L}\left\{u(x, t)\right\} = \int_0^\infty e^{-st} e^{-3t} \sin(\pi x)\, dt = \sin(\pi x) \lim_{b \to \infty} \int_0^b e^{-(s+3)t}\, dt$$

$$= \sin(\pi x) \lim_{b \to \infty} \left[\frac{-1}{s+3} e^{-(s+3)t} \right]_0^b$$

$$= \frac{-\sin(\pi x)}{s+3} \lim_{b \to \infty} \left[e^{-(s+3)b} - 1 \right] = \frac{\sin(\pi x)}{s+3}. \qquad \blacksquare$$

EXAMPLE 11.7. Find the Laplace transform of $\cos(x+t)$. By Definition (11.56), we have

$$\mathcal{L}\left\{u(x, t)\right\} = \int_0^\infty e^{-st} \cos(x+t)\, dt$$

$$= \lim_{b \to \infty} \int_0^b e^{-st} (\cos x \cos t - \sin x \sin t)\, dt$$

$$= \lim_{b \to \infty} \left[\int_0^b e^{-st} \cos x \cos t\, dt - \int_0^b e^{-st} \sin x \sin t\, dt \right]$$

$$= \cos x \lim_{b \to \infty} \int_0^b e^{-st} \cos t\, dt - \sin x \lim_{b \to \infty} \int_0^b e^{-st} \sin t\, dt$$

$$= \cos x \lim_{b \to \infty} \left[\frac{e^{-st}}{s^2+1} (-s \cos t + \sin t) \right]_0^b$$

$$- \sin x \lim_{b \to \infty} \left[\frac{e^{-st}}{s^2+1} (-s \sin t - \cos t) \right]_0^b$$

$$= \frac{\cos x}{s^2+1} \lim_{b \to \infty} \left[e^{-sb} (-s \cos b + \sin b) + s \right]$$

$$-\frac{\sin x}{s^2+1}\lim_{b\to\infty}\left[e^{-sb}\left(-s\sin b-\cos b\right)+1\right]$$

$$=\frac{s\cos x}{s^2+1}-\frac{\sin x}{s^2+1}=\frac{s\cos x-\sin x}{s^2+1}.\qquad\blacksquare$$

Examples (11.6 and 11.7) demonstrate the Laplace transform of a function of two variables where one variable is a parameter. We now move on to the Laplace transform solution method of a PDE.

If we assume Theorem (60) holds true for the Laplace transform of a PDE, which it does, then we have for $\dfrac{\partial u(x,t)}{\partial t}$, $\dfrac{\partial^2 u(x,t)}{\partial t^2}$, and $\dfrac{\partial^2 u(x,t)}{\partial x^2}$ the Laplace transforms;

$$\mathcal{L}\left\{\frac{\partial u(x,t)}{\partial t}\right\}=s\mathcal{L}\left\{u(x,t)\right\}-u(x,0)=sU(x,s)-u(x,0),\qquad(11.57)$$

$$\mathcal{L}\left\{\frac{\partial^2 u(x,t)}{\partial t^2}\right\}=s^2\mathcal{L}\left\{u(x,t)\right\}-su(x,0)-\frac{\partial u(x,0)}{\partial t}$$

$$=s^2U(x,s)-su(x,0)-\frac{\partial u(x,0)}{\partial t},\qquad(11.58)$$

and

$$\mathcal{L}\left\{\frac{\partial^2 u(x,t)}{\partial x^2}\right\}=\int_0^\infty e^{-st}\frac{\partial^2 u(x,t)}{\partial x^2}\,dt$$

$$=\int_0^\infty \frac{\partial^2}{\partial x^2}\left[e^{-st}u(x,t)\right]\,dt$$

$$=\frac{d^2}{dx^2}\int_0^\infty\left[e^{-st}u(x,t)\right]\,dt=\frac{d^2U(x,s)}{dx^2},\qquad(11.59)$$

respectively.

A close examination of Equations (11.57 and 11.58) indicates that the Laplace transform is ideally suited for PDEs with initial conditions, in particular, the heat and wave equations in one spatial dimension where the spatial variable has either finite or semi-infinite BCs. Also, Equation (11.59) informs us that the Laplace transform of a linear second-order PDE is a linear second-order ODE. The first two examples demonstrate this important concept.

EXAMPLE 11.8. Suppose that we have a very thin rod with perfect lateral insulation of length L. Further suppose that the rod has an initial temperature distribution of $\alpha\sin\left(\dfrac{\pi x}{L}\right)$ and the ends are held in a $0°$ bath. Find the time-dependent solution for $u(x, t)$.

The physical model may be described by the mathematical equations

$$\frac{\partial u(x, t)}{\partial t} = \frac{\partial^2 u(x, t)}{\partial x^2},$$

subject to

$$u(0, t) = 0 \text{ and } u(L, t) = 0$$

and

$$u(x, 0) = \alpha \sin\left(\frac{\pi x}{L}\right).$$

First, we find the Laplace transforms of all equations. We have, by Equations (11.57 and 11.59),

$$\mathcal{L}\left\{\frac{\partial u(x, t)}{\partial t}\right\} = s\mathcal{L}\left\{u(x, t)\right\} - u(x, 0) = sU(x, s) - u(x, 0)$$

$$= sU(x, s) - \alpha \sin\left(\frac{\pi x}{L}\right) \tag{11.60}$$

and

$$\mathcal{L}\left\{\frac{\partial^2 u(x, t)}{\partial x^2}\right\} = \frac{d^2 U(x, s)}{dx^2}. \tag{11.61}$$

Also, the BCs become

$$\mathcal{L}\left\{u(0, t)\right\} = U(0, s) = 0 \text{ and } \mathcal{L}\left\{u(L, t)\right\} = U(L, s) = 0. \tag{11.62}$$

Next, we combine Equations (11.60 and 11.61). This yields a nonhomogeneous linear second-order ODE, which is

$$\frac{d^2 U(x, s)}{dx^2} - sU(x, s) = -\alpha \sin\left(\frac{\pi x}{L}\right). \tag{11.63}$$

We solve Equation (11.63) by breaking it into homogeneous and particular parts. The homogeneous part is

$$\frac{d^2 U_h(x, s)}{dx^2} - sU_h(x, s) = 0,$$

which has the solution

$$U_h(x, s) = c_1 \sinh(\sqrt{s}x) + c_2 \cosh(\sqrt{s}x). \tag{11.64}$$

For the particular part, we use the method of undetermined coefficients. This method may be found in Appendix C. The solution for the particular part is

$$U_p(x, s) = \frac{\alpha L^2}{sL^2 + \pi^2} \sin\left(\frac{\pi x}{L}\right). \tag{11.65}$$

Combining Equations (11.64 and 11.65) yields

$$U(x, s) = U_h(x, s) + U_p(x, s)$$

$$= c_1 \sinh(\sqrt{s}x) + c_2 \cosh(\sqrt{s}x) + \frac{\alpha L^2}{sL^2 + \pi^2} \sin\left(\frac{\pi x}{L}\right). \tag{11.66}$$

Applying the BCs, Equation (11.62), implies $c_1 = c_2 = 0$. Therefore,

$$U(x, s) = \frac{\alpha L^2}{sL^2 + \pi^2} \sin\left(\frac{\pi x}{L}\right). \tag{11.67}$$

Finally, we find the inverse Laplace transform of Equation (11.67), which yields the time-dependent solution $u(x, t)$. *Note:* A table of inverse Laplace transforms may be found in Appendix F. We have

$$u(x, t) = \mathcal{L}^{-1}\{U(x, s)\} = \mathcal{L}^{-1}\left\{\frac{\alpha L^2}{sL^2 + \pi^2} \sin\left(\frac{\pi x}{L}\right)\right\}$$

$$= \alpha e^{-\frac{\pi^2 t}{L^2}} \sin\left(\frac{\pi x}{L}\right). \tag{11.68}$$

Equation (11.68) is exactly what we should expect as the answer. ■

EXAMPLE 11.9. Consider a perfectly flexible string of length L with the ends attached to frictionless sleeves, which move vertically up and down. Further, suppose the string has no initial displacement. However, the initial velocity is given as $\cos\left(\frac{\pi x}{L}\right)$. Find the displacement $u(x, t)$.

The physical model may be described by the mathematical equations

$$\frac{\partial^2 u(x, t)}{\partial t^2} = \frac{\partial^2 u(x, t)}{\partial x^2},$$

subject to

$$\frac{\partial u(0, t)}{\partial x} = 0 \text{ and } \frac{\partial u(L, t)}{\partial x} = 0$$

and

$$u(x, 0) = 0 \text{ and } \frac{\partial u(x, 0)}{\partial t} = \cos\left(\frac{\pi x}{L}\right).$$

First, we find the Laplace transforms of all equations. We have, by Equations (11.58

and 11.59),

$$\mathcal{L}\left\{\frac{\partial^2 u(x,t)}{\partial t^2}\right\} = s^2 \mathcal{L}\left\{u(x,t)\right\} - su(x,0) - \frac{\partial u(x,0)}{\partial t}$$

$$= s^2 U(x,s) - su(x,0) - \frac{\partial u(x,0)}{\partial t}$$

$$= s^2 U(x,s) - \cos\left(\frac{\pi x}{L}\right) \qquad (11.69)$$

and

$$\mathcal{L}\left\{\frac{\partial^2 u(x,t)}{\partial x^2}\right\} = \frac{d^2 U(x,s)}{dx^2}. \qquad (11.70)$$

Also, the BCs become

$$\mathcal{L}\left\{\frac{\partial u(0,t)}{\partial x}\right\} = \frac{dU(0,s)}{dx} = 0 \text{ and } \mathcal{L}\left\{\frac{\partial u(L,t)}{\partial x}\right\} = \frac{dU(L,s)}{dx} = 0. \qquad (11.71)$$

Next, we combine Equations (11.69 and 11.70). This yields a nonhomogeneous linear second-order ODE, which is

$$\frac{d^2 U(x,s)}{dx^2} - s^2 U(x,s) = -\cos\left(\frac{\pi x}{L}\right). \qquad (11.72)$$

We solve Equation (11.72) by breaking it into homogeneous and particular parts. The homogeneous part is

$$\frac{d^2 U_h(x,s)}{dx^2} - s^2 U_h(x,s) = 0,$$

which has the solution

$$U_h(x,s) = c_1 \sinh(sx) + c_2 \cosh(sx). \qquad (11.73)$$

For the particular part, we use the method of undetermined coefficients. This method may be found in Appendix C. The solution for the particular part is

$$U_p(x,s) = \frac{L^2}{s^2 L^2 + \pi^2} \cos\left(\frac{\pi x}{L}\right). \qquad (11.74)$$

Combining Equations (11.73 and 11.74) yields

$$U(x,s) = U_h(x,s) + U_p(x,s)$$

$$= c_1 \sinh(sx) + c_2 \cosh(sx) + \frac{L^2}{s^2 L^2 + \pi^2} \cos\left(\frac{\pi x}{L}\right). \qquad (11.75)$$

Applying the BCs, Equation (11.71), implies $c_1 = c_2 = 0$. Therefore,

$$U(x, s) = \frac{L^2}{s^2 L^2 + \pi^2} \cos\left(\frac{\pi x}{L}\right). \tag{11.76}$$

Finally, we find the inverse Laplace transform of Equation (11.76), which yields the time-dependent solution $u(x, t)$. We have

$$u(x, t) = \mathcal{L}^{-1}\left\{U(x, s)\right\} = \mathcal{L}^{-1}\left\{\frac{L^2}{s^2 L^2 + \pi^2} \cos\left(\frac{\pi x}{L}\right)\right\}$$

$$= \frac{L}{\pi} \sin\left(\frac{\pi t}{L}\right) \cos\left(\frac{\pi x}{L}\right). \tag{11.77}$$

Again, Equation (11.77) is exactly what we should expect as the answer. ∎

In the next example, we solve a linear second-order PDE where the spatial variable has semi-infinite boundaries. A semi-infinite boundary has one boundary at $x = 0$. The other boundary is at infinity.

EXAMPLE 11.10. Consider a semi-infinite solid. The temperature distribution is determined from the heat equation if the flow is one-dimensional. Suppose that this is the case, and the initial temperature of the solid is $0°$. Further suppose that the left boundary, $x = 0$, has a constant flux of heat into the solid of $-\alpha$. Solve for the temperature distribution, $u(x, t)$ in the solid.

The physical model may be described by the mathematical equations

$$\frac{\partial u(x, t)}{\partial t} = k \frac{\partial^2 u(x, t)}{\partial x^2},$$

subject to

$$\frac{\partial u(0, t)}{\partial x} = -\alpha \text{ and } \lim_{x \to \infty} u(x, t) = 0$$

and

$$u(x, 0) = 0$$

First, we find the Laplace transforms of all equations. We have, by Equations (11.57 and 11.59),

$$\mathcal{L}\left\{\frac{\partial u(x, t)}{\partial t}\right\} = s\mathcal{L}\left\{u(x, t)\right\} - u(x, 0) = sU(x, s) - u(x, 0)$$

$$= sU(x, s) \tag{11.78}$$

and

$$\mathcal{L}\left\{k\frac{\partial^2 u(x, t)}{\partial x^2}\right\} = k\frac{d^2 U(x, s)}{dx^2}. \tag{11.79}$$

Also, the BCs become

$$\mathcal{L}\left\{\frac{\partial u(0,t)}{\partial x}\right\} = \frac{dU(0,s)}{dx} = -\frac{\alpha}{s} \text{ and } \mathcal{L}\left\{\lim_{x\to\infty} u(x,t)\right\} = \lim_{x\to\infty} U(x,s) = 0. \quad (11.80)$$

Next, we combine Equations (11.78 and 11.79). This yields a homogeneous linear second-order ODE, which is

$$k\frac{d^2 U(x,s)}{dx^2} - sU(x,s) = 0. \quad (11.81)$$

We usually write the solution of Equation (11.81) in terms of hyperbolic sine and hyperbolic cosine functions. However, in this case, it is easier to write the solution in terms of exponential functions. Therefore, we have

$$U(x,s) = c_1 e^{-x\sqrt{s}/\sqrt{k}} + c_2 e^{-x\sqrt{s}/\sqrt{k}}. \quad (11.82)$$

Applying the first BC, $\dfrac{dU(0,s)}{dx} = -\dfrac{\alpha}{s}$, from Equation (11.80) to Equation (11.82) yields

$$\frac{\alpha\sqrt{k}}{s\sqrt{s}} + c_2 = c_1.$$

Thus, $U(x,s)$ may be written as

$$U(x,s) = \frac{\alpha\sqrt{k}}{s\sqrt{s}} e^{-x\sqrt{s}/\sqrt{k}} + c_2\left(e^{-x\sqrt{s}/\sqrt{k}} + e^{x\sqrt{s}/\sqrt{k}}\right), \quad (11.83)$$

Applying the second BC, $\lim_{x\to\infty} U(x,s) = 0$, from Equation (11.80) to Equation (11.82) implies $c_2 = 0$. Thus, the solution is

$$U(x,s) = \frac{\alpha\sqrt{k}}{s\sqrt{s}} e^{-x\sqrt{s}/\sqrt{k}}. \quad (11.84)$$

Finally, we find the inverse Laplace transform of Equation (11.84), which yields the time-dependent solution $u(x,t)$. We have

$$u(x,t) = \mathcal{L}^{-1}\left\{U(x,s)\right\} = \mathcal{L}^{-1}\left\{\alpha\sqrt{k}\frac{e^{-x\sqrt{s}/\sqrt{k}}}{s\sqrt{s}}\right\}$$

$$= \alpha\left[2\sqrt{\frac{kt}{\pi}} e^{\frac{-x^2}{4t}} - x\left\{erfc\left(\frac{x}{2\sqrt{kt}}\right)\right\}\right] \quad (11.85)$$

where the inverse Laplace transform as found in Appendix F.   ∎

The last two sections of this chapter deal with the Fourier transform. In the next section, we discuss the Fourier transform and the properties of the Fourier transform, which are similar to those of the Laplace transform. In the final section, we use Fourier transforms to solve PDEs.

EXERCISES 11.3

11.3.1. Find the Laplace transform of the following ODEs:

(1) $u''(x) - u'(x) - 4u(x) = 0$; $u(0) = 1$, $u'(0) = -1$.

(2) $u''(x) - 3u'(x) + u(x) = 0$; $u(0) = 1$, $u'(0) = 0$.

(3) $u''(x) + 2u'(x) - 5u(x) = 0$; $u(0) = 3$, $u'(0) = 2$.

(4) $u''(x) - 4u'(x) + 4u(x) = 0$; $u(0) = 1$, $u'(0) = 1$.

(5) $u''(x) - 2u'(x) + 4u(x) = e^{-x}$; $u(0) = 0$, $u'(0) = 2$.

11.3.2. Show that

$$\frac{e^{-a\sqrt{s}}}{s}$$

is the Laplace transform of

$$\text{erfc}\left(\frac{a}{2\sqrt{t}}\right).$$

11.3.3. Show that

$$\frac{e^{-a\sqrt{s}}}{s\sqrt{s}}$$

is Laplace transform of

$$2\sqrt{\frac{t}{\pi}}e^{\frac{-a^2}{4t}} - a\left\{\text{erfc}\left(\frac{a}{2\sqrt{t}}\right)\right\}.$$

11.3.4. Show that

$$\frac{e^{-a\sqrt{s}}}{\sqrt{s}\left(\sqrt{s} + b\right)}$$

is the Laplace transform of

$$e^{ab}e^{b^2 t}\left\{\text{erfc}\left(b\sqrt{t} + \frac{a}{2\sqrt{t}}\right)\right\}.$$

11.3.5. Show that

$$\frac{be^{-a\sqrt{s}}}{s\left(\sqrt{s}+b\right)}$$

is the Laplace transform of

$$-e^{ab}e^{b^2t}\left\{\text{erfc}\left(b\sqrt{t}+\frac{a}{2\sqrt{t}}\right)\right\}+\text{erfc}\left(\frac{a}{2\sqrt{t}}\right).$$

11.3.6. Prove the following properties about the error function:

(1) $\text{erf}\left(-x\right)=-\text{erf}\left(x\right)$.

(2) $\dfrac{d}{dx}\left(\text{erf}\left(x\right)\right)=\dfrac{2e^{-x^2}}{\sqrt{\pi}}$.

(3) $\displaystyle\int \text{erf}\left(x\right)dx=x\text{erf}\left(x\right)+\dfrac{e^{-x^2}}{\sqrt{\pi}}+c$.

(4) $\text{erf}\left(x\right)=\dfrac{2}{\sqrt{\pi}}\sum_{n=0}^{\infty}\dfrac{(-1)^n x^{2n+1}}{n!(2n+1)}$.

11.3.7. Solve the following boundary value problems

(1) $\dfrac{\partial u(x,t)}{\partial t}=4\dfrac{\partial^2 u(x,t)}{\partial x^2}$, $x>0$, $t>0$,
subject to $u(0,t)=4$; $\lim\limits_{x\to\infty}u(x,t)=0$ and $u(x,0)=10$.

(2) $\dfrac{\partial u(x,t)}{\partial t}=k_0\dfrac{\partial^2 u(x,t)}{\partial x^2}$, $x>0$, $t>0$,
subject to $u(0,t)=\alpha$; $\lim\limits_{x\to\infty}u(x,t)=\beta$ and $u(x,0)=\gamma$.

(3) $\dfrac{\partial u(x,t)}{\partial t}=2\dfrac{\partial^2 u(x,t)}{\partial x^2}$, $x>0$, $t>0$,
subject to $\dfrac{\partial u(0,t)}{\partial x}=0$; $\lim\limits_{x\to\infty}u(x,t)=0$ and $u(x,0)=-40$.

(4) $\dfrac{\partial u(x,t)}{\partial t}=k_0\dfrac{\partial^2 u(x,t)}{\partial x^2}$, $x>0$, $t>0$,
subject to $\dfrac{\partial u(0,t)}{\partial x}=\alpha$; $\lim\limits_{x\to\infty}u(x,t)=\beta$ and $u(x,0)=\gamma$.

(5) $\dfrac{\partial u(x,t)}{\partial t}=\dfrac{\partial^2 u(x,t)}{\partial x^2}$, $x>0$, $t>0$,
subject to $\dfrac{\partial u(0,t)}{\partial x}=0$; $\lim\limits_{x\to\infty}\dfrac{\partial u(x,t)}{\partial x}=0$ and $u(x,0)=25$.

(6) $\dfrac{\partial u(x,t)}{\partial t} = k_0 \dfrac{\partial^2 u(x,t)}{\partial x^2}$, $x > 0$, $t > 0$,

subject to $\dfrac{\partial u(0,t)}{\partial x} = \alpha$; $\lim\limits_{x \to \infty} \dfrac{\partial u(x,t)}{\partial x} = \beta$ and $u(x,0) = \gamma$.

(7) $\dfrac{\partial u(x,t)}{\partial t} = .22 \dfrac{\partial^2 u(x,t)}{\partial x^2}$, $x > 0$, $t > 0$,

subject to $u(0,t) = 0$; $\lim\limits_{x \to \infty} \dfrac{\partial u(x,t)}{\partial x} = -10$ and $u(x,0) = \gamma$.

(8) $\dfrac{\partial u(x,t)}{\partial t} = k_0 \dfrac{\partial^2 u(x,t)}{\partial x^2}$, $x > 0$, $t > 0$,

subject to $u(0,t) = \alpha$; $\lim\limits_{x \to \infty} \dfrac{\partial u(x,t)}{\partial x} = \beta$ and $u(x,0) = \gamma$.

11.3.8. Solve the following boundary value problems:

(1) $\dfrac{\partial^2 u(x,t)}{\partial t^2} = 4 \dfrac{\partial^2 u(x,t)}{\partial x^2}$, $x > 0$, $t > 0$,

subject to $u(0,t) = 0$; $\lim\limits_{x \to \infty} \dfrac{\partial u(x,t)}{\partial x} = 0$ and $u(x,0) = 0$;

$\dfrac{\partial u(x,0)}{\partial t} = -24$.

(2) $\dfrac{\partial^2 u(x,t)}{\partial t^2} = c^2 \dfrac{\partial^2 u(x,t)}{\partial x^2}$, $x > 0$, $t > 0$,

subject to $u(0,t) = \alpha$; $\lim\limits_{x \to \infty} \dfrac{\partial u(x,t)}{\partial x} = \beta$ and $u(x,0) = 0$;

$\dfrac{\partial u(x,0)}{\partial t} = -v_0$.

(3) $\dfrac{\partial^2 u(x,t)}{\partial t^2} = 9 \dfrac{\partial^2 u(x,t)}{\partial x^2}$, $x > 0$, $t > 0$,

subject to $\dfrac{\partial u(0,t)}{\partial x} = 0$; $\lim\limits_{x \to \infty} \dfrac{\partial u(x,t)}{\partial x} = 0$ and $u(x,0) = 0$;

$\dfrac{\partial u(x,0)}{\partial t} = -\sin(\pi x)$.

(4) $\dfrac{\partial^2 u(x,t)}{\partial t^2} = c^2 \dfrac{\partial^2 u(x,t)}{\partial x^2}$, $x > 0$, $t > 0$,

subject to $\dfrac{\partial u(0,t)}{\partial x} = \alpha$; $\lim\limits_{x \to \infty} \dfrac{\partial u(x,t)}{\partial x} = \beta$ and $u(x,0) = 0$;

$\dfrac{\partial u(x,0)}{\partial t} = -v_0$.

(5) $\dfrac{\partial^2 u(x,t)}{\partial t^2} = 16 \dfrac{\partial^2 u(x,t)}{\partial x^2}$, $x > 0$, $t > 0$,

subject to $\dfrac{\partial u(0,t)}{\partial x} = 0$; $\lim_{x \to \infty} u(x,t) = 0$ and $u(x,0) = 0$;

$$\dfrac{\partial u(x,0)}{\partial t} = -\cos(\pi x).$$

(6) $\dfrac{\partial^2 u(x,t)}{\partial t^2} = c^2 \dfrac{\partial^2 u(x,t)}{\partial x^2}$, $x > 0$, $t > 0$,

subject to $\dfrac{\partial u(0,t)}{\partial x} = \alpha$; $\lim_{x \to \infty} u(x,t) = \beta$ and $u(x,0) = 0$;

$$\dfrac{\partial u(x,0)}{\partial t} = -v_0 4.$$

(7) $\dfrac{\partial^2 u(x,t)}{\partial t^2} = 2\dfrac{\partial^2 u(x,t)}{\partial x^2}$, $x > 0$, $t > 0$,

subject to $\dfrac{\partial u(0,t)}{\partial x} = 0$; $\lim_{x \to \infty} u(x,t) = 0$ and $u(x,0) = xe^{-x}$;

$$\dfrac{\partial u(x,0)}{\partial t} = 0.$$

(8) $\dfrac{\partial^2 u(x,t)}{\partial t^2} = c^2 \dfrac{\partial^2 u(x,t)}{\partial x^2}$, $x > 0$, $t > 0$,

subject to $\dfrac{\partial u(0,t)}{\partial x} = \alpha$ $\lim_{x \to \infty} u(x,t) = \beta$ and $u(x,0) = f(x)$;

$$\dfrac{\partial u(x,0)}{\partial t} = g(x).$$

11.3.9. Suppose you have a uniform tightly stretched string from $x = 0$ to $x = L$. Find the displacement of the string if the initial displacement is $A\sin\left(\dfrac{\pi x}{L}\right)$ and the initial velocity is 0.

11.3.10. A uniform bar is clamped at $x = 0$ and is initially at rest. If a constant force of 10 lbs is applied to the free end at $x = 2\pi$, the longitudinal displacement $u(x,t)$ of a cross section of the bar is determined from

$$\dfrac{\partial^2 u(x,t)}{\partial t^2} = \dfrac{\partial^2 u(x,t)}{\partial x^2}, \quad 0 < x < 2\pi, \ t > 0,$$

subject to

$$u(0,t) = 0; \quad \dfrac{\partial u(2\pi,t)}{\partial x} = 10 \text{ and } u(x,0) = 0; \quad \dfrac{\partial u(x,0)}{\partial t} = 0.$$

Solve for $u(x,t)$ using Laplace transforms.

11.3.11. An infinite porous slab of width 2 is immersed in a solution of constant concentration C_0. A dissolved substance in the solution diffuses into the slab at a constant diffusion rate of D. Determine the concentration $C(x,t)$ in the slab.

11.3.12. A flash burn at $x = 0$ on an uniform rod of semi-infinte length may be described mathematically as

$$\frac{\partial u(0, t)}{\partial x} = -A\delta(t),$$

where $\delta(t)$ is the Dirac delta function. Suppose you have an uniform steel rod of semi-infinte length, which has perfect lateral insulation, and there is a flash burn of $1200°C$ (approximate temperature of burning magnesium mixed with white gun powder) at the end $x = 0$. Find the time-dependent solution.

11.3.13. Given a semi-infinite perfectly elastic uniform string where the string is initially at rest. Determine the displacement, $u(x, t)$, if the end $x = 0$ has a boundary condition of

$$f(t) = \begin{cases} \sin(\pi t), & 0 \le t \le 1 \\ 0, & t > 1. \end{cases}$$

11.3.14. Solve the boundary value problem

$$\frac{\partial u(x, t)}{\partial t} = \frac{\partial^2 u(x, t)}{\partial x^2},$$

subject to

$$\frac{\partial u(0, t)}{\partial x} = 100 - u(0, t); \ \lim_{x \to \infty} u(x, t) = 0 \text{ and } u(x, 0) = 0.$$

11.3.15. Consider a one-dimensional uniform rod of length L, which has no lateral insulation in a medium of temperature u_0. Determine the temperature distribution in the rod, $u(x, t)$ if the initial temperature in the rod is a constant u_1.

11.3.16. Consider a very long telephone transmission line initially at a constant potential u_0. If the line is grounded at $x = 0$ and insulated at the far end, find the potential $u(x, t)$ for any time t. *Hint:* See Equation (3.15) where $L = 0$.

11.3.17. PROJECT: A semi-infinite uniform tightly-stretched string is initially at rest. Suppose the end at $x = 0$ is fixed at 0 and a concentrated load of magnitude F_0 moves along the string with a constant velocity v, starting at the point $x = 0$ at $t = 0$.

(1) Find the displacement of the string for any time t if the force is acting positive y-direction where $v \ne c$. *Hint:* This is a nonhomogeneous wave equation where the where the external force is give as

$$F_0 \delta\left(t - \frac{x}{v}\right).$$

(2) Find the displacement of the string for any time t of the force is acting negative y-direction where $v \neq c$.

(3) Repeat part (1) and (2) where $v = c$.

(4) Suppose $F_0 = 1$. Use mathematical software of your choice to graph the solution when $v = \frac{1}{2}c$; when $v = c$; when $v = \frac{3}{2}c$.

11.3.18. PROJECT: Find the steady state temperature distribution in a thin uniformly insulated sheet which corresponds with the upper half plane where the temperature is maintained at

$$
u(x,0) = \begin{cases} 100 & |x| < \pi \\ 0 & |x| > \pi \end{cases}
$$

and $u(x,y) \to 0$ as $y \to \infty$. *Hint:* Use Laplace's equation. Also, there is not an infinite set of eigenvalues, instead there is a continuous family of solutions.

11.3.19. PROJECT:

(1) Suppose you have a homogeneous uniform one-dimensional rod with perfect lateral insulation and its ends, at $x = 0$ and $x = L$, in 0 degree baths. Further suppose that the initial temperature of the rod is αx, for α a real constant. Show that the temperature distribution in the rod, $u(x,t)$ may be expressed both by a Fourier series and by using Laplace transforms. *Hint:* The Laplace transform should be the infinite sum of the difference of two error functions, erf.

(2) In a one page paper, discuss which solution you think will converge more quickly for a short period of time and for a long period of time.

(3) Suppose your rod is made of cast iron and is 10 cm long, calculate the temperature of the midpoint of the rod when

  (a) $\alpha = 50$, $t = 10$ seconds, $t = 1$ min, $t = 10$ min, and $t = 50$ min.

  (b) $\alpha = -50$, $t = 10$ sec, $t = 1$ min, $t = 10$ min, and $t = 50$ min.

  In each case, determine if your suspected convergence in Part (2) is correct.

(4) Suppose your rod is made of cast iron and is 10 cm long, calculate the temperature of the midpoint of the rod when the initial condition is given as

$$
u(x,0) = \begin{cases} x, & 0 \leq x < 5 \\ 10 - x, & 5 \leq x < 10 \end{cases}
$$

11.3.20. PROJECT:[1]

(1) Find the solution of the semi-infinite cable equation for a neuron

$$\frac{\partial v}{\partial t} = \frac{\partial^2 v}{\partial x^2} - v, \tag{11.86}$$

subject to the clamped voltage IC

$$v(x, 0) = 0 \tag{11.87}$$

with current injection at $x = 0$

$$\frac{\partial v(0, t)}{\partial x} = -R_i \sigma_m I_0 H(t).$$

where t, is given in microseconds, R_i is the intracellular resistance, σ_m is in cm, and $I_0 H(t)$ is the current injection where $H(t)$ is the heavysided function.

Hint: Use the identity

$$\frac{2}{s\sqrt{s+1}} = \frac{1}{s+1-\sqrt{s+1}} - \frac{1}{s+1+\sqrt{s+1}}.$$

(2) Show that as $t \to \infty$

$$v(x, t) \to R_i \sigma_m I_0 e^{-x}$$

11.4 THE FOURIER TRANSFORM

In the previous section, we mentioned that transforms come in pairs. For instance, the Laplace transform of a function, $f(x)$, is

$$\mathcal{L}\{f(t)\} = F(s) = \int_0^\infty e^{-st} f(t)\, dt, \tag{11.88}$$

and the Inverse Laplace transform of $F(s)$ is

$$f(t) = \mathcal{L}^{-1}\{F(s)\} \frac{1}{2\pi i} \int_{\alpha - i\infty}^{\alpha + i\infty} F(s) e^{st}\, ds. \tag{11.89}$$

The pair concept is very true for Fourier transforms. However, the Fourier transform pair is somewhat more obvious than the Laplace transform pair. If we consider the Fourier integral form in Equation (11.32), which is restated here

$$f(x) = \frac{1}{2\pi} \int_{-\infty}^{\infty} e^{i\theta x}\, d\theta \int_{-\infty}^{\infty} f(\xi) e^{-i\theta\xi}\, d\xi, \tag{11.90}$$

---

[1] Adapted from James Keener and James Sneyd, Mathematical Physiology, ©1998 by Springer-Verlag, New York, pp. 261-262. Reprinted by permission.

then the integral

$$\int_{-\infty}^{\infty} f(\xi)e^{-i\theta\xi}\, d\xi$$

is a function of θ and is known as the Fourier transform of $f(x)$. Thus, the Fourier transform is stated as in the following definition:

Definition 61.

$$\mathcal{F}\{f(x)\} = F(\theta) = \int_{-\infty}^{\infty} f(\xi)e^{-i\theta\xi}\, d\xi \qquad (11.91)$$

Since Equation (11.90) produces the Fourier integral of $f(x)$, the inverse Fourier transform is given by Definition (62)

Definition 62.

$$f(x) = \mathcal{F}(\theta)^{-1} = \frac{1}{2\pi}\int_{-\infty}^{\infty} F(\theta)e^{i\theta x}\, d\theta. \qquad (11.92)$$

Just as in the Laplace transform pair, finding the Fourier transform of a function, $f(x)$, is much easier than finding the inverse Fourier transform, which involves contour integration. Contour integration is a standard topic in a complex variables course and will not be discussed here. However, we do include several examples where we find and graph the Fourier transform of a function.

EXAMPLE 11.11. Suppose

$$f(x) = \begin{cases} 1, & |x| < 5 \\ \\ 0, & |x| > 5. \end{cases} \qquad (11.93)$$

Then, the Fourier transform is

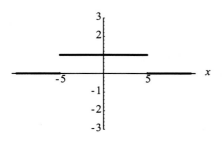

Figure 11.14: The graph of $f(x)$.

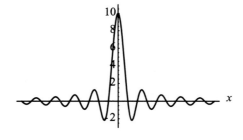

Figure 11.15: The Fourier transform, $F(\theta)$ of $f(x)$.

$$\mathcal{F}\{f(x)\} = F(\theta) \quad = \quad \int_{-\infty}^{\infty} f(\xi) e^{-i\theta\xi}\, d\xi$$

$$= \quad \int_{-5}^{5} e^{-i\theta\xi}\, d\xi = \left. \frac{e^{-i\theta\xi}}{-i\theta} \right|_{-5}^{5}$$

$$= \quad \frac{e^{i5\theta} - e^{-i5\theta}}{i\theta} = 2\frac{\sin 5\theta}{\theta}, \qquad (11.94)$$

where $\theta \neq 0$. For $\theta = 0$, $F(\theta) = 10$. Figure (11.14) is the graph of the function $f(x)$. Figure (11.15) is the graph of the Fourier transform, $F(\theta)$, of the function. ■

In general, if

$$f(x) = \begin{cases} 1, & |x| < a \\ 0, & |x| > a, \end{cases} \qquad (11.95)$$

then, the Fourier transform is

$$\mathcal{F}\{f(x)\} = \frac{e^{ia\theta} - e^{-ia\theta}}{i\theta} = 2\frac{\sin a\theta}{\theta}, \qquad (11.96)$$

where $\theta \neq 0$. For $\theta = 0$, $F(\theta) = 2\alpha$.

EXAMPLE 11.12. Consider the function

$$f(x) = \begin{cases} e^{-x}, & |x| < a \\ 0, & |x| > a. \end{cases} \qquad (11.97)$$

Then, the Fourier transform of $f(x)$ is

$$\mathcal{F}\{f(x)\} = F(\theta) \quad = \quad \int_{-\infty}^{\infty} f(\xi) e^{-i\theta\xi}\, d\xi$$

$$= \quad \int_{-a}^{a} e^{-\xi} e^{-i\theta\xi}\, d\xi$$

$$= \quad \int_{-a}^{a} e^{-(1+i\theta)\xi}\, d\xi$$

$$= \quad \left. \frac{-e^{-(1+i\theta)\xi}}{1+i\theta} \right|_{-a}^{a}$$

$$= \quad \frac{e^{(1+i\theta)a} - e^{-(1+i\theta)a}}{1+i\theta}$$

$$= \quad \frac{e^{i(\theta-i)a} - e^{-i(\theta-i)a}}{i(\theta-i)} = 2\frac{\sin(a(\theta-i))}{\theta-i}. \quad ■ \qquad (11.98)$$

From the previous example, we see that the Fourier transform of a relatively common function quickly becomes complicated. For your convenience, there is a table of Fourier transforms in Appendix F. We now consider Fourier cosine and sine transform pairs.

11.4.1   Fourier Cosine and Sine Transforms

We start by stating the Fourier cosine and sine transform pairs. Then, we use them on the same function in an example.

From the Fourier cosine integral,

$$f(x) = \frac{2}{\pi} \int_0^\infty \cos(\theta x)\, d\theta \int_0^\infty f(\xi) \cos(\theta \xi)\, d\xi, \tag{11.99}$$

we see that the integral

$$\int_0^\infty f(\xi) \cos(\theta \xi)\, d\xi$$

is a function of θ. This integral is known as the Fourier cosine transform of $f(x)$. Equation (11.100) is the manner in which it is usually stated as

$$\mathcal{F}_C\{f(x)\} = F(\theta) = \int_0^\infty f(\xi) \cos(\theta \xi)\, d\xi. \tag{11.100}$$

Note: In Equation (11.100), the variable is changed from u to x, and we use $\mathcal{F}_C$ to indicate that we are taking the Fourier cosine transform. Since Equation (11.99) produces the Fourier cosine integral of $f(x)$, the inverse Fourier cosine transform is given by

$$f(x) = \mathcal{F}_C^{-1}\{F(\theta)\} = \frac{2}{\pi} \int_0^\infty F(\theta) \cos(\theta x)\, d\theta. \tag{11.101}$$

Similarly, the Fourier sine transform is derived from the Fourier sine integral,

$$f(x) = \frac{2}{\pi} \int_0^\infty \sin(\theta x)\, d\theta \int_0^\infty f(\xi) \sin(\theta \xi)\, d\xi. \tag{11.102}$$

Again, the integral

$$\int_0^\infty f(\xi) \sin(\theta \xi)\, d\xi$$

is a function of θ, and it is known as the Fourier sine transform. Its usual form is

$$\mathcal{F}_S\{f(x)\} = F(\theta) = \int_0^\infty f(\xi) \sin(\theta \xi)\, d\xi. \tag{11.103}$$

The inverse Fourier sine transform is given by

$$f(x) = \mathcal{F}_S^{-1}\{F(\theta)\} = \frac{2}{\pi} \int_0^\infty F(\theta) \sin(\theta x)\, d\theta. \tag{11.104}$$

EXAMPLE 11.13. Find the Fourier cosine and sine transform of the function, as shown in Figure (11.16),

$$f(x) = \begin{cases} 1, & 0 < x < a \\ 0, & a < x. \end{cases} \qquad (11.105)$$

We start with the Fourier cosine transform. We have

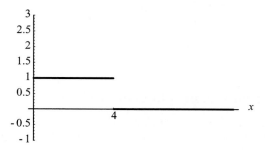

Figure 11.16: Graph of $f(x)$.

$$\begin{aligned} \mathcal{F}_C\{f(x)\} = F(\theta) &= \int_0^\infty f(\xi)\cos(\theta\xi)\,d\xi \\ &= \int_0^a \cos(\theta\xi)\,d\xi \\ &= \frac{\sin(\theta\xi)}{\theta}\bigg|_0^a = \frac{\sin(a\theta)}{\theta}. \qquad (11.106) \end{aligned}$$

For the Fourier sine transform, we have

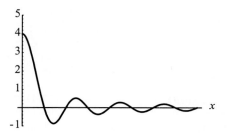

Figure 11.17: The Fourier cosine transform, $F(\theta)$ of $f(x)$.

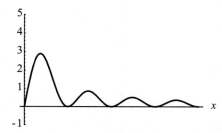

Figure 11.18: The Fourier sine transform, $F(\theta)$ of $f(x)$.

$$\mathcal{F}_S\{f(x)\} = F(\theta) \quad = \quad \int_0^\infty f(\xi)\sin(\theta\xi)\,d\xi$$

$$= \quad \int_0^a \sin(\theta\xi)\,d\xi$$

$$= \quad -\left.\frac{\cos(\theta\xi)}{\theta}\right|_0^a = \frac{1-\cos(a\theta)}{\theta}. \qquad (11.107)$$

Figure (11.16) is the graph of the function $f(x)$ for $a = 4$, Figure (11.17) is the graph of the Fourier cosine transform of $f(x)$ for $a = 4$, and Figure (11.18) is the graph of the Fourier sine transform of $f(x)$ for $a = 4$. ∎

We now develop some necessary theorems about Fourier transforms, which we need when solving PDEs.

11.4.2   Fourier Transform Theorems

Before we can solve any PDEs with Fourier transforms, we must understand certain related theorems. For instance, what is the Fourier transform of a first, second, or even n^{th} derivative? Also, do the same properties hold true for Fourier transforms that hold true for Laplace transforms? These are important questions, and we start by stating a theorem.

Theorem 63. *Given functions $f(x)$ and $g(x)$ whose Fourier transforms exist and any constants, c_1 and c_2 we have*

$$\mathcal{F}\{c_1 f(x) + c_2 g(x)\} = c_1 \mathcal{F}\{f(x)\} + c_2 \mathcal{F}\{g(x)\},$$

which means Fourier transforms are a linear operator.

Proof.

$$\mathcal{F}\{c_1 f(x) + c_2 g(x)\} = \int_{-\infty}^\infty (c_1 f(\xi) + c_2 g(\xi))\, e^{-i\theta x}\,d\xi. \qquad (11.108)$$

Since integration is a linear operator, the right side of Equation (11.88) becomes

$$c_1 \int_{-\infty}^\infty f(\xi)e^{-i\theta\xi}\,d\xi + c_2 \int_{-\infty}^\infty g(\xi)e^{-i\theta\xi}\,d\xi,$$

which equals

$$c_1 \mathcal{F}\{f(x)\} + c_2 \mathcal{F}\{g(x)\}.$$

□

Fourier transforms like Laplace transforms are a linear operator. Since Fourier transforms are going to be used to solve PDEs, it would seem reasonable to hope that they transform a first, second, and n^{th} derivatives similar to Laplace transforms. We find this out in the next theorem.

Theorem 64. *Suppose $f(x)$ is a continuous function on the $x-$axis and $f(x) \to 0$ as $|x| \to \infty$. Also, suppose $f'(x)$ is absolutely integrable on the $x-$axis. Then,*

$$\mathcal{F}\{f'(x)\} = i\theta \mathcal{F}\{f(x)\}. \tag{11.109}$$

Proof. Since $f'(x)$ is absolutely integrable, we know that a Fourier transform exists. Therefore, we have

$$\mathcal{F}\{f'(x)\} = \int_{-\infty}^{\infty} f'(\xi)e^{-i\theta\xi}\, d\xi.$$

Integrating by parts yields

$$\mathcal{F}\{f'(x)\} = \left[f(\xi)e^{-i\theta\xi}\big|_{-\infty}^{\infty} - (-i\theta)\int_{-\infty}^{\infty} f(\xi)e^{-i\theta\xi}\, d\xi.$$

$f(x)$ is continuous and $f(x) \to 0$ as $|x| \to \infty$ implies that

$$\left[f(\xi)e^{-i\theta\xi}\big|_{-\infty}^{\infty} = 0.$$

Therefore,

$$\mathcal{F}\{f'(x)\} = i\theta \int_{-\infty}^{\infty} f(\xi)e^{-i\theta\xi}\, d\xi = i\theta \mathcal{F}\{f(x)\}.$$

$\square$

For second derivatives, we use Theorem (64) twice. This yields

$$\mathcal{F}\{f''(x)\} = i\theta \mathcal{F}\{f'(x)\} = -\theta^2 \mathcal{F}\{f(x)\}. \tag{11.110}$$

In general, if $f(x)$ and its $n-1$ derivatives, $f^k(x)$, $k = 1, 2, 3, \ldots, n-1$ are continuous functions on the $x-$axis, and $f(x)$ and its $n-1$ derivatives go to 0 as $|x| \to \infty$, then, if $f^n(x)$ is absolutely integrable on the $x-$axis, we have

$$\mathcal{F}\{f^n(x)\} = (i\theta)^n \mathcal{F}\{f(x)\}. \tag{11.111}$$

We next state the convolution theorem for Fourier transforms.

Theorem 65. *If $f(x)$ and $g(x)$ are piecewise-smooth on every finite interval on the $x-$axis and absolutely integrable, then*

$$\mathcal{F}\{f * g\} = \mathcal{F}\{f\}\mathcal{F}\{g\}. \tag{11.112}$$

Proof. For our purposes a convolution $f * g$ of the function f and g is defined by

$$h(x) = (f * g)(x) = \int_{-\infty}^{\infty} f(y)g(x-y)\, dy = \int_{-\infty}^{\infty} f(x-y)g(y)\, dy.$$

Thus, by the definition of a Fourier transform, we have

$$\mathcal{F}\{f * g\} = \int_{-\infty}^{\infty} \int_{-\infty}^{\infty} f(y)g(\xi - y)e^{-i\theta\xi}\, dy d\xi. \tag{11.113}$$

Interchanging the order of integration yields

$$\mathcal{F}\{f * g\} = \int_{-\infty}^{\infty} \int_{-\infty}^{\infty} f(y)g(\xi - y)e^{-i\theta\xi}\, d\xi dy. \tag{11.114}$$

Letting $s = \xi - y$, we have $\xi = s + y$ and Equation (11.114) becomes

$$\mathcal{F}\{f * g\} = \int_{-\infty}^{\infty} \int_{-\infty}^{\infty} f(y)g(s)e^{-i\theta(s+y)}\, ds dy,$$

which can be stated as

$$\mathcal{F}\{f * g\} = \int_{-\infty}^{\infty} f(y)e^{-i\theta y}\, dy \int_{-\infty}^{\infty} g(s)e^{-i\theta s}\, ds.$$

Thus,

$$\mathcal{F}\{f * g\} = \mathcal{F}\{f\}\mathcal{F}\{g\},$$

which completes the proof. □

Note: The method of convolution is used extensively when solving the heat equation in an infinite spatial variable domain. If we let $\mathcal{F}\{f\} = \hat{f}(\theta)$ and $\mathcal{F}\{g\} = \hat{g}(\theta)$, then we have

$$\mathcal{F}\{f * g\} = \hat{f}(\theta)\hat{g}(\theta).$$

Taking the inverse transform of both sides yields

$$(f * g)(x) = \frac{1}{2\pi} \int_{-\infty}^{\infty} \hat{f}(\theta)\hat{g}(\theta)e^{i\theta x}\, d\theta.$$

Theorem (64) is similar for Fourier cosine and sine transforms, which we state here for future reference. For the Fourier cosine transform, we have

$$\mathcal{F}_C\{f'(x)\} = \theta\mathcal{F}_S\{f(x)\} - f(0) \tag{11.115}$$

and

$$\mathcal{F}_C\{f''(x)\} = -\theta^2\mathcal{F}_C\{f(x)\} - f'(0). \tag{11.116}$$

For the Fourier sine transform, we have

$$\mathcal{F}_S\{f'(x)\} = -\theta\mathcal{F}_C\{f(x)\} \tag{11.117}$$

and

$$\mathcal{F}_S\left\{f''(x)\right\} = -\theta^2 \mathcal{F}_C\left\{f(x)\right\} + \theta f(0). \qquad (11.118)$$

In the next section, we investigate solving PDEs using Fourier transforms.

EXERCISES 11.4

11.4.1. Compute and graph the Fourier transforms of the following functions:

(1)
$$f(x) = \begin{cases} 2, & |x| < 5 \\ 0, & 5 < x. \end{cases}$$

(2)
$$f(x) = \begin{cases} e^{-2|x|}, & |x| < 1 \\ 0, & \text{otherwise.} \end{cases}$$

(3)
$$f(x) = \begin{cases} e^{-x}, & x > 0 \\ 0, & \text{otherwise.} \end{cases}$$

(4)
$$f(x) = \begin{cases} e^{x}, & x < 0 \\ 0, & \text{otherwise.} \end{cases}$$

(5)
$$f(x) = \begin{cases} xe^{-2x}, & x > 0 \\ 0, & \text{otherwise.} \end{cases}$$

(6)
$$f(x) = \begin{cases} -1, & -1 < x < 0 \\ 1, & 0 < x < 1 \\ 0, & \text{otherwise.} \end{cases}$$

(7)
$$f(x) = \frac{1}{x^2 + 25}.$$

(8)
$$f(x) = \begin{cases} x, & 0 < x < 2 \\ 2x - 4, & 2 < x < 4 \\ 0, & \text{otherwise.} \end{cases}$$

11.4.2. Compute and graph the Fourier cosine transforms for the following functions:

(1)
$$f(x) = \begin{cases} 1, & 0 < x < 2 \\ 0, & \text{otherwise.} \end{cases}$$

(2)
$$f(x) = e^{-x^2}.$$

(3)
$$f(x) = x^{-1/2}.$$

(4)
$$f(x) = x^{-4}.$$

(5)
$$f(x) = \frac{3}{x^2 + 9}.$$

(6)
$$f(x) = \begin{cases} x, & 0 < x < 1 \\ 0, & \text{otherwise.} \end{cases}$$

(7)
$$f(x) = \begin{cases} x, & 0 < x < 2 \\ 2x - 4, & 2 < x < 4 \\ 0, & \text{otherwise.} \end{cases}$$

(8)
$$f(x) = \begin{cases} -1, & 0 < x < 1 \\ 1, & 1 < x < 2 \\ 0, & \text{otherwise.} \end{cases}$$

11.4.3. Compute and graph the Fourier sine transforms for the following functions:

(1)
$$f(x) = \begin{cases} 3, & 0 < x < 2 \\ 0, & \text{otherwise.} \end{cases}$$

(2)
$$f(x) = e^{-x^2}.$$

(3)
$$f(x) = x^{-1/2}.$$

(4)
$$f(x) = x^{-4}.$$

(5)
$$f(x) = \frac{x}{x^2 + 9}.$$

(6)
$$f(x) = \begin{cases} x, & 0 < x < 1 \\ 0, & \text{otherwise.} \end{cases}$$

(7)
$$f(x) = \begin{cases} x, & 0 < x < 2 \\ 2x - 4, & 2 < x < 4 \\ 0, & \text{otherwise.} \end{cases}$$

(8)
$$f(x) = \begin{cases} -1, & 0 < x < 1 \\ 1, & 1 < x < 2 \\ 0, & \text{otherwise.} \end{cases}$$

11.4.4. This exercise develops the Fourier transform of a shift of a function on the x−axis. Show that if $f(x)$ has a Fourier transform, then $f(x - a)$ for some $a \in \mathbb{R}$, and that

$$\mathcal{F}\{f(x - a)\} = e^{-ia\theta} \mathcal{F}\{f(x)\}.$$

11.4.5. This exercise develops a shift on the θ−axis. Show that if $\hat{f}(\theta)$ is the Fourier transform of $f(x)$, then $\hat{f}(\theta - a)$ is the Fourier transform of $e^{iax} f(x)$.

11.4.6. This exercise develops the basic properties of convolutions. Prove the following properties of convolutions:

(1) Commutativity of convolutions: $f * g = g * f$.
(2) Associativity of convolutions: $f * (g * h) = (f * g) * h$.
(3) Commutation with translation: Given $a \in \mathbb{R}$, let $f_a(x) = f(x - a)$. Show that $f_a * g = f * g_a = (f * g)_a$.

11.4.7. Let $f(x) = e^{-x^2/2}$ and $g(x) = e^{-|x|}$.

(1) Find the Fourier transform of $f(x)$ and $g(x)$.
(2) Find the Fourier transform of the convolution of f and g.
(3) Let $a = 2$. Find the Fourier transform of $f_a * g$. *Hint:* Reference the previous exercise.

11.4.8. Another definition of the Fourier transform is

$$\hat{f}(\theta) = \mathcal{F}\{f(x)\} = \frac{1}{\sqrt{2\pi}} \int_{-\infty}^{\infty} f(\xi) e^{-i\theta\xi} \, d\xi, \tag{11.119}$$

and of the inverse Fourier transform is

$$f(x) = \mathcal{F}^{-1}\left\{\hat{f}(\theta)\right\} = \frac{1}{\sqrt{2\pi}} \int_{-\infty}^{\infty} \hat{f}(\theta) e^{i\theta x}\, d\theta. \qquad (11.120)$$

(1) Show that the above two definitions are equivalent to Definitions (61 and 62).

(2) Using Definition (11.119) find the Fourier transform of e^{-ax^2} for $a > 0$. Compare your answer with that in the table of Fourier transforms in Appendix F.

(3) Using Definition (11.119) find the Fourier transform of

$$f(x) = \begin{cases} 1, & |x| < 1 \\ \\ 0, & |x| > 1. \end{cases}$$

Compare your answer with that in the table of Fourier transforms in Appendix F. Explain the differences.

11.4.9. Show that the inverse Fourier transform is a linear operator.

11.5 FOURIER TRANSFORM SOLUTION METHOD OF PDES

This section deals with PDEs, which are functions of two variables, x and t. Again, we note that a Fourier transform is designed to work with one variable. However, we learned in Chapter 6, The Calculus of Fourier series, that we could take the derivative of a Fourier series with respect to a parameter t. We will use that same type of technique here, and again the parameter is t. Thus, when using Definition (11.91), the definition of the Fourier transform of a function of two variables, $u(x, t)$ becomes

$$\mathcal{F}\left\{u(x, t)\right\} = \hat{u}(\theta, t) = \int_{-\infty}^{\infty} u(\xi, t) e^{-i\theta\xi}\, d\xi.$$

We demonstrate this new definition with an example.

EXAMPLE 11.14. Suppose

$$u(x, t) = \begin{cases} e^t \sin x, & |x| < \pi \\ \\ 0, & \text{otherwise.} \end{cases}$$

Then, the Fourier transform is

$$\hat{u}(\theta, t) = \mathcal{F}\{u(x, t)\} \quad = \quad \mathcal{F}\{e^t \sin x\}$$

$$= \quad \int_{-\infty}^{\infty} e^t \sin \xi e^{-i\theta \xi} \, d\xi$$

$$= \quad e^t \int_{-\infty}^{\infty} \sin \xi e^{-i\theta \xi} \, d\xi$$

$$= \quad e^t \int_{-\pi}^{\pi} \sin \xi e^{-i\theta \xi} \, d\xi = 2ie^t \frac{\sin(\pi\theta)}{\theta^2 - 1}.$$

Thus, $\hat{u}(\theta, t) = 2ie^t \dfrac{\sin(\pi\theta)}{\theta^2 - 1}$. ∎

If we assume Theorem (64) holds true for the Fourier transform of a PDE, which it does, then for $\dfrac{\partial u(x, t)}{\partial t}$, $\dfrac{\partial^2 u(x, t)}{\partial t^2}$, and $\dfrac{\partial^2 u(x, t)}{\partial x^2}$ where $-\infty < x < \infty$ and $t > 0$ we have the Fourier transforms;

$$\mathcal{F}\left\{\frac{\partial u(x, t)}{\partial t}\right\} = \frac{d}{dt}\hat{u}(\theta, t), \tag{11.121}$$

$$\mathcal{F}\left\{\frac{\partial^2 u(x, t)}{\partial t^2}\right\} = \frac{d^2}{dt^2}\hat{u}(\theta, t), \tag{11.122}$$

and

$$\mathcal{F}\left\{\frac{\partial^2 u(x, t)}{\partial x^2}\right\} = (i\theta)^2 \hat{u}(\theta, t), \tag{11.123}$$

respectively. Also, we have

$$\mathcal{F}\left\{\frac{\partial^n u(x, t)}{\partial x^n}\right\} = (i\theta)^n \hat{u}(\theta, t). \tag{11.124}$$

The method for solving PDEs by Fourier transforms is quite similar to that of Laplace transforms. We will use the notation $\hat{u}(\theta, t)$ introduced in this section to simplify the equations. Thus, we have the following steps:

(1) Given a PDE with infinite boundaries and initial conditions, take the Fourier transform of the infinite boundary value problem while holding t as a parameter. You will get an ODE in the variable t, with function $\hat{u}(\theta, t)$. *Note:* The ODE only depends on the variable t.

(2) Solve the ODE and find $\hat{u}(\theta, t)$.

(3) Take the inverse Fourier transform of $\hat{u}(\theta, t)$ to find $u(x, t)$.

Example (11.15) illustrates the use of the previous steps.

EXAMPLE 11.15. Given a uniform infinite rod with perfect lateral insulation, thermal diffusivity of 1, and initial temperature distribution of e^{-x^2}, solve the following problem using Fourier transforms.
First, we must state the mathematical formulation of the problem. We have

$$\frac{\partial u(x,t)}{\partial t} = \frac{\partial^2 u(x,t)}{\partial x^2}, \quad -\infty < x < \infty \text{ and } t > 0, \tag{11.125}$$

subject to

$$u(x,0) = f(x) = e^{-x^2} \tag{11.126}$$

Step 1. Take the Fourier transform of the entire problem, which is from Equations (11.121 and 11.123) and Equation (11.91),

$$\frac{d\hat{u}(\theta,t)}{dt} = -\theta^2 \hat{u}(\theta,t), \tag{11.127}$$

subject to

$$\mathcal{F}\{u(x,0)\} = \hat{u}(\theta,0)$$

$$= \mathcal{F}\{e^{-\xi^2}\} = \int_{-\infty}^{\infty} e^{-\xi^2} e^{-i\theta\xi} \, d\xi$$

$$= \frac{1}{\sqrt{4\pi}} e^{-\theta^2/4} = \hat{f}(\theta), \tag{11.128}$$

where the Fourier transform of the IC was found in Appendix F. Although the Fourier transform of the IC is $\dfrac{1}{\sqrt{4\pi}} e^{-\theta^2/4}$, the preferred form is $\hat{f}(\theta)$. This form becomes important when we must do the inverse Fourier transform.
Step 2. Equation (11.127) is a first-order ODE in t. Its solution is

$$\hat{u}(\theta,t) = A(\theta)e^{-\theta^2 t}. \tag{11.129}$$

Applying the IC, Equation (11.128), to Equation (11.129) yields

$$\hat{u}(\theta,0) = \frac{1}{\sqrt{4\pi}} e^{-\theta^2/4} = \hat{f}(\theta) = A(\theta). \tag{11.130}$$

Therefore, our solution is

$$\hat{u}(\theta,t) = \frac{1}{\sqrt{4\pi}} e^{-\theta^2/4} e^{-\theta^2 t} = \hat{f}(\theta)e^{-\theta^2 t}. \tag{11.131}$$

Step 3. Find the inverse Fourier transform of Equation (11.131), which is, by Equation (11.92),

$$u(x,t) = \mathcal{F}(\theta)^{-1} \quad = \quad \frac{1}{2\pi} \int_{-\infty}^{\infty} \hat{u}(\theta,t) e^{i\theta x} \, d\theta$$

$$= \quad \frac{1}{2\pi} \int_{-\infty}^{\infty} \frac{1}{\sqrt{4\pi}} e^{-\theta^2/4} e^{-\theta^2 t} e^{i\theta x} \, d\theta$$

$$= \quad \frac{1}{2\pi} \int_{-\infty}^{\infty} \hat{f}(\theta) e^{-\theta^2 t} e^{i\theta x} \, d\theta. \tag{11.132}$$

If we let $\hat{g}(\theta) = e^{-\theta^2 t}$ in Equation (11.132), then we have

$$u(x,t) = \frac{1}{2\pi} \int_{-\infty}^{\infty} \hat{f}(\theta) \hat{g}(\theta) e^{i\theta x} \, d\theta. \tag{11.133}$$

Equation (11.133) may be recognizable as a form of Equation (11.119), which is

$$(f * g)(x) = \frac{1}{2\pi} \int_{-\infty}^{\infty} \hat{f}(\theta) \hat{g}(\theta) e^{i\theta x} \, d\theta.$$

By the definition of convolution,

$$(f * g)(x) = \int_{-\infty}^{\infty} f(y) g(x - y) \, dy, \tag{11.134}$$

we must find the inverse Fourier transform of $\hat{g}(\theta)$. Note that

$$\mathcal{F}\left\{e^{-ax^2}\right\} = \frac{1}{\sqrt{4\pi a}} e^{-\theta^2/4a}.$$

If we let $t = \dfrac{1}{4a}$, then $a = \dfrac{1}{4t}$. Then, we can solve for the inverse Fourier transform of $\hat{g}(\theta)$. We have

$$\mathcal{F}\left\{e^{-x^2/4t}\right\} = \sqrt{\frac{t}{\pi}} e^{-\theta^2 t} = \sqrt{\frac{t}{\pi}} \hat{g}(\theta).$$

Thus, the inverse Fourier transform of $\hat{g}(\theta)$ is

$$\sqrt{\frac{\pi}{t}} e^{-x^2/4t}.$$

Replacing x with $x - y$ in Equation (11.134), we get the final answer of

$$u(x,t) = (f * g)(x) \quad = \quad \frac{1}{2\pi} \int_{-\infty}^{\infty} f(y) \sqrt{\frac{\pi}{t}} e^{-(x-y)^2/4t} \, dy$$

$$= \quad \frac{1}{2\sqrt{\pi t}} \int_{-\infty}^{\infty} f(y) e^{-(x-y)^2/4t} \, dy$$

$$= \quad \frac{1}{2\sqrt{\pi t}} \int_{-\infty}^{\infty} e^{-y^2} e^{-(x-y)^2/4t} \, dy. \quad \blacksquare \tag{11.135}$$

In integral equation theory, $\hat{g}(\theta) = e^{-\theta^2 t}$, in Example (11.15) is known as a kernel. In this case, it is typically known as the heat kernel or Gauss's kernel. The Gaussian is the typical bell-shaped curve well known in statistics and probability. In general, the solution of the heat equation on an infinite domain, $-\infty < x < \infty$, with an initial temperature distribution, $f(x)$, is the convolution of the heat kernel, $\hat{g}(\theta) = e^{-\theta^2 t}$ with the initial temperature distribution, $f(x)$. Thus, given the general heat equation

$$\frac{\partial u(x, y)}{\partial t} = k_0 \frac{\partial^2 u(x, t)}{\partial x^2},$$

subject to the initial temperature distribution of

$$u(x, 0) = f(x),$$

our solution is

$$u(x, t) = \frac{1}{2\sqrt{\pi k_0 t}} \int_{-\infty}^{\infty} f(y) e^{-(x-y)^2/4k_0 t} \, dy. \tag{11.136}$$

Our next example will make use of Equation (11.136) and the error function, which was introduced in Section 11.3, Laplace transforms.

EXAMPLE 11.16. Consider a uniform infinite rod with perfect lateral insulation, thermal diffusivity of iron, and initial temperature distribution of 30°C for $|x| < 1$ and 0 otherwise. Solve the problem using Fourier transforms and the error function. First, we must state the mathematical formulation of the problem. We have

$$\frac{\partial u(x, t)}{\partial t} = .22 \frac{\partial^2 u(x, t)}{\partial x^2}, \quad -\infty < x < \infty \text{ and } t > 0, \tag{11.137}$$

subject to

$$u(x, 0) = f(x) = \begin{cases} 30, & |x| < 1 \\ 0, & |x| > 1. \end{cases} \tag{11.138}$$

Using Equation (11.136),

$$u(x, t) = \frac{1}{2\sqrt{\pi k_0 t}} \int_{-\infty}^{\infty} f(y) e^{-(x-y)^2/4k_0 t} \, dy,$$

we know that the solution as an inverse Fourier transform is

$$\begin{aligned}
u(x, t) &= \frac{1}{2\sqrt{\pi .22 t}} \int_{-\infty}^{\infty} 30 e^{-(x-y)^2/.88t} \, dy \\
&= \frac{15}{\sqrt{\pi .22 t}} \int_{-1}^{1} e^{-(x-y)^2/.88t} \, dy.
\end{aligned} \tag{11.139}$$

Note: The integral is from -1 to 1 since the integral is 0 for $|x| > 1$.

If we let $z = \dfrac{x-y}{\sqrt{.88t}}$, then $dz = \dfrac{-1}{\sqrt{.88t}}\,dy = \dfrac{-1}{2\sqrt{.22t}}\,dy$. Also, the limits of integration change to $\dfrac{x-1}{2\sqrt{.22t}}$ and $\dfrac{x+1}{2\sqrt{.22t}}$. Thus, Equation (11.139) becomes

$$
\begin{aligned}
u(x,t) &= \frac{15}{\sqrt{\pi.22t}} \int_{\frac{x+1}{2\sqrt{.22t}}}^{\frac{x-1}{2\sqrt{.22t}}} e^{-z^2}\left(-2\sqrt{.22t}\right) dz \\[2mm]
&= \frac{30}{\sqrt{\pi}} \int_{\frac{x-1}{2\sqrt{.22t}}}^{\frac{x+1}{2\sqrt{.22t}}} e^{-z^2}\, dz \\[2mm]
&= \frac{30}{\sqrt{\pi}} \left(\int_{0}^{\frac{x+1}{2\sqrt{.22t}}} e^{-z^2}\, dz - \int_{0}^{\frac{x-1}{2\sqrt{.22t}}} e^{-z^2}\, dz \right) \\[2mm]
&= 15\left(\operatorname{erf}\left[\frac{x+1}{2\sqrt{.22t}}\right] - \operatorname{erf}\left[\frac{x-1}{2\sqrt{.22t}}\right]\right). \qquad (11.140)
\end{aligned}
$$

Figure (11.19) graphs the temperature distribution, $u(x,t)$ for several values of t.∎

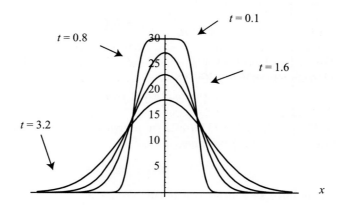

Figure 11.19: Temperature distribution, $u(x,t)$, for several values of t.

Our last example for this section is the wave equation.

EXAMPLE 11.17. Consider a tightly stretched uniform one-dimensional string of infinite length. Suppose the initial displacement is $f(x) = \dfrac{1}{1+x^2}$ and the initial velocity is $g(x) = 0$. Determine the solution as an inverse Fourier transform. The mathematical formulation of the problem is

$$
\frac{\partial^2 u(x,t)}{\partial t^2} = \frac{\partial^2 u(x,t)}{\partial x^2}, \quad -\infty < x < \infty,\ t > 0 \qquad (11.141)
$$

subject to

$$u(x,0) = \frac{1}{1+x^2}$$

$$\frac{\partial u(x,0)}{\partial t} = 0 \tag{11.142}$$

Step 1. Find the Fourier transform of Equations (11.141 and 11.142). We have

$$\frac{d^2 \hat{u}(\theta, t)}{dt^2} = -\theta^2 \hat{u}(\theta, t), \tag{11.143}$$

subject to

$$\mathcal{F}\{u(x,0)\} = \hat{u}(\theta, 0)$$

$$= \mathcal{F}\left\{\frac{1}{1+x^2}\right\} = \int_{-\infty}^{\infty} \frac{1}{1+\xi^2} e^{-i\theta\xi} \, d\xi$$

$$= \pi e^{-\theta} = \hat{f}(\theta). \tag{11.144}$$

and

$$\mathcal{F}\left\{\frac{\partial u(x,0)}{\partial t}\right\} 0 = \hat{g}(\theta). \tag{11.145}$$

Step 2. Solve the ODE in t subject to the ICs, $\hat{f}(\theta)$ and $\hat{g}(\theta)$. Equation (11.143) is a second-order ODE that has the solution

$$\hat{u}(\theta, t) = A(\theta)\cos(\theta t) + B(\theta)\sin(\theta t). \tag{11.146}$$

Applying the ICs to Equation (11.146), we arrive at $A(\theta) = \hat{f}(\theta)$ and $B(\theta) = 0$. Therefore, Equation (11.146) becomes

$$\hat{u}(\theta, t) = \hat{f}(\theta)\cos(\theta t) = \pi e^{-\theta}\cos(\theta t). \tag{11.147}$$

Step 3. Find the inverse Fourier transform. We have

$$u(x,t) = \mathcal{F}(\theta)^{-1} = \frac{1}{2\pi} \int_{-\infty}^{\infty} \hat{u}(\theta, t) e^{i\theta x} \, d\theta$$

$$= \frac{1}{2\pi} \int_{-\infty}^{\infty} \pi e^{-\theta}\cos(\theta t) e^{i\theta x} \, d\theta$$

$$= \frac{1}{2} \int_{-\infty}^{\infty} e^{-\theta}\cos(\theta t) e^{i\theta x} \, d\theta. \tag{11.148}$$

Equation (11.148) expresses the solution in terms of an inverse Fourier transform. Many times these expressions can be evaluated. However, the inverse Fourier transform solution of wave equation with infinite boundaries may be manipulated into d'Alembert's solution. This is left as an exercise. ∎

EXERCISES 11.5

11.5.1. Solve the following heat problems using Fourier transform method:

(1) $\dfrac{\partial u(x,t)}{\partial t} = \dfrac{\partial^2 u(x,t)}{\partial x^2}$, $-\infty < x < \infty$ and $t > 0$, subject to

$$u(x,0) = f(x) = e^{-|x|}.$$

(2) $\dfrac{\partial u(x,t)}{\partial t} = \dfrac{\partial^2 u(x,t)}{\partial x^2}$, $-\infty < x < \infty$ and $t > 0$, subject to

$$u(x,0) = f(x) = \begin{cases} 0, & x < -1 \\[2mm] -25, & -1 < x < 0 \\[2mm] 25, & 0, x < 1 \\[2mm] 0, & 1 < x. \end{cases}$$

(3) $\dfrac{\partial u(x,t)}{\partial t} = \dfrac{\partial^2 u(x,t)}{\partial x^2}$, $-\infty < x < \infty$ and $t > 0$, subject to

$$u(x,0) = f(x) = \begin{cases} 0, & x < -1 \\[2mm] x, & -1 < x < 1 \\[2mm] 0, & 1 < x. \end{cases}$$

(4) $\dfrac{\partial u(x,t)}{\partial t} = 4\dfrac{\partial^2 u(x,t)}{\partial x^2}$, $-\infty < x < \infty$ and $t > 0$, subject to

$$u(x,0) = f(x) = \begin{cases} 0, & x < -1 \\[2mm] -x, & -1 < x < 1 \\[2mm] 0, & 1 < x. \end{cases}$$

11.5.2. Solve the following wave problems using Fourier transforms:

(1) $\dfrac{\partial^2 u(x,t)}{\partial t^2} = \dfrac{\partial^2 u(x,t)}{\partial x^2}$, $-\infty < x < \infty$, $t > 0$,

subject to $u(x,0) = 0$ and $\dfrac{\partial u(x,0)}{\partial t} = \dfrac{1}{1+x^2}$.

(2) $\dfrac{\partial^2 u(x,t)}{\partial t^2} = \dfrac{\partial^2 u(x,t)}{\partial x^2}$, $-\infty < x < \infty$, $t > 0$,

subject to $u(x,0) = \dfrac{\sin(\pi x)}{x}$ and $\dfrac{\partial u(x,0)}{\partial t} = 0$.

(3) $\dfrac{\partial^2 u(x,t)}{\partial t^2} = \dfrac{\partial^2 u(x,t)}{\partial x^2}$, $-\infty < x < \infty$, $t > 0$,

subject to $u(x,0) = 0$ and $\dfrac{\partial u(x,0)}{\partial t} = e^{-2x^2}$.

(4) $\dfrac{\partial^2 u(x,t)}{\partial t^2} = \dfrac{\partial^2 u(x,t)}{\partial x^2}$, $-\infty < x < \infty$, $t > 0$,

subject to $u(x,0) = \dfrac{1}{1+x^2}$ and $\dfrac{\partial u(x,0)}{\partial t} = e^{-2x^2}$.

11.5.3. Consider

$$\frac{\partial^2 u(x,t)}{\partial t^2} = c^2 \frac{\partial^2 u(x,t)}{\partial x^2}, \quad -\infty < x < \infty, \, t > 0,$$

subject to

$$u(x,0) = f(x) \text{ and } \frac{\partial u(x,0)}{\partial t} = g(x).$$

(1) Solve for $u(x,t)$ using the Fourier transform method.

(2) d'Alembert's solution may be derived from the answer to part (1).

11.5.4. Solve the diffusion equation with convection. That is, solve

$$\frac{\partial u(x,t)}{\partial t} = k_0 \frac{\partial^2 u(x,t)}{\partial x^2} + h \frac{\partial u(x,t)}{\partial x}, \quad -\infty < x < \infty,$$

subject to

$$u(x,0) = f(x).$$

11.5.5. Solve the linearized Korteweg-deVries equation. That is, solve

$$\frac{\partial u(x,t)}{\partial t} = \frac{\partial^3 u(x,t)}{\partial x^3}, \quad -\infty < x < \infty,$$

subject to

$$u(x,0) = \begin{cases} 1, & |x| < 2 \\ 0, & |x| > 2. \end{cases}$$

11.5.6. Solve by Fourier transforms and graph the solution of

$$\frac{\partial^2 u(x,t)}{\partial t \partial x} = \frac{\partial^2 u(x,t)}{\partial x^2}, \quad -\infty < x < \infty,$$

subject to

$$u(x,0) = e^{-|x|}.$$

11.5.7. Consider

$$\frac{\partial^2 u(x,t)}{\partial t^2} = \frac{\partial^4 u(x,t)}{\partial x^4}, \quad -\infty < x < \infty,$$

subject to

$$u(x,0) = \frac{1}{1+x^2}.$$

(1) Find the Fourier transform of the boundary value problem.

(2) Solve the resulting ODE.

(3) Justify a reasonable assumption about the boundedness of the solution of the ODE.

(4) Apply the IC.

(5) Give $u(x,t)$ in the form of an inverse Fourier transform form.

11.5.8. Solve

$$\frac{\partial^2 u(x,t)}{\partial t^2} = a^2 \frac{\partial^3 u(x,t)}{\partial t \partial^2 x} - b \frac{\partial^4 u(x,t)}{\partial x^4}, \quad -\infty < x < \infty$$

subject to

$$u(x,0) = f(x) \text{ and } \frac{\partial u(x,0)}{\partial t} = g(x)$$

for $u(x,t)$ and leave the answer in inverse Fourier transform form.

11.5.9. Given

$$\frac{\partial u(x,t)}{\partial t} = k_0 \frac{\partial^2 u(x,t)}{\partial x^2}, \quad -\infty < x < \infty \text{ and } t > 0,$$

subject to

$$u(x,0) = f(x) = e^{-ax^2}$$

show that the solution is

$$u(x,t) = \frac{1}{\sqrt{g(t)}} e^{-ax^2/g(t)} \text{ where } g(t) = 4ak_0 t + 1.$$

11.5.10. Solve the following variable-coefficient PDEs using the Fourier transform method:

(1) $\dfrac{\partial u(x,t)}{\partial t} = -t\dfrac{\partial u(x,t)}{\partial x}$, subject to $u(x,0) = \begin{cases} e^{-x}, & x > 0 \\ 0, & x < 0. \end{cases}$

(2) $\dfrac{\partial u(x,t)}{\partial t} = t\dfrac{\partial u(x,t)}{\partial x}$, subject to $u(x,0) = \cos x^2$.

(3) $\dfrac{\partial u(x,t)}{\partial t} = t\dfrac{\partial u(x,t)}{\partial x}$, subject to $u(x,0) = \sin x^2$.

(4) $\dfrac{\partial u(x,t)}{\partial t} = \sin t\dfrac{\partial u(x,t)}{\partial x}$, subject to $u(x,0) = \sin x^2$.

(5) $\dfrac{\partial u(x,t)}{\partial t} = \cos t\dfrac{\partial u(x,t)}{\partial x}$, subject to $u(x,0) = \cos x^2$.

(6) $\dfrac{\partial u(x,t)}{\partial t} = e^{-t}\dfrac{\partial^u(x,t)}{\partial x^2}$,
  subject to $u(x,0) = 256$.

11.5.11. PROJECT: Solve the Klein-Gordon equation in one dimension. That is, solve

$$\frac{\partial^2 u(x,t)}{\partial t^2} = c^2\frac{\partial^2 u(x,t)}{\partial x^2} - m^2 u(x,t), \quad -\infty < x < \infty,$$

subject to

$$u(x,0) = 0 \text{ and } \frac{\partial u(x,0)}{\partial t} = 1.$$

Note: The Klein-Gordon equation is one of a number of equations which model elementary particles; electrons, mesons, quarks, etc. where m is the mass of the particle

(1) Find the Fourier transforms of $\dfrac{\partial^2 u(x,t)}{\partial t^2}$, $\dfrac{\partial^2 u(x,t)}{\partial x^2}$, $u(x,t)$, and the ICs.
    Note: The Fourier transform of $u(x,t)$ is $\hat{u}(\theta,t)$.

(2) Solve the resulting ODE and apply the ICs. *Hint:* $\hat{u}(\theta,t) = \dfrac{\sin\left(t\sqrt{c^2\theta^2 + m^2}\right)}{\sqrt{c^2\theta^2 + m^2}}$.

(3) Setup the inverse Fourier Transform.

(4) From the theory of cylindrical functions it is known that

$$\frac{\sin r}{r} = \frac{1}{2}\int_0^\pi J_0(r\sin\phi\sin\tau)e^{ir\cos\phi\cos\tau}\sin\tau\,d\tau$$

(5) Make the following substitutions:

$$r^2 = t^2(c^2\theta^2 + m^2), \; r\sin\phi = imt, \text{ and } r\cos\phi = c\theta t$$

(6) After working this part of the problem, you should arrive at the following answer:

$$u(x,t) = \frac{1}{2c} J_o\left(m\sqrt{t^2 - \frac{x^2}{c^2}}\right) \text{ for } |x| < ct.$$

11.5.12. PROJECT: Solve the nonhomogeneous Klein-Gordon equation. That is, solve

$$\frac{\partial^2 u(x,t)}{\partial t^2} = c^2\frac{\partial^2 u(x,t)}{\partial x^2} - m^2 u(x,t) + xt, \; -\infty < x < \infty,$$

subject to

$$u(x,0) = 0 \text{ and } \frac{\partial u(x,0)}{\partial t} = 0.$$

Hint: Follow the steps of the previous problem.

Appendix A

Summary of the Spatial Problem

In this appendix, we give a summary of the solutions of the spatial problem for homogeneous Dirichlet and Neumann boundary conditions. The eigenvalues and corresponding eigenfunctions are given for each of the four possible sets of boundary conditions.

Given

$$\frac{\partial^2 \varphi(x)}{\partial x^2} = -\lambda \varphi(x),$$

we have the following results:

Boundary Conditions	Eigenvalues	Eigenfunctions
$\begin{cases} \varphi(0) = 0 \\ \varphi(L) = 0 \end{cases}$	$\lambda_n = \left(\dfrac{n\pi}{L}\right)^2$ $n = 1, \ 2, \ 3, \ldots$	$\varphi_n(x) = \sin \dfrac{n\pi x}{L}$
$\begin{cases} \varphi(0) = 0 \\ \dfrac{\partial \varphi(L)}{\partial x} = 0 \end{cases}$	$\lambda_n = \left(\dfrac{(2n-1)\pi}{2L}\right)^2$ $n = 1, 2, 3, \ldots$	$\varphi_n(x) = \sin \dfrac{(2n-1)\pi x}{2L}$
$\begin{cases} \dfrac{\partial \varphi(0)}{\partial x} = 0 \\ \varphi(L) = 0 \end{cases}$	$\lambda_n = \left(\dfrac{(2n-1)\pi}{2L}\right)^2$ $n = 1, \ 2, \ 3, \ldots$	$\varphi_n(x) = \cos \dfrac{(2n-1)\pi x}{2L}$
$\begin{cases} \dfrac{\partial \varphi(0)}{\partial x} = 0 \\ \dfrac{\partial \varphi(L)}{\partial x} = 0 \end{cases}$	$\lambda_0 = 0$ $\lambda_n = \left(\dfrac{n\pi}{L}\right)^2$ $n = 1, \ 2, \ 3, \ldots$	$\varphi_0(x) = c_0$ $\varphi_n(x) = \cos \dfrac{n\pi x}{L}$
$\begin{cases} \varphi(-L) = \varphi(L) \\ \dfrac{\partial \varphi(-L)}{\partial x} = \dfrac{\partial \varphi(L)}{\partial x} \end{cases}$	$\lambda_0 = 0$ $\lambda_n = \left(\dfrac{n\pi}{L}\right)^2$ $n = 1, \ 2, \ 3, \ldots$	$\varphi_0(x) = c_0$ $\begin{cases} \varphi_n(x) = \cos \dfrac{n\pi x}{L} \\ \varphi_n(x) = \sin \dfrac{n\pi x}{L} \end{cases}$

Appendix B

Proofs of Related Theorems

Basically, the flow of this appendix is the statement of the theorem and then the proof, with only a few comments. Any assumptions about your prior knowledge of these theorems are given before the statement of the theorem to be proved.

There are five theorems in the text that have proofs in this appendix. Two are from Chapter 2, two are from Chapter 4, and one is from Chapter 10. The appendix is sectioned according to chapter and sectioned within chapter according to theorem.

B.1 THEOREMS FROM CHAPTER 2

B.1.1 Leibniz's Formula

For the proof of Leibniz's formula, I assume that you are familiar with various theorems from calculus, namely, the fundamental theorem of integral calculus from Calculus I, and the integral interchange theorem from Calculus III. Also, I assume that you understand the concept of integration depending on a parameter, which you may have encountered in Calculus III.

Theorem 66. *Suppose $f(x,t)$ and the partial derivative $\dfrac{\partial f(x,t)}{\partial t}$ are continuous in some region of the xt-plane where $a \le x \le b$ and $c \le t \le d$, then*

$$\frac{d}{dt}\left[\int_a^b f(x,t)\ dx\right] = \int_a^b \frac{\partial f(x,t)}{\partial t}\ dx.$$

Proof. Let

$$g(t) = \int_a^b \frac{\partial f(x,t)}{\partial t}\ dx,\ c \le t \le d.$$

Since $\dfrac{\partial f(x,t)}{\partial t}$ is continuous, we know that $g(t)$ is continuous for $c \le t \le d$. Thus,

465

for $c \le t \le d$, we have

$$\int_c^d g(t) \, dt = \int_c^d \int_a^b \frac{\partial f(x,t)}{\partial t} \, dx \, dt. \tag{B.1}$$

Therefore, by the integral interchange theorem, we may switch the order of integration in Equation (B.1). We have

$$
\begin{aligned}
\int_c^d g(t) \, dt &= \int_c^d \int_a^b \frac{\partial f(x,t)}{\partial t} \, dx \, dt = \int_a^b \int_c^d \frac{\partial f(x,t)}{\partial t} \, dt \, dx \\
&= \int_a^b [f(x,d) - f(x,c)] \, dx = \int_a^b f(x,d) \, dx - \int_a^b f(x,c) \, dx \\
&= F(d) - F(c),
\end{aligned}
$$

where $F(d)$ and $F(c)$ are integrations depending on a parameter. If we let d be a variable t, we have

$$F(t) - F(c) = \int_c^t g(u) \, du. \tag{B.2}$$

Equation (B.2) can now be differentiated with respect to t. Hence, by the fundamental theorem of integral calculus we have

$$F'(t) = g(t) = \int_a^b \frac{\partial f(x,t)}{\partial t} \, dx.$$

Note:

$$F'(t) = \frac{d}{dt} \int_a^b f(x,t) \, dx.$$

Thus, Liebnitz's formula is proved. $\square$

B.1.2   Maximum–Minimum Theorem

Suppose we are given

$$\frac{\partial u(x,t)}{\partial t} = k \frac{\partial^2 u(x,t)}{\partial x^2}, \tag{B.3}$$

subject to Dirichlet BCs

$$u(0,t) = g(t)$$
$$u(L,t) = h(t) \tag{B.4}$$

and IC

$$u(x,0) = f(x). \tag{B.5}$$

Equation (B.1) tells us that for $0 < x < L$ and $0 < t$

$$\frac{\partial u(x,t)}{\partial t} - k\frac{\partial^2 u(x,t)}{\partial x^2} = 0. \tag{B.6}$$

We want to show that the maximum-minimum value occurs on the sides of the rectangle, R, given by

$$0 \leq x \leq L \text{ and } 0 \leq t \leq T$$

as shown in Figure B.1. Since k is a positive constant in Equation (B.6), we will

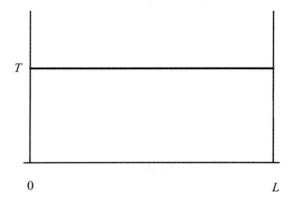

Figure B.1: $u(x,t)$ in the rectangle $0 \leq x \leq L$ and $0 \leq t \leq T$.

assume that it has the value of 1. Therefore, Equation (B.6) becomes

$$\frac{\partial u(x,t)}{\partial t} - \frac{\partial^2 u(x,t)}{\partial x^2} = 0. \tag{B.7}$$

The proof of the maximum–minimum theorem is done by contradiction. Also, the proof of the minimum part is provided. The maximum part is similar.

Theorem 67. *(Maximum–Minimum Theorem) Let T be an element of the real numbers such that $T > 0$. Suppose the function $u(x,t)$ is continuous in a rectangle, R, given by*

$$0 \leq x \leq L \text{ and } 0 \leq t \leq T,$$

as shown in Figure (B.1), and satisfies the heat equation given in Equation (B.3) in the interior of the rectangle. (Note: T is considered in the interior of the rectangle, since any $T_1 > T$ would imply T is in the interior.) Then, $u(x,t)$ attains its maximum or minimum on the base of the rectangle $t = 0$ or on the vertical sides of the rectangle $x = 0$ or $x = L$.

Proof. Let m be the minimum value of $u(x,t)$ in R and suppose that the minimum value of $u(x,t)$ on the lower base and vertical sides of R is $m + \epsilon$ where $\epsilon > 0$. Let

(x_0, t_0), where $0 < x_0 < L$ and $0 < t_0$, be a point where $u(x,t)$ attains minimum in R. Thus, $u(x_0, t_0) = m$. Consider the function

$$\mu(x,t) = u(x,t) - \frac{\epsilon}{4L^2}(x - x_0)^2. \tag{B.8}$$

Note:

$$\frac{\partial \mu(x,t)}{\partial t} = \frac{\partial u(x,t)}{\partial t} \tag{B.9}$$

and

$$\frac{\partial^2 \mu(x,t)}{\partial x^2} = \frac{\partial^2 u(x,t)}{\partial x^2} - \frac{\epsilon}{2L^2}. \tag{B.10}$$

On the lower base and vertical sides of R,

$$\mu(x,t) \geq m + \epsilon - \frac{\epsilon}{4} = m + \frac{3\epsilon}{4}, \tag{B.11}$$

whereas $\mu(x_0, t_0) = m$. Therefore, the minimum value of μ in R is not achieved on the lower base and vertical sides of R. Let (x_1, t_1), where $0 < x_1 < L$ and $0 < t_1 \leq T$, be a point where μ attains its minimum. At (x_1, t_1), μ must satisfy the necessary conditions for a minimum, that is,

$$\frac{\partial \mu(x_1, t_1)}{\partial t} = 0,$$

if $t_1 < T$, or

$$\frac{\partial \mu(x_1, t_1)}{\partial t} \leq 0,$$

if $t_1 = T$, and

$$\frac{\partial^2 \mu(x_1, t_1)}{\partial x^2} \geq 0.$$

Hence, at (x_1, t_1),

$$\frac{\partial \mu(x_1, t_1)}{\partial t} - \frac{\partial^2 \mu(x_1, t_1)}{\partial x^2} \leq 0.$$

However,

$$\frac{\partial \mu(x_1, t_1)}{\partial t} - \frac{\partial^2 \mu(x_1, t_1)}{\partial x^2} = \frac{\partial \mu(x,t)}{\partial t}$$

$$= \frac{\partial u(x,t)}{\partial t} - \frac{\partial^2 \mu(x,t)}{\partial x^2}$$

$$= \frac{\partial^2 u(x,t)}{\partial x^2} + \frac{\epsilon}{2L^2} > 0.$$

Thus, we have a contradiction. Therefore the minimum value of $u(x,t)$ must be attained on the lower base or vertical sides of the rectangle R. $\square$

B.2   THEOREMS FROM CHAPTER 4

You may not have previously encountered the next theorems in this form, in a linear algebra course. In your ODE course, you may have seen something similar. However, this material, in an ODE course, is concentrated on solving systems of linear ODEs.

B.2.1   Eigenvectors of Distinct Eigenvalues Are Linearly Independent

The proof of Theorem (68) is done by induction. For those unfamiliar with this type of proof, I suggest you study a text on logic and proof.

Theorem 68. *Let λ_1, λ_2, $\lambda_3, \ldots, \lambda_n$ be distinct eigenvalues of an $n \times n$ matrix. Then the corresponding eigenvectors $\overrightarrow{x_1}$, $\overrightarrow{x_2}$, $\overrightarrow{x_3}, \ldots \overrightarrow{x_n}$ form a linearly independent set of vectors. That is, $c_1\overrightarrow{x_1} + c_2\overrightarrow{x_2} + c_3\overrightarrow{x_3} + \ldots + c_n\overrightarrow{x_n} = 0$ if and only if $c_i = 0$ for $1 \leq i \leq n$.*

Proof. Let λ_1, λ_2, $\lambda_3, \ldots, \lambda_n$ be distinct eigenvalues of an $n \times n$ matrix. We know that for each eigenvalue, there exists a corresponding eigenvector, usually expressed as $\overrightarrow{x_1}$, $\overrightarrow{x_2}$, $\overrightarrow{x_3}, \ldots \overrightarrow{x_n}$. For $n = 1$, we have one eigenvalue and one eigenvector and the result follows from normal algebra on the real numbers $\mathbb{R}$. Now assume that every set of $(n - 1)$ eigenvectors, which correspond to the $(n - 1)$ distinct eigenvalues of a given square matrix, A, are linearly independent. Let $\overrightarrow{x_1}$, $\overrightarrow{x_2}$, $\overrightarrow{x_3}, \ldots, \overrightarrow{x_{n-1}}$ be the eigenvectors of A corresponding to the distinct eigenvalues λ_1, λ_2, $\lambda_3, \ldots, \lambda_{n-1}$. Thus, $\lambda_i \neq \lambda_j$ for $i \neq j$. Suppose that $\overrightarrow{x_1}$, $\overrightarrow{x_2}$, $\overrightarrow{x_3}, \ldots, \overrightarrow{x_{n-1}}$, $\overrightarrow{x_n}$ are linearly dependent. Thus, there exist constants c_i for $1 \leq i \leq n$ that are not all equal to 0, such that

$$c_1\overrightarrow{x_1} + c_2\overrightarrow{x_2} + c_3\overrightarrow{x_3} + \ldots + c_n\overrightarrow{x_n} = 0. \tag{B.12}$$

We assume that $c_1 \neq 0$. Now,

$$
\begin{aligned}
0 &= (A - \lambda_1 I_n)\left(c_1\overrightarrow{x_1} + c_2\overrightarrow{x_2} + c_3\overrightarrow{x_3} + \ldots + c_n\overrightarrow{x_n}\right) \\[2mm]
&= c_1\left(A\overrightarrow{x_1} - \lambda_1\overrightarrow{x_1}\right) + c_2\left(A\overrightarrow{x_2} - \lambda_1\overrightarrow{x_2}\right) + \ldots + c_n\left(A\overrightarrow{x_n} - \lambda_1\overrightarrow{x_n}\right) \\[2mm]
&= c_1\left(\lambda_1\overrightarrow{x_1} - \lambda_1\overrightarrow{x_1}\right) + c_2\left(A\overrightarrow{x_2} - \lambda_1\overrightarrow{x_2}\right) + \ldots + c_n\left(A\overrightarrow{x_n} - \lambda_1\overrightarrow{x_n}\right).
\end{aligned}
$$

Since

$$c_1\left(A\overrightarrow{x_1} - \lambda_1\overrightarrow{x_1}\right) = c_1\left(\lambda_1\overrightarrow{x_1} - \lambda_1\overrightarrow{x_1}\right)$$

by Definition (15), we have

$$0 = c_2\left(A\overrightarrow{x_2} - \lambda_1\overrightarrow{x_2}\right) + \ldots + c_n\left(A\overrightarrow{x_n} - \lambda_1\overrightarrow{x_n}\right).$$

But λ_2 through λ_n are $(n-1)$ distinct eigenvalues, so by the induction hypothesis, $\vec{x_2}, \vec{x_3}, \ldots, \vec{x_n}$ are linearly independent. Therefore,

$$c_2 \left(\lambda_1 - \lambda_2\right) = \ldots = c_n \left(\lambda_1 - \lambda_2\right).$$

But each $\lambda_i - \Lambda_j \neq 0$. Hence, $c_2 = c_3 = \ldots = c_n = 0$. Thus, Equation (B.12) becomes

$$c_1 \vec{x_1} = 0,$$

which implies $\vec{x_1} = 0$ since $c_1 \neq 0$. This is a contradiction. Therefore, the eigenvectors $\vec{x_1}$, $\vec{x_2}$, $\vec{x_3}, \ldots, \vec{x_n}$ are linearly independent. $\square$

B.2.2 Eigenvectors of Distinct Eigenvalues of an n by n Matrix Form a Basis for $\mathbb{R}^n$

Generally, Theorem (69) is considered a corollary or lemma of Theorem (68). As such it has a very short proof.

Theorem 69. *If an $n \times n$ matrix A has n distinct eigenvalues, then the corresponding eigenvectors form a basis for $\mathbb{R}^n$.*

Proof. Let $\vec{x_1}$, $\vec{x_2}$, $\vec{x_3}, \ldots \vec{x_n}$ be the corresponding eigenvectors of the distinct eigenvalues λ_1, λ_2, $\lambda_3, \ldots \lambda_n$ of the $n \times n$ matrix A. By Theorem (68), we know that the eigenvectors are linearly independent. We must show that they span $\mathbb{R}^n$. Let $\vec{y}$ be any vector in $\mathbb{R}^n$. Then, the set of vectors $\{\vec{x_1}, \vec{x_2}, \vec{x_3}, \ldots \vec{x_n}, \vec{y}\}$ is linearly dependent because each vector is a linear combination of any basis vectors of $\mathbb{R}^n$. Thus, the constants c_1, c_2, $c_3, \ldots, c_n$, and c exist where at least one is not 0, such that

$$c_1 \vec{x_1} + c_2 \vec{x_2} + c_3 \vec{x_3} + \ldots + c_n \vec{x_n} + c \vec{y} = 0. \tag{B.13}$$

Since $\vec{x_1}$, $\vec{x_2}$, $\vec{x_3}, \ldots, \vec{x_n}$ are linearly independent, we must have $c \neq 0$ in Equation (B.13). Therefore,

$$c_1 \vec{x_1} + c_2 \vec{x_2} + c_3 \vec{x_3} + \ldots + c_n \vec{x_n} = -c \vec{y}.$$

Since $\vec{y}$ was an arbitrary vector of $\mathbb{R}^n$, $\vec{x_1}$, $\vec{x_2}$, $\vec{x_3}, \ldots, \vec{x_n}$ span $\mathbb{R}^n$. Hence, $\vec{x_1}$, $\vec{x_2}$, $\vec{x_3}, \ldots, \vec{x_n}$ is a basis of $\mathbb{R}^n$. $\square$

B.3 THEOREM FROM CHAPTER 5

Since there is only one theorem from Chapter 5, we will not separate this section into subsections.

The maximum principle has two parts–maximum value occuring on the boundary or a minimum value occuring on the boundary. We prove the minimum value occuring on the boundary. The proof of the maximum value occuring on the boundary is similar.

Laplace's equation in two dimensions is

$$\frac{\partial^2 u}{\partial x^2} + \frac{\partial^2 u}{\partial y^2} = 0. \tag{B.14}$$

Theorem 70. *(Maximum Principle) Let Ω be a bounded set in $\mathbb{R}^2$. Let $u(x,y)$ be a harmonic function in Ω, while $u(x,y)$ must be continuous in the union of Ω and the boundary of Ω, denoted $\partial\Omega$. Then, the maximum and minimum values of $u(x,y)$ are attained on the $\partial\Omega$, unless $u(x,y)$ is identically equal to a constant.*

Proof. Let $\epsilon > 0$ be given. Let $\mu(x,y) = u(x,y) - \epsilon\left(x^2 + y^2\right)$. Then,

$$\frac{\partial^2 \mu(x,y)}{\partial x^2} + \frac{\partial^2 \mu(x,y)}{\partial y^2} = \frac{\partial^2 u(x,y)}{\partial x^2} + \frac{\partial^2 u(x,y)}{\partial y^2} - 4\epsilon = 0 - r\epsilon < 0.$$

But $\dfrac{\partial^2 \mu(x,y)}{\partial x^2} + \dfrac{\partial^2 \mu(x,y)}{\partial y^2} \geq 0$ at an interior minimum point, by the second derivative test from calculus. Thus, $\mu(x,y)$ has no interior minimum in Ω. Since $\mu(x,y)$ is a continuous function it must have a minimum in $\Omega \bigcup \partial\Omega$. Let the minimum of $\mu(x,y)$ be attained at (x_0, y_0) on $\partial\Omega$. Then, for all (x,y) in Ω we have

$$u(x,y) > \mu(x,y) > \mu(x_0, y_0) = u(x_0, y_0) + \epsilon(x^2 + y^2) \geq \min_{\partial\Omega} u + \epsilon r^2,$$

where r is the radius of the circle centered at the origin that contains $\Omega \bigcup \partial\Omega$. Since this is true for all $\epsilon > 0$, we have

$$u(x,y) > \min_{\partial\Omega} u$$

for all (x,y) in Ω. $\qquad\square$

Appendix C

Basics from Ordinary Differential Equations

C.1 SOME SOLUTION METHODS FOR FIRST-ORDER ODES

A first-order ordinary differential equation (ODE) has the form

$$S(t)\frac{du(t)}{dt} + K(t)u(t) = F(t). \tag{C.1}$$

If $F(t) = 0$, then the equation is homogeneous. For the homogeneous case, $u(t) \equiv 0$ (where $\equiv$ means identically equal to) is the trivial solution. For a solution to exist, it is important that $S(t)$, $K(t)$, and $F(t)$ be continuous functions on some interval $a < t < b$ where the ODE is defined.

If $S(t)$ is not equal to 0 anywhere on the interval $a < t < b$, then Equation (C.1) may be written in the more familiar form as Equation (C.2),

$$u'(t) + k(t)u(t) = f(t). \tag{C.2}$$

Equation (C.2) is said to be linear since $u'(t)$ and $u(t)$ are linear. Therefore, let us first consider Equation (C.2) where $k(t)$ is a constant and then complete this section by discussing Equation (C.2) where $k(t)$ is a more interesting function.

C.1.1 First-Order ODE Where k(t) Is a Constant

The following are two examples of commonly occurring first-order ODEs. These particular ODEs are very important when solving PDEs where $k(t)$ is a constant.

EXAMPLE C.1. Consider

$$u'(t) = \lambda u(t) \tag{C.3}$$

where λ is a constant. *Note:* In Equation (C.3), the function, $f(t)$, is the constant 0. Assuming $u(t) \neq 0$ for any t, we can rewrite Equation (C.3) as

$$\frac{u'(t)}{u(t)} = \lambda. \tag{C.4}$$

You arrive at the solution by remembering that

$$\frac{u'(t)}{u(t)} = \frac{\frac{du(t)}{dt}}{u(t)} = \lambda,$$

which can be written as

$$\frac{du(t)}{u(t)} = \lambda dt.$$

Assuming that $u(t) > 0$, we can integrate. The solution is

$$\ln[u(t)] = \lambda t + \alpha.$$

Now remembering some basic calculus, we arrive at the more compact and useful solution to Equation (C.4), which is

$$u(t) = ce^{\lambda t}. \qquad\blacksquare$$

EXAMPLE C.2. Consider

$$u'(t) + au(t) = \lambda. \tag{C.5}$$

Note: In Equation (C.5), the function $f(t)$ is the constant λ, and a is a nonzero multiplicative constant of $u(t)$. We solve Equation (C.5) using the usual technique. That is, we find an integrating factor. Therefore, we must determine a function $g(t)$ such that when you multiply Equation (C.5) by $g(t)$, the left side is recognized as a derivative of a product. Many times you can quickly discover the function $g(t)$ by remembering your calculus. In this case, if we choose $g(t) = e^{at}$, we can form the product $e^{at}u(t)$. When we then take the derivative of this product, we have

$$\frac{d}{dt}\left[e^{at}u(t)\right] = e^{at}u'(t) + ae^{at}u(t) = e^{at}\left[u'(t) + au(t)\right]. \tag{C.6}$$

The left side of Equation (C.5) and the contents of the brackets on the right side of Equation (C.6) are equal. This means that we have the correct integrating factor. Thus, we can multiply both sides of Equation (C.5) by e^{at} and get

$$e^{at}[u'(t) + au(t)] = \lambda e^{at}.$$

Making use of the work we did to produce Equation (C.6), we have

$$e^{at}[u'(t) + au(t)] = [u(t)e^{at}]'$$

resulting in

$$[u(t)e^{at}]' = \lambda e^{at}.$$

Integrating both sides with respect to t yields

$$u(t)e^{at} = \frac{\lambda e^{at}}{a} + c.$$

And, finally, solving for $u(t)$, we find that

$$u(t) = ce^{-at} + \frac{\lambda}{a}. \qquad \blacksquare$$

In both examples we have an arbitrary constant c. When c is unknown, we have a general solution to the ODE and there exists a family of curves as solutions. For example, if we let $\lambda = 1$ and $a = \frac{1}{2}$ in Example C.2, then the following graph (Figure (C.1)) depicts some members of the family of curves that solve Equation (C.5).

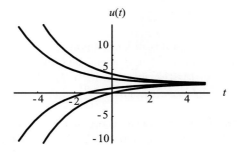

Figure C.1: Plot of $ce^{\frac{-t}{2}} + 2$ for $c = -2, -1, 1, 2$.

In most applications, the solution $u(t)$ is known for some value of $t = t_0$ where $u(t_0) = u_0$. This is generally called an initial condition (IC). The IC actually describes a point (t_0, u_0) through which the solution curve passes, resulting in a specific solution. For example, if in $u(t) = ce^{-at} + \frac{\lambda}{a}$ (Example C.2) we had $\lambda = 1$, $a = \frac{1}{2}$, and an IC of $u(0) = 1$, then our solution would become $u(t) = -e^{(\frac{-t}{2})} + 2$. This is graphed in Figure (C.2).

C.1.2   First-Order ODE Where k(t) Is a Function

In this section, we consider solution methods for the first-order ODE Equation (C.2) where $k(t)$ is a continuous function on some interval $a < t < b$ where the ODE is defined.

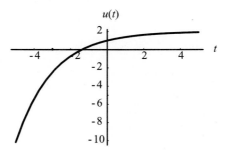

Figure C.2: Plot of $-e^{\frac{-t}{2}} + 2$.

EXAMPLE C.3. Consider the following ODE subject to the stated IC:

$$u'(t) - 2tu(t) = t, \quad u(0) = 1. \tag{C.7}$$

There are three ways of solving this ODE. First, we could use separation of variables and solve the related problem

$$\frac{u'(t)}{1 + 2u(t)} = t.$$

A second method is to treat the original problem as nonhomogeneous. In this case, we solve the homogeneous equation $u_h(t)$ and then solve for a particular solution $u_p(t)$. Adding $u_h(t)$ and $u_p(t)$ yields the solution for $u(t)$. Although this method works, I do not recommend it, because we solve the same problem twice and do twice the amount of work.

The third method is the integrating factor method. In Equation (C.7), we have $k(t) = -2t$, which suggests that an appropriate integrating factor is $f(t) = e^{-t^2}$. To prove this, consider

$$\frac{d}{dt}\left[e^{-t^2}u(t)\right] = e^{-t^2}u'(t) - 2te^{-t^2}u(t) = e^{-t^2}\left[u'(t) - 2tu(t)\right].$$

As in the previous subsection, we see that $f(t) = e^{-t^2}$ is the desired integrating factor. Now following our process used in Example C.2, we form

$$e^{-t^2}[u'(t) - 2tu(t)] = te^{-t^2},$$

which becomes

$$\left[e^{-t^2}u(t)\right]' = te^{-t^2}.$$

Integrating both sides, we find

$$u(t)e^{-t^2} = -\frac{1}{2}e^{-t^2} + c.$$

And then solving for $u(t)$ yields

$$u(t) = -\frac{1}{2} + ce^{t^2},$$

the general solution. Using the given IC $u(0) = 1$, we find $c = \frac{3}{2}$. Therefore, the specific solution of Equation (C.7) is

$$u(t) = \frac{3}{2}e^{t^2} - \frac{1}{2},$$

which is graphed in Figure (C.3). ■

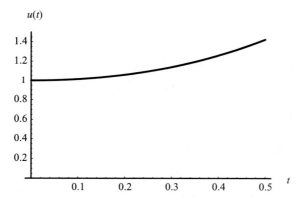

Figure C.3: The graph of $u(t) = \frac{3}{2}e^{t^2} - \frac{1}{2}$ for $0 \le t \le 0.5$.

The ability to solve first-order ODEs is an interesting mental exercise, but the real benefit comes with applying the ability. Applications occur in engineering, biology, chemistry, environmental studies, and many other disciplines. Later, we give an example of an application in elementary mechanics.

In elementary mechanics, we assume motion of a rigid body is along a straight line. Thus, Newton's law of motion applies, giving us "mass times acceleration equals the sum of the external forces." In mathematical terms, this translates to

$$F = ma.$$

Knowing this fact, we proceed to our example.

EXAMPLE C.4. Suppose a body of mass 10 kgs is projected upward with an initial velocity of 100 m/s. If we assume the gravitational attraction of the earth is constant, and we neglect all other forces acting on the body, we can find the maximum height attained by the body, the time at which the maximum height is reached, and the time it takes the body to return to its starting point.
Solution: We begin by assuming the positive direction is upward and the origin is the surface of the earth. We have, from Newton's law, $F = ma$, (force equals the

mass of the object times acceleration due to gravity), $(10kg)(-9.8 \ m/s^2)$. Force also equals the mass of the object times acceleration, which is $(10 \ kg)\dfrac{dv}{dt}$. Thus, we must solve

$$(10 \ kg)\frac{dv}{dt} = (10 \ kg)(-9.8 \ m/s^2),$$

which can be written as

$$\frac{dv}{dt} = (-9.8 \ m/s^2).$$

This means

$$v(t) = (-9.8 \ m/s^2)t + c.$$

Knowing the initial velocity is 100 m/s, we find

$$v(t) = (-9.8 \ m/s^2)t + 100 \ m/s.$$

Now realizing velocity is the rate of change of position with respect to time, we have

$$\frac{dx}{dt} = v(t) = (-9.8 \ m/s^2)t + 100 \ m/s.$$

This means the maximum height will occur when $\dfrac{dx}{dt} = 0$. This occurs when $t = \dfrac{100}{9.8} \ s$, which answers the second question—when does the maximum height occur? Solving for $x(t)$ yields

$$x(t) = \frac{(-9.8 \ m/s^2)t^2}{2} + 100t \ m/s.$$

Note: Since initial distance is at the surface of the earth, implying $x(0) = 0$, the constant of integration, c, equals 0. Applying $t = \dfrac{100}{9.8} \ s$, yields

$$x_{\max} = \frac{100^2}{2(9.8)} \ m.$$

To answer the third question, the time it takes the body to return to its starting point, we multiply the time it takes to reach maximum height by two, giving us $2t = \dfrac{200}{9.8} \ s.$ ■

As the above example shows, using basic laws to acquire an answer to a problem often results in solving a first-order ODE. This usually means that the problem is described by a first-order ODE. For instance, consider problems describing radioactive decay. Here the ODE is $Q'(t) = kQ(t)$, where k is the constant of proportionality. Other examples of problems that are described by first-order ODEs are determining compound interest, studying epidemics or population growth, solving optimization problems, and modeling electrical circuits.

Not all physical systems are conveniently modeled by first-order ODEs. For instance, electrical and mechanical vibrations, which are also governed by Newton's law, $F = ma$, are more easily modeled by a second-order ODE. We take up the task of reviewing second-order ODEs in the next section.

C.2   SOME SOLUTION METHODS OF SECOND-ORDER ODEs

A second-order ODE has the form

$$S(t)\frac{d^2u(t)}{dt^2} + K(t)\frac{du(t)}{dt} + H(t)u(t) = F(t). \tag{C.8}$$

If $F(t) = 0$, then the equation is homogeneous. The homogeneous case, $u(t) \equiv 0$, is known as the trivial solution. As in first-order ODEs, for the solution to exist it is important that $S(t)$, $K(t)$, $G(t)$, and $F(t)$ be continuous functions on some interval $\alpha < t < \beta$. Also, if we assume that $S(t)$ does not equal 0 anywhere on the interval $\alpha < t < \beta$, we can convert Equation (C.8) to the familiar form of

$$u''(t) + k(t)u'(t) + h(t)u(t) = f(t). \tag{C.9}$$

Solutions for Equation (C.9) depend on the nature of $k(t)$, $h(t)$, and $f(t)$; I will limit this section to solutions of second-order linear-constant coefficient ($k(t)$ and $h(t)$ are constants) ODEs,

$$u''(t) + bu'(t) + cu(t) = f(t). \tag{C.10}$$

C.2.1   Second-Order Linear Homogeneous ODEs

From your previous study in linear homogeneous second-order constant-coefficient ODEs, two independent solutions, $u_1(t)$ and $u_2(t)$, are expected. By the principle of superposition, any arbitrary linear combination of the two independent solutions ($c_1u_1(t) + c_2u_2(t)$) is also a solution. Given the equation

$$u''(t) + bu'(t) + cu(t) = 0, \tag{C.11}$$

we assume that the solution is of the form $u(t) = e^{rt}$. Finding both the first and second derivatives of $u(t)$ and plugging them into Equation (C.11) yields

$$r^2e^{rt} + bre^{rt} + ce^{rt} = 0. \tag{C.12}$$

Factoring out e^{rt} and multiplying Equation (C.12) by e^{-rt} gives us

$$r^2 + br + c = 0. \tag{C.13}$$

Equation (C.13) is known as the **characteristic equation** of Equation (C.12). We solve Equation (C.13) for roots r_1 and r_2 by using the quadratic formula. This yields

$$r_1 = \frac{-b + (b^2 - 4(1)(c))^{\frac{1}{2}}}{2(1)} = \frac{-b + (b^2 - 4c)^{\frac{1}{2}}}{2}$$

and

$$r_2 = \frac{-b - (b^2 - 4(1)(c))^{\frac{1}{2}}}{2(1)} = \frac{-b - (b^2 - 4c)^{\frac{1}{2}}}{2}.$$

The discriminant $(b^2 - 4c)$ determines the solution type. We have three possible solution types:

- real and unequal roots: $b^2 - 4c > 0$;
- real and equal roots: $b^2 - 4c = 0$; and
- complex roots: $b^2 - 4c < 0$.

Real and Unequal Roots

For real and unequal roots, $r_1 \neq r_2$. Therefore, the solution of Equation (C.11) is

$$u(t) = c_1 e^{r_1 t} + c_2 e^{r_2 t}. \tag{C.14}$$

However, there are other ways of doing the linear combination. For instance, consider the form of r_1 and r_2 wherein

$$r_1 = \frac{-b + (b^2 - 4c)^{\frac{1}{2}}}{2} = \frac{-b}{2} + \frac{(b^2 - 4c)^{\frac{1}{2}}}{2}$$

and

$$r_2 = \frac{-b - (b^2 - 4c)^{\frac{1}{2}}}{2} = \frac{-b}{2} - \frac{(b^2 - 4c)^{\frac{1}{2}}}{2}.$$

Then, Equation (C.14) could be written as

$$u(t) = e^{-\frac{bt}{2}} \left[\frac{d_1 e^{\gamma t} + d_2 e^{-\gamma t}}{2} \right] \text{ where } \gamma = \frac{(b^2 - 4c)^{\frac{1}{2}}}{2}$$

or

$$u(t) = e^{-\frac{bt}{2}} \left[\frac{m_1 e^{\gamma t} - m_2 e^{-\gamma t}}{2} \right] \text{ where } \gamma = \frac{(b^2 - 4c)^{\frac{1}{2}}}{2}.$$

If $d_1 = d_2 = m_1 = m_2 = 1$ in the above two linear combinations, we immediately notice that we can use forms of Euler's equations, which are

$$\cosh ax = \frac{e^{ax} + e^{-ax}}{2}$$

and

$$\sinh ax = \frac{e^{ax} - e^{-ax}}{2},$$

called the hyperbolic sine and hyperbolic cosine functions. Doing so allows us to form the new linear combination

$$u(t) = e^{\frac{-bt}{2}} \left[\alpha \cosh \gamma t + \beta \sinh \gamma t \right]; \, \gamma = \frac{(b^2 - 4c)^{\frac{1}{2}}}{2},$$

in which α and β are arbitrary constants. We use this form of the solution to Equation (C.11) with real roots, because it is much more advantageous in our study of PDEs. The advantage is discussed in both Chapters 4 and 5. *Note:* If the term $b^2 - 4c$, is a perfect square, then we will not have the hyperbolic sine and hyperbolic cosine as part of the solution.

EXAMPLE C.5. Solve

$$u''(t) + 3u'(t) + u(t) = 0.$$

Assuming $u(t) = e^{rt}$, we get the characteristic equation

$$r^2 + 3r + 1 = 0.$$

The quadratic formula yields the following roots:

$$r_1 = \frac{-3 + (9 - 4)^{\frac{1}{2}}}{2} = \frac{-3 + \sqrt{5}}{2}$$

and

$$r_2 = \frac{-3 - (9 - 4)^{\frac{1}{2}}}{2} = \frac{-3 - \sqrt{5}}{2} = -\frac{3 + \sqrt{5}}{2}.$$

Letting $\eta = \frac{\sqrt{5}}{2}$ and $\xi = \frac{-3}{2}$, $r_1 = \xi + \eta$ and $r_2 = \xi - \eta$. This means that the solution is

$$u(t) = c_1 e^{(\xi + \eta)t} + c_2 e^{(\xi - \eta)t} = e^{\xi t} \left(c_1 e^{\eta t} + c_2 e^{-\eta t} \right) \qquad \text{(C.15)}$$

or

$$u(t) = e^{\frac{-3}{2}t} \left[\alpha \cosh \left(\frac{\sqrt{5}}{2} t \right) + \beta \sinh \left(\frac{\sqrt{5}}{2} t \right) \right],$$

which is the preferred form. ∎

Real and Equal Roots

We have real and equal roots whenever the discriminant $b^2 - 4c = 0$ applies. Then, $r_1 = r_2 = r$. To have a complete solution, we must develop a second solution. We do this by using the reduction of order method. For a full discussion on the reduction of order method, please refer to *Elementary Differential Equations,* by Boyce and DiPrima. However, since this is not a course on ODEs, we jump to the general solution of Equation (C.11), which is

$$u(t) = \alpha e^{rt} + \beta t e^{rt}.$$

Note: This is the form of the solution only for ODEs with constant coefficients.

EXAMPLE C.6. Solve

$$u''(t) + 4u'(t) + 4u(t) = 0.$$

Assuming $u(t) = e^{rt}$, we develop the characteristic equation

$$r^2 + 4r + 4 = 0,$$

which factors into

$$(r + 2)^2 = 0.$$

Thus, we have real and equal roots

$$r = -2,$$

which yield one solution

$$e^{-2t}.$$

So our second solution is

$$te^{-2t}.$$

Remember, our second solution depends on using the reduction of order method and the ODE having constant coefficients. Combining the two solutions yields

$$u(t) = \alpha e^{-2t} + \beta t e^{-2t}. \qquad \blacksquare$$

Complex Roots

For complex roots, both r_1 and r_2 are complex numbers. That is, each root has the form $A + Bi$, where A is the real and Bi is the complex part. *Note:* A and B are real constants, thus, the roots to the characteristic equation, Equation (C.12) are

$$r_1 = \frac{-b + |b^2 - 4c|^{\frac{1}{2}} i}{2} = \frac{-b}{2} + \frac{|b^2 - 4c|^{\frac{1}{2}}}{2} i$$

and

$$r_2 = \frac{-b - |b^2 - 4c|^{\frac{1}{2}} i}{2} = \frac{-b}{2} - \frac{|b^2 - 4c|^{\frac{1}{2}}}{2} i.$$

Thus, we know the two solutions of Equation (C.11) with complex roots are

$$u_1(t) = e^{-\frac{bt}{2}} e^{\gamma it} \text{ and } u_2(t) = e^{-\frac{bt}{2}} e^{-\gamma it} \text{ where } \gamma = \frac{|b^2 - 4c|^{\frac{1}{2}}}{2}.$$

Therefore, one possible solution of Equation (C.11) with complex roots is

$$u(t) = c_1 e^{-\frac{bt}{2}} e^{\gamma it} + c_2 e^{-\frac{bt}{2}} e^{-\gamma it}, \text{ where } \gamma = \frac{|b^2 - 4c|^{\frac{1}{2}}}{2}. \tag{C.16}$$

However, there are other ways of doing the linear combination. For instance, you could do the linear combination as

$$u(t) = e^{-\frac{bt}{2}} \left[\frac{d_1 e^{\gamma it} + d_2 e^{-\gamma it}}{2} \right] \text{ where } \gamma = \frac{|b^2 - 4c|^{\frac{1}{2}}}{2}$$

or

$$u(t) = e^{-\frac{bt}{2}} \left[\frac{m_1 e^{\gamma it} - m_2 e^{-\gamma it}}{2i} \right] \text{ where } \gamma = \frac{|b^2 - 4c|^{\frac{1}{2}}}{2}.$$

If we let $d_1 = d_2 = m_1 = m_2 = 1$ in the above two linear combinations, we immediately notice that we can use Euler's equations, which are

$$\cos ax = \frac{e^{iax} + e^{-iax}}{2}$$

and

$$\sin ax = \frac{e^{iax} - e^{-iax}}{2i}.$$

Doing so forms the new linear combination

$$u(t) = e^{\frac{-bt}{2}} \left[\alpha \cos \gamma t + \beta \sin \gamma t \right]; \gamma = \frac{|b^2 - 4c|^{\frac{1}{2}}}{2},$$

in which α and β are arbitrary constants. We will use this form of the solution to Equation (C.11) with complex roots, because it is much more advantageous in our study of PDEs. Again, the advantage is described in Chapters 4 and 5.

EXAMPLE C.7. Solve

$$u''(t) + 6u'(t) + 10u(t) = 0.$$

Assuming $u(t) = e^{rt}$, the characteristic equation is

$$r^2 + 6r + 10 = 0,$$

which has roots

$$r_1 = -3 + i$$

and

$$r_2 = -3 - i.$$

These roots yield the general solution

$$u(t) = e^{-3t} \left[\alpha \cos(t) + \beta \sin(t) \right]. \qquad \blacksquare$$

We have covered all three possible general solutions to Equation (C.11). For a specific solution, we must determine the coefficients. Since the basic idea behind the solution to a linear homogeneous second-order ODE is two integrations, you can see that we have two constants of integration to solve for. These constants of integration have become, through various manipulations, the coefficients of our general solution (α and β in the previous). Thus, a well-defined solution needs two initial conditions. If the ODE has time, t, as an independent variable, then the first IC is usually the initial position of the physical body you are modeling. The second IC is generally the initial velocity of the body and, as such, is the derivative of the function describing the position of the body. For instance, $u(t_0) = a$ and $\frac{du(t_0)}{dt} = b$. Notice that a and b are constants and not functions.

Another way of describing the beginning state of a body with a second-order ODE without using initial conditions is to describe its boundaries. In this case, the problem is called a boundary value problem (BVP). Boundary conditions (BCs) are generally given when the equation is restricted to an interval and the endpoints are known. Like initial conditions, we need two BCs. Examples of BCs for $0 < t < L$ are the following:

$$u(0) = a, \quad u(L) = b;$$
$$u'(0) = a, \quad u'(L) = b;$$
$$u(0) = a, \quad u'(L) = b;$$

or

$$u'(0) = a, \quad u(L) = b.$$

As in first-order ODEs, the ability to solve second-order ODEs is an interesting mental exercise, but the real benefit comes with applying the ability. Two problems are mechanical and electrical oscillations. Mechanical oscillations can further be broken into two areas: those with, and those without, external force. Those that have no external force are called free vibrations. A free vibration equation is homogeneous. Those that involve an external force are not homogeneous and are called forced vibrations or forced motion. In the following example, we describe a free vibration application in mechanics.

EXAMPLE C.8. Consider a weight of 16 lbs attached to a steel spring that has a natural length of 2 feet. The mass stretches the spring 0.1 ft. Suppose the system is started in motion by stretching the spring an additional 0.25 ft downward; then it's released. Determine and solve the resulting equation of motion, neglecting air resistance.

Solution: Our basic equation is derived from Newton's law of motion, $F = ma$. If we let $u(t)$ represent the displacement of the mass from equilibrium, we have $ma = m\dfrac{d^2u}{dt^2}$, where $m = \dfrac{\text{weight}}{\text{gravity}} = \dfrac{16 \text{ lbs}}{32 \text{ (ft/s}^2)} = 0.5$ lbs-s^2/ft. Though we have no external driving force, we do have a force due to the spring. This force we call F_s, and it always acts to restore the spring to equilibrium. The force F_s is proportional to the displacement of the spring, u; this is called Hooke's law. $F_s = ku$, where k is the spring constant. We determine the "spring modulus" k by dividing the weight by the distance the weight stretches the spring. This gives $k = \dfrac{16 \text{ lbs}}{0.1 \text{ ft}} = 160\dfrac{\text{lbs}}{\text{ft}}$. Putting it all together results in

$$0.5\frac{d^2u}{dt^2} \quad = \quad -160u$$

$$\text{lbs-s}^2/\text{ft} \qquad \text{lbs/ft}$$

or

$$\frac{d^2u}{dt^2} + 320u/\text{s}^2 = 0,$$

which has the solution

$$u(t) = c_1 \cos(8\sqrt{5}t) + c_2 \sin(8\sqrt{5}t) \text{ ft}.$$

We know the initial velocity is 0, thus $c_2 = 0$. The initial displacement is 0.25 feet, which means $c_1 = 0.25$. Therefore,

$$u(t) = 0.25 \cos(8\sqrt{5}t) \text{ ft}$$

is the final answer. ∎

We complete our review of second-order linear constant-coefficient ODEs with a short discussion of the nonhomogeneous problem.

C.2.2 Second-Order Linear Nonhomogeneous ODEs

All second-order linear constant-coefficient nonhomogeneous ODEs are solved using a basic premise. This premise is that the ODE may be separated into a homogeneous part, $u_h(t)$, and a particular part, $u_p(t)$. Consider Equation (C.17),

$$u''(t) + bu'(t) + cu(t) = f(t). \tag{C.17}$$

The homogeneous part of Equation (C.17) is

$$u_h''(t) + bu_h'(t) + cu_h(t) = 0,$$

where $u_h(t)$ refers to the homogeneous part of the nonhomogeneous ODE. We discovered in the previous section that the solution to the homogeneous equation is

$$u_h(t) = c_1 u_{h_1}(t) + c_2 u_{h_2}(t),$$

where $u_{h_1}(t)$ and $u_{h_2}(t)$ are two linearly independent solutions. The particular part of Equation (C.17) really refers to any solution, $u_p(t)$, of the nonhomogeneous equation, Equation (C.17), which is different from $u_{h_1}(t)$ and $u_{h_2}(t)$. Thus, the complete solution to Equation (C.17) is

$$u(t) = u_p(t) + u_h(t) = u_p(t) + c_1 u_{h_1}(t) + c_2 u_{h_2}(t).$$

Hence, we must develop a methodology for obtaining $u_p(t)$.

There are two major solution methods to consider when solving for $u_p(t)$. The first is the method of variation of parameters, which is a general method. It may be used to solve for $u_p(t)$ in any second-order linear constant-coefficient nonhomogeneous ODE. The second is the method of undetermined coefficients, a specific method that uses an educated guess of the form of the solution $u_p(t)$ based on $f(t)$ in Equation (C.17). This method will only work for a select few second-order linear constant-coefficient nonhomogeneous ODEs. In this course, I discuss only the method of undetermined coefficients. Also, the discussion is example based.

Consider the following examples:

EXAMPLE C.9. Suppose

$$u''(t) + 3u'(t) + 2u(t) = 3t^2 + 2t - 1. \tag{C.18}$$

We solve Equation (C.18) by first solving the homogeneous part, which is

$$u_h''(t) + 3u_h'(t) + 2u_h(t) = 0, \tag{C.19}$$

where $u_h(t)$ refers to the homogeneous solution. Assuming $u_h(t) = e^{rt}$, we form the characteristic equation

$$r^2 + 3r + 2 = 0,$$

which factors into

$$(r + 2)(r + 1) = 0. \tag{C.20}$$

Equation (C.20) has real and unequal roots of $r = -2$ and $r = -1$. Therefore, we know that $u_{h_1}(t) = e^{-2t}$ and $u_{h_2}(t) = e^{-t}$. Thus,

$$u_h(t) = c_1 u_{h_1}(t) + c_2 u_{h_2}(t) = c_1 e^{-2t} + c_2 e^{-t}. \tag{C.21}$$

(*Note:* Equation (C.21) is not in the form of hyperbolic sine and hyperbolic cosine functions because $b^2 - 4c$ is a perfect square).

We must now determine the particular solution, $u_p(t)$. First, we notice that $f(t)$ is a polynomial in the independent variable t, $f(t) = 3t^2 + 2t - 1$. This suggests that a polynomial may be the form of the particular solution, $u_p(t)$. Thus, we let

$$u_p(t) = At^2 + Bt + C. \tag{C.22}$$

We then find the first and second derivatives of Equation (C.22), which are

$$u'_p(t) = 2At + B \tag{C.23}$$

and

$$u''_p(t) = 2A. \tag{C.24}$$

Then, we replace $u'(t)$ with $u'_p(t)$ and $u''(t)$ with $u''_p(t)$ in Equation (C.18), resulting in

$$u''_p(t) + 3u'_p(t) + 2u_p(t) = 3t^2 + 2t - 1,$$

which becomes

$$2A + 3\left(2At + B\right) + 2\left(At^2 + Bt + C\right) = 3t^2 + 2t - 1. \tag{C.25}$$

We solve for the constants A, B, and C in Equation (C.25). By matching the coefficients of similar terms of the variable t on both sides of the equation, we have

$$2A + 3B + 2C = -1, \tag{C.26}$$
$$6A + 2B = 2 \tag{C.27}$$

and

$$2A = 3. \tag{C.28}$$

From Equation (C.28), we know $A = \dfrac{3}{2}$. Thus, by replacing A by $\dfrac{3}{2}$ in Equation (C.27), we find $B = \dfrac{-7}{2}$. Knowing A and B and using Equation (C.26) we find $C = \dfrac{13}{4}$. Therefore,

$$u_p(t) = \frac{3}{2}t^2 - \frac{7}{2}t + \frac{13}{4}. \tag{C.29}$$

Combining Equation (C.21) and Equation (C.29), we get

$$u(t) = u_p(t) + u_h(t) = \frac{3}{2}t^2 - \frac{7}{2}t + \frac{13}{4} + c_1 e^{-2t} + c_2 e^{-t}. \qquad \blacksquare$$

EXAMPLE C.10. Suppose

$$u''(t) + 3u'(t) + 2u(t) = \sin t. \tag{C.30}$$

We know from the previous example that

$$u_h(t) = c_1 u_{h_1}(t) + c_2 u_{h_2}(t) = c_1 e^{-2t} + c_2 e^{-t}.$$

Since $f(t) = \sin t$, we assume $u_p(t) = A \cos t + B \sin t$. Thus,

$$u_p'(t) = -A \sin t + B \cos t \tag{C.31}$$

and

$$u_p''(t) = -A \cos t - B \sin t. \tag{C.32}$$

Replacing $u''(t)$ with $u_p''(t)$, $u'(t)$ with $u_p'(t)$, and $u(t)$ with $u_p(t)$ in Equation (C.30) yields

$$u_p''(t) + 3u_p'(t) + 2u_p(t) = \sin t,$$

which becomes

$$-A \cos t - B \sin t + 3\left(-A \sin t + B \cos t\right) + 2\left(A \cos t + B \sin t\right) = \sin t. \tag{C.33}$$

We solve for the constants A and B in Equation (C.33). By matching the coefficients of similar terms of the function $\sin t$ and $\cos t$ on both sides of the equation, we have

$$-B - 3A + 2B = B - 3A = 1 \tag{C.34}$$

and

$$-A + 3B + 2A = A + 3B = 0. \tag{C.35}$$

Solving Equations (C.35 and C.34) simultaneously yields $A = \dfrac{-3}{10}$ and $B = \dfrac{1}{10}$. Therefore, the solution to Equation (C.30) is

$$u(t) = u_p(t) + u_h(t) = \frac{-3}{10} \cos t + \frac{1}{10} \sin t + c_1 e^{-2t} + c_2 e^{-t}.$$

Note: Using $u_p(t) = A \sin t$ would not have given you the correct answer for $u(t)$. ∎

This completes our review of ODEs. For an extensive review of this topic, I suggest *Elementary Differential Equations*, by Boyce and DiPrima.

Appendix D

Mathematical Notation

Throughout this text, we use mathematical notation that may be new to you, such as $\mathbb{R}$ to represent the real number line, commonly called the Reals. Also, in your study of mathematics, you may have thought mathematicians are notation happy. However, this is not the case. Notation, and knowing the correct definition of notation, is very important. Another example that shows the importance of notation is

$$\int_0^1 x \ dx.$$

In English, this notation says integrate the function x on the interval 0 to 1 with respect to x. I think the notation is much clearer than the English version and certainly easier to write. I hope you agree. Thus, you easily see it is important in your mathematical education to learn some basic notation, particularly notation found in many text books, such as names of spaces and mapping notation.

The real number line is known by the symbol $\mathbb{R}$ and is called one-space. The natural numbers, a subset of the real numbers, use the symbol $\mathbb{N}$. The xy-coordinate plane is commonly known as $\mathbb{R}^2$ and is called two-space. Three-dimensional space, the xyz-coordinate system, is known as $\mathbb{R}^3$ and is called three-space. Both $\mathbb{R}^2$ and $\mathbb{R}^3$ are easily identified. They are formed by taking copies of the real number line, then intersecting them at right angles to each other. Since we live in three dimensional space, it is hard to imagine what $\mathbb{R}^4$, $\mathbb{R}^5$, or $\mathbb{R}^n$ where $n \in \mathbb{N}$, meaning n is any natural number, look like. But mathematically they exist, and they are quite useful. Also, the spaces that are formed by copies of the real number line, such as $\mathbb{R}^2$, $\mathbb{R}^3$ or $\mathbb{R}^n$, are known as finite-dimensional spaces. One reason they are considered finite is the number in the superscript; it indicates a finite number of copies of the real number line. The reason for the superscript (hence, why they are finite dimensional) will be addressed in the next section.

The other notation that is important to know is mapping notation. Mapping notation is very important. It is closely tied to functions, and is easily understood. For instance, when you add the number 2 to the number 3, $2 + 3$, you are actually dealing with a function. The function is addition. It would be silly to write mapping notation for the addition of 2 and 3. However, if you wanted to indicate addition of all real numbers to each other, then mapping notation is much easier and more compact. The mapping notation indicating that you can add all real numbers to each other is

$$+ : \mathbb{R} \times \mathbb{R} \longrightarrow \mathbb{R}.$$

It says that you are adding a real number from one copy of $\mathbb{R}$ to a real number from another copy of $\mathbb{R}$, and you expect the answer to be a third copy of $\mathbb{R}$.

Generally, we do not use mapping notation for addition. We reserve it for other things. For example, suppose you wanted to discuss, in general terms, all the functions that are in $\mathbb{R}^2$. Mapping notation is a great help. The notation would simply be

$$f : \mathbb{R} \longrightarrow \mathbb{R}.$$

Notice the mapping has two single copies of $\mathbb{R}$. Adding their superscripts together indicates the graphs of the functions are living in two-space. What would the mapping look like for functions in $\mathbb{R}^3$? One might naturally imagine it is

$$f : \mathbb{R} \times \mathbb{R} \longrightarrow \mathbb{R}.$$

This indicates functions that take a point from the xy-plane and map it to a value on the z-axis giving us graphs in the normal $\mathbb{R}^3$ space. But we could have a mapping which looks like

$$f : \mathbb{R} \longrightarrow \mathbb{R} \times \mathbb{R}.$$

This mapping indicates functions which take a value on the z-axis to a point on the xy-plane. It is another copy of $\mathbb{R}^3$, but with a slightly different twist. The important thing is knowing where the mapping originates and where it terminates.

Appendix E

Summary of Thermal Diffusivity of Common Materials

In this appendix, a summary of thermal diffusivity, k, for common materials is given. Remember, thermal diffusivity is given by

$$k = \frac{K_0}{c\rho},$$

where K_0 is the thermal conductivity, c is the specific heat, and ρ is the mass density of the given material.

Thermal conductivity, K_0, is considered temperature independent provided the temperature variation is not too great. If the temperature variation is great enough, then the formula for calculating the thermal conductivity is given by

$$K_0' = \frac{1}{T_2 - T_1} \int_{T_1}^{T_2} K_0 \, dx,$$

where K_0' is the new thermal conductivity. However, for most problems in this text, we will assume that thermal conductivity is temperature independent.

Thermal Diffusivity of Common Materials

Material	k in ft^2/hr	k in cm^2/sec
Metals:		
Aluminum	3.33	.852
Copper	4.42	1.12
Gold	4.68	1.18
Iron, cast	0.66	0.12
Iron, pure	0.7	0.22
Lead	0.95	0.246
Mercury	0.172	0.044
Nickel	0.6	0.141
Silver	6.6	1.616
Steel, mild	0.48	0.124
Tungsten	2.39	0.725
Zinc	1.6	0.403
Non-metals:		
Asbestos	0.01	0.003
Brick, fire clay	0.02	0.005
Cork	0.006	0.002
Glass	0.023	0.006
Granite	0.05	0.014
Ice	0.046	0.011
Oak, across grain	0.0062	0.002
Pine, across grain	0.0059	0.002
Quartz Sand, dry	0.008	0.002
Rubber	0.003	0.001
Water	0.005	0.001

Appendix F

Tables of Fourier and Laplace Transforms

In this appendix, we provide common Fourier and Laplace transforms. The transforms on the following pages may have a slightly different form than those that you have seen before. It is an interesting exercise to show that both forms are correct.

F.1   TABLES OF FOURIER, FOURIER COSINE, AND FOURIER SINE TRANSFORMS

Table of Fourier Transforms

	$f(x)$	$\mathcal{F}\{f(x)\} = F(\theta)$
1.	$\begin{cases} 1, & \lvert x \rvert < a \\ 0, & a < \lvert x \rvert \end{cases}$	$\dfrac{1}{\pi}\dfrac{\sin(a\theta)}{\theta}$
2.	$\dfrac{1}{x^2 + a^2},\ a > 0$	$\dfrac{\pi e^{-a\theta}}{a}$
3.	$\begin{cases} e^{-ax}, & a > 0,\, x > 0 \\ 0, & \text{otherwise} \end{cases}$	$\dfrac{1}{a + i\theta}$
4.	$\begin{cases} e^{-ax}, & a > 0,\, x < 0 \\ 0, & \text{otherwise} \end{cases}$	$\dfrac{1}{a - i\theta}$
5.	$e^{-ax^2},\ a > 0$	$\dfrac{1}{\sqrt{4\pi a}}e^{-\theta^2/4a}$
6.	$e^{-a\lvert x \rvert},\ a > 0$	$\dfrac{2a}{a^2 + \theta^2}$
7.	$\dfrac{x}{x^2 + a^2},\ a > 0$	$i\pi e^{-a\theta}$
8.	$\dfrac{2a}{x^2 + a^2},\ a > 0$	$e^{-a\lvert\theta\rvert}$

Table of Fourier Transforms (continued)

	$\mathbf{f(x)}$	$\mathcal{F}\{\mathbf{f(x)}\} = \mathbf{F}(\theta)$
9.	$\dfrac{\sin(ax)}{x},\ a > 0$	$\begin{cases} \pi, & |\theta| < a \\[2mm] \dfrac{\pi}{2}, & |\theta| = a \\[2mm] 0, & |\theta| > a \end{cases}$
10.	$\cos(ax^2),\ a > 0$	$\sqrt{\dfrac{\pi}{a}}\cos\left\{\dfrac{\theta^2}{4a} - \dfrac{\pi}{4}\right\}$
11.	$\sin(ax^2),\ a > 0$	$\sqrt{\dfrac{\pi}{a}}\cos\left\{\dfrac{\theta^2}{4a} + \dfrac{\pi}{4}\right\}$

Table of Fourier Cosine Transforms

	$\mathbf{f(x)}$	$\mathcal{F_C}\{\mathbf{f(x)}\} = \mathbf{F}(\theta)$
1.	$\begin{cases} 1, & 0 < x < a \\[2mm] 0, & a < x \end{cases}$	$\dfrac{\sin(a\theta)}{\theta}$
2.	$\dfrac{a}{x^2 + a^2},\ a > 0$	$\dfrac{\pi e^{-a\theta}}{2a}$
3.	$e^{-ax},\ a > 0$	$\dfrac{a}{a^2 + \theta^2}$

Table of Fourier Cosine Transforms (continued)

	$\mathbf{f(x)}$	$\mathcal{F}_\mathbf{C}\left\{\mathbf{f(x)}\right\} = \mathbf{F}(\theta)$
4.	xe^{-ax}, $a > 0$	$\dfrac{a^2 - \theta^2}{\left(a^2 + \theta^2\right)^2}$
5.	e^{-ax^2}, $a > 0$	$\sqrt{\dfrac{\pi}{4a}}\,e^{-\theta^2/4a}$
6.	$x^{n-1}e^{-ax}$, $a > 0$	$\dfrac{\Gamma(n)\cos\left(n\tan^{-1}(\theta/a)\right)}{(a^2 + \theta^2)^{n/2}}$
7.	$\cos(ax)e^{-ax}$, $a > 0$	$\dfrac{a\theta^2 + 2a^3}{4a^4 + \theta^4}$
8.	$\sin(ax)e^{-ax}$, $a > 0$	$\dfrac{2a^3 - a\theta^2}{4a^4 + \theta^4}$
9.	$\begin{cases} \cos x, & 0 < x < a \\ \\ 0, & a < x \end{cases}$	$\dfrac{1}{2}\left[\dfrac{\sin(a(1-\theta))}{1-\theta} + \dfrac{\sin(a(1+\theta))}{1+\theta}\right]$
10.	$\dfrac{\sin ax}{x}$, $a > 0$	$\begin{cases} \pi/2, & \theta < a \\ \pi/4, & \theta = a \\ 0, & \theta > a \end{cases}$
11.	$\sin(ax^2)$, $a > 0$	$\sqrt{\dfrac{\pi}{8a}}\left[\cos\dfrac{\theta^2}{4a} - \sin\dfrac{\theta^2}{4a}\right]$
12.	$\cos(ax^2)$, $a > 0$	$\sqrt{\dfrac{\pi}{8a}}\left[\cos\dfrac{\theta^2}{4a} + \sin\dfrac{\theta^2}{4a}\right]$

Table of Fourier Sine Transforms

	$\mathbf{f(x)}$	$\mathcal{F}_S\left\{\mathbf{f(x)}\right\} = \mathbf{F}(\theta)$
1.	$\begin{cases} 1, & 0 < x < a \\ 0, & a < x \end{cases}$	$\dfrac{1 - \cos(a\theta)}{\theta}$
2.	$\dfrac{x}{x^2 + a^2},\ a > 0$	$\dfrac{\pi e^{-a\theta}}{2}$
3.	$e^{-ax},\ a > 0$	$\dfrac{\theta}{a^2 + \theta^2}$
4.	$xe^{-ax},\ a > 0$	$\dfrac{2a\theta}{\left(a^2 + \theta^2\right)^2}$
5.	$xe^{-ax^2},\ a > 0$	$\dfrac{\sqrt{\pi}}{4a^{3/2}}\theta e^{-\theta^2/4a}$
6.	$x^{n-1}e^{-ax},\ a > 0$	$\dfrac{\Gamma(n)\sin\left(n\tan^{-1}(\theta/a)\right)}{(a^2 + \theta^2)^{n/2}}$
7.	$\cos(ax)e^{-ax},\ a > 0$	$\dfrac{\theta^3}{4a^4 + \theta^4}$
8.	$\sin(ax)e^{-ax},\ a > 0$	$\dfrac{2a^2\theta}{4a^4 + \theta^4}$
9.	$\begin{cases} \sin x, & 0 < x < a \\ 0, & a < x \end{cases}$	$\dfrac{1}{2}\left[\dfrac{\sin(a(1-\theta))}{1-\theta} + \dfrac{\sin(a(1+\theta))}{1+\theta}\right]$

Table of Fourier Sine Transforms (continued)

	$\mathbf{f(x)}$	$\mathcal{F}_S\{\mathbf{f(x)}\} = \mathbf{F}(\theta)$
10.	$\dfrac{\sin ax}{x},\ a > 0$	$\dfrac{1}{2}\ln\left[\dfrac{\theta + a}{\theta - a}\right]$
11.	$\dfrac{\cos(ax)}{x},\ a > 0$	$\begin{cases} 0, & \theta < a \\ \pi/4, & \theta = a \\ \pi/2, & \theta > a \end{cases}$
12.	$\dfrac{e^{-ax}}{x},\ a > 0$	$\tan^{-1}\left(\dfrac{\theta}{a}\right)$

F.2   TABLE OF LAPLACE TRANSFORMS

Table of Laplace Transforms

	$\mathbf{f(t)}$	$\mathcal{L}\{\mathbf{f(t)}\} = \mathbf{F(s)}$
1.	1	$\dfrac{1}{s},\ s > 0$
2.	t	$\dfrac{1}{s^2}$
3.	$t^n,\ n \in \mathbb{N}$	$\dfrac{n!}{s^n},\ s > 0$
4.	$\dfrac{1}{\sqrt{\pi t}}$	$\dfrac{1}{\sqrt{s}}$

Table of Laplace Transforms (continued)

	$\mathbf{f(t)}$	$\mathcal{L}\{\mathbf{f(t)}\} = \mathbf{F(s)}$
5.	$e^{\alpha t}$	$\dfrac{1}{s-\alpha},\ s > \alpha$
6.	$\sin(\alpha t)$	$\dfrac{\alpha}{s^2 + \alpha^2},\ s > 0$
7.	$t^p,\ p > -1$	$\dfrac{\Gamma(p+1)}{s^{p+1}},\ s > 0$
8.	$\cos(\alpha t)$	$\dfrac{s}{s^2 + \alpha^2},\ s > 0$
9.	$\sinh(\alpha t)$	$\dfrac{\alpha}{s^2 - \alpha^2},\ s > |\alpha|$
10.	$\cosh(\alpha t)$	$\dfrac{s}{s^2 - \alpha^2},\ s > |\alpha|$
11.	$e^{\alpha t}\sin(\beta t)$	$\dfrac{\beta}{(s-\alpha)^2 + \beta^2},\ s > \alpha$
12.	$e^{\alpha t}\cos(\beta t)$	$\dfrac{s-\alpha}{(s-\alpha)^2 + \beta^2},\ s > \alpha$
13.	$t^n e^{\alpha t},\ n \in \mathbb{N}$	$\dfrac{n!}{(s-\alpha)^{n+1}},\ s > \alpha$
14.	$u(t-\alpha)$, unit step function	$\dfrac{e^{-\alpha s}}{s},\ s > 0\text{x}$
15.	$u(t-\alpha)f(t-\alpha)$	$e^{-\alpha s}F(s)$

Table of Laplace Transforms (continued)

	$\mathbf{f(t)}$	$\mathcal{L}\{\mathbf{f(t)}\} = \mathbf{F(s)}$
16.	$\delta(t - \alpha)$, Dirac delta function	$e^{-\alpha s}$
17.	$e^{\alpha t} f(t)$	$F(s - \alpha)$
18.	$\displaystyle\int_0^t f(t - u)g(u)\,du$, Convolution Integral	$F(s)G(s)$
19.	$\dfrac{1}{\sqrt{\pi t}} e^{\frac{-a^2}{4t}}$	$\dfrac{e^{-a\sqrt{s}}}{\sqrt{s}}$
20.	$\dfrac{a}{2\sqrt{\pi t^3}} e^{\frac{-a^2}{4t}}$	$e^{-a\sqrt{s}}$
21.	$erfc\left(\dfrac{a}{2\sqrt{t}}\right)$, Complementary Error Function	$\dfrac{e^{-a\sqrt{s}}}{s}$
22.	$2\sqrt{\dfrac{t}{\pi}} e^{\frac{-a^2}{4t}} - a\left\{erfc\left(\dfrac{a}{2\sqrt{t}}\right)\right\}$	$\dfrac{e^{-a\sqrt{s}}}{s\sqrt{s}}$
23.	$e^{ab}e^{b^2 t}\left\{erfc\left(b\sqrt{t} + \dfrac{a}{2\sqrt{t}}\right)\right\}$	$\dfrac{e^{-a\sqrt{s}}}{\sqrt{s}(\sqrt{s} + b)}$
24.	$-e^{ab}e^{b^2 t}\left\{erfc\left(b\sqrt{t} + \dfrac{a}{2\sqrt{t}}\right)\right\} + erfc\left(\dfrac{a}{2\sqrt{t}}\right)$	$\dfrac{be^{-a\sqrt{s}}}{s(\sqrt{s} + b)}$
25.	$H(t - a)$	$\dfrac{e^{-as}}{s}$

Bibliography

1. Asmar, Nakhle H. *Partial Differential Equations and Boundary Value Problems*. Englewood Cliffs: Prentice-Hall, 2000.

2. Bartle, Robert G. *The Elements of Real Analysis*, 2nd ed. New York: Wiley, 1976.

3. Boyce, William E. and DiPrima, Richard C. *Elementary Differential Equations*, 2nd ed. New York: Wiley, 1969.

4. Boyer, Carl B. *A History of Mathematics*, 2nd ed. New York: Wiley, 1991

5. Byron, Frederick W. Jr. and Fuller, Robert W. *Mathematics of Classical and Quantum Physics*. New York: Dover, 1992.

6. Budak, B. M., Samarskii, A. A., and Tikhonov, A. N. *a Collection of Problems of Mathematical Physics*, New York: Dover, 1988.

7. Churchill, Ruel V. *Fourier Series and Boundary Value Problems*, $2^{n}d$ ed. New York: McGraw-Hill, 1969.

8. Courant, R. and Hilbert D. *Methods of Mathematical Physics*. New York: Interscience, 1953.

9. CRC *Standard Mathematical Tables*, 18th ed. Cleveland: CRC Press, 1970.

10. Dettman, John W. *Mathematical Methods in Physics and Engineering*. New York: Dover, 1988.

11. Duchateau, Paul and Zachmann, David W. *Partial Differential Equations*. New York: McGraw-Hill, 1986.

12. Farlow, Stanley J. *Partial Differential Equations for Scientists and Engineers*. New York: Dover, 1993.

13. Feynman, Richard P., Leighton, Robert B., and Sands, Matthew *the Feynman Lectures on Physics*. Reading, MA: Addison-Wesley, 1966

14. Haberman, Richard *Elementary Applied Partial Differential Equation with Fourier Series and Boundary Value Problems* 2nd ed. Englewood Cliffs: Prentice-Hall, 1987.

15. Hayt, William H. Jr. *Engineering Electromagnetics*, 5th ed. New York, McGraw-Hill, 1989.

16. Keener, James and Sneyd, James *Mathematical Physiology*. New York: Springer-Verlag, 1998.

17. Körner, T. W. *Fourier Analysis*. Cambridge, England: Cambridge University Press, 1995.

18. Kreyszig, Erwin *Advanced Engineering Mathematics*, 7th ed. New York: Wiley, 1993.

19. Kreyszig, Erwin *Introductory Functional Analysis with Applications*. New York: Wiley, 1978.

20. Lay, Steven R. *Analysis with an Introduction to Proof*, 2nd. New Jersey: Prentice Hall, 1990.

21. O'Neil, Peter V. *Advanced Engineering Mathematics*, 4th ed. Boston: PWS-Kent, 1995.

22. Özisik, M. Necati *Heat Conduction*. New York: Wiley, 1980.

23. Powers, David L. *Boundary Value Problems*, 2nd ed. New York: Academic Press, 1979.

24. Rainville, Earl D. *Infinite Series*. New York: Macmillian, 1967.

25. Rainville, Earl D. *Special Functions*. New York: Macmillian, 1960.

26. Renardy, Michael and Rogers, Robert C. *An Introduction to Partial Differential Equations*. New York: Springer-Verlag, 1993.

27. Sagen, Hans *Boundary and Eigenvalue Problems in Mathematical Physics*. New York: Dover, 1989.

28. Spiegel, Murray R. *Theory and Problems of Fourier Analysis with Applications to Boundary Value Problems*. New York: McGraw-Hill, 1974.

29. Spiegel, Murray R. and Liu, John *Mathematical Handbook of Formulas and Tables*, 2nd ed. New York: McGraw-Hill,1999.

30. Strauss, Walter A. *Partial Differential Equations An Introduction*. New York: Wiley, 1992.

31. Taylor, A. B. *Mathematical Models in Applied Mechanics*. Oxford, England: Clarendon Press, 1986.

32. Tolstov, Georgi P. *Fourier Series*. New York: Dover, 1976.

33. Trench, William F. *Elementary Differential Equations*. Pacific Grove: Brooks/Cole, 2000.

34. Trim, Donald W. *Applied Partial Differential Equations.* Boston: PWS-Kent, 1990.

35. Walker, P. L. *The Theory of Fourier Series and Integrals.* New York: Wiley-Interscience, 1986.

36. Wolfram, Stephen *The Mathematica Book*, 3rd ed. Champaign, IL.: Wolfram Media and Cambridge University Press, 1996.

37. Wylie, Ray C. and Barrett, Louis C. *Advanced Engineering Mathematics*, 6th ed. New York: McGraw-Hill, 1995.

38. Zachmanoglou, E. C. and Thoe, Dale W. *Introduction to Partial Differential Equations with Applications.* New York: Dover, 1986.

39. Zill, Dennis G. and Cullen, Michael R. *Advanced Engineering Mathematics.* Sudbury: Jones and Barlett, 2000.

Index